Atlas of Yellowstone

# Atlas of Yellowstone

Senior Editor
## W. Andrew Marcus
*University of Oregon*

Cartographic Editor
## James E. Meacham
*University of Oregon*

Yellowstone Editor
## Ann W. Rodman
*Yellowstone National Park*

Production Manager
## Alethea Y. Steingisser
*University of Oregon*

Consulting Editor
## Stuart Allan

Text Editor
## Ross West

University of California Press, one of the most distinguished university presses in the United States, enriches lives around the world by advancing scholarship in the humanities, social sciences, and natural sciences. Its activities are supported by the UC Press Foundation and by philanthropic contributions from individuals and institutions. For more information, visit www.ucpress.edu.

University of California Press
Berkeley and Los Angeles, California

University of California Press, Ltd.
London, England

Library of Congress Cataloging-in-Publication Data

Atlas of Yellowstone / senior editor, W. Andrew Marcus ; cartographic editor, James E. Meacham ; Yellowstone editor, Ann W. Rodman ; production manager, Alethea Y. Steingisser ; text editor, Ross West ; contributing editor, Stuart Allan.
    p. cm.
Includes bibliographical references and index.
 ISBN 978-0-520-27155-5 (cloth : alk. paper)
 1.  Yellowstone National Park—Maps. 2.  Physical geography—Yellowstone National Park—Maps. 3.  Human geography—Yellowstone National Park—Maps. 4.  Wildlife refuges—Yellowstone National Park—Maps.  I. Marcus, W. Andrew. II. Meacham, James E. III. Rodman, Ann Winne, 1958–

G1477.Y4A8 2012
 912.09787'52—dc23
              2011037533

The *Atlas of Yellowstone* is dedicated to

William (Bill) G. Loy
*who devoted so much of his life to cartography*

John Varley
*who has devoted so much of his life to Yellowstone*

and to the many individuals who have
worked to preserve the spirit of Yellowstone

# Contents

University of Oregon   xi

Montana State University   xiii

University of Wyoming   xv

Acknowledgments   xviii

Preface   xx

## Geographic Setting

Yellowstone in the World   2

Yellowstone in the Region   4

Greater Yellowstone Detail   6

The World's First National Park   8

Political Boundaries   10

## Human Geography

Archaeology   14

American Indians   16

Sheep Eaters   18

Catlin and the American Indian   20

Exploration   22

Early Maps, 1808–1814   24

Early Maps, 1836–1865   26

Early Maps, 1869–1872   28

Jackson and Moran   30

Yellowstone Art   32

Early Science History   34

Science History   36

Road History   38

Development at Old Faithful   40

Roads and Trails   42

Traffic   44

Park Visitation   46

Tetons Climbing History   48

The Economy   50

Labor and Employment   52

Income   54

Agriculture   56

Market Access   58

Wildland Economies   60

Protected Areas   62

Land Ownership   64

Population   66

County Population   68

City Population   70

Education   72

Race and Ethnicity   74

Religion and Politics   76

## Physical Geography

Elevation   80

Cross Sections   82

Landforms

    National Parks   84

    Park Headquarters   86

    Canyons and Domes   88

    Lava Flows and Glacial Erosion   90

    Overthrust Belt and Glacial Features   92

Grand Teton Geology   94

Yellowstone Geology   96

Yellowstone Volcano   98

Geothermal Activity   100

Hydrothermal Areas   102

Geysers   104

Earthquakes   106

Glaciers   108

Yellowstone Lake   110

Drainage Basins   112

Rivers   114

Streamflow   116

Flow Regimes   118

Waterfalls   120

Precipitation   122

| | |
|---|---|
| Temperature | 124 |
| Climate Change | 126 |
| Wetlands | 128 |
| Soils | 130 |
| Ecoregions | 132 |
| Vegetation | 134 |
| Landscape Change | 140 |
| Fire History | 142 |
| 1988 Fires | 144 |

## Wildlife

| | |
|---|---|
| Grizzly Bears | 148 |
| Wolves | 150 |
| Coyotes | 152 |
| Bison | 154 |
| Bison Movement | 156 |
| Elk | 158 |
| Fish | 160 |
| Potential Wildlife Habitats | 162 |
| Sagebrush-Steppe Habitat | 164 |
| Thermophiles | 166 |
| Dinosaurs | 168 |
| Vertebrate Species | 170 |

## Reference Maps

| | |
|---|---|
| Greater Yellowstone Reference Maps | 178 |
| Bozeman | 180 |
| Billings | 182 |
| Rexburg | 184 |
| Cody | 186 |
| Pocatello | 188 |
| Lander | 190 |
| National Park Reference Maps | 192 |
| Electric Peak | 194 |

| | |
|---|---|
| Tower Junction | 196 |
| Silver Gate | 198 |
| West Yellowstone | 200 |
| Canyon Village | 202 |
| Lamar Valley | 204 |
| Old Faithful | 206 |
| Lake Village | 208 |
| East Entrance | 210 |
| Bechler Meadows | 212 |
| Lewis Lake | 214 |
| Thorofare | 216 |
| Flagg Ranch | 218 |
| Grand Teton | 220 |
| Moose | 222 |
| Gazetteer | 224 |
| USGS Map Index | 232 |
| Counties | 236 |
| Place Names | |
| Greater Yellowstone Cultural Names | 238 |
| Greater Yellowstone Physical Names | 240 |
| Yellowstone National Park | 242 |
| Grand Teton National Park | 248 |

## Afterword, Sources and Sponsors

| | |
|---|---|
| Afterword | 252 |
| Sponsoring Institutions | |
| University of Oregon | 256 |
| Montana State University | 257 |
| Yellowstone Park Foundation | 258 |
| University of Wyoming | 259 |
| Contributing Organizations | 260 |
| Sources | 262 |
| Index | 268 |

# University of Oregon
**A World-Class Teaching and Research University**

Research and teaching go hand in hand at a major public university. This *Atlas of Yellowstone* is a perfect example.

The atlas was originally conceived of as a class project. Students in an advanced cartography course developed the first maps for the atlas—see the pages on glaciers, climate, fire history, and wildlife as just a few impressive examples. Working closely with University of Oregon scholars, our students explored the vast treasure that is Yellowstone research, selected from it the most telling details, and, using an array of sophisticated data-presentation techniques, fashioned those discrete bits of knowledge into an organized, meaningful whole. Such work is the essence of our mission as a research university.

The *Atlas of Yellowstone* is also an example of how the University of Oregon integrates research and teaching to serve the state, the nation, and the international community. This is the first major atlas of a national park. Fittingly, its subject is the extraordinary place that inspired the very idea of a national park, the place where that idea became reality—and in doing so forever changed the world.

This atlas is a remarkable work of collaborative scholarship, providing a synthesis and visualization of data that, in many cases, has never before been portrayed. At the same time, the atlas is a major teaching tool. Whether you are a tourist to Yellowstone and Grand Teton national parks, a schoolteacher, a park interpreter, or a resource manager, the atlas portrays the often complex stories of the Yellowstone region in ways that are both initially accessible and yet deep enough to reward extended or repeated investigation.

In carrying out this research and teaching mission, the atlas provides a service to our schools, the citizens of our nation, and the global community of those interested in the natural world and our place in it. As you turn the pages of this atlas, please remember how these stories and graphics came to be—through the vision and hard work of students and faculty members collaborating on research and learning. It is the core of what we do.

On behalf of the University of Oregon, I hope you enjoy and learn from this wonderful resource. It is a tremendous accomplishment and a fitting tribute to everyone who helped make it possible.

Richard W. Lariviere, President

# Montana State University

From the time explorers first ventured into what is now Yellowstone National Park and observed its wonders, Yellowstone has inspired contemplation, exploration, and awe.

America's oldest national park, which is the model for conservation efforts worldwide, continues to lure explorers, but today those explorers include scientists in addition to adventurers and tourists, and the wonders of the region have inspired research and discoveries that are changing our world.

It is this exploration that we celebrate with the *Atlas of Yellowstone.*

Montana State University is honored to be a part of this project, as it is proud to play a key role in this scholarly exploration of Yellowstone National Park.

Located just ninety miles from Yellowstone's northernmost border, Montana State University is the sole university located in the Greater Yellowstone Ecosystem. Because Yellowstone National Park is our backyard, it serves as our largest laboratory, providing hands-on learning opportunities for virtually every college and department in the university. As a result, MSU has received more National Science Foundation grants based in Yellowstone than any other university in the country, earning MSU the official distinction as The University of the Yellowstone®.

Our location within the Greater Yellowstone Ecosystem informs teaching, learning, and research at our university. It enriches the experiences of our students, faculty, staff, and community with scores of programs and opportunities. For example, scientists from the Thermal Biology Institute and Center for Biofilm Engineering, both based at MSU, study microbes that thrive in the extreme temperatures of the park's geothermal features for possible innovations ranging from alternate fuel sources to tools to fight cancer. MSU ecologists have conducted landmark studies of the park's diverse mammal, bird, and fish species, including the interaction of gray wolves with elk and bison. The Interagency Grizzly Bear Study Team based at MSU monitors and studies grizzlies, helping ensure the existence of the Ursus arctos horribilis in the Greater Yellowstone Ecosystem. And, MSU earth scientists study how climate changes have affected the ecosystem in the past and how those changes will likely affect the area in the future.

MSU artists, photographers, and filmmakers find the park to be a glorious studio as well as an inspiring subject. Scholars of Yellowstone's history utilize collections housed at MSU's Renne Library. Even young explorers come to MSU's Museum of the Rockies to an interactive learning experience geared to children, the *Explore Yellowstone* exhibit; and our School of Architecture is helping National Park Service planners envision what the park should look like in the future.

Montana State University is proud of these partnerships with Yellowstone National Park as well as the opportunity to help unlock the mysteries of our dynamic landscape. It is likewise honored to be affiliated with other universities in our region on this impressive project that shines new light on the crown jewel of America's park system.

Waded Cruzado, President

# University of Wyoming

As the first national park in the world, Yellowstone National Park is one of our nation's finest achievements in public lands. It is a natural, cultural, and historical treasure, not only for all Americans but also for all people who experience this park and its glories. It is a place that provides us with a deep appreciation of the history of the United States and a deep understanding of nature and our own well-being.

Yellowstone National Park provides a rich natural laboratory for University of Wyoming faculty members and students. The sole four-year research university in the state, the University of Wyoming is committed to sustained and increasing prominence in the area of environment and natural resources. Yellowstone National Park is part of the physical landscape that has distinguished the university's scholarship in this important arena.

The University of Wyoming has long served as a key partner with the park. Our faculty members have applied their interdisciplinary teaching and research to many environmental questions while exploring critical ecological issues such as geologic and geothermal hazards and resources, threatened and endangered species, ecosystem integrity, invasive species, migration corridor protection, and fire management.

The challenges facing resource managers of our national parks have grown increasingly complex, and effective conservation of Yellowstone's natural and cultural treasures requires a reliance on interdisciplinary scholarship and the ability to adapt to a changing world. UW offers strong scientific expertise in biology, hydrology, geology, and geography to further explore and investigate ecosystem change, as well as a rich history of interfacing science and management.

Over the years, our faculty members have conducted important scientific research that has informed management of Yellowstone and Grand Teton national parks. Our researchers and graduate students were on the leading edge of forest fire ecology studies in Yellowstone, including the wildfires of 1988. We are engaged in studying the unprecedented impact of bark beetles on lodgepole pine forests, conducting inventories and restoration efforts for cultural and historical resources, studying the critical cross-boundary issues related to large mammals in the park, and evaluating the dynamics of invasive species such as the New Zealand mud snail, lake trout, and Canada thistle. Our geophysicists have imaged the mantle plume beneath the park, an important contribution to geologic hazard assessment. Our scientists have also contributed new knowledge on the geology relating to the Earth's crust in the Tetons, the earliest record for Himalayan-style tectonics in the world. The University of Wyoming is also a repository of cultural and historical resources on Yellowstone, all available to the public and scholars at the American Heritage Center and the UW libraries. Both contain original and unique collections on the history of the park.

Yellowstone National Park provides a natural laboratory where the University of Wyoming engages in state-of-the-art research, exposes students to natural resource challenges, and collaborates with other institutions and organizations to apply scientific knowledge to management of the park. Yellowstone National Park represents one of our greatest achievements in preserving our national heritage, for the benefit and enjoyment of our residents and for people from around the world who come to Wyoming to visit this national treasure.

Tom Buchaan, President

# Supporters of the *Atlas of Yellowstone*

The individuals and institutions listed below have made the creation of the atlas possible. We gratefully acknowledge their contributions.

## Sponsoring Institutions

University of Oregon                    Montana State University
University of Wyoming                   Yellowstone Park Foundation

## Contributing Organizations

Allan Cartography                       Buffalo Bill Historical Center
Headwaters Economics                    Yellowstone Ecological Research Center

## Friends of the *Atlas of Yellowstone*

Individual contributions of $1,000 or more

Dr. and Mrs. John L. Allen
The Anders Foundation
Valerie E. Anders and William A. Anders
Josie Berry
Mila Berry
The Taylor Fithian family
Mark A. Fonstad
William F. Gary and Jane C. Gary
The Gold family
Geoffrey M. Jacquez and Lauren Wagner
William G. Loy Memorial Scholarship Fund in Geography
Ann Gerlinger Lyman and Ronald G. Lyman
W. Andrew Marcus and Lisa N. Marcus
Richard A. Marston
James E. Meacham
Jacqueline J. Shinker and Thomas Minkley
Peter K. Simpson and Lynne L. Simpson
Ross West and Barbara West
Carolyn Miller Younger and Ralph W. Younger

# Acknowledgments

Hundreds of individuals helped make the *Atlas of Yellowstone*. It is impossible in this limited space to adequately acknowledge their contributions of time, resources, and expertise. The *Afterword* at the end of this volume reviews how the atlas was developed over a period of ten years and describes the many ways in which people contributed to its creation; these acknowledgements list the individuals who contributed. Many individuals helped the effort in more than one way, so their names are listed in several sections below. The large majority of participants volunteered their time. Dedicated salaried production staff members put in long hours beyond the standard working day and gave of their talents above and beyond the call of duty.

Names below are listed in alphabetical order within groupings based on the nature of their contributions. Full names are given, but titles are not.

## Contributing Experts

Two or more topic experts typically contributed to each subject covered in the *Atlas of Yellowstone*. Sometimes the editors took the lead on topics within their areas of expertise, but the large majority of topics were developed in collaboration with other scholars. Experts' professional affiliations are not listed below because many have changed their home institutions or retired over the ten-year production period. The large majority (but not all) of the experts listed below were or are presently affiliated with the following: Yellowstone and Grand Teton national parks, the University of Oregon, Montana State University, the University of Wyoming, the Buffalo Bill Historical Center, Headwaters Economics, the Museum of the Rockies, the Yellowstone Ecological Research Center, the U.S. Geological Survey, or the U.S. Forest Service.

The "Sources" section in the back of the atlas lists the specific experts who contributed to each particular topic. In more general terms, scholars who contributed their time to the "Geographic Setting" portion of the atlas were Paul Schullery, John Varley, and William K. Wyckoff. The "Human Geography" portion of the atlas is based on expert guidance from C. Melvin Aikens, Stuart Allan, Bruce A. Blonigen, Sarah Bone, Christine Brindza, Story Clark, Colleen Curry, Brian W. Dippie, Daniel H. Eakin, Dana Fernbach, Anne Fiefield, Elaine Skinner Hale, Susan Hardwick, Christie Hendrix, Donald G. Holtgrieve, Ann Johnson, Lawrence L. Loendorf, Gene Martin, Alexander B. Murphy, Tom Olliff, Zehra Osman, Ray Rasker, Bill Resor, John Sacklin, Paul Schullery, Peter K. Simpson, Rosemary Sucec, Alexander M. Tait, John Varley, James Walker, Rick Wallen , Katie White, Lee H. Whittlesey, and William K. Wyckoff.

Experts contributing to the "Physical Geography" section of the atlas were Miles C. Barger, Patrick J. Bartlein, Peter Bengeyfield, Pat Bigelow, Paul A. Caffrey, Stephan G. Custer, Don G. Despain, Mark A. Fonstad, Daniel G. Gavin, Carrie Guiles, Henry P. Heasler, Mary Hektner, Cheryl Jaworowski, David R. Lageson, Richard A. Marston, Grant A. Meyer, Lisa A. Morgan, James M. Omernik, Greg Pederson, Kenneth L. Pierce, Mitchell J. Power, Roy Renkin, Paul Rubenstein, Sarah L. Shafer, Patrick Shanks, Jacqueline J. Shinker, Ramesh Sivanpillai, Mike Stevens, Mark Story, Erik R. Strandhagen, Kathryn E. Watts, Jennifer Whipple, Cathy Whitlock, and Lee H. Whittlesey.

Scientists who guided the content of the "Wildlife" section were Mark D. Andersen, Gary P. Beauvais, Tami Blackford, Robert L. Crabtree, Mary Ann Franke, Jim Garry, Chris Geremia, Jeffrey K. Gillan, Kerry Guenther, William P. Inskeep, Lynn Kaeding, Todd M. Koel, Timothy R. McDermott, Mary Meagher, Glenn Plumb, Charles R. Preston, James G. Schmidt, Charles C. Schwartz, J.W. Sheldon, Douglas W. Smith, Cristina Takacs-Vesbach, Kelli Trujillo, Rick Wallen, David M. Ward, Molly Ward, and P.J. White.

Stuart Allan provided text and cartographic oversight for maps in the "Reference" section, all of which were proofed by staff members at Yellowstone and Grand Teton national parks. Lee H. Whittlesey

provided the text for the place-name descriptions.

Experts often led us to data sets that required access to and guidance on use of Yellowstone geospatial data sets. Geographic information system (GIS) specialists at various agencies who provided data to the atlas project were Agnes Badin de Montjoie, Steven M. Bassett, Sarah Bone, Michela Bongiorno, Chiara Dorati, Elizabeth D. Fano, Jeffrey Gillan, Allison Klein, Eric J. Meyer, Julie Rose, Shannon Savage, Hilary Smith, and Brandt Winkelman. We especially wish to thank Carrie Guiles, who—from the very inception of the project—contributed an immense amount of time to compiling GIS data sets for the atlas.

## Cartography and Editing

A veritable army of cartographers contributed materials to this atlas. In the first two years of the project, geography graduate student Erik R. Strandhagen played a major role in developing page pairs. A large portion of the cartographic content was finished in the final two years of the project when student cartographers Miles C. Barger and Eric R. Stipe and staff cartographer Steven M. Bassett worked under production manager Alethea Y. Steingisser. We are immensely indebted to them for the quality of their work, for their willingness to suffer the slings and arrows of constant artistic and scientific critique, and for their good cheer in working under tight deadlines.

The *Atlas of Yellowstone* benefited from guidance received from some of the world's top cartographers. Stuart Allan was the project's cartographic mentor from its inception, providing insights on issues ranging from overall content and topic sequences to the critically important minutiae of labeling features on maps, as well as being the lead expert on several topics. Tom Patterson's guidance, particularly with regard to shaded relief representation, significantly improved many pages in the atlas. The atlas also benefited from conversations with Neil H. Allen, Lawrence J. Andreas, Gene Martin, and William McNulty. Alexander M. Tait took the lead on producing graphics and content for the dramatic "Tetons Climbing History" pages. We are indebted to many of our colleagues in the North American Cartographic Information Society (NACIS) for their critical input on the atlas as it developed over the years.

The original idea for the atlas derived from a class project in Advanced Cartography, taught by James E. Meacham in the Department of Geography at the University of Oregon in the winter quarter of 2004. Each student's final class project consisted of a topic page pair for the then hypothetical *Atlas of Yellowstone*. Student projects on climate, Yellowstone park geology, transportation, glaciers, vegetation, and wolves later evolved into content included in the present atlas. Undergraduates students in the class were Jacob M. Blair, Grace A. Burgwyn, Tory G. Caputo, Andrew R. Case, Justin R. Cooley, Berend L. Diderich, Joseph A. Fagliano, Bowen A. Garner, Michael J. Geffel, Jordan F. Glubka, Jay P. Grayson, Bonnie L. Hoskinson-Wiebe, Derek B. Kellenbeck, Rita L. Pick, Shun Ho Sham, and Justin T. Thomas. Graduate students in the class were Jonathan W. Day, Adrianna L. Hirtler, Jonathon L. McConnel, and Alethea Y. Steingisser.

Many University of Oregon students who learned about the atlas through this first course or subsequent Advanced Cartography courses went on to work on the project as paid cartographers in the InfoGraphics Lab, Department of Geography. Students who took the course as undergraduates and later worked on the project included Steven M. Bassett, Brook S. Eastman, Julia J. Giebultowicz, Jesse L. Nett, Kristen M. Phelan, Eric R. Stipe, and Lauren F. Thompson. Graduate students who worked on the atlas in the InfoGraphics Lab were Miles C. Barger, Matthew Derrick, Nick Martinelli, Jonathon L. McConnel, Benjamin W. Metcalfe, Alethea Y. Steingisser, and Erik R. Strandhagen. All of the students above made significant contributions to the atlas content.

The constant procession of editors, experts, and cartographers through the InfoGraphics Lab meant that at one time or another every employee of the lab was conscripted to provide expert advice on handing geospatial data, software use, or logistics. We appreciate their patience with our interruptions and their willingness to help with problems, both big and small. In addition to the cartographers already listed above, the members of the InfoGraphics Lab staff were Blake E. Andrew, Jacob D. Bartruff, Michael Engelmann, Kenneth S. Kato, Nicholas P. Kohler, and Erik B. Steiner.

The 1:100,000 and 1:500,000 reference maps and the gazetteer at the back of the atlas were produced by Allan Cartography. Employees of Allan Cartography who worked on the atlas were Stuart Allan, Neil H. Allen, Lawrence J. Andreas, Thaddeus A. Lenker, and Eric J. Meyer.

Ross West edited all text throughout the atlas to ensure consistency of voice and style; he also identified issues with graphics that were near completion. Barbara West provided invaluable assistance in this work. Scott Skelton proofread the entire atlas with coordinating assistance from John Crosiar. We are indebted to them for the meticulous attention and devotion to quality they brought to the editing process. We are equally grateful for their good humor in receiving so many materials at the last minute in the face of tight deadlines—their graciousness in these circumstances made our work far more pleasurable.

## Support

The willingness of administrative national park staff to work with us and provide information was essential to completing the atlas. In addition to their contributions to individual topics that we note above and at the back of the atlas, we wish to extend our thanks to the following individuals for making us welcome within the hallways of the national park system: Lisa Baril, Pat Bigelow, Tami Blackford, Rob Daley, Chris Geremia, Mary Hektner, Christie Hendrix, Kathy Mellander, Tom Olliff, Tom Patterson, Glenn Plumb, Lindsay Robb, John Sacklin, John Varley, and Sue Wolff.

Our advisory board of Mark Berry, Charles R. Preston, Bill Resor, Peter K. Simpson, and John Varley provided sage advice on fundraising and atlas content, as well as hosting us in their homes and introducing us to their worlds of Yellowstone. We look forward to more such visits in the future. Jennifer Slater and Jim Slater introduced us to board member Simpson—we thank them for opening such a welcoming door for us.

At the University of Oregon, we wish to thank Andrea Heid, Sandra Knauber, Mary Milo, and Sonja Anthone for their unflagging willingness to process employee contracts, handle accounting, and help with our many travel plans and bills. Publishing, marketing, developing a publishing contract, and fundraising all required institutional support—guidance on these topics was provided by Sarah Bungum, Guy Maynard, Chris Michel, Colin Miller, Susan Thelen, and Kathrin Walsch. We wish to especially thank Charles R. Williams for his help in developing a contract with the University of California Press, and Jane C. Gary, whose friendship, endless belief in the atlas project, and fundraising expertise helped support both the atlas and the morale of the atlas team.

The atlas never would have launched without strong internal funding from the University of Oregon. This initial funding, as well as support for the duration of the atlas project, was made possible because of the stalwart support from many members of the University of Oregon's senior administration, including Scott Coltrane, Lorraine Davis, Frances Dyke, Rich Linton, Marianne Nicols, Larry Singell, and Priscilla Southwell, who provided the first seed grant for the atlas through the College of Arts and Sciences. We also wish to thank David M. Dooley and John Varley at Montana State University and Chamois Lynnette Andersen, Ingrid C. Burke, Jeff Hamerlinck, and Jacqueline J. Shinker at the University of Wyoming, all of whom championed the cause of their home institutions providing funding for the *Atlas of Yellowstone*. Funding from these three universities and a Canon U.S.A.

grant funded through the Yellowstone Park Foundation provided the core financing necessary to construct the atlas.

Contributions from donors have also made a major difference in our ability to compile data, pay cartographers, and take the time necessary to produce high-quality maps. We extend our thanks to John L. Allen, Mrs. John L. Allen, the Anders Foundation of Valerie E. Anders and William A. Anders, Josie Berry, Mila Berry, Barbara Bailey Bergeron and Lester Bergeron, Madeleine and Robert Caufield, the Taylor Fithian Family, Mark A. Fonstad, William F. Gary and Jane C. Gary, Diane Opdenweyer Hazen and Herbert Hazen, Geoffrey M. Jacquez, the William G. Loy Memorial Scholarship Fund in Geography, Ann Gerlinger Lyman and Ronald G. Lyman, W. Andrew Marcus and Lisa N. Marcus, Richard A. Marston, James E. Meacham, Jacqueline J. Shinker and Thomas Minkley, Peter K. Simpson and Lynne L. Simpson, James Stembridge Jr., Lauren Wagner, James Walker, Ross and Barbara West, and Carolyn Miller Younger and Ralph W. Younger.

It was our pleasure to work closely with several nonprofit institutions that provided immense amounts of expertise, data, and other in-kind support. For their belief in this project and their willingness to commit time and energy to it, we are indebted to the Buffalo Bill Historical Center, Headwaters Economics, and the Yellowstone Ecological Research Center.

Dan Kaveney provided critical advice to us in our early phases of seeking a publisher. We are fortunate in having the University of California Press as our publisher for this volume. Their tremendous experience in and commitment to publishing atlases has provided us a safe haven during a turbulent time in the world of print books. We are grateful to Lester Rowntree for first introducing us to UC Press and to Anthony D. Barnosky and Richard A. Marston who reviewed the atlas at an early stage of completion. We also owe thanks to Jenny Wapner, who set up the publishing contract, and to Kim Robinson, who has shepherded us through the review and prepress process that has allowed us to reach the point of publication.

We are deeply indebted to all of the individuals and institutions listed above. This atlas exists because of their generosity of spirit, their belief in the power of explaining the world through maps, and their commitment to telling the many tales of Yellowstone.

*W. Andrew Marcus*
*James E. Meacham*
*Ann W. Rodman*
*Alethea Y. Steingisser*

*June 30, 2011*

# Preface

## Why create an *Atlas of Yellowstone*?

In the early 1970s, famed British author and commentator Alistair Cooke was asked his views on the United States. He had traveled the country for two years and had put together a series of television shows called, simply, *America*. What he found left many people, including myself, disturbed.

He observed that it is almost impossible to instantly identify a place upon exiting an airplane in America—for what you see is just miles of motels, second-hand car lots, hamburger stands, refuse, and litter. In short, what he discovered almost fifty years ago was that any individuality of place had disappeared and every place looked alike. He too was disturbed and made the point that it's not only a question of recovering what is lost but it's also a question of saving what is left. Reflecting, we could rightfully ask, "So how are we doing; is more being lost and anything being saved?" The *Atlas of Yellowstone* is a unique attempt to begin anew answering these questions and restarting this important dialogue. Happily, our readers will have to make their own judgments and draw their own conclusions.

I don't know if Alistair Cooke had the opportunity to visit America's national parks and other significant wildlands during his two-year tour. I argue that if he had left his airplane at the gate of, say, one of our national parks he could very well have reached the opposite conclusion. Clearly, these areas are distinctively American, and, by definition and law, unique. Each has beauty and character beyond comparison. I suggest it is our nation's wildlands where visitors—foreign and domestic—most readily and dramatically encounter and are enriched by the qualities of our nation.

How very fortunate we are that we have places like the Greater Yellowstone with its mix of publicly owned park, forest, and refuge lands that impart such powerful feelings to its visitors. For one proof of this, if indeed any is needed, we can turn to economic science and conclude that over time, for at least a century and a half, visitors continue to be drawn to the Greater Yellowstone in ever greater numbers. By economic benchmarks, this supports the contention that people are attracted to beauty, character, and uniqueness.

Naturally, regional businesspeople love this demographic trend and have established a myriad of goods and services to take advantage of the cash cow the Greater Yellowstone provides. Estimates of the value of the annual visitor onslaught vary widely but generally have a range of $1–4 billion annually, an impressive sum by almost any standard.

Students of the Greater Yellowstone economy marvel about several aspects of this reality because it is reminiscent of the old fable about the goose that laid the golden eggs. One such facet is that the original price tag of this financial wonder was minuscule because the parks and forests were simply being moved from the federally owned public domain to federal parks, forests, and refuges. Later, additional lands were added to the Greater Yellowstone estate by private donations or purchased with monies from the Land and Water Conservation Fund, which is supported largely by royalties paid to the federal government from oil and gas leases on public lands. Of course, the people also have to sustain annual funding to manage and maintain these lands, and these funds are paid for by the U.S. Department of the Treasury along with user and concession fees.

These publicly available figures invite financial effectiveness valuations, the best known of which is the cost-benefit analysis. At its simplest, and because the costs of originating the Greater Yellowstone were minimal, the annual upkeep and management costs of the government units of the Greater Yellowstone form the significant costs to society. In recent times, I estimate these costs are around $300 million per year, which are far, far less than the estimated $1–4 billion of annual benefits. Thus, it appears that the Greater Yellowstone has a very attractive cost-benefit ratio, is an excellent value for the taxpayer-owners, has obviously been sustainable in the past, and should continue for the foreseeable future. Fans of the Greater Yellowstone should be cautiously comfortable with this news because financial sustainability has been and remains a vital concept in current dialogues over the role of government in our society.

As important as a favorable cost-benefit ratio and sustainable operations are to politicians and resource managers, they certainly aren't the only measures that are important. A wise wildland scientist once said to me, "The only thing the national parks and other wildlands offer humans is education, inspiration, and wonder." I am pretty sure he used the word only as a tongue-in-cheek remark, for what else of any importance is there? It's a wonderful ideal for any land or resource manager to have and fairly easy for the public to embrace.

For many of us, knowledge is the principle that helps feed, stimulate, and nurture education, inspiration, and wonder. It is also generally acknowledged that science produces the finest knowledge. I think we all know deep within ourselves that knowledge is inherently valuable to humankind, but it is also exceedingly difficult to try to value.

Throughout its history, the Greater Yellowstone has produced a wealth of fine scientific knowledge, and in fact it is one of the few large wildlands of the world that enjoys the lofty status of being data-rich. But being rich in scientific data doesn't mean that the information is accessible to anyone besides other scientists, so a small cottage industry has started up in the Greater Yellowstone to help make that knowledge accessible. The *Atlas of Yellowstone* is surely the finest manifestation of scientific translation produced thus far by this movement because a worthy atlas requires a data-rich environment to even be contemplated. Add to that a sizable cadre of able science translators to produce the atlas and the result is a synthesis equally useful to the public and scientists alike.

This allows us to be more sensitive to the multiple perspectives that the visitors to and people of the Greater Yellowstone bring to all conversations around the heritage and resources of this region. Often those conversations become acrimonious because of divergent understanding regarding past events. The *Atlas of Yellowstone* provides an important tool for moderating those disagreements by providing a baseline of data from which conversations can logically proceed. Indeed, some of the work from the atlas, such as the first-time maps of changes in bison migrations over time, have already been used by citizens and resource managers to provide a common basis for discussions at public meetings and before Congress.

Much of the value of the atlas will derive from the methods that the editors are following to develop its content. They have strived to reach out to many different groups in the Yellowstone region, including the National Park Service and other federal agencies, major museums and universities, and citizens of the region. The atlas content is thus as wide-ranging as the people doing research in the region. At the same time, limiting each story (for example, the history of fires in Yellowstone) to a pair of pages maintains a tight focus on each story line and avoids the kind of mishmash that can result if too many voices speak at once. The tight focus and quality assurance are further maintained by the assignment of two experts to each story, insuring that the reporting and data meet high scholarly standards.

I happened to have the good fortune to spend more than three decades in the science business of Yellowstone National Park. My manna was Yellowstone's natural world, and its delivery vehicle was discovery science. I admit to always considering myself a mediocre scientist, but Greater Yellowstone had around 300 to 400 annually permitted researchers, many of them truly outstanding scientists, thus I compensated my shortcomings by talking with them regularly and in depth. In the field or over lunch, each of these scientists could tell an interesting, even exciting story about their work, but for a variety of good and not-so-good reasons, not often in their scientific papers. For science junkies like me, this was not only sad but it also deprived park managers of information they needed to manage well. Perhaps more important, it also deprived a substantial portion of the public the ability to be better educated about the Greater Yellowstone; it left them less inspired and disadvantaged many of an enhanced sense of wonder.

I am confident and pleased that publishing the *Atlas of Yellowstone* corrects a longstanding deficiency in science translation and synthesis, and that all of us who enjoy perusing fabulous maps and poring over charts and graphs can anticipate hours of enjoyment and edification ahead of us.

John Varley

# Geographic Setting

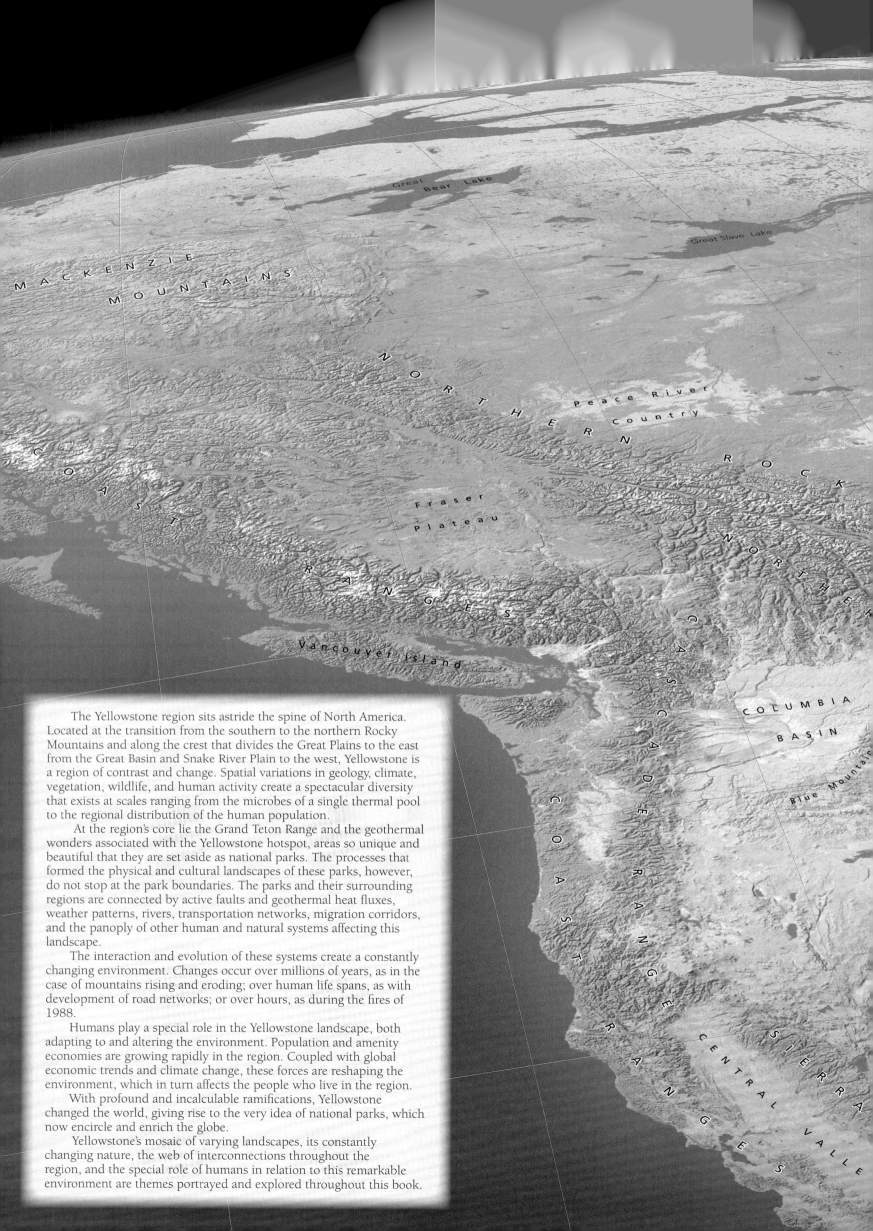

The Yellowstone region sits astride the spine of North America. Located at the transition from the southern to the northern Rocky Mountains and along the crest that divides the Great Plains to the east from the Great Basin and Snake River Plain to the west, Yellowstone is a region of contrast and change. Spatial variations in geology, climate, vegetation, wildlife, and human activity create a spectacular diversity that exists at scales ranging from the microbes of a single thermal pool to the regional distribution of the human population.

At the region's core lie the Grand Teton Range and the geothermal wonders associated with the Yellowstone hotspot, areas so unique and beautiful that they are set aside as national parks. The processes that formed the physical and cultural landscapes of these parks, however, do not stop at the park boundaries. The parks and their surrounding regions are connected by active faults and geothermal heat fluxes, weather patterns, rivers, transportation networks, migration corridors, and the panoply of other human and natural systems affecting this landscape.

The interaction and evolution of these systems create a constantly changing environment. Changes occur over millions of years, as in the case of mountains rising and eroding; over human life spans, as with development of road networks; or over hours, as during the fires of 1988.

Humans play a special role in the Yellowstone landscape, both adapting to and altering the environment. Population and amenity economies are growing rapidly in the region. Coupled with global economic trends and climate change, these forces are reshaping the environment, which in turn affects the people who live in the region.

With profound and incalculable ramifications, Yellowstone changed the world, giving rise to the very idea of national parks, which now encircle and enrich the globe.

Yellowstone's mosaic of varying landscapes, its constantly changing nature, the web of interconnections throughout the region, and the special role of humans in relation to this remarkable environment are themes portrayed and explored throughout this book.

The *Atlas of Yellowstone* tells many stories about Yellowstone across many different scales. At global extents, the tale of the first national park—Yellowstone—and its influence on the creation of other national parks requires maps of the entire Earth. At regional extents, maps follow the track of the Yellowstone hotspot from Oregon and Nevada along the Snake River Plain into Wyoming. In contrast, some stories require zooming to microscopic perspectives, as with the display of thermophiles, the microbes that inhabit specific portions of individual hot springs.

The location and scale of maps in this atlas are determined by the information necessary to tell a story *and* by available expertise and geographic data. These two requirements lead to some common map extents, such as one that displays Yellowstone and Grand Teton national parks and the surrounding national forests. This region captures a range of common ecosystem characteristics and is an area for which data are available from federal agencies. It is also one of the various ways in which the Greater Yellowstone Area is defined. Alternatively, economic maps often display the twenty counties around Yellowstone National Park. Key economic data are collected at the county scale, making this a useful extent and resolution for displaying the interaction of the park and surrounding economies.

The variations within Yellowstone and its connections to many different parts of the world make it impossible to portray its varied tales with a single regional extent. The many maps, images, and stories in this atlas do share one element in common, however—they are all connected to the region's heart, Yellowstone and Grand Teton national parks.

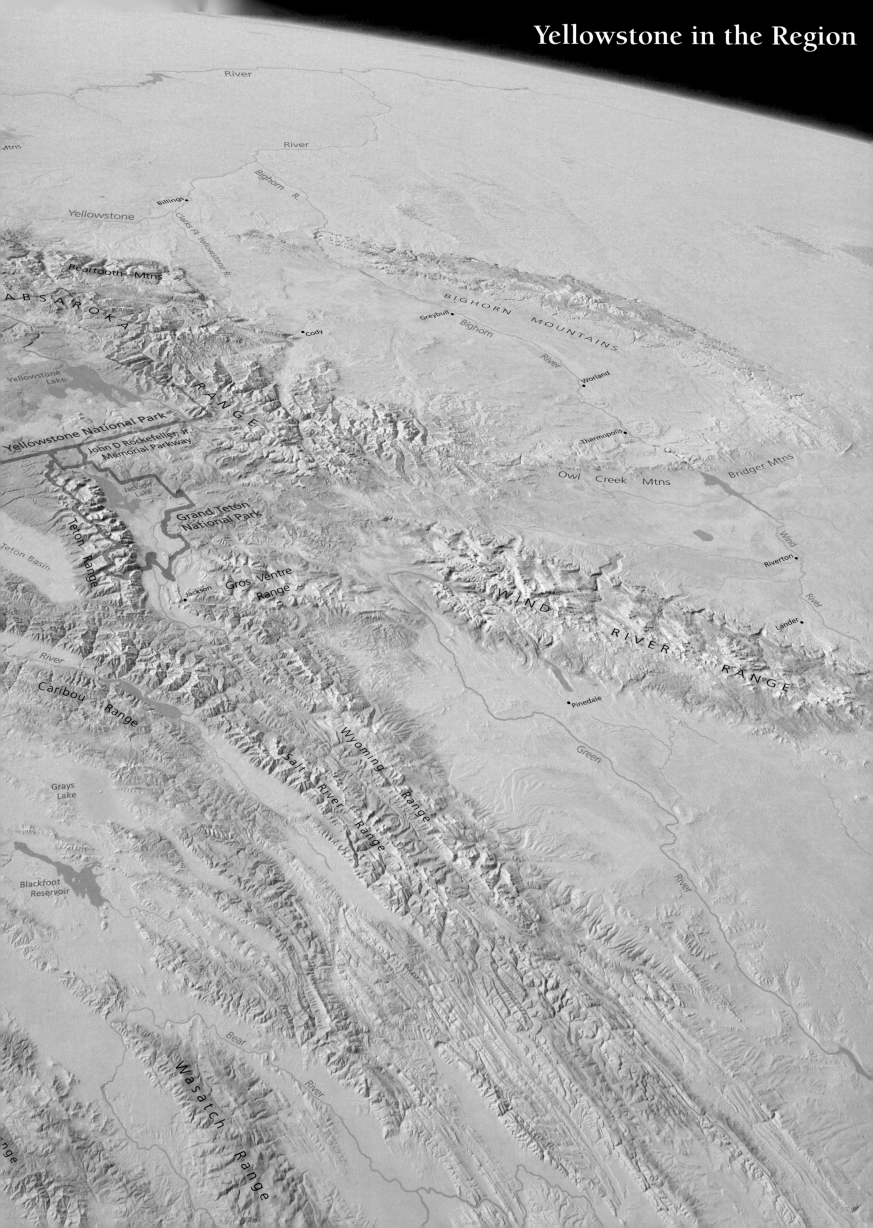

# Greater Yellowstone Detail

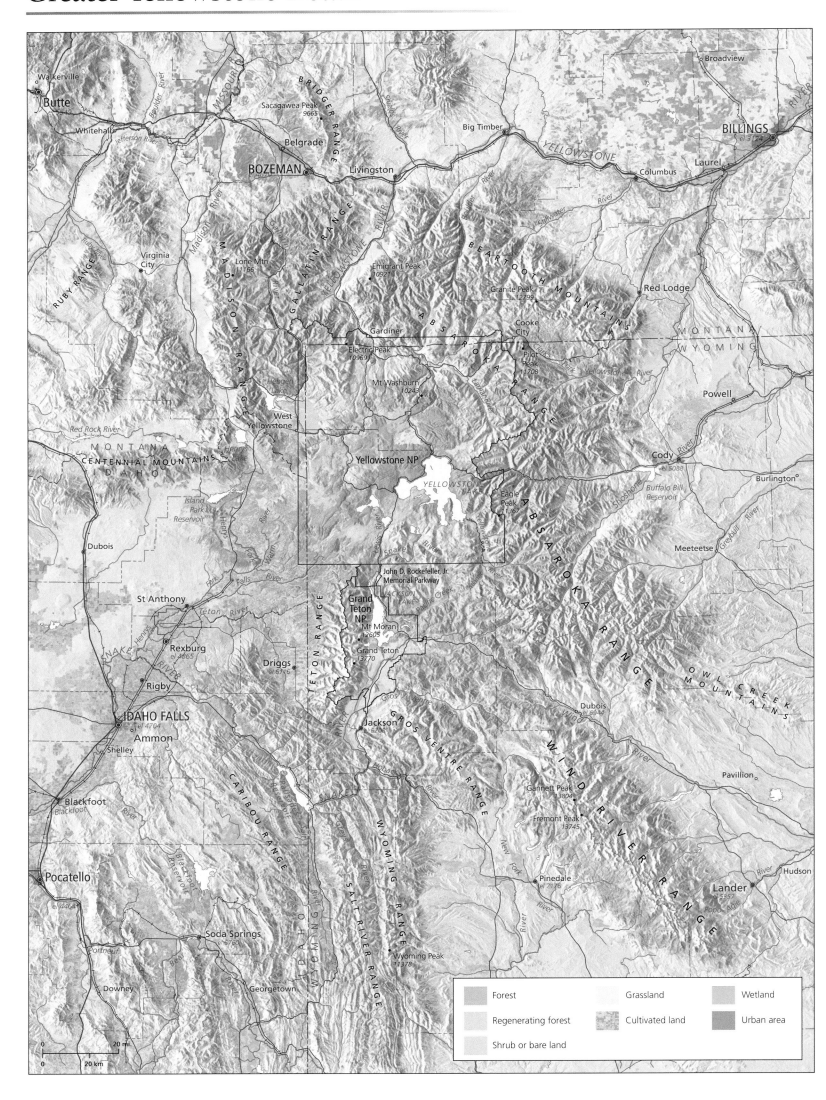

Walkerville
Butte
el 5548
Whitehall
Jefferson River
Boulder River
MISSOURI River
BRIDGER RANGE
Sacagawea Peak
9665
Big Timber
Broadview
YELLOWSTONE RIVER
BILLINGS
el 3124
Belgrade
BOZEMAN
Livingston
Shields River
Laurel
Columbus
Stillwater River
Madison River
GALLATIN RANGE
YELLOWSTONE RIVER
BEARTOOTH MOUNTAINS
Virginia City
RUBY RANGE
MADISON RANGE
Lone Mtn
11166
Emigrant Peak
10921
Granite Peak
12799
Red Lodge
Gardiner
ABSAROKA RANGE
Cooke City
MONTANA
WYOMING
Red Rock River
Hebgen Lake
West Yellowstone
Electric Peak
10969
Mt Washburn
10243
Pilot Peak
11708
Clarks Fork
Yellowstone River
Powell
MONTANA
IDAHO
CENTENNIAL MOUNTAINS
Henrys Lake
Yellowstone NP
Lamar River
Cody
el 5088
Burlington
Island Park Reservoir
Dubois
Henrys Fork
Falls River
Warm River
YELLOWSTONE LAKE
Eagle Peak
11367
ABSAROKA RANGE
Shoshone River
Buffalo Bill Reservoir
Meeteetse
Greybull River
St Anthony
Teton River
TETON RANGE
Lewis River
Snake River
John D. Rockefeller, Jr. Memorial Parkway
Grand Teton NP
Jackson Lake
Pacific Creek
OWL CREEK MOUNTAINS
Rexburg
el 4865
Driggs
el 6116
Mt Moran
12605
Grand Teton
13770
SNAKE RIVER
Rigby
Gros Ventre River
Dubois
el 6944
Wind River
IDAHO FALLS
el 4704
Ammon
Jackson
el 6209
GROS VENTRE RANGE
Shelley
Hoback River
Pavillion
CARIBOU RANGE
WYOMING RANGE
Gannett Peak
13804
Fremont Peak
13745
WIND RIVER RANGE
Blackfoot
Blackfoot River
Greys River
New Fork River
Pocatello
el 4464
Blackfoot Reservoir
SALT RIVER RANGE
Pinedale
el 7176
Lander
el 5357
Hudson
Soda Springs
el 5760
IDAHO WYOMING
Popo Agie
Portneuf River
Downey
Georgetown
Bear River
Wyoming Peak
11378

| | Forest | | Grassland | | Wetland |
|---|---|---|---|---|---|
| | Regenerating forest | | Cultivated land | | Urban area |
| | Shrub or bare land | | | | |

0        20 mi.
0    20 km

## Cody, Wyoming

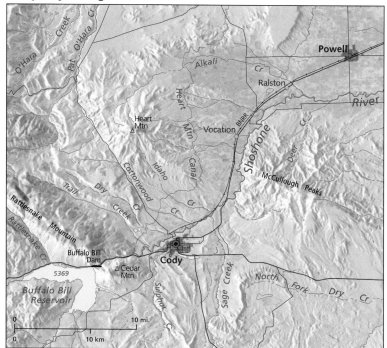

## Jackson, Wyoming

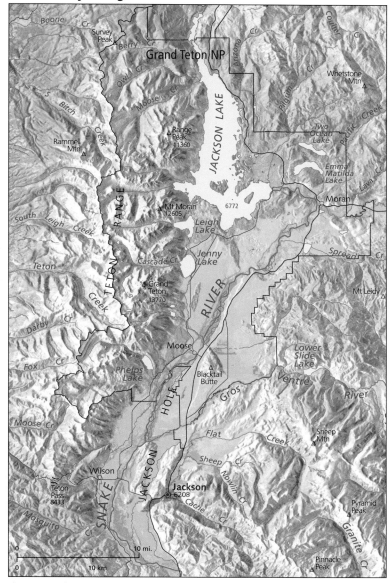

## Bozeman, Livingston, and Gardiner, Montana

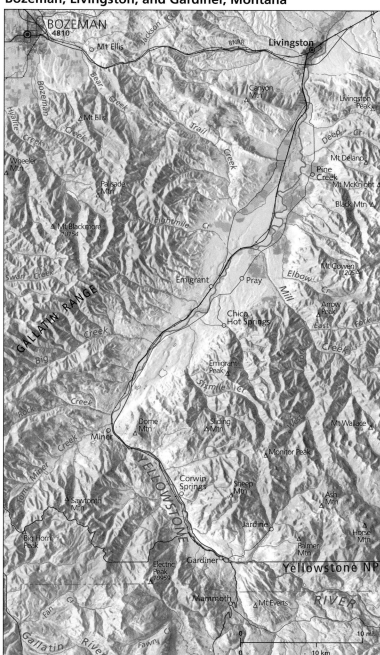

## West Yellowstone, Montana

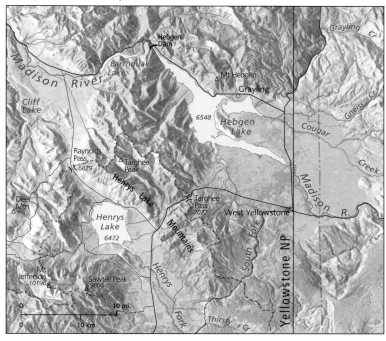

The maps on these pages are derived from the 2001 National Land Cover Database (NLCD). This data set was developed by a consortium of nine federal agencies and is housed at the U.S. Geological Survey. The maps represent classifications based on Landsat satellite imagery (thirty-meter resolution) from three seasons combined with data on slope, aspect, and slope position for each ground location. The NLCD has twenty-nine classes of land cover; these have been generalized here into six categories. These divisions accurately portray overall patterns of land cover, although the map may be incorrect at individual sites.

# The World's First National Park

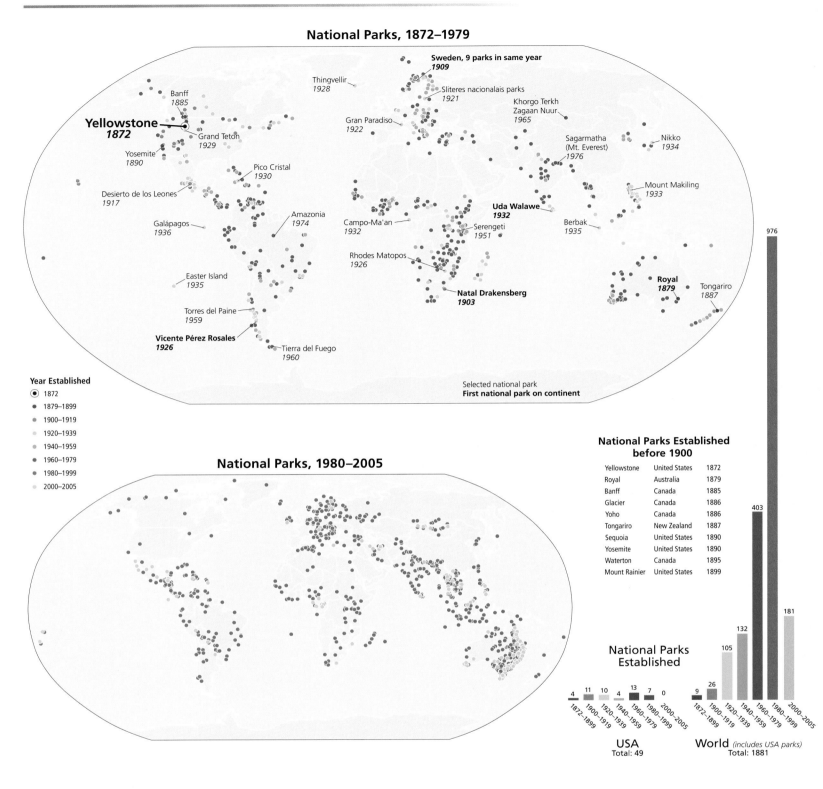

**National Parks, 1872–1979**

**National Parks, 1980–2005**

National Parks Established before 1900

| Yellowstone | United States | 1872 |
|---|---|---|
| Royal | Australia | 1879 |
| Banff | Canada | 1885 |
| Glacier | Canada | 1886 |
| Yoho | Canada | 1886 |
| Tongariro | New Zealand | 1887 |
| Sequoia | United States | 1890 |
| Yosemite | United States | 1890 |
| Waterton | Canada | 1895 |
| Mount Rainier | United States | 1899 |

National Parks Established

USA Total: 49

World (includes USA parks) Total: 1881

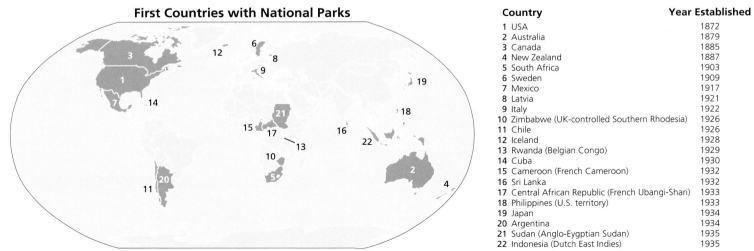

**First Countries with National Parks**

| Country | Year Established |
|---|---|
| 1 USA | 1872 |
| 2 Australia | 1879 |
| 3 Canada | 1885 |
| 4 New Zealand | 1887 |
| 5 South Africa | 1903 |
| 6 Sweden | 1909 |
| 7 Mexico | 1917 |
| 8 Latvia | 1921 |
| 9 Italy | 1922 |
| 10 Zimbabwe (UK-controlled Southern Rhodesia) | 1926 |
| 11 Chile | 1926 |
| 12 Iceland | 1928 |
| 13 Rwanda (Belgian Congo) | 1929 |
| 14 Cuba | 1930 |
| 15 Cameroon (French Cameroon) | 1932 |
| 16 Sri Lanka | 1932 |
| 17 Central African Republic (French Ubangi-Shari) | 1933 |
| 18 Philippines (U.S. territory) | 1933 |
| 19 Japan | 1934 |
| 20 Argentina | 1934 |
| 21 Sudan (Anglo-Eygptian Sudan) | 1935 |
| 22 Indonesia (Dutch East Indies) | 1935 |

The international press took little notice when the U.S. Congress created the world's first national park in 1872—nor was there much fanfare in the United States. But despite lengthy growing pains, the national park movement flourished beyond the imaginations of its founders. By 2010, 58 national parks, plus an additional 335 national monuments, seashores, battlefields, and other special sites, celebrated the American landscape and experience.

The 1872 Yellowstone Act called for basic land and resource preservation and the creation of "a public park or pleasuring-ground for the benefit and enjoyment of the people." The Organic Act of 1916 established the National Park Service and required that national parks, monuments, and reservations be preserved "unimpaired for the enjoyment of future generations." This radical notion of perpetual conservation posed extraordinary challenges for preserve supervisors (since no templates for this kind of management existed) and forced park administrators to learn by doing. Today, park management continues to evolve and remains a hopeful blend of art, science, and politics. Still, as American writer and conservationist Wallace Stegner observed, "National parks are the best idea we ever had. Absolutely American, absolutely democratic, they reflect us at our best." The worldwide popularity of national parks buttresses Stegner's assertion. Today 180 countries have nearly 1,900 parks or equivalent reserves.

## U.S. National Parks

| Park Name | Year Established | Acres 2009 | Visitors 2009 | First Protected as | Year |
|---|---|---|---|---|---|
| 1 Yellowstone National Park | 1872 | 2,219,791 | 3,295,187 | Same | 1872 |
| 2 Sequoia and King's Canyon National Parks* | 1890/1940 | 859,901 | 1,574,466 | Sequoia and General Grant National Parks | 1890 |
| 3 Yosemite National Park | 1890 | 761,267 | 3,737,472 | Federal grant to California | 1864 |
| 4 Mount Rainier National Park | 1899 | 236,381 | 1,151,654 | Same | 1899 |
| 5 Crater Lake National Park | 1902 | 183,224 | 446,516 | Same | 1902 |
| 6 Wind Cave National Park | 1903 | 28,295 | 587,868 | Same | 1903 |
| 7 Mesa Verde National Park | 1906 | 52,485 | 550,377 | Same | 1906 |
| 8 Glacier National Park | 1910 | 1,013,322 | 2,031,348 | Same | 1910 |
| 9 Rocky Mountain National Park | 1915 | 265,758 | 2,822,325 | Same | 1915 |
| 10 Hawaii Volcanoes National Park | 1916 | 323,431 | 1,233,105 | Same | 1916 |
| 11 Lassen Volcanic National Park | 1916 | 106,372 | 365,639 | National Monument | 1907 |
| 12 Denali National Park & Preserve | 1917 | 6,075,030 | 358,041 | Mt. McKinley National Park | 1917 |
| 13 Acadia National Park | 1919 | 47,389 | 2,227,698 | Sieur de Monts National Monument | 1916 |
| 14 Grand Canyon National Park | 1919 | 1,217,403 | 4,348,068 | National Monument | 1908 |
| 15 Zion National Park | 1919 | 146,597 | 2,735,402 | Mukuntuweap National Monument | 1909 |
| 16 Hot Springs National Park | 1921 | 5,550 | 1,284,707 | Reservation | 1832 |
| 17 Shenandoah National Park | 1926 | 199,099 | 1,120,981 | Same | 1926 |
| 18 Bryce Canyon National Park | 1928 | 35,835 | 1,216,377 | National Monument | 1923 |
| 19 Grand Teton National Park | 1929 | 310,044 | 2,580,081 | Same | 1929 |
| 20 Carlsbad Caverns National Park | 1930 | 46,766 | 432,639 | National Monument | 1923 |
| 21 Great Smoky Mountains National Park | 1934 | 522,051 | 9,491,437 | Same | 1934 |
| 22 Mammoth Cave National Park | 1934 | 52,830 | 503,856 | Same | 1941 |
| 23 Olympic National Park | 1938 | 922,650 | 3,276,459 | Same | 1938 |
| 24 Isle Royale National Park | 1940 | 571,790 | 14,653 | Same | 1940 |
| 25 Big Bend National Park | 1944 | 801,163 | 363,905 | Texas Canyons State Park | 1933 |
| 26 Everglades National Park | 1947 | 1,509,127 | 900,882 | Same | 1947 |
| 27 Virgin Islands National Park | 1956 | 14,737 | 415,847 | Same | 1956 |
| 28 Haleakala National Park | 1961 | 33,222 | 1,109,104 | Hawaii National Park-Haleakala Section | 1916 |
| 29 Petrified Forest National Park | 1962 | 221,621 | 631,613 | National Monument | 1906 |
| 30 Canyonlands National Park | 1964 | 337,598 | 436,241 | Same | 1964 |
| 31 Guadalupe Mountains National Park | 1966 | 86,416 | 198,882 | Same | 1966 |
| 32 North Cascades National Park | 1968 | 504,781 | 26,972 | Same | 1968 |
| 33 Redwood National Park | 1968 | 112,582 | 444,426 | Same | 1968 |
| 34 Arches National Park | 1971 | 76,679 | 996,312 | National Monument | 1929 |
| 35 Capitol Reef National Park | 1971 | 241,904 | 617,208 | National Monument | 1937 |
| 36 Voyageurs National Park | 1971 | 218,210 | 222,429 | Same | 1971 |
| 37 Badlands National Park | 1978 | 242,756 | 933,918 | National Monument | 1939 |
| 38 Theodore Roosevelt National Park | 1978 | 70,447 | 586,928 | National Memorial Park | 1947 |
| 39 Biscayne National Park | 1980 | 172,971 | 437,745 | National Monument | 1968 |
| 40 Channel Islands National Park | 1980 | 249,561 | 348,745 | National Monument | 1938 |
| 41 Gates of the Arctic National Park & Preserve | 1980 | 8,472,506 | 9,975 | National Monument | 1978 |
| 42 Glacier Bay National Park & Preserve | 1980 | 3,281,789 | 438,361 | National Monument | 1925 |
| 43 Katmai National Park & Preserve | 1980 | 4,093,078 | 43,035 | National Monument | 1918 |
| 44 Kenai Fjords National Park | 1980 | 669,984 | 218,358 | Same | 1980 |
| 45 Kobuk Valley National Park | 1980 | 1,750,717 | 1,879 | National Monument | 1978 |
| 46 Lake Clark National Park & Preserve | 1980 | 4,030,025 | 9,711 | National Monument | 1978 |
| 47 Wrangell-St Elias National Park & Preserve | 1980 | 13,175,901 | 59,966 | National Monument | 1978 |
| 48 Great Basin National Park | 1986 | 77,180 | 84,974 | National Monument | 1922 |
| 49 National Park of American Samoa | 1988 | 10,500 | 3,242 | Same | 1988 |
| 50 Dry Tortugas National Park | 1992 | 64,701 | 52,011 | Fort Jefferson National Monument | 1935 |
| 51 Death Valley National Park | 1994 | 3,373,042 | 828,574 | National Monument | 1933 |
| 52 Joshua Tree National Park | 1994 | 790,636 | 1,304,471 | National Monument | 1936 |
| 53 Saguaro National Park | 1994 | 91,440 | 665,234 | National Monument | 1933 |
| 54 Black Canyon of the Gunnison National Park | 1999 | 30,750 | 178,495 | National Monument | 1933 |
| 55 Cuyahoga Valley National Park | 2000 | 32,855 | 2,589,288 | National Recreation Area | 1974 |
| 56 Great Sand Dunes National Park & Preserve | 2000 | 85,932 | 289,955 | National Monument | 1932 |
| 57 Wolf Trap National Park for the Performing Arts | 2002 | 130 | 466,752 | Wolf Trap Farm Park for the Performing Arts | 1966 |
| 58 Congaree National Park | 2003 | 26,546 | 122,970 | National Monument | 1976 |

*King's Canyon was designated in 1940 and added onto Sequoia National Park

**Protected Areas of the United States**

■ National Parks
☐ All other federally protected areas
(includes all other public protected land areas such as BLM and USFS lands and non-national park National Park Service units)

North Cascades 1968
Olympic 1938
Mount Rainier 1899
Glacier 1910
Voyageurs 1971
Isle Royale 1940
Acadia 1919
Theodore Roosevelt 1978
Yellowstone 1872
Crater Lake 1902
Grand Teton 1929
Redwood 1968
Wind Cave 1903
Badlands 1978
Lassen Volcanic 1916
Rocky Mountain 1915
Cuyahoga Valley 2000
Yosemite 1890
Great Basin 1986
Arches 1971
Black Canyon of the Gunnison 1999
Wolf Trap 2002
Shenandoah 1926
Capitol Reef 1971
Sequoia & Kings Canyon 1890
Zion 1919
Canyonlands 1964
Great Sand Dunes 2000
Death Valley 1994
Bryce Canyon 1928
Mesa Verde 1906
Mammoth Cave 1934
Channel Islands 1980
Grand Canyon 1919
Great Smoky Mountains 1934
Joshua Tree 1994
Petrified Forest 1962
Congaree 2003
Saguaro 1994
Hot Springs 1921
Carlsbad Caverns 1930
Guadalupe Mountains 1966
Big Bend 1944
Biscayne 1980
Everglades 1947
Dry Tortugas 1992

ALASKA
Kobuk Valley 1980
Gates of the Arctic 1980
Denali 1917
Wrangell-St. Elias 1980
Lake Clark 1980
Katmai 1980
Kenai Fjords 1980
Glacier Bay 1980
BERING SEA
GULF OF ALASKA
PACIFIC OCEAN
400 miles
400 kilometers

HAWAII
PACIFIC OCEAN
MAUI
Haleakala 1916
HAWAII
Hawaii Volcanoes 1916

AMERICAN SAMOA
PACIFIC OCEAN
National Park of American Samoa 1988

Scale for all areas except Alaska
0    200 miles
0    200 kilometers

PUERTO RICO and VIRGIN ISLANDS
Virgin Islands 1956
CARIBBEAN SEA

# Political Boundaries

Yellowstone National Park's political geography has evolved within a larger context of changing national, regional, and local boundaries. Before 1840, European Americans had divided much of the West into large, sparsely settled political units—both the British and the Americans claimed the Oregon Country, Spain and Mexico had a presence in the Southwest, and America had acquired from France the vast tracts of the Louisiana Purchase.

After 1850, America's continental consolidation was largely complete, having secured the Oregon Country in 1846 and Mexican Cession lands in 1848. As the number and size of western settlements grew, authorities gradually created new political constituencies by subdividing large territories into their now-familiar shapes.

When Yellowstone National Park was established in 1872, it included a sizable portion of Wyoming Territory as well as smaller pieces of Idaho and Montana territories. The regional capitals—Cheyenne City, Boise City, and Virginia City (later Helena)—were distant from the park, which complicated legal matters arising within Yellowstone. In 1894, a federal statute upheld the validity of all state laws within their respective portions of the park, but specified that the judicial district of Wyoming would be responsible for prosecuting criminal offenses within park boundaries.

County boundaries also changed over time. When the park was created, only a few large, loosely governed counties sprawled across the area. Populations in the area increased greatly between 1890 and 1930 and residents demanded—and got—more political control and access to a nearby county seat. Thus, for example, what was eastern Idaho's Bingham County in 1890 now includes portions of six different counties.

## State Boundary Development

### 1820
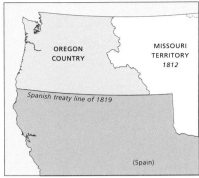

### 1850

### 1860

### 1870
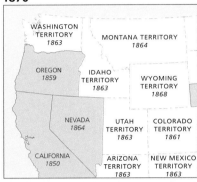

### 1900
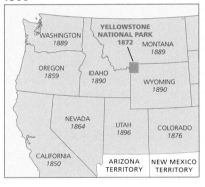

Shifting national, territorial, and state boundaries continually reconfigured the West's political geography during the nineteenth century. Before 1850, large, loosely organized units defined much of the interior West. By late in the century, however, continuing population growth in the region led to full statehood for Montana (1889), Idaho (1890), and Wyoming (1890).

## Greater Yellowstone County Boundaries

### 1870

### 1890

### 1910
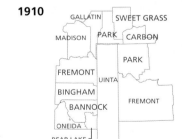

### 1930
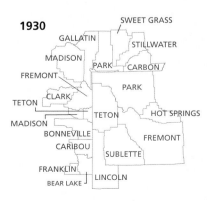

### 2010

BEAR LAKE — Existing county with no boundary changes
SWEETWATER — Newly-established county

The elaborate subdivision of the six original counties in the Greater Yellowstone Area has produced more than twenty counties today—a dramatic fragmentation and multiplication of local political jurisdictions.

The original boundaries of Yellowstone National Park were drawn with a primary goal of including all the geothermal and geological wonders. The effort succeeded almost completely in capturing geothermal surface features, but park proponents found even the apparently generous rectangle of wilderness—roughly 55 by 65 miles—insufficient for the growing number of other purposes the park would be asked to serve.

During the park's first half-century, numerous expansion proposals were seriously debated. Some people sought to extend park boundaries to the southeast to include the entire headwater drainage system of the Yellowstone River; others wanted to extend the park far eastward, across the Absaroka Mountains toward the Bighorn Basin, to embrace important timberlands and wildlife ranges. Still others aimed for a southern extension to bring Jackson Hole and the Teton Range into the park. Though they failed, these proposals reveal the extent to which conservationists as early as the 1880s were thinking in terms of ecosystems and hoping to establish park boundaries more ecologically meaningful than the original "box."

In an important sense, these early efforts did bear fruit. Though the park was not substantially enlarged to include these many desired landscapes, alternative means of protecting much of the acreage at issue were found—"forest reserves" were created east (1891) and south (1897) of the park, and later to the north and west, all as part of the national forest system. Similarly, the Teton Range and Jackson Hole, after many years of bitter political dispute, became Grand Teton National Park, established in 1929. At first the park included only the mountains. It reached nearly its current size in 1950.

Yellowstone's geographical and institutional integrity was in peril for half a century. While some early park advocates were campaigning for expansion, other interest groups were working just as ardently to lop off various portions of the park or to build small and large dams on several park streams, including on the Yellowstone River itself. The stakes were high, and the fate of the park was up for grabs well into the 1920s.

Yellowstone National Park boundaries were not revised until 1929. That year, an important area of petrified forest in the Specimen Creek drainage was added to the western end of the north boundary; the headwaters of Pebble Creek were added to the park's northeastern corner; and much of the east boundary of the park was realigned (for a gain of 78 square miles of park area) along a crest of the Absaroka Mountains.

Other changes followed. In 1932, 7,600 acres of ungulate winter range were added northwest of the North Entrance, on the west side of the Yellowstone River. A symbolically important change in land jurisdiction occurred in 1972 when the John D. Rockefeller Memorial Parkway was created, closing the gap between Grand Teton and Yellowstone national parks.

## Yellowstone and Grand Teton National Parks

**1872**

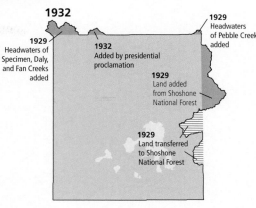

YELLOWSTONE
NATIONAL PARK
*est. 1872*

**1932**

1929
Headwaters of
Specimen, Daly,
and Fan Creeks
added

1932
Added by presidential
proclamation

1929
Headwaters
of Pebble Creek
added

1929
Land added
from Shoshone
National Forest

1929
Land transferred
to Shoshone
National Forest

GRAND TETON
NATIONAL PARK
*est. 1929*

**1950**

1950
Land transferred
to Teton
National Forest

1950
Land transferred
to National Elk Refuge

**2010**

JOHN D. ROCKEFELLER
MEMORIAL PARKWAY
*est. 1972*

- ▨ Park additions
- ▤ Removals
- ▢ Existing park land
- ▨ Inholdings

## Forests and Parks

**1902**

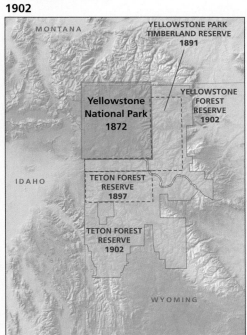

MONTANA

YELLOWSTONE PARK
TIMBERLAND RESERVE
1891

Yellowstone
National Park
1872

YELLOWSTONE
FOREST
RESERVE
1902

IDAHO

TETON FOREST
RESERVE
1897

TETON FOREST
RESERVE
1902

WYOMING

**2010**

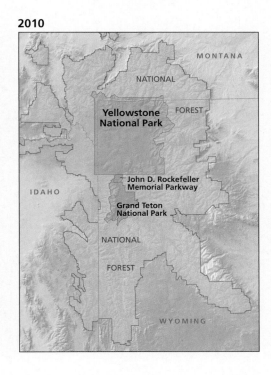

MONTANA

NATIONAL

FOREST

Yellowstone
National Park

John D. Rockefeller
Memorial Parkway

Grand Teton
National Park

IDAHO

NATIONAL

FOREST

WYOMING

11

# Human Geography

# Archaeology

Archaeology is the study of the remains left behind by earlier people. Scientists in this field seek to understand what these people did and when, how they adapted to their environment, and what factors influenced their decisions. The most recent glacial ice was gone from greater Yellowstone valleys by 15,000 years ago, opening the area to habitation by the earliest local people.

The time since then falls into four general periods preceding European contact, each based on human adaptations to the changing environment. During the Early Precontact period, people used large unnotched projectile points attached to throwing spears to hunt now-extinct animals. After glacial times, as the environment warmed and animals became smaller, technology changed with points becoming smaller, the addition of notches for attachment to shafts, and the use of the atlatl or throwing stick. The bow and arrow mark the beginning of the Late Precontact period (around the year 200), and arrow points become even smaller. Pottery appeared in the region about 400, but not in what would become Yellowstone park until roughly 1400 to 1550, when local pottery is assumed to have been made by relatives of the modern Shoshone. There are also changes in the locations of campsites and intensity of use of the park.

The earliest people of the region were pedestrian big-game hunters, who in the late Pleistocene and early Holocene sought at first the now-extinct *Bison antiquus* and later the modern form of *Bison bison*. The practice of hunting bison and other game continued during the warmer middle Holocene, but smaller game, such as mountain sheep and pronghorn antelope, and plant foods were increasingly important, especially in the higher western Plains. In late Holocene times, many Plains groups took up the cultivation of maize and other plants of Mesoamerican origin and formed villages but also continued the older hunting pattern to varying degrees. Throughout the 1700s, a new Plains cultural pattern developed rapidly with the adoption of the horse after its introduction by the Spanish. This was a highly mobile and rich way of life that included extensive travel, trade, and warfare, both among rival tribes and against invading European Americans. The period of contact is the shortest—only 200 years—and contains

## Hopewell Culture Archaeologic Sites with Obsidian from Greater Yellowstone

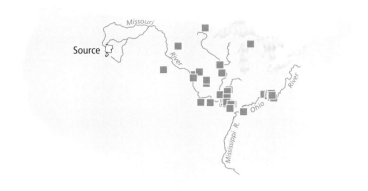

American Indian sites as well as those of explorers, the military, and the park administration.

A compilation of radiocarbon dates shows the intensity of park use has varied through time. A large number of radiocarbon dates between 900 and 1,800 years ago represent roasting pits and other cooking features in sites exposed by Yellowstone River terrace erosion. Further research should add detail to what is known of early park use patterns.

The Greater Yellowstone Area is rich in resources needed by people to make a living: animals, edible and medicinal plants, and sources for stone tools. Obsidian was particularly important because it was easily flaked, was available in unlimited quantities from many sources, and was widely used. Obsidian is also a telling marker for researchers because each source of the volcanic glass has a unique combination of trace elements, which permit accurate identification. Obsidian Cliff is the most notable obsidian source for artifacts; stone from this area was taken from the park, sometimes to distant places.

## Yellowstone Archaeology Sites

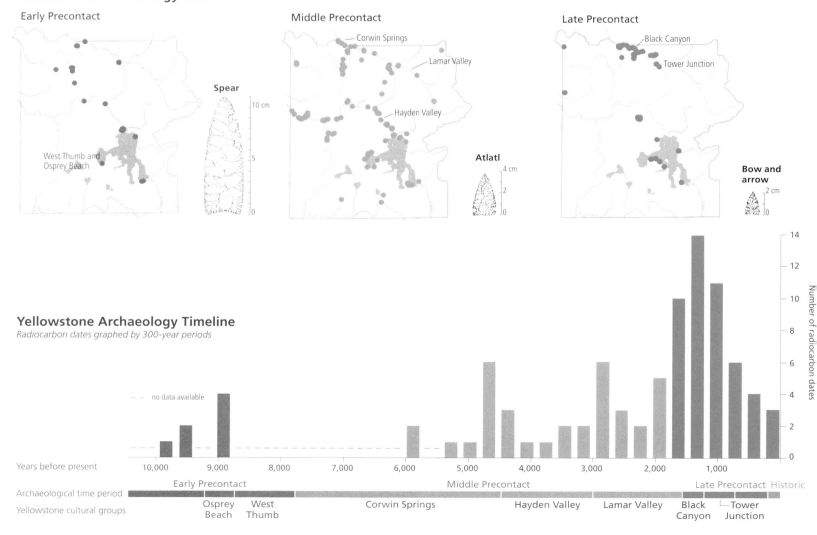

## Seasonal Movement of Early Precontact Cody Complex Peoples

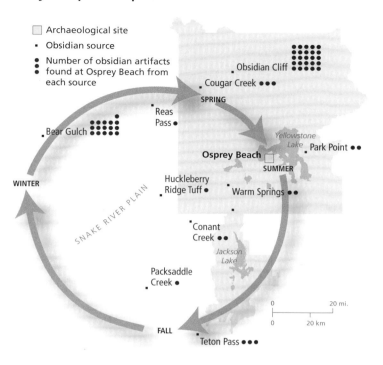

- ☐ Archaeological site
- ▪ Obsidian source
- ⋮ Number of obsidian artifacts found at Osprey Beach from each source

Obsidian Cliff
Cougar Creek ●●●
SPRING
Reas Pass ▪
Bear Gulch
*Yellowstone Lake*  ▪ Park Point ●●
Osprey Beach ☐
SUMMER
Huckleberry Ridge Tuff ▪
Warm Springs ●●
WINTER
SNAKE RIVER PLAIN
Conant Creek ●●
*Jackson Lake*
Packsaddle Creek ▪
FALL
Teton Pass ●●●

0     20 mi.
0     20 km

Material from another significant site, Bear Gulch, was brought into the Yellowstone area from Idaho.

One illuminating contact between the park and the outside world comes from the inclusion of Obsidian Cliff and Bear Gulch obsidian in burial mounds and villages of the Hopewell culture in the Midwest. This occurred between 1,800 and 2,200 years ago when Pelican Lake peoples of the Middle Precontact period were in the park. The volume of obsidian in these distant sites shows that it was intentionally procured and transported more than a thousand miles. Large blocks of obsidian in the Ohio sites show that this was a well-organized trade and that significant expeditions were involved, not merely casual hand-to-hand exchanges.

An Early Precontact site on Yellowstone Lake provides insight into human activity of the period. Osprey Beach is assigned to the Cody Complex (8,500 to 9,600 years ago). These people hunted rabbit, deer, bighorn sheep, bear, and *Bison antiquus*. Obsidian was a preferred tool stone and was collected during yearly travels. At 7,000 feet of elevation, Yellowstone Lake has seventy or fewer frost-free days per year and deep winter snow, so people were most likely at the site during the summer. A hypothetical seasonal travel route shows how Cody Complex people might have moved around the greater Yellowstone ecosystem during the course of a year obtaining raw material for tools from various obsidian sources. Later groups (Pelican Lake and McKean Complex of the Middle Precontact period) had different seasonal travel patterns that brought them into the park from the north and south.

Researchers are in the early stages of understanding a precise timeline of human habitation. The use of various areas fluctuated through time, but these locations are found throughout the park, including near geysers and thermal areas. Clearly, the myth that Indians were afraid of the geysers is not true, but details of how these people went about their lives are usually unknown or unclear. Future research should sample those places underrepresented by previous inventories, salvage sites being lost to erosion, continue the obsidian artifact sourcing to determine movements and preferences of different peoples, seek evidence for the use of plants and animals through time, clarify travel routes, and correlate cultural changes with environmental changes such as the Little Ice Age. Few historic Indian sites have been found and none can be assigned to a historic tribal group. It would be important to find clues to the ethnicity of their creators.

Yellowstone National Park is at the core of the Greater Yellowstone Area and connects to human groups in all directions. Cultural groups traveled through the park and procured resources necessary to make a living. Understanding Yellowstone park's archaeology illuminates historic and precontact use of the entire Yellowstone region.

## Mummy Cave

Mummy Cave is one of the region's foremost archaeological sites. The cave's many cultural layers reflect nearly continuous use by humans for more than 9,000 years and are extremely helpful in dating other less well-preserved sites. People camped here about every 300 years, leaving a record of the plants, animals and artifacts they used in making make their livings. This record shows human adaptations to a changing climate. It is notable that the site was abandoned after about 1600, which mirrors the decline in radiocarbon dates available for the park.

Mummy Cave was dry, offering excellent protection for organic materials such as clothing, basketry, feathers, cordage, wood, animal bones, plant remains, and other perishable artifacts. Organic remains are rare in this region where usually only stone artifacts remain. Charcoal from hearths provides radiocarbon for dating discovered objects.

Bighorn sheep were the most common source of animal remains in every layer. Mule deer were also represented, but there was an almost total absence of bison and surprisingly few elk. The only human remains found in the cave were those of an adult male, wrapped in a sheepskin robe and lying in a flat area outlined with rocks.

Late Precontact artifacts are dominated by the dense remains in Layer 36, which contained more than 100 Rose Springs projectile points, including one still in the arrow shaft.

The deepest strata in the Early Precontact period contained prismatic stone blades, known to be Paleo-Indian tools. Leaf-shaped projectile points from the lowest levels are identified as Angostura points.

Mummy Cave offers the best long-term cultural sequence known for the High Plains in the vicinity of Yellowstone. Its artifacts show clear cultural connections with the Great Plains in general but also indicate linkages to the Great Basin region west of the Rockies. They reflect the hunting patterns suited to the higher, colder environment of the region, where deer and mountain sheep could rival bison as a food source.

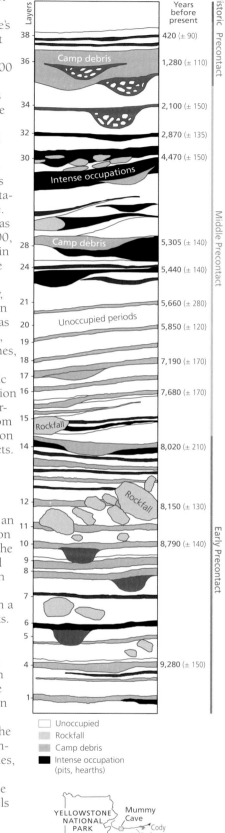

| Layers | Years before present |
|---|---|
| 38 | 420 (± 90) |
| 36 Camp debris | 1,280 (± 110) |
| 34 | 2,100 (± 150) |
| 32 | 2,870 (± 135) |
| 30 | 4,470 (± 150) |
| Intense occupations | |
| 28 Camp debris | 5,305 (± 140) |
| 24 | 5,440 (± 140) |
| 21 | 5,660 (± 280) |
| 20 Unoccupied periods | 5,850 (± 120) |
| 19 | |
| 18 | 7,190 (± 170) |
| 17 | |
| 16 | 7,680 (± 170) |
| 15 Rockfall | |
| 14 | 8,020 (± 210) |
| 12 Rockfall | 8,150 (± 130) |
| 11 | |
| 10 | 8,790 (± 140) |
| 9 | |
| 8 | |
| 7 | |
| 6 | |
| 5 | |
| 4 | 9,280 (± 150) |
| 1 | |

Historic · Late Precontact · Middle Precontact · Early Precontact

- ☐ Unoccupied
- ▨ Rockfall
- ▨ Camp debris
- ■ Intense occupation (pits, hearths)

YELLOWSTONE NATIONAL PARK — Mummy Cave — Cody
N. Fork Shoshone River

# American Indians

American Indians have had a widespread presence in and around Yellowstone for the past 12,000 years. American Indian place names for Yellowstone and its features precede European American settlement and indicate the importance of Yellowstone to tribes of the region. Oral histories, archeological evidence, and historical accounts further demonstrate the persistent use of Yellowstone by American Indians. Hundreds of generations used Yellowstone for residences, hunting, resource procurement, ceremonies, and corridors of travel. Battles, burials, and intercultural gatherings also took place inside the present park boundaries.

## American Indian Names for the Yellowstone Area

| Associated Tribe | Name | Translation |
|---|---|---|
| Assiniboine and Sioux | Pahaska | *White Mountain country* |
| Blackfoot | Aisitsi | *Many smoke* |
| Bannock | Panaiti-Toiai'l | *Yellowstone country* |
| Comanche | Ohatiipi | *Yellow rock* |
| Crow | Aw' Pawishe | *Land of steam* |
| Nez Perce | Me-mut-nee-spah | *Boiling earth* |
| Nez Perce | Kuuseyn'eyéekt | *Buffalo expedition* |
| Salish-Kootenai | K ali ssens | No translation available |
| Shoshone | pa'nd | *Up high* |
| Shoshone | Gooch-a-moonk-be-heah | *The buffalo heart* |

*Tribal names shown do not represent all tribes associated with Yellowstone.*

## American Indian Place Names

### Key to Tribal Names
**Present day feature name**

| Crow | Shoshone |
|---|---|
| Kiowa | Shoshone-Bannock* |
| Nez Perce | Salish-Kootenai |

*Translation* (color indicates tribal association)

- ■ Indian campsite
- — — Bannock trail
- — – Nez Perce trail
- ⋯⋯ Major Indian trails

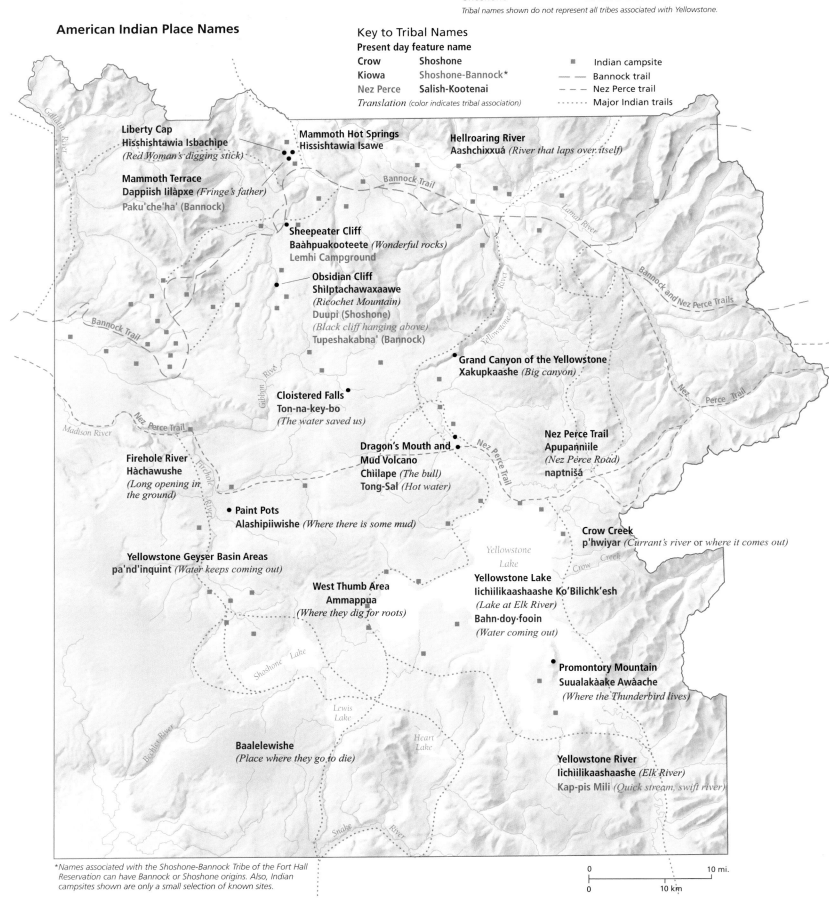

**Liberty Cap**
Hisshishtawia Isbachipe
*(Red Woman's digging stick)*

**Mammoth Terrace**
Dappiish Iilàpxe *(Fringe's father)*
Paku'che'ha' (Bannock)

**Mammoth Hot Springs**
Hissishtawia Isawe

**Hellroaring River**
Aashchixxuá *(River that laps over itself)*

**Sheepeater Cliff**
Baàhpuakooteete *(Wonderful rocks)*
Lemhi Campground

**Obsidian Cliff**
Shilptachawaxaawe
*(Ricochet Mountain)*
Duupi (Shoshone)
*(Black cliff hanging above)*
Tupeshakabna' (Bannock)

**Grand Canyon of the Yellowstone**
Xakupkaashe *(Big canyon)*

**Cloistered Falls**
Ton-na-key-bo
*(The water saved us)*

**Dragon's Mouth and Mud Volcano**
Chiilape *(The bull)*
Tong-Sal *(Hot water)*

**Nez Perce Trail**
Apupanniile
*(Nez Perce Road)*
naptnišá

**Firehole River**
Hàchawushe
*(Long opening in the ground)*

**Paint Pots**
Alashipiiwishe *(Where there is some mud)*

**Crow Creek**
p'hwiyar *(Currant's river* or *where it comes out)*

**Yellowstone Geyser Basin Areas**
pa'nd'inquint *(Water keeps coming out)*

**West Thumb Area**
Ammappua
*(Where they dig for roots)*

**Yellowstone Lake**
Iichiilikaashaashe Ko'Bilichk'esh
*(Lake at Elk River)*
Bahn·doy·fooin
*(Water coming out)*

**Promontory Mountain**
Suualakàake Awàache
*(Where the Thunderbird lives)*

**Baalelewishe**
*(Place where they go to die)*

**Yellowstone River**
Iichiilikaashaashe *(Elk River)*
Kap-pis Mili *(Quick stream, swift river)*

*Gallatin River*
*Bannock Trail*
*Lamar River*
*Bannock and Nez Perce Trails*
*Yellowstone River*
*Nez Perce Trail*
*Gibbon River*
*Nez Perce Trail*
*Madison River*
*Firehole River*
*Yellowstone Lake*
*Crow Creek*
*Shoshone Lake*
*Lewis Lake*
*Heart Lake*
*Bechler River*
*Snake River*

*Names associated with the Shoshone-Bannock Tribe of the Fort Hall Reservation can have Bannock or Shoshone origins. Also, Indian campsites shown are only a small selection of known sites.*

| 0 | | 10 mi. |
| 0 | 10 km | |

American Indians have maintained their relationships with Yellowstone into contemporary times. Tribes received elk and bison carcasses from 1930 to 1967 when park managers actively reduced surplus game herds. Tribal members have been employed in the park since the 1920s. More recently, tribes have collaborated on park management strategies; eighty tribes contributed to park plans for Yellowstone bison. Various native groups conduct traditional ceremonies in Yellowstone to commemorate past events, to bless future endeavors, or to protect the ecosystem and its inhabitants.

American Indian tribes that occupied the Yellowstone region in the past were all highly mobile, their constituent bands ranging over extensive territories both within and outside of the modern park area in the course of their regular movements. The Sheep Eater Shoshone were most closely centered on the area of the modern park, but many other groups commonly spent time there and drew on its important natural and spiritual resources. Seven tribes have treaties with the United States acknowledging title to lands within Yellowstone: Shoshone-Bannock, Eastern Shoshone, Nez Perce, Confederated Salish and Kootenai, Blackfeet, Crow, and Gros Ventre. Ethnohistoric records, legal determinations by the Indian Claims Commission and U.S. Court of Claims, and modern ethnographic consultations have established that in all, twenty-six still-existing tribes were "traditionally associated" with the area.

It is important to note that contact period treaties and other government actions put together many Indian groups originally separate as "consolidated" or "confederated" tribes. The contemporary "judicially established Indian lands" reflect this process and do not faithfully replicate the aboriginal distributions of these groups.

The relationship between American Indians and Yellowstone has changed through the years. The United States system of Indian reservations, initiated in the 1830s, enforced boundaries that often had little relation to tribal homelands. Many American Indians crossed these boundaries to hunt and travel on their traditional lands, some of which included the park area. At the time of Yellowstone National Park's inception in 1872, Indians in the western United States were commonly seen by European American settlers as "marauding Indians" who presented "a serious danger . . . to the game and forests of a portion of the Yellowstone National Park." Thus, park managers were charged with keeping American Indians out of the Yellowstone area to encourage and guarantee the "safety" of tourists. In the modern era, Yellowstone park managers engage in systematic consultation and collaboration with representatives of tribes that have been confirmed to have longstanding traditional connections with the park and active interest in sustaining the ecological and cultural values associated with it.

**Approximate Tribal Distributions in 1850**

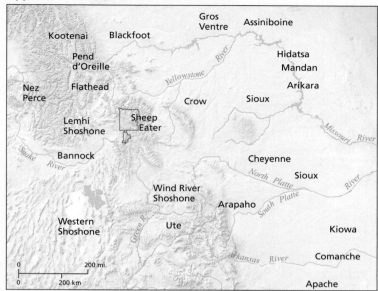

**Traditional Indian Territories at Mid-Nineteenth Century, as Judicially Established in 1978**

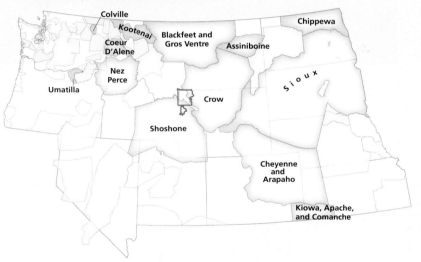

**The Twenty-six Associated Tribes of Yellowstone National Park, 2005**

# Sheep Eaters

**Shoshone Camp**

This painting by David Joaquin shows a Sheep Eater camp as it might have been in 1650 in the Lamar Valley of Yellowstone National Park. Joaquin shows a second group arriving to trade.

This interpretation of Sheep Eater use of Yellowstone is based on archaeological and ethnographic research. When Yellowstone was dedicated as a national park, the Indians living in the region were principally the Tukadika or Sheep Eater Shoshone. The Sheep Eaters were one of a large group of regional Indians that spoke a Numic or Shoshone language. The Agaidika or Salmon Eaters, who lived along the river bottoms of Idaho, were close neighbors. Often, nicknames and epithets are considered to be derogatory, but the name Sheep Eater actually denoted a people who lived in the mountains among the powerful spirits and had ample resources for their livelihood.

Like other Shoshone bands, the Sheep Eaters were migratory, spending the warmer months in today's Yellowstone park and moving off the mountains for the winter season. These movements are easy to trace because the Sheep Eaters followed the migratory routes of the bighorn sheep, off the Absaroka Mountains to the east and south into areas providing prime winter and spring habitat.

While traveling through this region in the summer of 1835, fur trapper Osborne Russell wrote in his journal, "On the north and West were towering rocks several thousand feet high which seem to overhang this little vale—Thousands of mountain Sheep were scattered up and down feeding on the short grass which grew among the cliffs and crevices."

Not surprisingly, this is the same area where the greatest numbers of ancient sheep traps are found. These structures are made of wooden fences that converge in a V-shape to a rectangular catch pen resembling a modern corral. Hunters drove the sheep into the corrals and then dispatched them with clubs and spears.

The Sheep Eaters also fished, collected pine nuts, and dug root vegetables. Plants (balsamroot, biscuit root, bitterroot, bistort, camas, sego lily, spring beauty, tobacco root, wild onion, and yampa) were dug and eaten raw or processed in large underground ovens. These plants likely made up half the Sheep Eater diet.

The Sheep Eaters were renowned for their exceptionally large dogs. Osborne Russell offers an early description of a Sheep Eater camp in the Lamar River valley where there were six men, seven women, eight to ten children, and about thirty dogs. These dogs, found among Indians across the upper Plains, were large, robust, wolf-like animals with unusually long legs and broad heads. Frederick Kurz noted in 1851 that these dogs "differ slightly from wolves, howl like them, do not bark, and not infrequently mate with them."

**Bighorn Sheep Sightings, 1835–1881, and Sheep Traps**

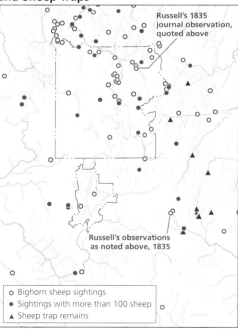

Russell's 1835 journal observation, quoted above

Russell's observations as noted above, 1835

- ○ Bighorn sheep sightings
- ● Sightings with more than 100 sheep
- ▲ Sheep trap remains

The remains of a bighorn sheep trap located in the Absaroka Mountains, southeast of Yellowstone National Park. The photograph shows collapsed converging walls leading to a catch pen. Superintendent Philetus Norris describes several of these drivelines and catch pens in his 1881 annual report on the condition of the park.

The Sheep Eaters used their dogs as pack animals, carrying up to fifty pounds or pulling a travois with a load of as much as seventy pounds. The dogs also assisted the Sheep Eaters in hunting bighorn sheep, driving the animals into traps. The Sheep Eaters loved their dogs to the extent that they fed them before taking food themselves and tied leather booties on the dogs' feet in winter to protect them from snow and ice. A dog was often sacrificed when its owner died—so the two could be buried together.

The Sheep Eaters made use of all parts of the bighorn sheep. They were known for the high-quality clothing they assembled from their ample supply of sheep hides. Their sheep-horn bows were perhaps the most ingenious of their artifacts. They were skillfully made from rams' horns soaked in hot water (which the Sheep Eaters had readily available in Yellowstone thermal areas) until they were pliable. Bow parts were cut from the horns and straightened by binding them between two pieces of wood. After considerable cutting, abrading, and smoothing, they were joined to form the two arms of the bow and covered on one side with layers of sinew affixed with hide glue. These horn bows had a great deal of elastic energy and were extremely efficient in the hands of a good archer.

Arrows were fashioned from the straight shafts of chokecherry and willow bushes. Arrow points were made from the obsidian, chert, and chalcedony found at several places in Yellowstone park.

Pots carved out of steatite, or soapstone, are another important Sheep Eater artifact. There are no sources of this stone in Yellowstone park, so the Sheep Eaters had to travel to locations in the Wind River Mountains to obtain it. Men made the vessels that were then owned by women and passed down to their daughters through the generations. Most of the pots were small, holding about a quart, with a few of larger size, holding as much as a gallon. The pots were set directly on a fire to boil food.

Contrary to a popular myth, the Sheep Eaters were not afraid of Yellowstone's geysers and hot springs. As noted above, they used the hot water for making bows and for therapeutic baths. The Sheep Eaters believed that the violent geysers and bubbling mud pots were caused by underwater spirits. They believed that by fasting and praying they could acquire one of these powerful underwater creatures as a guardian spirit.

Most Sheep Eater men and some women sought the guidance of a spiritual force through a vision quest ritual. A common way for Sheep Eaters to seek such power was to purify with a sweat bath and then go to a petroglyph site for several days of prayer. The spiritual deities that they encountered existed in the rocks behind the petroglyphs.

**Section of William Clark's 1814 Map**

A portion of William Clark's 1814 map showing the location of the Ne-Moy and Yeppe bands of Shoshone Indians in the immediate Yellowstone National Park region. Ne-Moy is quite likely a derivation of *Newe*, a name the Sheep Eaters frequently used for themselves. Yeppe is more obscure, but may refer to the Shoshone word *yap* that was used for a root vegetable and indicates a band of Root Eaters. "Lake Eustis" was an early name for Yellowstone Lake. "Colter's route in 1807" refers to the mountain trek made by John Colter at the request of Manuel Lisa.

## Hot Springs and Petroglyphs

The Sheep Eaters believed that the thermal areas of Yellowstone were connected by an underground passage to the hot springs at Thermopolis, Wyoming. Powerful underwater spirits were frequently encountered along this passageway. The area to the northwest of Thermopolis is covered with petroglyph images of water spirits. Most of these sites are in the sandstone outcrops below the mountains. Northwest of Thermopolis, Legend Rock is one of the many petroglyph sites in this corridor. The water spirits at some of these sites are shown beneath lines to demonstrate their underwater location. These spirits, or *pan dzoavits* (*pan* for water and *dzoavits* for spirits in the Shoshoni language), were known for their large hands that could reach out and grab people. A special class of water spirits bear female attributes. These water spirit women were especially feared and able to offer invulnerability in a time of war. If a Sheep Eater could successfully obtain her as a spirit helper, he was recognized as a very significant person. For example, Togwotee, a noted Sheep Eater, had the power of the water spirit woman.

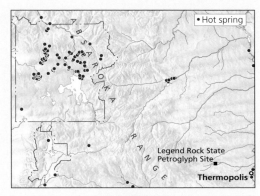

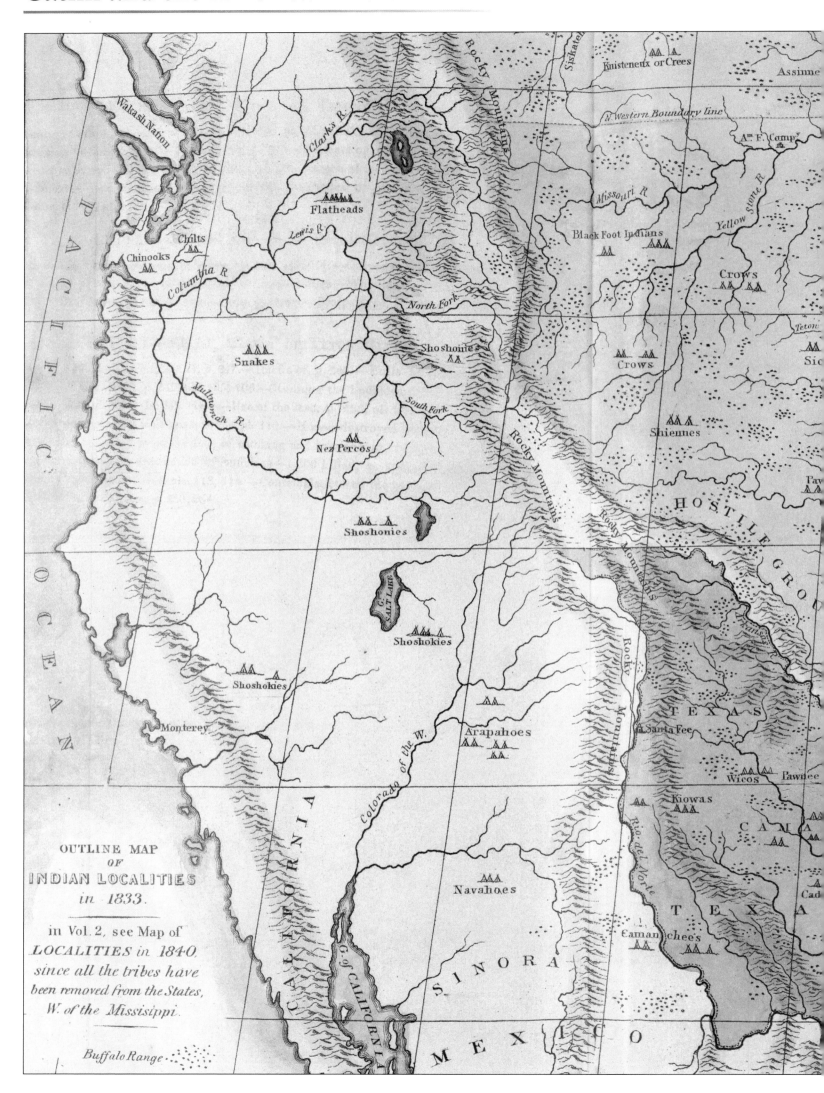

OUTLINE MAP
OF
INDIAN LOCALITIES
in 1833.

in Vol. 2, see Map of
*LOCALITIES in 1840,*
*since all the tribes have*
*been removed from the States,*
*W. of the Missisippi.*

*Buffalo Range*

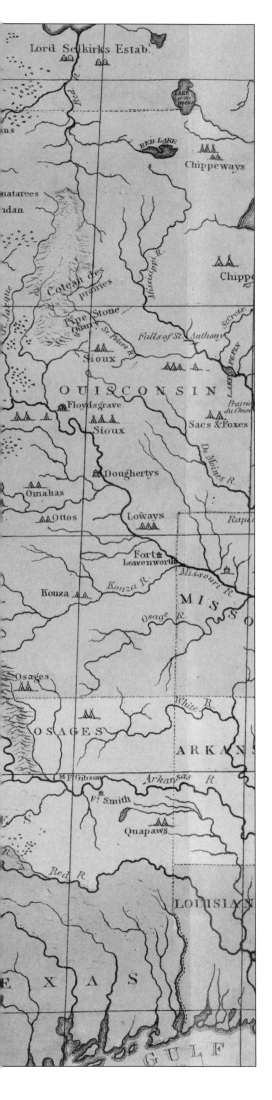

Traditional historical scholarship tended to celebrate the westward movement of European American settlers as a triumphant pageant of civilization, but more recently scholars describe "the winning of the West" as a violent series of conquests, crushing and disrupting the ancient cultures. Only a few people with the tools to document those cultures arrived early enough on the western scene to do so. One of those pioneer cultural observers was American artist George Catlin, who produced hundreds of portraits of American Indians and a wealth of paintings and drawings of native life and landscapes. Between 1830 and 1836, Catlin made a series of trips to various parts of the West, including an 1832 passage up the Missouri River as far as the mouth of the Yellowstone River at Fort Union (his closest approach to present Yellowstone National Park). His extensive travels resulted in his milestone *Letters and Notes on the Manners, Customs, and Condition of the North American Indians* (1841). Among the book's treasures is a comprehensive map (left) of tribal distribution yet—made just in time, as political, social, and economic forces were about to scramble the entire western cultural landscape.

It was also during his Missouri River trip that Catlin was inspired to propose the most sweeping conservation initiative in North American history. He urged, in 1833 and later, that the heart of the continent, from the Mexican border to Lake Winnipeg, be set aside as a "magnificent park," which would include Indians "preserved in their pristine beauty and wildness." Though today we are troubled by his perception of so many dynamic human cultures as suitable subjects for static exhibition, Catlin's pronouncement rightly earned him enduring stature as a founder of the park movement, which would eventually bring Yellowstone National Park into existence.

**Stu-mick-o-súcks, Buffalo Bull's Back Fat, Head Chief, Blood Tribe**

**Peh-tó-pe-kiss, Eagle's Ribs, a Piegan Chief**

**Buffalo Chase, Bull Protecting a Cow and Calf**

# Exploration

## First Contact, 1806–1813

The first known European Americans to enter the Greater Yellowstone Area were U.S. citizens under the command of Capt. William Clark, returning from their journey to the mouth of the Columbia River. They traveled eastward up the Gallatin Valley, across Bozeman Pass, and down the Yellowstone River to the Missouri, approaching within about fifty miles of present-day Yellowstone National Park. Party member John Colter returned to the region in 1807 (through 1808), becoming the first known white man to enter the park area and witness its geothermal features. In 1811 and 1812, members of the fur-trading Astorians passed along the southern edge of the ecosystem on journeys to and from Oregon. The Astorians or Colter may have been the first nonnatives to see the Teton Range.

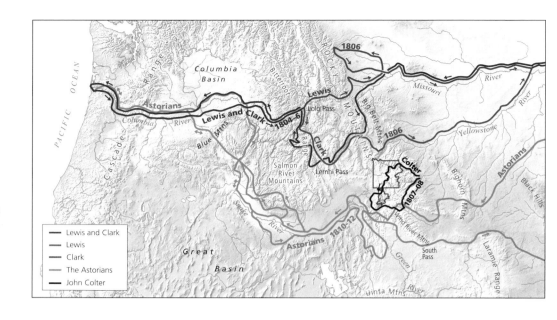

## Journeys of a Trapper, 1835–1839

Osborne Russell, a literate Maine-born trapper, traveled extensively through the northern Rocky Mountains in the closing years of the fur trade. Repeatedly crisscrossing the region in and around Yellowstone, Russell traveled with or encountered some of the era's best-known professional trappers and traders, including Jim Bridger and Nathaniel Wyeth. His journal provides essential background on life in the fur trade—from the extraordinary perils that were a matter of daily routine to the staggering demands of the trapper's work.

Russell's journal also documents dramatic and even violent encounters with Native Americans. Despite this, he wrote insightful and even sympathetic depictions of these longtime local inhabitants. For example, his 1835 report on Sheep Eater Indians of the Lamar Valley is a thoughtful rendering of a people who, though obviously prospering, would for many years in numerous other accounts be misjudged as stunted and miserable.

A careful and keen observer of the natural world, Russell wrote significant accounts of geothermal features, but his reports on wildlife are irreplaceable. His observations of animals in settings throughout the Yellowstone ecosystem offer modern environmental scholars an unsurpassed view of the ecological community of the Northern Rockies more than 170 years ago.

Russell's vivid, matter-of-fact accounts—of a trapper's life, native populations and brutal battles with them, the risks of wilderness travel, the sublime beauty of the landscape—still bring that era to life for modern readers and make his journal the first enduring classic of Yellowstone literature.

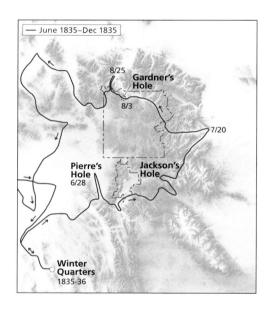

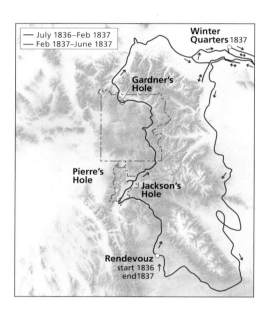

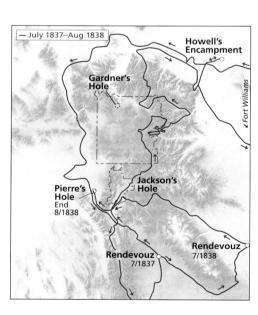

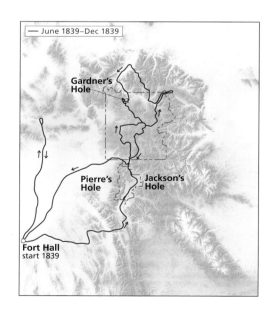

## Expeditions, 1860–1870

The first official exploration party to Yellowstone since Lewis and Clark was headed by Capt. William Raynolds. The expedition skirted the central ecosystem but failed to reach the upper Yellowstone River because of deep snows.

Then in 1869, almost on a lark, Montanans David Folsom, Charles Cook, and William Peterson traveled to many of the area's most amazing sites. Though their additions to knowledge of the region were invaluable to later parties, many people doubted their stories.

It was left to the Washburn party of 1870 to follow a similar route and bring back "definitive knowledge" of Yellowstone. Thus, European American exploration had been under way for more than sixty years before "discovery" became official.

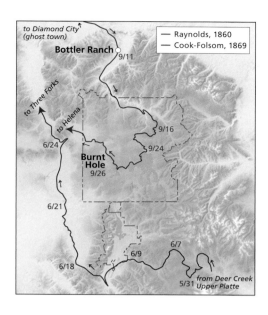

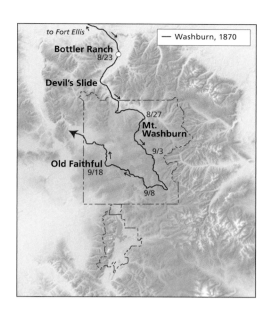

## The Search for Gold, 1863–1870

Gold strikes in Idaho and Montana brought prospectors to the region in the 1860s. While many prospectors passed through the present-day park hunting for gold or traveling to more promising fields, the area never yielded meaningful finds.

In 1863, Walter DeLacy led a party of dozens of prospectors through Yellowstone, adding to the geographic knowledge gradually being accumulated by many such wanderers. That same year, prospectors George Huston and H.W. Wyant explored the northeastern and east-central portions of the present park, naming prominent landmarks and learning still more about the area. In the late-1860s and early-1870s, A. Bart Henderson routinely traveled in the region, helping to pioneer the New World Mining District near the northeast entrance of the present-day park.

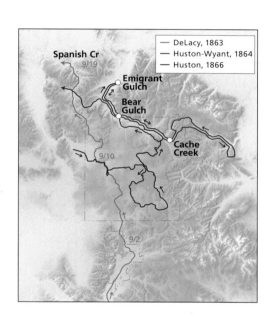

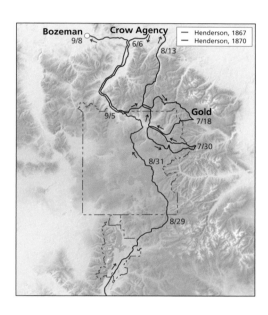

## Surveys, 1871–1872

On the heels of the Washburn party came veteran explorer Ferdinand Hayden and the U.S. Geological Survey. Hayden and U.S. Army engineers Barlow and Heap, following routes of many European American predecessors, applied scientific rigor to the investigation of Yellowstone's geographic mysteries.

Yet for all its achievement, the Hayden survey's wisest move may have been including a photographer, William Henry Jackson, and an artist, Thomas Moran. Between them, Jackson and Moran brilliantly captured Yellowstone's weirdness, wonder, and beauty, settling once and for all any doubts about the existence of this singular landscape and helping to persuade Congress that here, indeed, was a place worth saving.

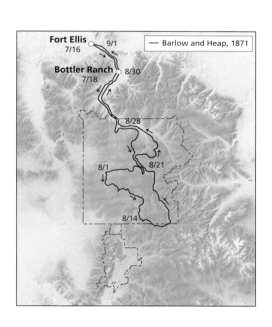

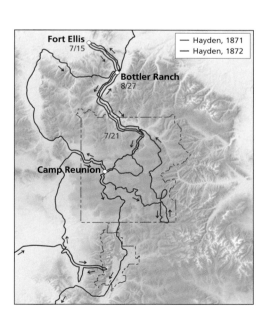

# Early Maps, 1808–1814

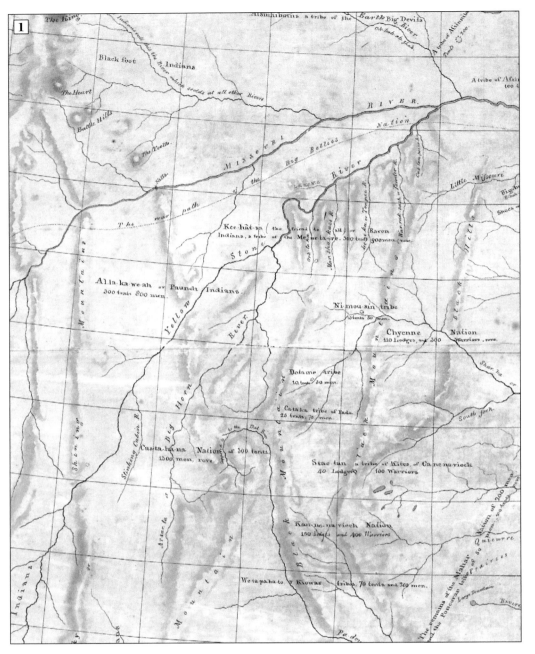

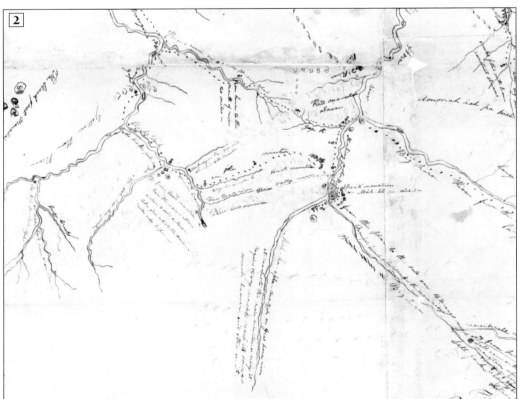

## 1. Clark and King's Map of 1805

William Clark used information from Indians and traders such as Hughes Heney to develop this map while the Lewis and Clark Expedition was at Fort Mandan in December 1804. He sent it east on a keelboat in April 1805. Nicolas King made two copies, which are known today as the State Department copy and the War Department copy (largely identical).

Rivers were crucial to early travelers on foot and horseback for three main reasons: they provided a pathway of sorts through uncharted lands; access to drinking water was vital; and animals—food—frequented waterways. Clark and King's 1805 map is one of the earliest to show the Bighorn and Yellowstone rivers and the Tongue and Powder rivers, tributaries to the Yellowstone. Intriguingly, it also shows "Stinking Cabin River" near the headwaters of the Yellowstone. This notation may represent today's Shoshone River. The "tar" and other hot springs on the river were shown on Jim Bridger's 1851 map as "Sulphur Springs" or "Colter's Hell." Colter's Hell was a term trappers used to refer to the thermal area at the forks of today's Shoshone River, not to present-day Yellowstone National Park.

## 2. Clark (Drouillard) Map of 1808

George Drouillard, one of the hunters for Lewis and Clark, returned from the West to St. Louis in September of 1808. He gave his map sketches to William Clark, who drew the Drouillard map that same year. The original sketches are now in the Library of Congress and the map is in the Missouri Historical Society's Drouillard collection.

This is the first map to depict the Yellowstone-Bighorn River area in any detail and contributed to the content of Clark's map of 1810. Although difficult to read, it shows the Yellowstone River, Bighorn River, Little Bighorn River, Stinking Water (today's Shoshone River), Clark's Fork River, Pryor's Fork, and Tongue River. It contains useful information for historians, including trade routes, seven Indian winter campsites (probably Crow), and Indian place names.

Clark made separate notes, based on Drouillard's information, that supplement this map. Drouillard therefore did not draw directly onto Clark's map but was a source for it. One of Clark's notes on this page may refer to the Yellowstone country, stating, "At the head of this river [Yellowstone] the natives give an account that there is frequently herd a loud noise, like Thunder, which makes the earth Tremble, they State that they seldom go there because their children cannot sleep—and conceive it possessed of spirits, who were averse that men Should be near them." This may have been partially responsible for the misunderstanding, long held by European Americans, that many Indian tribes avoided Yellowstone's geyser areas for fear of dangerous spirits.

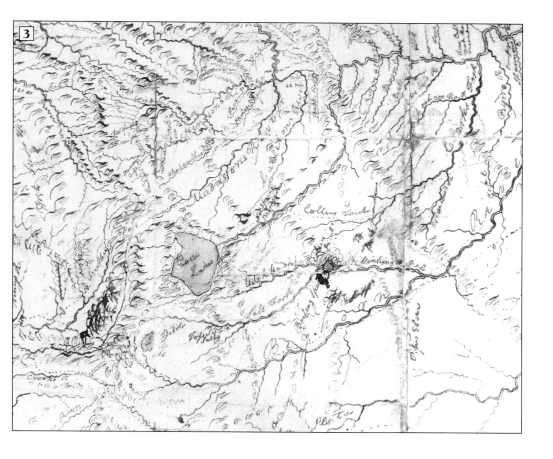

### 3. Clark's Manuscript Map of 1806–1811 (or Clark's Map of 1810)

William Clark prepared this map in 1810. It represents one version of a series of modifications he made, based largely on information he received from George Drouillard, continually from 1806 to 1811. It shows the track of the Lewis and Clark Expedition across the American West. The map was forwarded to printer Nicolas Biddle in 1810, and it became the basis for Lewis and Clark's map of 1814.

The map shows "Eustis Lake," which is present-day Yellowstone Lake. To the south, "L. Biddle," later incorrectly transcribed as "Lake Riddle," probably represents Jackson Lake. "Madisons River," "Jeffersons River," and "Gallitons River" are all shown to the north of Eustis Lake across a range of mountains. This is likely the first time these features appear on a map compiled by European Americans. Also shown is "hot spring brimstone" near "Colters rout" where it crossed the Yellowstone River. Some historians believe this refers to hot springs near Tower Fall where Colter may have crossed the river. The term "Colters rout" is written so it tracked east from what was probably the Yellowstone River; Clark likely obtained this information from John Colter. To the east, "Boiling Springs" are shown on the "Stinking" river. These probably represent present-day Tar Spring or DeMaris Springs on the Shoshone River. To the northwest is "hot spring" near the "Wisdom" river, probably a reference to one of the commercial hot springs in that part of Montana today.

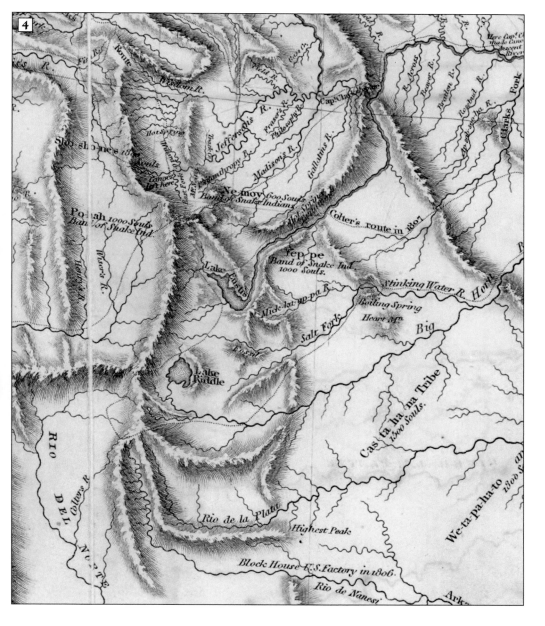

### 4. Lewis and Clark's Map of 1814 (Samuel Lewis's engraving)

Samuel Lewis prepared an engraving from the Clark's Map of 1810. This new version was published in 1814 in Nicolas Biddle's *History of the Expedition Under the Command of Captains Lewis and Clark*. Lake Biddle, probably present-day Jackson Lake, became "Lake Riddle," confusing map readers for more than half a century. Probably the most important contribution this map makes to Yellowstone National Park history is its detailed portrayal of "Colter's Route in 1807." Although the exact route is still generalized, he clearly followed the Yellowstone River and crossed it near "Hot Spring Brimstone," perhaps hot springs near Tower Fall. Locations of several bands of Indians are noted, including the "Yep-pe," in or just east of Yellowstone Park. The Madison, Gallatin, and Jefferson rivers are also clearly shown, the headwaters of the first two being in Yellowstone National Park.

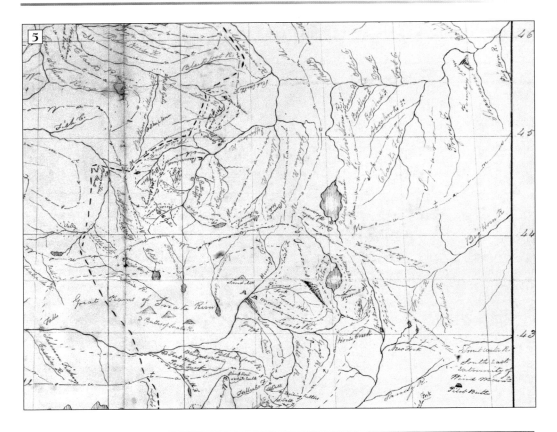

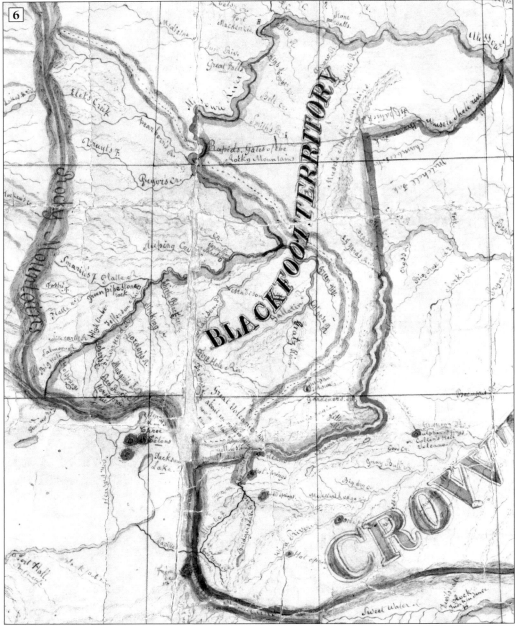

### 5. Ferris's Manuscript Map of 1836

Lost for more than a century, Warren Ferris's map would have exerted incredible influences upon the geographers of its day had it been known to persons other than its creator. Instead, it remained in a family trunk until discovered and published in 1940 by Paul Phillips in his *Life in the Rocky Mountains*. Today the original is at Brigham Young University.

Warren Ferris was an independent fur trapper. His map portrays an extensive and generally accurate knowledge of the northern Rocky Mountains. Yellowstone Lake, the Yellowstone River, and the Madison-Gallatin-Jefferson systems are clearly shown and labeled with their modern names. The map shows "Boiling water Volcanoes" on the southwest shore of Yellowstone Lake and "spouting fountains" in the headwaters of the Madison River, perhaps a reference to the Old Faithful area. The map does, however, have some inaccuracies. "Burnt Hole," for example, is shown to the east of the Madison River headwaters, although Yellowstone historians believe the actual location of this feature is east of Henry's Lake and west of present West Yellowstone, Montana.

### 6. Father DeSmet's Map of 1851

While at the 1851 treaty council at Fort Laramie, Jim Bridger drew a sketch map for Father Pierre-Jean DeSmet. Yellowstone historian Aubrey Haines called this amazingly accurate sketch the "Bridger-DeSmet manuscript map." DeSmet transferred most of Bridger's information to a formal map, the DeSmet Map of 1851. These two maps are probably the most important of the pre-1860s maps of the upper Yellowstone River country. They now repose at St. Louis University.

The DeSmet map is notable for its improved accuracy and detail relative to previous maps. Yellowstone Lake is named and labeled "60 miles by 7." "Madison R.", "Gallatin R.", "Gardeners R.", and Atlantic and Pacific creeks are shown and named. Also of importance is the crinkled outline of today's Grand Canyon of the Yellowstone with "Falls 250 feet" at its upper end and Colter Ford near Tower Fall indicated by a heavy pen stroke across the river. This makes it clear that Bridger knew of that important crossing. The map portrays many thermal features, including "Hot Springs" on the east shore of Yellowstone Lake, a "Volcano" at the outlet of the lake, and "Sulphur Mountain" (Mammoth Hot Springs). At the head of the Firehole River, the map notes a "Great Volcanic Region about 100 miles in extent now in state of eruption." The 100 miles extent suggests this refers to a broad area that includes the Upper, Midway, Lower, Shoshone, West Thumb, and Norris geyser basins.

Of particular importance east of the Yellowstone plateau is the location of "Sulphur Springs or Colter's Hell." Thus Bridger and DeSmet correctly pinpointed the location of Colter's Hell. The sulfur springs correspond to either present DeMaris Springs or the "tar spring" documented by fur trappers (now inundated by Buffalo Bill Reservoir).

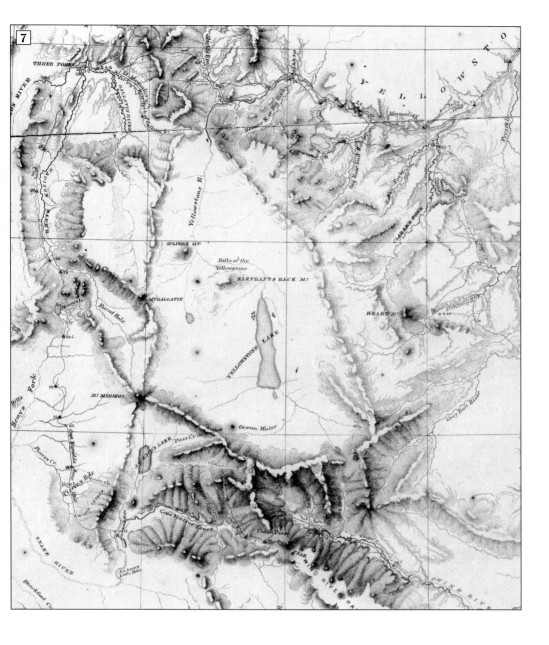

On his newer map, DeSmet formalized the name "Fire Hole Riv.", added "Little Falls" (Tower Fall), added "Bridger's Lake and Riv.", mislabeled the Lamar River as "Beaver Creek" (it was "Meadow" on Bridger's sketch), and omitted Bridger's lines showing the Slough and Hellroaring drainages.

## 7. Raynolds Map of 1860

Army Capt. William F. Raynolds conducted a government survey in 1859–60 of the country south and west of present-day Yellowstone National Park. He was accompanied by guide Jim Bridger and geologist F.V. Hayden. Raynolds produced a map of the surrounding area that left its central portion, the park itself, largely blank.

The map shows a surprising lack of knowledge—at least among Raynolds and his men—of the Yellowstone Park area, even at this late date, with, for example, an elongated Yellowstone Lake and the "Yellowstone R." running north from it. Near the "Falls of the Yellowstone," "Elephants Back Mt." probably represents present Mount Washburn. Below the falls in the apparent location of Mammoth Hot Springs is "Sulpher Mt.", while a "Mt. Gallatin" appears in the location of present Mount Holmes. This map also confirms the location of the valley of "Burnt Hole" as being west of Mount Holmes and east of Targhee Pass on the Madison River and "The Tetons" as being southwest of "Jackson's Lake."

Raynolds stated "that at no very distant day the mysteries of this region will be fully revealed, and though small in extent, I regard the valley of the upper Yellowstone as the most interesting unexplored district in our widely expanded country."

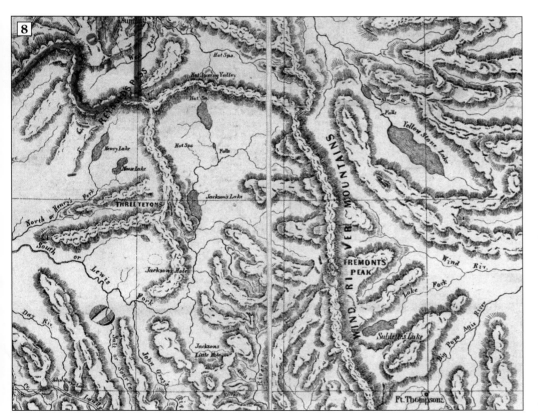

## 8. DeLacy Map of 1865

Walter W. DeLacy was a prospector who entered the present Yellowstone Park area in 1863 with a party of forty men and was the first European American to discover Shoshone Lake. At the time, he failed to get credit for discovery of the upper Yellowstone country because he did not fully publish a report until 1876. He did, however, produce a map in 1865 that the First Legislative Assembly of Montana used to lay out the original counties in Montana. It reposes today in the Montana Historical Society at Helena. Because of this map, Carl Wheat called DeLacy "Montana's great cartographic pioneer."

DeLacy's map was the first to show Shoshone and Lewis lakes, although unnamed, as flowing south to the Snake River system. It also shows the Upper and probably Lower geyser basins as "Hot Spring Valley" and shows "Hot Springs" on the Snake River where the Lewis River flows into it. Much of this map is essentially empty where Yellowstone National Park is, although it does show Yellowstone Lake flowing northwest into the Yellowstone River, with falls near the outlet.

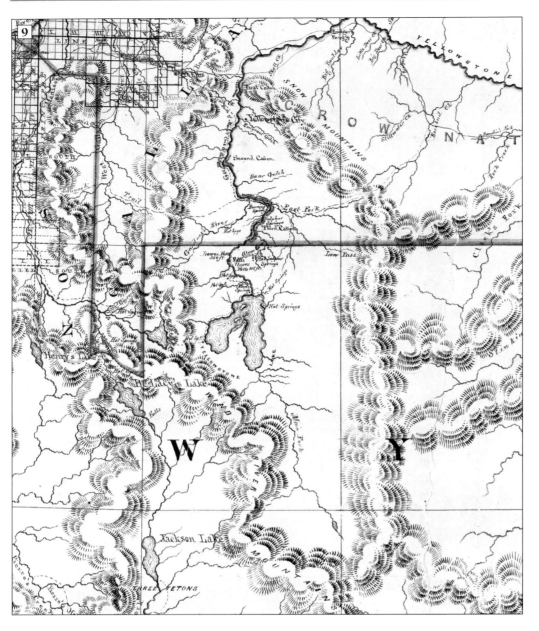

## 9. General Land Office Map, Henry Washburn, 1869

Henry Dana Washburn was surveyor general of Montana Territory. In late 1869, he worked with civil engineer and Yellowstone explorer Walter DeLacy to produce a map based on the explorations of the Folsom party of that same summer. This map is stored in the National Archives.

The map reveals the Yellowstone region—at long last—in reasonably accurate fashion. Yellowstone Lake, for example, is properly shaped for the first time. The map also displays greater detail than ever seen before, including features such as the "route of Messrs Cook & Folsom 1869," "Gardner's River" with its "Hot Spgs" (Mammoth Hot Springs), the "East Fork" (Lamar River) with "Burning Spring" (Calcite Springs) near its mouth, "Alum Creek" and thermal features on Broad Creek, the falls of the Yellowstone, "Hot Spgs" noted at Crater Hills and on Trout Creek, a "Mud Spring" near the outlet of Yellowstone Lake (Mud Volcano), "Hot Spgs" on the western shore of the lake, and three unnamed islands in the lake. Also shown is a triangular "Madison Lake" (Shoshone Lake) west of Yellowstone Lake, "Hot Springs" at the head of the Madison River, and both "Hot Springs" and "Geysers" where the east branch of the Madison joins the Firehole River, a possible reference to present Lower Geyser Basin.

Awareness of Native American presence in the region is indicated by the "Crow Nation" and "Bannock Trail" notations. The Bannock Trail was one of a series of Indian trails crossing the present park; the map indicates that some of these trails were known to European American explorers in 1869.

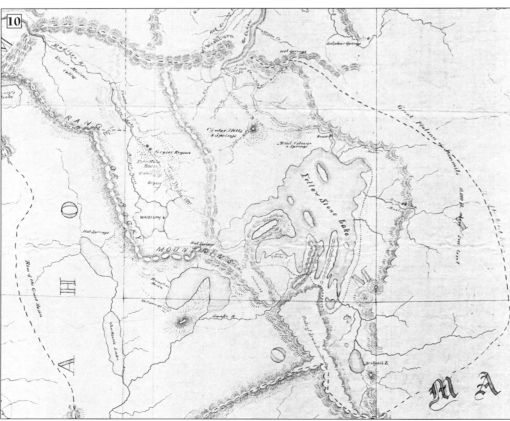

## 10. Doane Map of 1870

First Lt. Gustavus C. Doane led the military escort for the 1870 Washburn expedition, the party that received credit for the European American discovery of the Yellowstone National Park area. Although previous individuals and groups had explored the region, the Washburn expedition was the first to widely document their findings. Doane produced an important report on that expedition and the first official and detailed description of a trip through present Yellowstone National Park. The Doane Map of 1870 accompanied his narrative. It reposes today in the National Archives.

Doane's map includes information from DeLacy's modified map of 1870 (not shown here), which in turn used information provided by the 1869 Folsom party. It also added the 1870 route followed by the Washburn party and, more accurately than any previous map, delineated Yellowstone Lake's various arms. It shows today's Heart Lake (with a reversed peninsula) unnamed on the Pacific side of the Continental Divide and a large "Madison L." at the head of the Firehole River on the Atlantic side. The Washburn party did not explore that area well enough to know that the large lake they

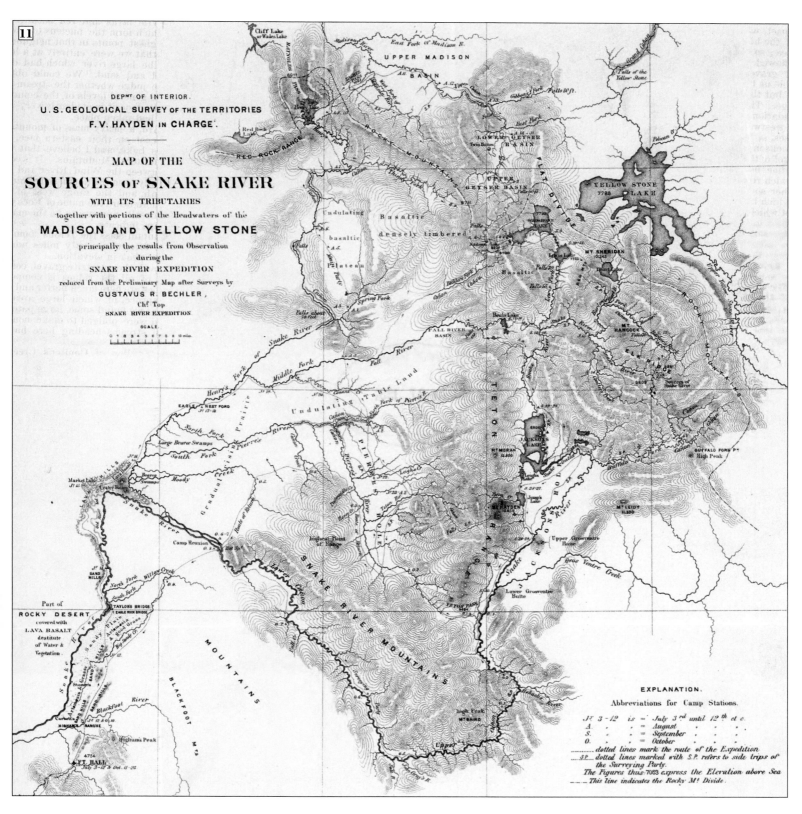

DEPMT OF INTERIOR.

U.S. GEOLOGICAL SURVEY OF THE TERRITORIES

F. V. HAYDEN IN CHARGE.

MAP OF THE

SOURCES OF SNAKE RIVER

WITH ITS TRIBUTARIES

together with portions of the Headwaters of the

MADISON AND YELLOW STONE

principally the results from Observation
during the

SNAKE RIVER EXPEDITION

reduced from the Preliminary Map after Surveys by

GUSTAVUS R. BECHLER,

Chf. Top
SNAKE RIVER EXPEDITION

SCALE

EXPLANATION.

Abbreviations for Camp Stations.

Jy 3 - 12   is   =   July 3rd until 12th et c.
A.      .      .   =   August      .      .   .
S.      .      .   =   September   .      .   .
O.      .      .   =   October     .      .   .
_____ dotted lines mark the route of the Expedition
..S.P.... dotted lines marked with S.P. refers to side trips of
                    the Surveying Party.
The Figures thus 7083 express the Elevation above Sea
_ _ _ _ This line indicates the Rocky Mt. Divide.

saw—today's Shoshone Lake—drained south to the Pacific rather than north through the Madison River to the Atlantic.

Doane's map also shows "Bridger's L." south of Yellowstone Lake, although much too large in size. Doane took this name from DeLacy's Map of 1870, which the party was carrying. Additionally, N.P. Langford stated, "Doane says that he thinks he has seen on an old map the name 'Bridger' given to some body of water near the Yellowstone." That map was probably the General Land Office Map of 1869 by Meredith (not shown here). In any case, the name Bridger Lake got onto today's maps in this way.

## 11. Hayden Survey Map by Bechler, 1872

This map, which appeared in F.V. Hayden's *Sixth Annual Report of the U.S. Geological and Geographical Survey of the Territories* in 1873, was drawn by Hayden Survey topographer Gustavus Bechler in 1872. It showed the upper portions of Snake River and southern Yellowstone National Park with accuracy unknown up to that time. Partially modeled on the Hayden Survey Map of 1871, this map added the country east and west of the Grand Teton Range. It purported to show the sources of the Snake River, although the real headwaters were still not known. The map mistakenly shows the

head branches of the Snake Buffalo and Buffalo Fork intermingling, rather than being separate.

This map was one of the first in the region to utilize modern triangulation techniques and equipment to locate features and map their shape. It heralded the end of the old mapping style. Subsequent maps are far more precise and accurate.

# Jackson and Moran

**Grand Canyon of the Yellowstone, 1872**

It was through the work of early artists and photographers that the public first viewed the geysers, canyons, and other sights of Yellowstone. Before and after the establishment of Yellowstone National Park, artists found inspiration from the area's unique and magnificent features. Their captivating works, in turn, inspired generations of tourists to visit the park.

In 1871, Ferdinand V. Hayden led the U.S. Geological Survey expedition to the area that would become the national park. Before this venture, there were few amateur drawings detailing physical characteristics of the remote territory. Photographic evidence or professional artwork did not exist to substantiate reports of this region that was veiled in myth and mystery. Congress appropriated $40,000 for the Hayden survey to explore "the sources of the Missouri and Yellowstone Rivers." The party numbered almost forty men, specialists in many fields—including several artists and photographers. The survey team explored the region for two months, stopping at many areas that are today major attractions, among them the Grand Canyon of the Yellowstone, Mammoth Hot Springs, and the Upper and Lower geyser basins.

While Henry Wood Elliott was the expedition's official artist (and a trained topographer), it was Thomas Moran, a young, unknown artist, whose contribution is far more widely remembered. He accompanied the expedition as a guest of the U.S. Geological Survey and as an employee of the Northern Pacific Railroad. The railroad saw the value of using artistic depictions of the exotic West to attract eastern tourists.

Moran sketched and painted Yellowstone scenes on site and took extensive notes about the area's features and their qualities. Upon seeing the Grand Canyon of the Yellowstone and admiring its brilliant color, he declared it "beyond the reach of human art." In spite of that assessment, he soon returned to his studio and worked on his large-format painting, *Grand Canyon of the Yellowstone*. He also created many watercolors, such as the *Great*

**Old Faithful, 1872**

30

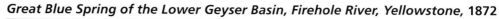

**Great Blue Spring of the Lower Geyser Basin, Firehole River, Yellowstone, 1872**

**Hot Springs on Gardner River, 1871**

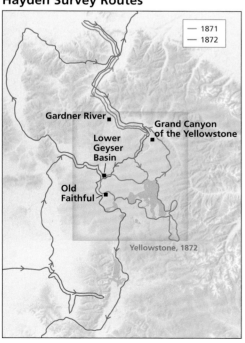

### Hayden Survey Routes

— 1871
— 1872

Gardner River

Lower
Geyser
Basin

Grand Canyon
of the Yellowstone

Old
Faithful

Yellowstone, 1872

### Hayden Survey, 1872

*Blue Spring of the Lower Geyser Basin, Firehole River, Yellowstone*. These works ultimately gave Moran recognition as a master painter. His idyllic images showed Americans that their country possessed the treasure trove of distinctive sights and geologic wonders that is Yellowstone.

The expedition's photographer, William Henry Jackson, captured the area's remarkable landscapes on film. Working side by side, Jackson photographed and Moran sketched hot springs, geysers, and canyon walls from similar vantage points. Moran often appeared in Jackson's photographs to provide scale. The photographs documented what the expedition members witnessed.

Jackson's images, along with Moran's field sketches, served as proof of the region's beauty and unique features. Hayden wrote an article on the expedition's findings, "The Wonders of the West," which appeared in *Scribner's Monthly* in February 1872. The article, Jackson's photographs, and Moran's drawings helped persuade Congress and President Ulysses S. Grant the following month to set aside Yellowstone as the nation's and world's first national park.

Jackson returned to the newly created park several times, including a trip in 1872 as a photographer for a second U.S. Geological Survey expedition to the area. Likewise, Moran returned two more times (1892 and 1900) and became known as "Thomas Yellowstone Moran."

# Yellowstone Art

**Geysers in Yellowstone, Albert Bierstadt, c. 1881**

Artistic representations of the "wonderland" of Yellowstone played a key role in establishing the national park, and this unique landscape has continued to inspire artists to the present day. Albert Bierstadt, a painter known for grand landscapes of the American West, visited the park in 1881. A rival of Moran, Bierstadt was in search of new material and saw an opportunity to create his own artistic interpretation of Yellowstone. He depicted the setting as unspoiled and tranquil, though even by this early date more and more people were making the challenging journey to the park. Fascinated by the geysers, Bierstadt created a series of paintings that included *Geysers in Yellowstone*. This painting hung in the White House during the presidency of Chester A. Arthur, who was inspired to see the park for himself, and did so on a visit in 1883.

Throughout the park's first decades, Yellowstone was proudly considered "the nation's art gallery" because of its captivating beauty and grandeur. Large numbers of visitors, including many artists, ventured into the area as road systems and transportation improved. While on a hunting trip to Wyoming in 1910, noted painter William R. Leigh took several excursions to the park and made hundreds of sketches. The Lower Falls and Grand Canyon of the Yellowstone—particularly, the play of light off the canyon walls—intrigued him, and in 1911 he completed *The Lower Falls of the Yellowstone*.

Throughout the twentieth century, Yellowstone acted as muse for American artists as well as those from other countries. Once at the park, however, it was difficult for those artists to avoid hoards of visitors gathering at the primary attractions. In the 1920s, German-born Carl Preussl painted city scenes with atmospheres of factory smoke and skyscrapers. His work *Old Faithful* departed from such subject matter by showing a natural occurrence, the geyser erupting, but one that paralleled the industrialization of the day. Preussl's painting shows tourists—women in fur coats and fashionably dressed men—with automobiles; they have temporarily traded in city life to encounter the great, all-natural engine, Old Faithful. No longer was the park viewed as

**Old Faithful, Carl Pruessl, 1929**

the untouched, pristine landscape of Moran or Bierstadt; humans are now depicted as involved with both nature and technology.

In addition to the famed geological features, wildlife also drew visitors to the park. The opportunity to see bears, bison, moose, and elk in a vast wilderness appealed to people from large, crowded cities. In the late 1940s and early 1950s, Olive Fell's work featuring Little Cub Bear appeared on postcards and posters sold in the park. In *Me and Old Faithful*, Fell combined wildlife and geysers, two iconic park attractions, into one simple composition.

The numbers of tourists traveling to the Yellowstone area continued to increase, with consequences for the region's natural features and wildlife. Today, conservation is a major concern, and some artists have incorporated this and related themes into their works. In her painting *Out to Lunch*, artist and environmental activist Anne Coe depicts bears at a picnic in Grand Teton National Park. Through this seemingly playful and cartoonish piece she compels viewers to consider the effects humans have on bears and the environment. Coe puts the bears in a ridiculous scene, breaking barriers between human and beast. She makes a statement, laced with humor, that people are taking advantage of the natural environment at the expense of the bears (and other wildlife), and that all species must share the national parks.

The art that emerged from before the park's inception up to the present day affirms Yellowstone's importance to the spirit of America and the West. Artists who experienced the park found something exciting and inspiring, something they wanted to share with the world—whether by capturing a distinctive natural feature or by commenting on an environmental issue. Art helped establish the park, and art has helped perpetuate its popularity.

**Me and Old Faithful**, Olive Fell, c. 1950

**Out to Lunch**, Anne Coe, 1990

# Early Science History

While national parks may be, as often claimed, America's best idea, most scholars agree that the origins of the national park *idea* are uncertain. Indisputably, the grand notion was the product of numerous minds, and many of those contributing minds were scientifically trained. A second question involves who was most responsible for the legislation that created Yellowstone National Park. Exactly what role scientific evidence played in establishing the park is unclear, but Yellowstone's early scientists demythologized the region and validated the reality of the wonders there so that even a skeptical Congress could see the urgency of preserving the area for future generations.

Science was rarely if ever invoked when it came to the opening of the American West and yet it was fairly common in the descriptions of the proposed and enacted Yellowstone park. For example, Yellowstone's second superintendent, P.W. Norris, wrote in 1878 that while the park was "destitute of valuable minerals, isolated and worthless of all else, [it is] matchless and invaluable as a field for scientists and a national health and pleasure resort of our people."

The scientific history of the Yellowstone area began after the era of fur trapping and near the close of the era of large, federally sponsored explorations of the western United States. These expeditions and their military escorts represent an extraordinary period in American empire building, when interdisciplinary scientific study was an essential element of the effort to define the national domain. The work, conducted by geologists, zoologists, botanists, and other scientists, was enriched by graphic and fine artists, photographers, and a level of literary accomplishment rare in the history of exploration. The published results of these expeditions stand as important accomplishments in the national biography and proved inspiring and persuasive for both lawmakers and the public.

In the Greater Yellowstone Area, the first great scientific venture since Lewis and Clark (who did not actually enter the area that would be the park) was the 1860 Raynolds expedition, members of which documented extensive portions of the region without entering present-day Yellowstone National Park. The pioneering studies of the park area itself were provided by Lt. Gustavus Doane's classic report on the Washburn Party's 1870 trip and the almost encyclopedic scientific chronicles of the Hayden surveys of 1871 and 1872. Hayden, a medical doctor with

## Hayden Survey of the Upper Geyser Basin, Firehole River, 1871

## Hague Geological Survey, 1904

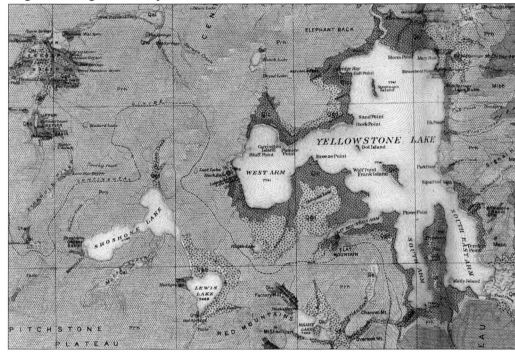

knowledge of geology, geography, ethnology, and entomology, was a significant force in the campaign for the park's creation and was probably the person who first drew the lines on a map that would become the park's original boundaries. Another early scientific milestone, Ludlow's expedition of 1875,

gathered much important information, including a comprehensive zoological report by noted scientist George Bird Grinnell.

Other explorations soon followed, together creating a baseline scientific understanding of the park. By 1900, researchers published findings on geology (e.g., Hayden,

## Yellowstone Science, 1865–1945

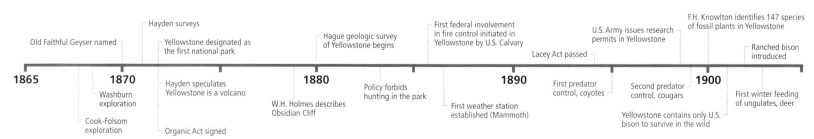

Holmes, Bradley), mineralogy (Peale), fossils (Knowlton), fish and aquatics (Forbes, Jordan, Evermann), parasitology (Linton), thermal microbes and algae (Weed, Setchell, Tildon), botany (Hague, Tweedy), and zoology (Grinnell, Merriam, Croker). While this list is not complete, it demonstrates that many of the finest minds in the science of the era were drawn to the Yellowstone region, as is also demonstrated by the timeline below.

Tentative first attempts at counting wildlife date to about 1881. Early infrastructure for environmental monitoring was established during the period of army control with water gauging sites (1889) and weather stations (1893). The first known permit to conduct scientific research was issued by the army to W.A. Setchell to collect and carry out research on life forms in hydrothermal features—this marked the beginning of the formalized process of managing scientific information about the park and greater Yellowstone. Today, the number of books, scientific papers, government reports, and other documents dealing with the natural and cultural resources of greater Yellowstone exceeds 25,000. It is quite likely that this is scientifically the best-known nature reserve in the world. But ask any scientist who works there and he or she will show you a long list of what still needs to be learned.

## Allen and Day Geothermal Survey, 1935

| Geyser | Height |
|---|---|
|  | *feet* |
| Old Faithful | 118 |
| Old Faithful | 128 |
| Old Faithful | 146 |
| Old Faithful | 150 |
| Old Faithful | 114 |
| Old Faithful | 140 |
| Old Faithful, 18 eruptions | 110–150 |
| Old Faithful, 20 eruptions | 118–160 |
| Imperial | 48–86 |
| Imperial | Max. 75 |
| White Dome | 18–29 |
| Great Fountain | 10–62 |
| Great Fountain, one splash | 90 |
| Oblong | 10–18 |
| Oblong | 14.5 |
| Daisy, oblique jet 75 ft. long | 67 |
| Solitary | 17–25 |
| Sawmill | 17–30 |
| Sawmill | 31.5 |
| Artemisia | 18–34 |
| Castle | 16.5 |
| Castle | 40 |
| Castle | 64 |
| Bead | 13 |
| Black Warrior or Steady | 21.5–26 |
| Jewel | 11.5–21.5 |
| Restless | 16.5 |
| Clepsydra | 5–24.5 |
| Narcissus | 2–16 |
| Grand | 71–166 |

## Preble's Map of Elk Winter Range, 1911

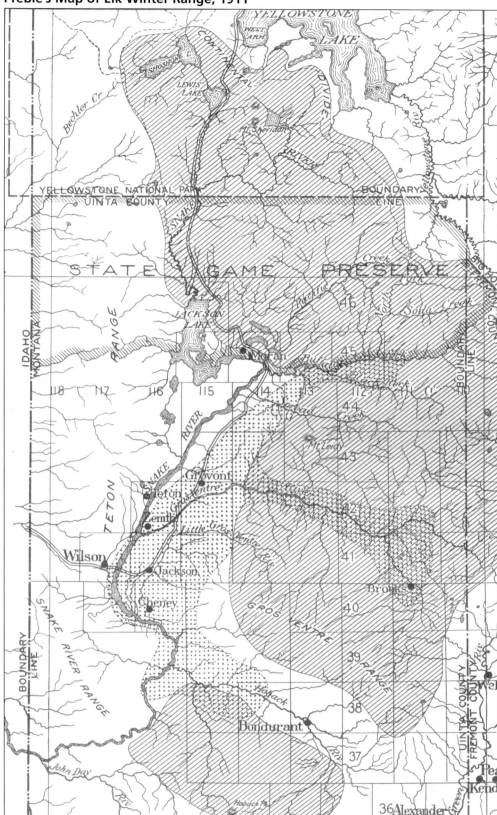

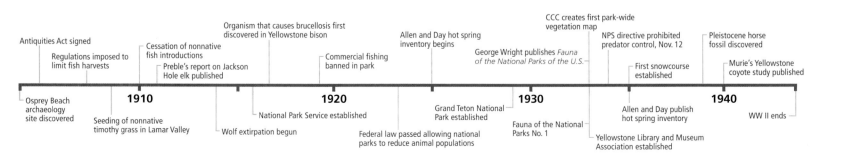

Antiquities Act signed

Regulations imposed to limit fish harvests

Cessation of nonnative fish introductions

Preble's report on Jackson Hole elk published

Organism that causes brucellosis first discovered in Yellowstone bison

Commercial fishing banned in park

Allen and Day hot spring inventory begins

George Wright publishes *Fauna of the National Parks of the U.S.*

CCC creates first park-wide vegetation map

NPS directive prohibited predator control, Nov. 12

First snowcourse established

Pleistocene horse fossil discovered

Murie's Yellowstone coyote study published

Osprey Beach archaeology site discovered

**1910**

Seeding of nonnative timothy grass in Lamar Valley

National Park Service established

Wolf extirpation begun

**1920**

Federal law passed allowing national parks to reduce animal populations

Grand Teton National Park established

Fauna of the National Parks No. 1

**1930**

Allen and Day publish hot spring inventory

Yellowstone Library and Museum Association established

**1940**

WW II ends

# Science History

The scientific enthusiasm of the nineteenth century was magnified in the twentieth century as more scientists pursued an ever-greater array of disciplines. Until the 1930s, most of the research in Yellowstone was foundational, meaning it was primarily descriptive in nature. In general, the examination of why the parks were a certain way and how natural processes worked did not enter the scientific conversation until the mid-to-late 1930s and did not become the norm until the 1960s. This change was a logical step as scientific understanding moved past traditional concepts that nature was static. The advent and flourishing of the ecological, geophysical, and geochemical sciences—aided by technological advances—gave scientists, land managers, and the public new and deeper insights into Yellowstone's complex natural systems and their ever-changing manifestations.

As a wider range of spatial and temporal scales was investigated, science focused less on the park as an isolated place and increasingly viewed it as a regional ecosystem, influenced by forces such as global climate change, human use and development, changing fire frequencies, the invasion of foreign organisms and human pollutants, and even the hiccupping of a pre-eruptive volcano.

Scientific exploration in the parks is broad in scope and varied in purpose. Basic research is without an immediate application, although such studies can lead to discoveries with tremendous implications, as with the realization that the Yellowstone landscape is part of a supervolcano. A great deal of park science is aimed at solving particular management problems. Public controversy also drives science, particularly in recent decades as resource managers have faced growing levels of public interest and skepticism. A maturing community of advocacy groups across the political and social spectra often uses data to dispute the official park perspective. In response, park managers have developed a deeper appreciation for the value of science in advancing legislative mandates and policy agendas—and even for defending their institutional identities. Since the park's establishment 140 years ago, Yellowstone institutional and ecological integrity have repeatedly been at risk. The park's rich scientific tradition helped see Yellowstone through many of these crises, largely due to the powerful influence of solid science in the federal courts. Science has also been known to sway Congress and the public. Jack Anderson, Yellowstone's superintendent in the late 1960s and early 1970s, once remarked that when walking into a federal

court he wanted to look over his shoulder and see a platoon of scientists following him, ready to lend support to his argument. The sentiment was prophetic; his tenure was a turning point in Yellowstone, with management moving beyond the subjective policy judgments of the past to using science as a basis for policy decisions.

Despite the importance of science to park management and the major contributions of the parks to the advancement of science, Congress did not formally add science to the statutory mission of the parks until 1998 with the passage of the National Parks Omnibus Management Act. This far-reaching legislation requires the National Park Service to identify, enumerate, and monitor the entirety of its natural resources and to use the best possible science when making resource decisions. As political process and public involvement continue to be drivers in the evolution of land and resource management, science informs the discussions that shape these changes.

## Wildlife Research

Researchers Frank and John Craighead "fitting" the first-ever satellite radio collar to Monique the elk, ca. 1970

## Microbial Research

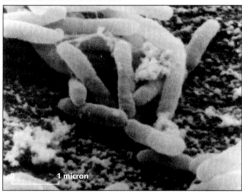

1 micron

Above is a scanning electron microscope photograph of a strain of the *Thermus aquaticus* YT-1 bacteria isolated by Thomas D. Brock and Hudson Freeze from Mushroom Hot Spring in 1966. The organism thrives at temperatures around 160°F. *Thermus aquaticus* produces an enzyme (Taq polymerase, also called Taq) that tremendously speeds up the copying of gene segments, a necessary step to have large enough samples for genetic analyses. Unlike other enzymes previously used for gene amplification, Taq is stable at the high temperatures required to split DNA into single strands, the first step in copying genes.

## Yellowstone Science, 1945–2010

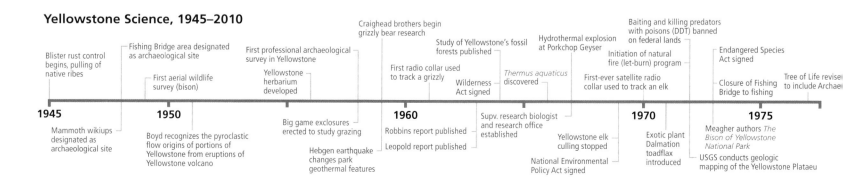

Blister rust control begins, pulling of native ribes

Fishing Bridge area designated as archaeological site

First aerial wildlife survey (bison)

First professional archaeological survey in Yellowstone

Yellowstone herbarium developed

Craighead brothers begin grizzly bear research

First radio collar used to track a grizzly

Study of Yellowstone's fossil forests published

Wilderness Act signed

*Thermus aquaticus* discovered

Hydrothermal explosion at Porkchop Geyser

Initiation of natural fire (let-burn) program

First-ever satellite radio collar used to track an elk

Baiting and killing predators with poisons (DDT) banned on federal lands

Endangered Species Act signed

Closure of Fishing Bridge to fishing

Tree of Life revised to include Archae

**1945** — **1950** — **1960** — **1970** — **1975**

Mammoth wikiups designated as archaeological site

Boyd recognizes the pyroclastic flow origins of portions of Yellowstone from eruptions of Yellowstone volcano

Big game exclosures erected to study grazing

Hebgen earthquake changes park geothermal features

Robbins report published

Leopold report published

Supv. research biologist and research office established

National Environmental Policy Act signed

Yellowstone elk culling stopped

Exotic plant Dalmation toadflax introduced

Meagher authors *The Bison of Yellowstone National Park*

USGS conducts geologic mapping of the Yellowstone Plateau

## Research Permits

Research activities that involve field-work and specimen collection within national parks require a permit. The early legislation that created Yellowstone (1872), established the National Park Service (1916), and protected antiquities (1906) directed park administrators to protect park resources and ensure that collections have a valid scientific purpose and are available to the public. Evidence from the park's archives suggests that Yellowstone started permitting collections as early as 1898.

Nationwide, more than 5,000 active research permits were granted by 266 park service units in 2007. Research activities within ten National Park Service units (Yellowstone, Everglades, Great Smoky Mountains, Yosemite, Hawaii Volcanoes, Death Valley, Big Bend, Point Reyes, Grand Canyon, and Rocky Mountain) accounted for more than a quarter of the permits issued that year. In any given year there are approximately 200 active permits in Yellowstone and seventy in Grand Teton. The number of permits in Yellowstone park grew dramatically in the 1980s when the National Park Service implemented more inclusive definitions of research that required a permit. The decline in total permits since 2000 is likely related to changes in the permitting system and declining research funding.

The majority of contemporary permit holders come from adjoining states, in part because researchers develop a heightened interest about their own "backyard." Many researchers, however, come from much farther away, including six countries outside the United States—evidence of the broad appeal of conducting science in Yellowstone.

Research in Yellowstone and Grand Teton national parks covers a wide range of subjects. In Yellowstone, examples of research topics from 2010 include interactions between fire and bark beetles, an inventory of prehistoric human use along road corridors, identifying microorganisms useful in biofuels production, the effect of climate change on native bees in the alpine zone, and the genetic adaptations of the yellow monkeyflower, a plant that survives in high temperature thermal areas and cool habitats alike. In a typical year, researchers publish roughly 100 journal articles related to their work in the park and every year several students earn master's or doctoral degrees for their Yellowstone studies.

### Research Permit Holders by State of Residence, 2001–2007

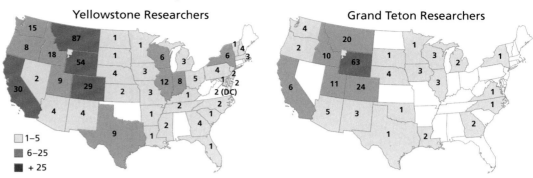

Yellowstone Researchers — Grand Teton Researchers

1–5
6–25
+ 25

### Yellowstone Research Permits

Permits, 1954–2010

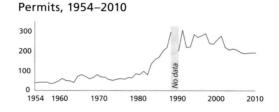

Foreign Research Permits, 2010

| Canada | 1 | New Zealand | 1 |
| Denmark | 1 | Portugal | 1 |
| Germany | 2 | United Kingdom | 1 |

Permit Categories, 2009

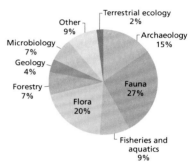

Terrestrial ecology 2%
Other 9%
Archaeology 15%
Microbiology 7%
Geology 4%
Forestry 7%
Fauna 27%
Flora 20%
Fisheries and aquatics 9%

### National Park Service Research Permits, 2007

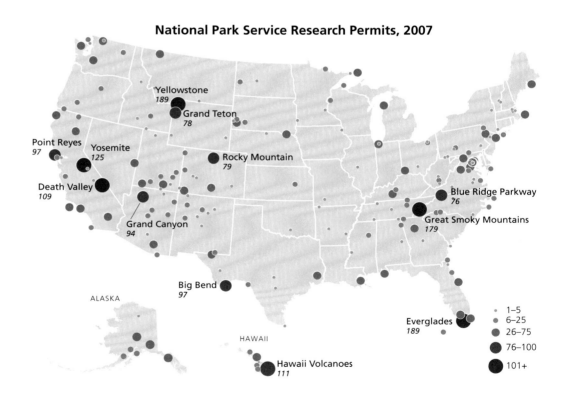

Yellowstone 189
Grand Teton 78
Point Reyes 97
Yosemite 125
Rocky Mountain 79
Death Valley 109
Grand Canyon 94
Big Bend 97
Blue Ridge Parkway 76
Great Smoky Mountains 179
Everglades 189
ALASKA
HAWAII
Hawaii Volcanoes 111

1–5
6–25
26–75
76–100
101+

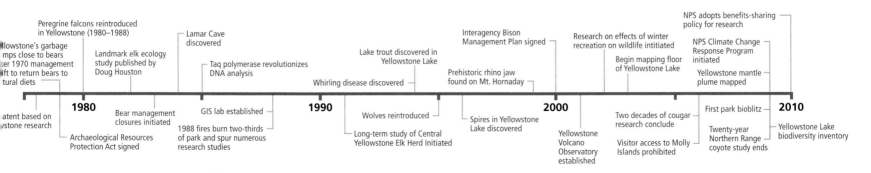

Peregrine falcons reintroduced in Yellowstone (1980–1988)

Yellowstone's garbage dumps close to bears after 1970 management shift to return bears to natural diets

Landmark elk ecology study published by Doug Houston

Lamar Cave discovered

Taq polymerase revolutionizes DNA analysis

Whirling disease discovered

Lake trout discovered in Yellowstone Lake

Interagency Bison Management Plan signed

Prehistoric rhino jaw found on Mt. Hornaday

Research on effects of winter recreation on wildlife intitiated

Begin mapping floor of Yellowstone Lake

NPS adopts benefits-sharing policy for research

NPS Climate Change Response Program initiated

Yellowstone mantle plume mapped

**1980**    **1990**    **2000**    **2010**

Patent based on Yellowstone research

Bear management closures initiated

Archaeological Resources Protection Act signed

GIS lab established

1988 fires burn two-thirds of park and spur numerous research studies

Wolves reintroduced

Long-term study of Central Yellowstone Elk Herd Initiated

Spires in Yellowstone Lake discovered

Yellowstone Volcano Observatory established

Two decades of cougar research conclude

Visitor access to Molly Islands prohibited

First park bioblitz

Twenty-year Northern Range coyote study ends

Yellowstone Lake biodiversity inventory

# Road History

**Chittenden Map, 1895**

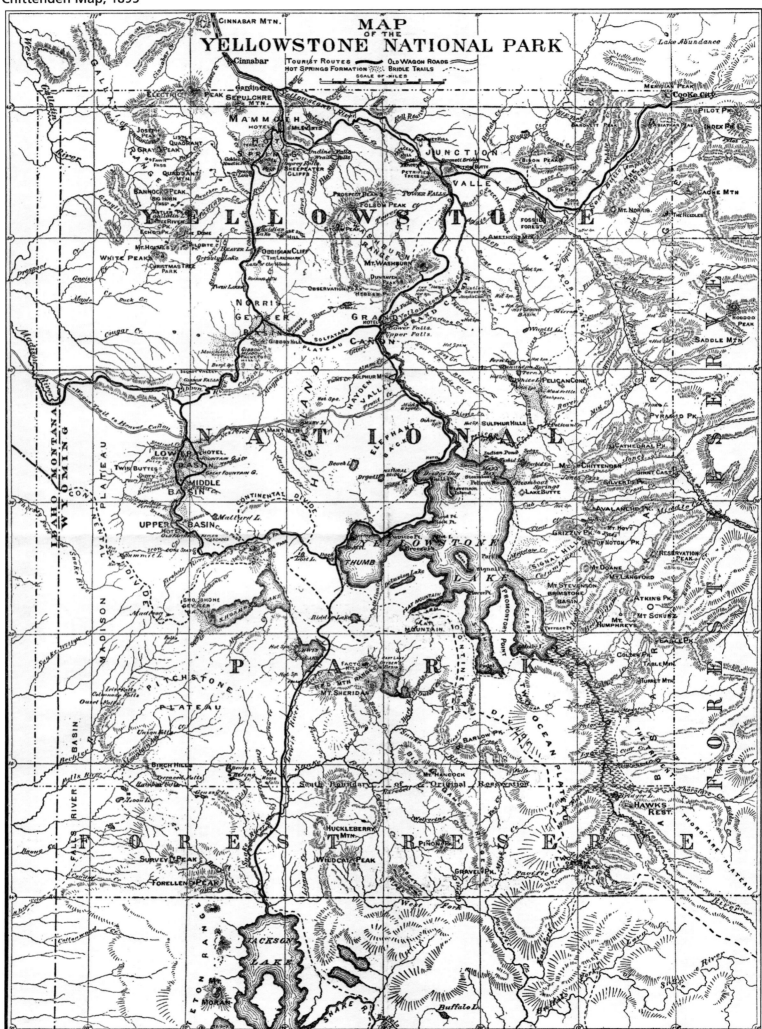

The current road system in Yellowstone National Park reflects a philosophy of blending with nature. The park's first superintendent, N.P. Langford, directed early road construction; the general quality of these roads was poor due to the lack of funds and challenging settings. The U.S. Army Corps of Engineers assumed responsibility for the park's roads in 1883. By the time Army engineer Lt. Hiram Chittenden left the park in 1906, Yellowstone had a good single-track wagon road system, the Grand Loop Road, and entrance roads.

After automobiles were officially permitted to enter the park in 1915, roads began a transformation from wagon ruts and dust to paved surfaces. The National Park Service (formed in 1916) continued its association with the Army Corp of Engineers and promoted the philosophy of harmonizing roads with their surrounding landscape. In 1926, the NPS and the Bureau of Public Roads agreed that the bureau would survey, construct, and improve the park's road system. The original Grand Loop and entrance roads proved to meet public needs, and managers decided to reconstruct the existing roads. By 1936, 200 miles of road had been improved and widened from the 18-foot Army standard to 28 feet, and nineteen major bridges had been constructed.

World War II stopped road construction activity; it was not until after 1956 (with the Mission 66 program) that improvements were again undertaken. In 1988, the Federal Highway Administration began a thirty-plus year road reconstruction program in the park.

## Development of Roads and Trails, 1877–1905

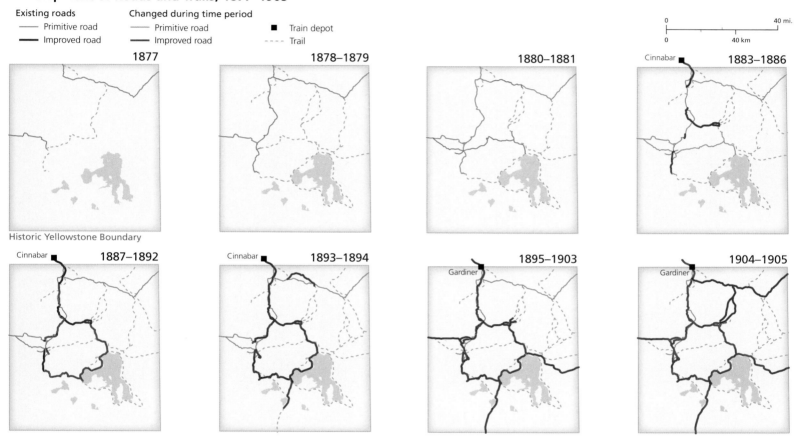

## Road Building in Yellowstone, 1877–1920

### Affiliation of Road Engineers

At the turn of the century, visitors traveling in the park were conveyed by horses, wagons, bicycles, or stagecoaches. Their common complaint about the roads was dust.
*Postcard by Frank J. Haynes.*

The arrival in 1883 of the Great Northern Railroad to Cinnabar and its extension south to Gardiner in 1903 significantly expanded access to the park. Until automobile travel became more common after 1915, most visitors arriving at the North Entrance (shown here) came by train.
*Postcard by Frank J. Haynes.*

In the 1920s, many of the park's roads remained narrow and unimproved, restricting traffic to one-way flow.
*Postcard by Frank J. Haynes.*

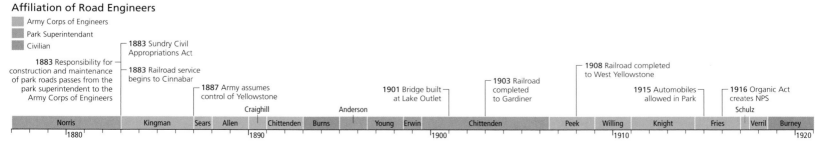

# Development at Old Faithful

The development of the Old Faithful area helps tell the story of parkwide development trends. Early trails and wagon roads allowed visitors access to individual park features and attractions. Improvements to these transportation routes influenced the location of tourist lodging and other support facilities. As numbers and demands of tourists changed, so did development within the park. Concessioners, the primary provider of tourist facilities, created much of the built environment. The National Park Service also influenced development as the agency attempted to balance the values of use and preservation.

At the time of the park's establishment in 1872, only trails existed throughout the Upper Geyser Basin. It was not long after the first narrow footbridge had been built across the Firehole River in 1877 that a system of trails and other footbridges ran through the area. Charged with protecting the park in 1886, the U.S. Army established a soldier station on the Firehole River to guard thermal features from vandalism. The U.S. Army Corps of Engineers improved the Upper Geyser Basin road system (as well as parkwide transportation networks) leading to the construction of facilities such as tent camps, a general store, and a photography studio. The Old Faithful Inn, one of the West's most important icons of park architecture, opened in 1904, accommodating wealthier tourists.

The advent of the automobile significantly altered tourism and development patterns within Yellowstone. Between 1916 and 1942,

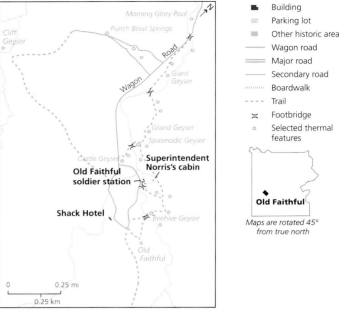

**1891**

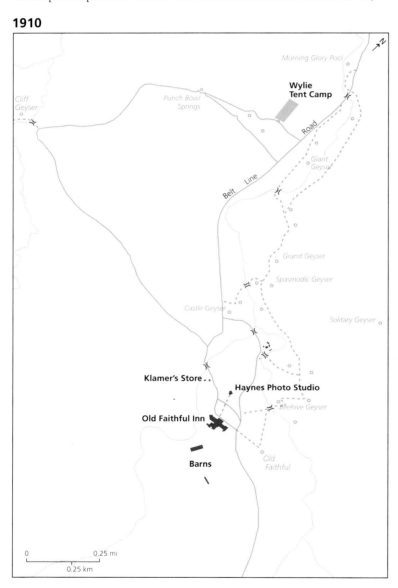

**1910**

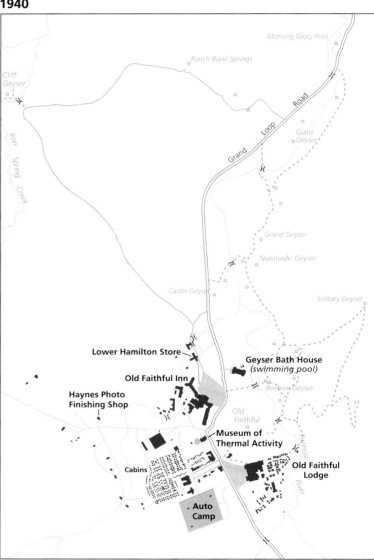

**1940**

## Parkwide Development and Major Events

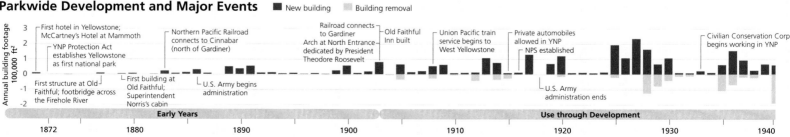

the annual number of visitors to the park rose from 36,000 to about 580,000; these visitors had widely varying tastes, schedules, and economic means—and, increasingly, their own cars. In response to this dramatically increasing influx, development of tourist services expanded greatly within the Old Faithful area. Overnight accommodations and meal services ranged from minimal to luxurious. Within the Upper Geyser Basin, the National Park Service built boardwalks and erected interpretive signs in an effort to protect thermal features from the onslaught of tourists.

The number of visitors to Yellowstone rose to one million in 1948. In 1956, the National Park Service responded to the national increase in park visitation by instituting Mission 66—a program to upgrade the nation's park infrastructure and facilities by 1966. Yellowstone's Mission 66 prospectus cautioned that development at Old Faithful (in particular the main access road that ran through the geyser basin) was encroaching on the thermal features and called for removal and relocation of the road. In 1963, the publication of the Leopold Report changed NPS philosophy to one of "natural regulation" with little manipulation of natural systems.

During the next decade, many of the buildings, older structures, and camping areas at Old Faithful were demolished. This made room for a new bypass road, an urban-style overpass, and two large parking lots (completed in 1972). These changes were intended to increase

protection of the area's resources by converting Old Faithful to a day-use area where visitors would come and go quickly and efficiently. The day-use plan was reversed in 1985.

Today, visitors can again stay overnight at Old Faithful, allowing time to explore the Upper Geyser Basin and wait for eruptions. An increasingly sophisticated understanding of human effects on hydrothermal processes now influences development decisions. A visitor education center was constructed in 2010, for example, using new technologies that are intended to protect the very hydrothermal processes the facility is meant to celebrate.

The development history near Old Faithful reflects similar changes throughout the park and provides a physical record of evolving philosophies of how to best balance tourism and preservation.

### Parkwide Total Building Footprint, 1871–2009

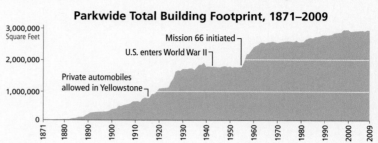

**1975**

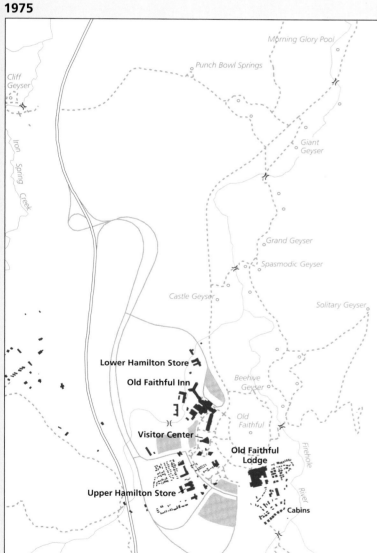

**2010**

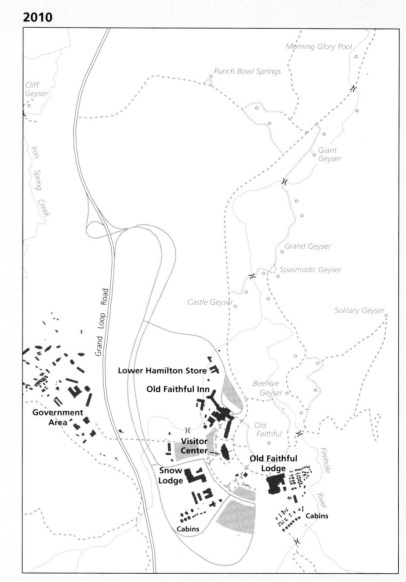

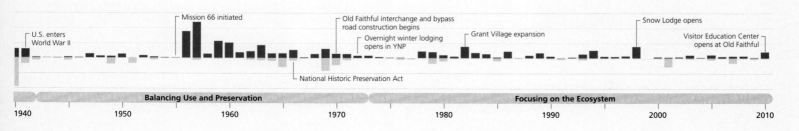

# Roads and Trails

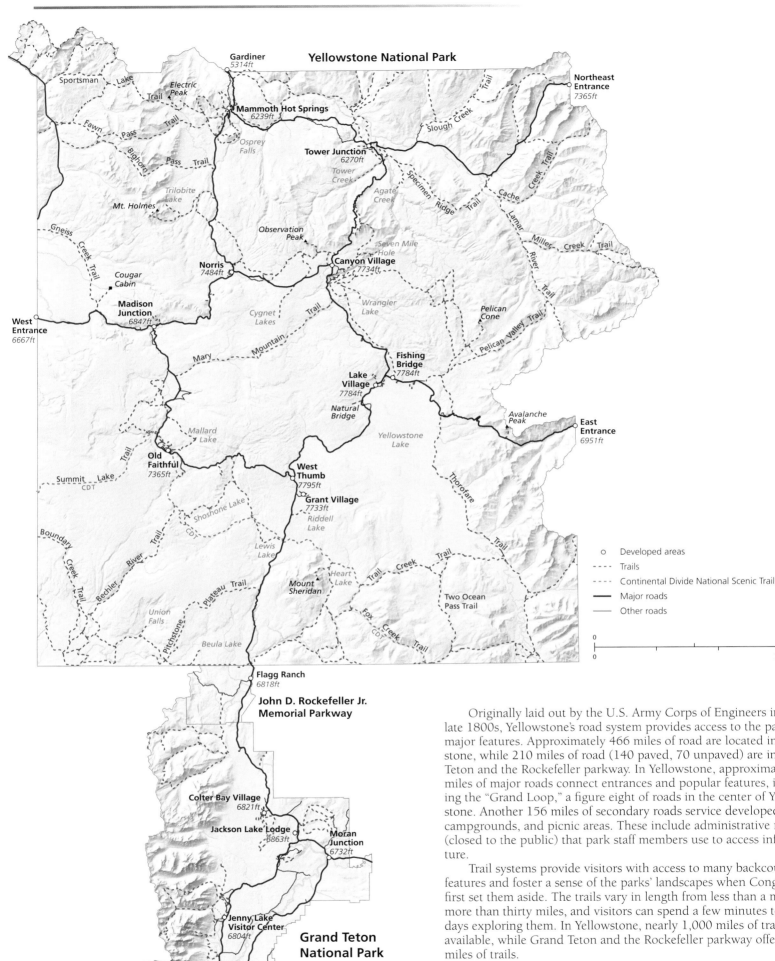

**Yellowstone National Park**

Gardiner
5314ft

Northeast
Entrance
7365ft

Sportsman Lake Trail

Electric
Peak

Mammoth Hot Springs
6239ft

Slough Creek Trail

Fawn Pass Trail

Osprey
Falls

Bighorn Pass Trail

Tower Junction
6270ft

Specimen Ridge Trail

Cache Creek Trail

Trilobite
Lake

Tower
Creek

Mt. Holmes

Agate
Creek

Lamar River Trail

Miller Creek Trail

Gneiss Creek Trail

Observation
Peak

Seven Mile
Hole

Cougar
Cabin

Norris
7484ft

Canyon Village
7734ft

West
Entrance
6667ft

Madison
Junction
6847ft

Cygnet
Lakes

Wrangler
Lake

Pelican
Cone

Mountain Trail

Mary

Pelican Valley Trail

Fishing
Bridge
7784ft

Lake
Village
7784ft

Avalanche
Peak

East
Entrance
6951ft

Natural
Bridge

Mallard
Lake

Yellowstone
Lake

Old
Faithful
7365ft

Trail

West
Thumb
7795ft

Summit Lake Trail

CDT

Grant Village
7733ft

Riddell
Lake

Shoshone Lake

CDT

Thorofare Trail

Boundary Creek Trail

River Trail

Lewis
Lake

Trail

Creek Trail

Plateau Trail

Mount
Sheridan

Heart
Lake

Trail

Two Ocean
Pass Trail

Union
Falls

Pitchstone Trail

Fox Creek Trail

CDT

Bechler

Beula Lake

Developed areas

- - - - Trails

- - - - - Continental Divide National Scenic Trail

——— Major roads

——— Other roads

0 ———————————— 20 mi.

0 ———————————— 30 km

Flagg Ranch
6818ft

**John D. Rockefeller Jr.
Memorial Parkway**

Colter Bay Village
6821ft

Jackson Lake Lodge
6863ft

Moran
Junction
6732ft

**Grand Teton
National Park**

Jenny Lake
Visitor Center
6804ft

Moose Junction
6509ft

Originally laid out by the U.S. Army Corps of Engineers in the late 1800s, Yellowstone's road system provides access to the park's major features. Approximately 466 miles of road are located in Yellowstone, while 210 miles of road (140 paved, 70 unpaved) are in Grand Teton and the Rockefeller parkway. In Yellowstone, approximately 310 miles of major roads connect entrances and popular features, including the "Grand Loop," a figure eight of roads in the center of Yellowstone. Another 156 miles of secondary roads service developed areas, campgrounds, and picnic areas. These include administrative roads (closed to the public) that park staff members use to access infrastructure.

Trail systems provide visitors with access to many backcountry features and foster a sense of the parks' landscapes when Congress first set them aside. The trails vary in length from less than a mile to more than thirty miles, and visitors can spend a few minutes to many days exploring them. In Yellowstone, nearly 1,000 miles of trails are available, while Grand Teton and the Rockefeller parkway offer 238 miles of trails.

The road and trail profiles illustrate the terrain of the parks. Many of the road segments follow river valleys, possibly leading to a misperception about the gentleness of Yellowstone's terrain. A review of profiles D, E, and F highlights the inclined shoulders of Mount Washburn and the Absaroka Range. Similarly, a trek from Cascade Canyon to Paintbrush Canyon (trail 4) in Grand Teton will tax all but the fittest hikers.

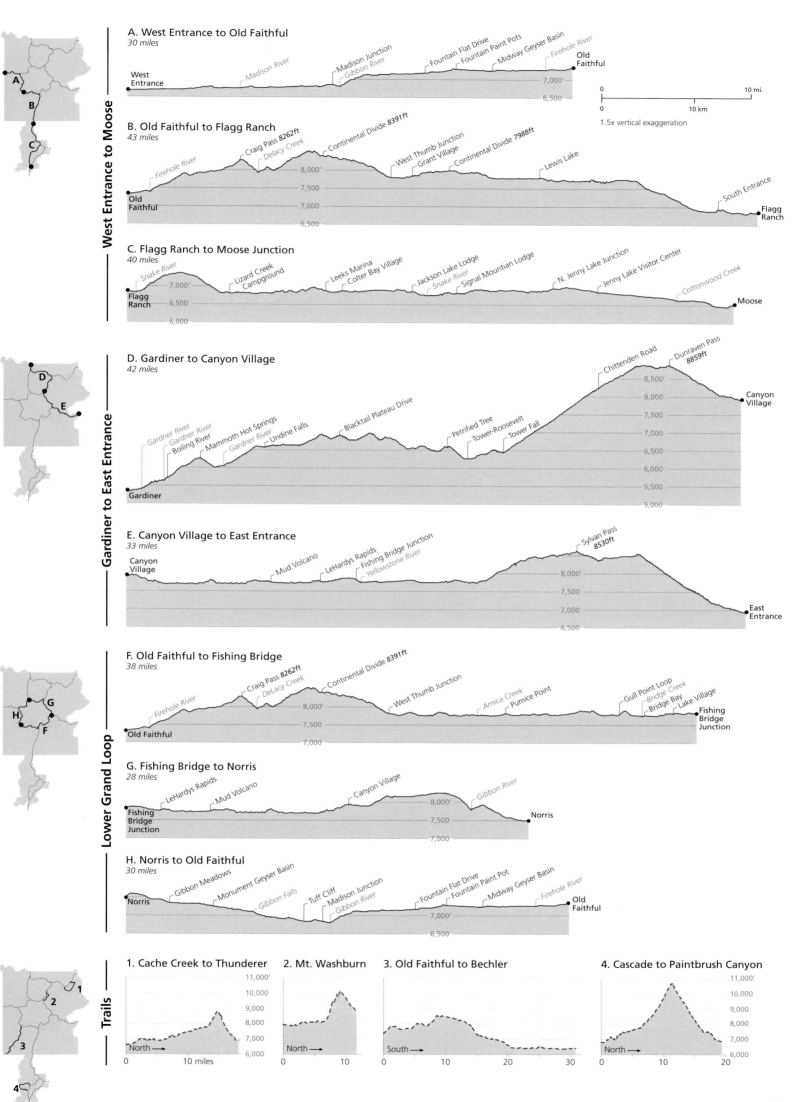

# Traffic

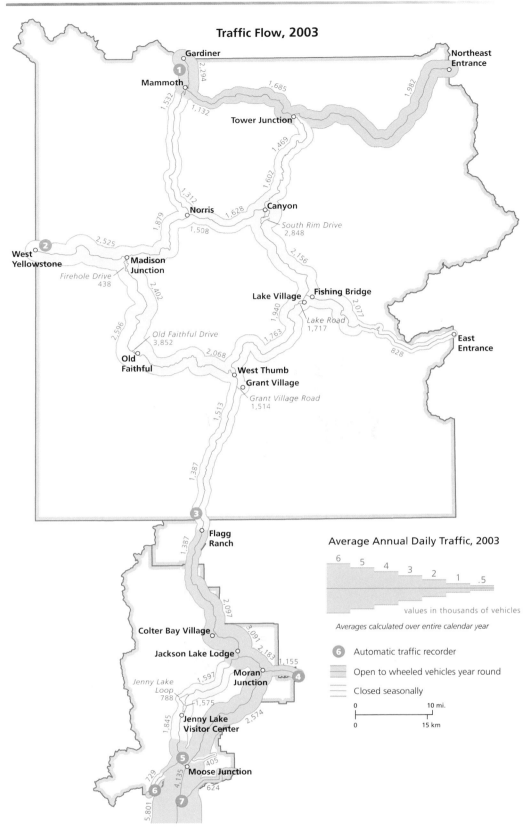

## Traffic Flow, 2003

Gardiner
Northeast Entrance
Mammoth
2,294
1,685
1,992
1,532
1,132
Tower Junction
1,469
1,602
Norris
1,312
1,628
Canyon
1,879
South Rim Drive 2,848
1,508
West Yellowstone
2,525
Madison Junction
Firehole Drive 438
2,402
2,156
Lake Village
Fishing Bridge
1,940
2,077
2,596
Lake Road 1,717
1,763
Old Faithful Drive 3,852
828
East Entrance
Old Faithful
2,068
West Thumb
Grant Village
Grant Village Road 1,514
1,513
1,387
Flagg Ranch
1,387
2,097
Colter Bay Village
3,091
2,183
1,155
Jackson Lake Lodge
1,597
Moran Junction
Jenny Lake Loop 788
1,575
2,574
1,845
Jenny Lake Visitor Center
405
729
4,135
Moose Junction
624
5,801

### Average Annual Daily Traffic, 2003

6   5   4   3   2   1   .5

values in thousands of vehicles

*Averages calculated over entire calendar year*

6 — Automatic traffic recorder
— Open to wheeled vehicles year round
— Closed seasonally

0 ——— 10 mi.
0 ——— 15 km

## Traffic Volume, 1990–2010

**Average Annual Daily Traffic Relative to 2010** — **Percent of Total Annual Traffic by Month**

**1 North Entrance, Yellowstone**
1992: 541 (80%)
2010: 679 (100%)

**2 West Entrance, Yellowstone***
1992: 936 (66%)
2010: 1,409 (100%)

**3 South Entrance, Yellowstone***
1992: 804 (111%)
2010: 727 (100%)

**4 U.S. 89, Inbound Traffic, Grand Teton**
1991: 677 (100%)
2010: 675 (100%)

**5 Moose Entrance, Grand Teton**
1991: 767 (88%)
2010: 873 (100%)

**6 Moose-Wilson Entrance, Grand Teton**
1991: 318 (80%)
2010: 395 (100%)

**7 Gros Ventre Junction, Grand Teton**
1991: 2,462 (70%)
2010: 3,507 (100%)

*\* Closed to car travel in winter months*

More than one million vehicles travel through Yellowstone and Grand Teton national parks in the summer months, often leading to congestion. Average daily traffic counts range from as many as 5,801 vehicles at Jackson to as few as 828 at the East Entrance road to Yellowstone National Park. Developed areas and popular destinations attract correspondingly higher traffic volumes. Variations in traffic flow can be relatively abrupt, as when visitors travel to Yellowstone Lake at Fishing Bridge but turn back rather than proceeding to the East Entrance. The parks' two-lane road system, lower speed limits, and design emphasis on helping visitors enjoy the parks reduces the traffic volume that roads can accommodate. Complicating traffic flows are the common, but difficult to predict, "animal jams" that occur when visitors stop to view wildlife. Miles-long traffic jams can result, leaving drivers further back in line fuming. Road construction, rapidly changing weather (it can snow any day of the year in the parks), and accidents can also affect traffic flow.

July and August have the highest traffic volumes in the parks, with relatively little traffic from November through April. Traffic volumes outside the parks are more evenly distributed through the year, although July and August remain peak months.

# Greater Yellowstone Traffic Flow, 2009

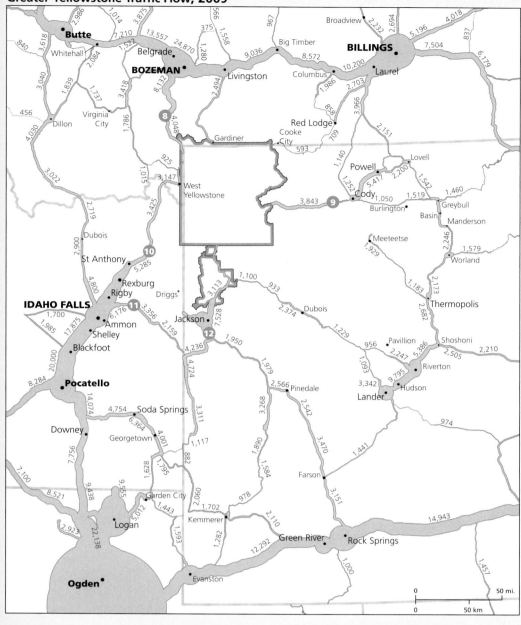

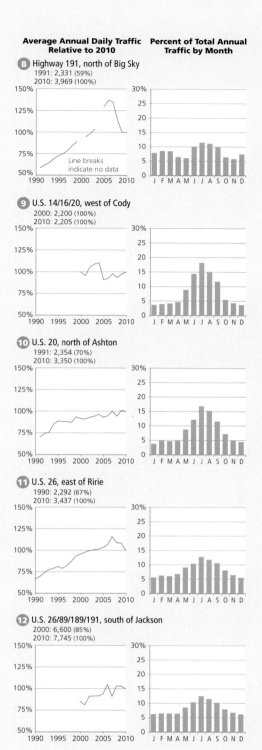

## Average Annual Daily Traffic Relative to 2010

## Percent of Total Annual Traffic by Month

**8 Highway 191, north of Big Sky**
1991: 2,331 (59%)
2010: 3,969 (100%)

Line breaks indicate no data

**9 U.S. 14/16/20, west of Cody**
2000: 2,200 (100%)
2010: 2,205 (100%)

**10 U.S. 20, north of Ashton**
1991: 2,354 (70%)
2010: 3,350 (100%)

**11 U.S. 26, east of Ririe**
1990: 2,292 (67%)
2010: 3,437 (100%)

**12 U.S. 26/89/189/191, south of Jackson**
2000: 6,600 (85%)
2010: 7,745 (100%)

Yellowstone and Grand Teton national parks are accessible by road. The nearest passenger train stop is across northern Montana. Bus service from nearby communities is limited. Visitors arriving at local airports typically rent cars to reach the parks. Commercial trucks are not allowed on park roads, except to support park and concessions operations. Outside the parks, large volumes of freight move along interstate highways. Railroads (the Montana Rail Link just north of Yellowstone and the BNSF Hi-Line in northern Montana) carry products such as grain and coal to the nation. The heavy rail traffic between Gillette and Casper, Wyoming, and points east largely reflects coal transport.

## Average Annual Daily Traffic, 2009

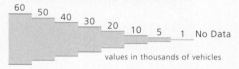

60  50  40  30  20  10  5  1  No Data

values in thousands of vehicles

**9** Automatic traffic recorder

## Airport Commercial Operations, 2008

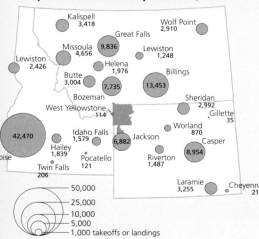

Kalispell 3,418
Wolf Point 2,910
Great Falls
Missoula 4,656
Lewiston 1,248
Lewiston 2,426
Helena 1,976
Butte 3,004
Billings 13,453
9,836
Bozeman 7,735
Sheridan 2,992
West Yellowstone 114
Gillette 35
Idaho Falls 1,579
Worland 870
Jackson 6,882
Casper 8,954
42,470
Hailey 1,839
Pocatello 121
Riverton 1,487
Twin Falls 206
Laramie 3,255
Cheyenne 212
Boise

50,000
25,000
10,000
5,000
1,000 takeoffs or landings

## Public Intercity Transportation, 2010

Shelby
Whitefish
Havre
Wolf Point
Great Falls
Lewiston
Missoula
Helena
Butte
Billings
Bozeman
Boise
Idaho Falls
Buffalo
Pocatello
Riverton
Gillette
Twin Falls
Casper
Rock Springs
Laramie
Cheyenne

• Bus stop
■ Amtrak stops
— Bus routes
— Passenger rail routes

## Annual Freight Tonnage, 2007

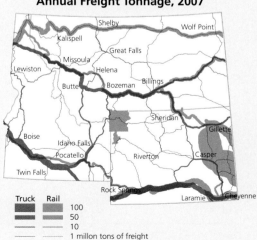

Shelby
Wolf Point
Kalispell
Great Falls
Missoula
Helena
Lewiston
Butte
Bozeman
Billings
Sheridan
Boise
Gillette
Idaho Falls
Pocatello
Riverton
Casper
Twin Falls
Rock Springs
Laramie
Cheyenne

Truck  Rail
100
50
10
1 million tons of freight

45

# Park Visitation

In Yellowstone's first decades, when the concept of a national park was still novel, managers realized that public support would be essential to the park's survival. They aggressively promoted the park and developed its infrastructure. These efforts, along with increasingly convenient automobile travel and other factors, led to rising visitor numbers through the 1920s and 1930s and an upsurge after World War II. Today, more than 150,000,000 visits to the park have been tallied. Thousands more "visit" Yellowstone and Grand Teton national parks through websites and webcams.

Visitation fluctuates from year to year in response to many influences—the economy, world wars, gas prices, and weather. The overall trend, however, has been upward: Yellowstone's estimated annual visitation first exceeded 10,000 in 1897, 100,000 in 1923, 1 million in 1948, 2 million in 1965, and 3 million in 1992.

Both parks are primarily summer destinations. Well over 3 million people visited Yellowstone between April and November of 2009, while fewer than 100,000 braved the park between December and March. More than half of the parks' annual visitation occurs in July and August. Summertime tourists often make the parks a stop on a longer road trip; three-quarters of Yellowstone's visitors come in one entrance and exit through another. Vacationers arrive from across the country and around the world—Canada and Europe provide the majority of international visitors.

Yellowstone is predominantly an automobile park. More than 95 percent of visitors in 2009 came in approximately 1,090,000 private vehicles (cars, trucks, recreational vehicles, and motorcycles), compared to fewer than 5,000 buses.

Yellowstone's visitors are trending toward shorter visits. In a 1987 survey, 31 percent of summer tourists spent one day or less in the

## Summer, 2009

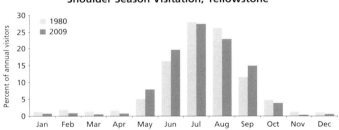

## Yellowstone Summer Visitors
### By Mode of Transport, 1993–2009

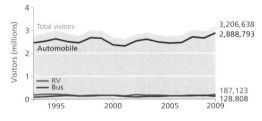

## State of Residence, 2006
*2,851 surveyed*

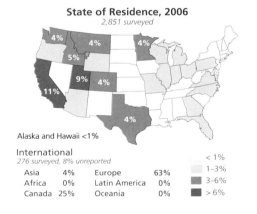

Alaska and Hawaii <1%

### International
*276 surveyed, 8% unreported*

| | | | |
|---|---|---|---|
| Asia | 4% | Europe | 63% |
| Africa | 0% | Latin America | 0% |
| Canada | 25% | Oceania | 0% |

Legend:
- < 1%
- 1–3%
- 3–6%
- > 6%

## Shoulder Season Visitation, Yellowstone

1980 / 2009

The "shoulder" seasons (spring and fall months) have been increasingly popular with visitors in recent decades. June and September are busy, with more than 1.1 million people enjoying Yellowstone during those combined months of 2009. Even May and October have risen in popularity—especially in years with good weather. In 2009, more than 385,000 people visited Yellowstone in those months. In the early days, Yellowstone opened on Memorial Day and closed Labor Day.

## Yellowstone Annual Visitation, 1872–2009

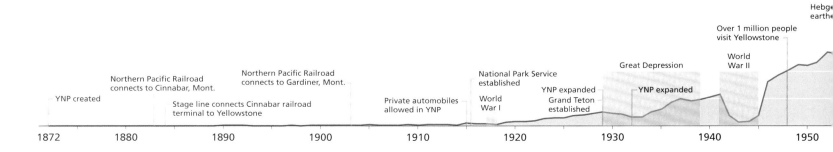

## Winter, 2009

North Entrance
47,259

Mammoth

West Entrance
26,830

East
Entrance
293

Old Faithful

South Entrance
12,402

Southbound traffic
14,178

Moran Entrance
40,514

Triangle X Ranch

Moose Entrance
29,616

Granite Canyon
18,216

Road Surface

Plowed
Groomed routes
Closed

## Yellowstone Winter Visitors
### By Mode of Transport, 1993–2009

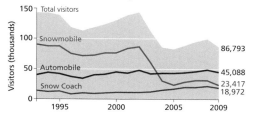

Total visitors

Snowmobile

Automobile

Snow Coach

86,793
45,088
23,417
18,972

Visitors (thousands)

1995   2000   2005   2009

### State of Residence, 2008
*384 surveyed*

16%
5%
5%
8%
5%
3%
5%
5%
3%

Alaska and Hawaii <1%

International
*18 surveyed*

| | | | |
|---|---|---|---|
| Asia | 0% | Europe | 61% |
| Africa | 0% | Latin America | 11% |
| Canada | 16% | Oceania | 11% |

< 1%
1–3%
3–6%
>6%

park. In summer 1996, 49 percent of visitors spent less than twenty-four hours in the park (the surveys' "length of stay" questions had different wording). In 1987, 68 percent of visitors spent two or more days in the park, while 49 percent did so in 2006. Interestingly, those spending four or more days went from 19 percent of summer visitors in 1987 to 22 percent in 2006.

The main gateway to Yellowstone is the West Entrance, through which 40 percent of summer visitors arrive. The entrance is adjacent to West Yellowstone, Montana, which has amenities for visitors staying overnight and making day trips to the park. In addition, the entrance is on routes from the population centers in California, Utah, Oregon, Washington, and Idaho, from which nearly a third of Yellowstone visitors come. The South Entrance provides access for 21 percent of visitors, reflecting proximity to Grand Teton National Park; Jackson, Wyoming; the Jackson Hole Airport; popular travel routes; and population centers in Utah and California. Seventeen percent of visitors come through the North Entrance.

Winter visitation is a fraction of summer activity at both parks. In 2009, less than 3 percent of visitors came to Yellowstone in the winter. More than half travel in wheeled vehicles, entering through Yellowstone's North Entrance and proceeding to Mammoth Hot Springs. Many continue on to the Lamar Valley and then to Silver Gate and Cooke City, Montana, just outside the Northeast Entrance. In winter, the road terminates just beyond Cooke City.

The balance of the park roads are closed to wheeled vehicles in the winter but opened for snowmobile and snow coach access. Since the mid-1950s, people have visited Old Faithful, the Grand Canyon of the Yellowstone, and other destinations in snow coaches or on snowmobiles. As of 2004, commercial guides are required for all snowmobile and snow coach visitors. Snowmobiles must be the cleanest and quietest machines available. Such rules have allowed visitors to enjoy the park while addressing the air quality, soundscape, wildlife, and safety concerns that arose in the 1980s and 1990s, an era of unlimited winter use.

## Backcountry, 2009

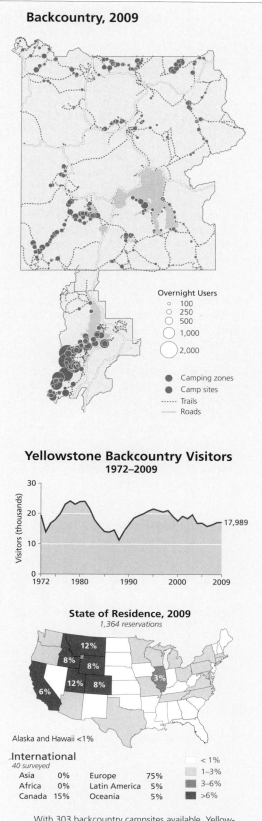

Overnight Users

100
250
500
1,000
2,000

Camping zones
Camp sites
Trails
Roads

## Yellowstone Backcountry Visitors
### 1972–2009

30

20

10

17,989

Visitors (thousands)

0
1972   1980   1990   2000   2009

### State of Residence, 2009
*1,364 reservations*

12%
8%
8%
12%
8%
3%
6%

Alaska and Hawaii <1%

International
*40 surveyed*

| | | | |
|---|---|---|---|
| Asia | 0% | Europe | 75% |
| Africa | 0% | Latin America | 5% |
| Canada | 15% | Oceania | 5% |

< 1%
1–3%
3–6%
>6%

With 303 backcountry campsites available, Yellowstone offers opportunities for visitors to get away from the crowds. Only about 18,000 of the 3.3 million visitors spent a night in the backcountry of Yellowstone in 2009. People access the remote regions of the park on foot (hikes ranging from one to thirty miles); with mules, llamas, or horses; and by boat. Yellowstone and Shoshone lakes have fifty four boat-accessible campsites along their shores.

In Grand Teton, backcountry opportunities abound, especially in the Teton Range. Both camping zones (visitors may camp anywhere; use of previously used locations is encouraged) and designated campsites are available. The parks emphasize using "leave no trace" principles and educate backcountry campers about bears. In Grand Teton for example, all persons camping below 10,000 feet are required to use bear-resistant food storage containers.

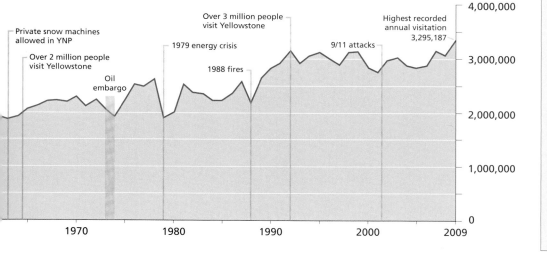

4,000,000

Over 3 million people
visit Yellowstone

Highest recorded
annual visitation
3,295,187

9/11 attacks

3,000,000

Private snow machines
allowed in YNP

1979 energy crisis

Over 2 million people
visit Yellowstone

1988 fires

2,000,000

Oil
embargo

1,000,000

0

1970   1980   1990   2000   2009

# Tetons Climbing History

The Teton Range has been central to the history of climbing in the United States. The profile of the range's lofty summits drew early explorers, and the steep ridges and alpine faces attracted experienced mountaineers. Tetons climbers of the late 1920s and 1930s established some of the most difficult alpine climbs in North America.

The geography of the region makes the Tetons well suited to climbing. The mountains thrust upward more than 7,000 feet from Jackson Hole along a sharp geological fault line. The short distance of the major peaks from the valley provides easy access to steep rock and ice. Solid crystalline rock cleaves into steep ridges and faces.

Early climbing in the Teton Range focused on the Grand Teton, the range's highest peak. Several expeditions from the 1870s to the early 1890s tried to reach the summit. James Stevenson and Nathaniel Langford, members of the 1872 Hayden geological survey, claimed to have reached the top but left no evidence. In 1898, William Owen and Franklin Spalding led a climbing party to the summit of the Grand to complete the first confirmed ascent of a major Teton peak.

The arrival of experienced climbers in the 1920s sparked the golden age of Tetons mountaineering. Climbers reached all the major summits and established new routes on soaring ridgelines and precipitous faces throughout the range. Albert Ellingwood of the Colorado Mountain Club made the first ascents of the Middle and South Tetons. Robert Underhill of the Appalachian Mountain Club climbed two new routes up the Grand. His North Ridge route (established with Fritiof Fryxell, 1931) was then the hardest alpine climb in America. Local Idahoan Paul Petzoldt was the first Tetons mountain guide and made many first ascents including the North Face of the Grand with his brother Eldon and Jack Durrance. Durrance established many additional hard classic lines on the Grand and other peaks.

In the decades that followed, new generations of climbers tested themselves on the Tetons peaks—on the classic lines and by establishing more difficult routes. They went on to tackle larger mountains in the great ranges of the world. The expedition that conducted the first American ascent of Everest in 1963 included four Tetons climbers.

## ❶ Central Tetons from the Southeast

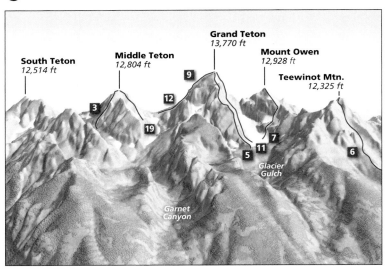

South Teton 12,514 ft
Middle Teton 12,804 ft
Grand Teton 13,770 ft
Mount Owen 12,928 ft
Teewinot Mtn. 12,325 ft

Garnet Canyon
Glacier Gulch

### Teton Panoramas

The view of the Teton Range from the east, as shown in the panorama below, is an iconic image of the Rocky Mountain west. The Tetons have become synonymous with rugged mountain peaks and appear in advertisements, movies, and other popular culture images. The Grand Teton and Mount Moran are the two most prominent summits in the range. Their ridges and faces contain the highest number of difficult alpine climbs.

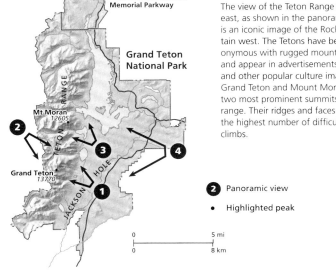

John D. Rockefeller, Jr. Memorial Parkway

Grand Teton National Park

Mt Moran 12605

TETON RANGE

JACKSON HOLE

Grand Teton 13770

❷ Panoramic view

• Highlighted peak

0 — 5 mi
0 — 8 km

## Difficulty of Teton Range First Ascents

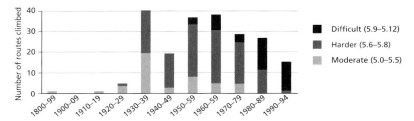

Number of routes climbed

1800–99, 1900–09, 1910–19, 1920–29, 1930–39, 1940–49, 1950–59, 1960–69, 1970–79, 1980–89, 1990–94

■ Difficult (5.9–5.12)
■ Harder (5.6–5.8)
■ Moderate (5.0–5.5)

## ❹ Tetons First Ascents

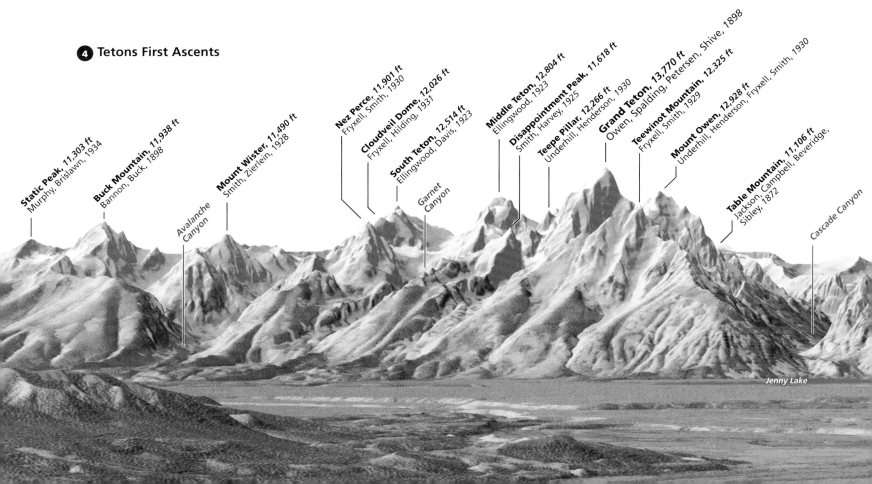

Static Peak, 11,303 ft
Murphy, Brislawn, 1934

Buck Mountain, 11,938 ft
Bannon, Buck, 1898

Avalanche Canyon

Mount Wister, 11,490 ft
Smith, Zierlein, 1928

Nez Perce, 11,901 ft
Fryxell, Smith, 1930

Cloudveil Dome, 12,026 ft
Fryxell, Hilding, 1931

South Teton, 12,514 ft
Ellingwood, Davis, 1923

Garnet Canyon

Middle Teton, 12,804 ft
Ellingwood, 1923

Disappointment Peak, 11,618 ft
Smith, Harvey, 1925

Teepe Pillar, 12,266 ft
Underhill, Henderson, 1930

Grand Teton, 13,770 ft
Owen, Spalding, Petersen, Shive, 1898

Teewinot Mountain, 12,325 ft
Fryxell, Smith, 1929

Mount Owen, 12,928 ft
Underhill, Henderson, Fryxell, Smith, 1930

Table Mountain, 11,106 ft
Jackson, Campbell, Beveridge, Sibley, 1872

Cascade Canyon

Jenny Lake

**2** Grand Teton from the Northwest

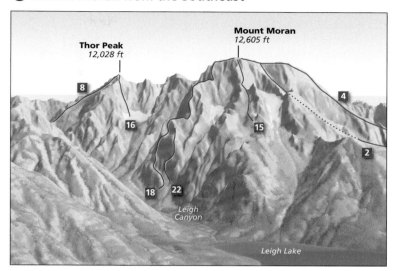

Teewinot Mtn.
12,325 ft

Mount Owen
12,928 ft

Grand Teton
13,770 ft

Enclosure

Middle Teton
12,804 ft

17  10  23  21  14  20  13  1

Valhalla
Canyon

Dartmouth
Basin

Cascade
Canyon

**3** Mount Moran from the Southeast

Thor Peak
12,028 ft

Mount Moran
12,605 ft

8  16  15  4  2

18  22

Leigh
Canyon

Leigh Lake

## Notable Climbs in the Teton Range

| | Year | Climb | Climbers |
|---|---|---|---|
| **1** | 1898 | Grand Teton, *Owen-Spalding* | Owen, Spalding, Petersen, Shive |
| | 1898 | Buck Mountain, *East Face* | Bannon, Buck |
| **2** | 1922 | Mount Moran, *Skillet Glacier* | Hardy, Rich, McNulty |
| **3** | 1923 | Middle Teton, *Ellingwood Couloir* | Ellingwood |
| | 1923 | South Teton, *Northwest Couloir* | Ellingwood, Davis |
| **4** | 1924 | Mount Moran, *Northeast Ridge* | Ellingwood |
| | 1925 | Disappointment Peak, *Lake Ledges* | Smith, Harvey |
| | 1928 | Mount Wister, *Northeast Couloir* | Smith, Zierlein |
| **5** | 1929 | Grand Teton, *East Ridge* | Underhill, Henderson |
| **6** | 1929 | Teewinot Mountain, *East Face* | Fryxell, Smith |
| | 1929 | Mount St. John, *West Ridge* | Fryxell, Smith |
| | 1929 | Symmetry Spire, *East Ridge* | Fryxell, Smith |
| | 1930 | Nez Perce Peak, *Northwest Couloirs* | Fryxell, Smith |
| **7** | 1930 | Mount Owen, *East Ridge* | Underhill, Henderson, Fryxell, Smith |
| | 1930 | Teepee Pillar, *Underhill-Henderson* | Underhill, Henderson |
| **8** | 1930 | Thor Peak, *South Slope (descent)* | Petzoldt, Strange |
| **9** | 1931 | Grand Teton, *Upper Exum Ridge* | Exum |
| | 1931 | Middle Teton, *North Ridge* | Underhill, Fryxell |
| **10** | 1931 | Grand Teton, *North Ridge* | Underhill, Fryxell |
| | 1931 | Cloudveil Dome, *East Ridge* | Fryxell, Hilding |
| | 1935 | Grand Teton, *Winter Ascent* | Brown, Petzoldt, Petzoldt |
| **11** | 1936 | Grand Teton, *North Face* | Durrance, Petzoldt, Petzoldt |
| **12** | 1936 | Grand Teton, *Lower Exum Ridge* | Durrance, Henderson |
| **13** | 1938 | Enclosure, *Northwest Ridge* | Durrance, Davis |
| **14** | 1940 | Grand Teton, *West Face* | Durrance, Coulter |
| **15** | 1941 | Mount Moran, *CMC Route* | Petzoldt, Hawkes, Clark, Plumley |
| | 1945 | Mount Owen, *North Face* | Kraus, Snyder |
| | 1950 | Cloudveil Dome, *North Face* | Pownall, Kenworthy |
| **16** | 1950 | Thor Peak, *East Face* | Exum, Brewer, Pownall |
| **17** | 1951 | Mount Owen, *North Ridge* | Emerson, Clayton |
| **18** | 1953 | Mount Moran, *Direct South Buttress* | Ortenburger, Emerson, Decker |
| | 1953 | Grand Teton, *North Face-Direct Finish* | Emerson, Ortenburger, Unsoeld |
| | 1954 | Nez Perce Peak, *South Ridge* | Merriam, Buckingham, Clark |
| | 1960 | Grand Teton, *Northwest Chimney* | Dornan, Ortenburger, Ortenburger |
| **19** | 1960 | Middle Teton, *Robbins-Fitschen* | Robbins, Fitschen |
| | 1961 | Teepe Pillar, *Northeast Face* | Robbins, Taylor |
| | 1961 | Middle Teton, *Taylor Route* | Robbins, Taylor |
| **20** | 1961 | Enclosure, *Black Ice Couloir* | Jacquot, Swedlund |
| | 1963 | Teewinot–Nez Perce, *Grand Traverse* | Steck, Long, Evans |
| | 1964 | Death Canyon, *Pillar of Death* | Medrick, Dornan |
| | 1964 | Death Canyon, *The Snaz* | Chouinard, Hempel |
| **21** | 1969 | Enclosure, *Lowe Route* | Lowe, Lowe |
| **22** | 1973 | Mount Moran, *South Buttress-Right* | Wunsch, Higbee |
| | 1977 | Enclosure, *High Route* | Fowler, Glenn |
| | 1979 | Grand Teton, *Route Canal* | Lowe, Fowler |
| | 1981 | Enclosure, *Visionquest Couloir* | Stern, Quinlan |
| **23** | 1981 | Grand Teton, *Loki's Tower* | Whiton, Stern |
| | 1983 | Grand Teton, *Speed Ascent (3:06)* | Thatcher |
| | 1985 | Enclosure, *Emotional Rescue* | Jackson, Rickert |
| | 1991 | Enclosure, *Lookin' For Trouble* | Beyer |
| | 1993 | Cloudveil Dome, *Nimbus Route* | Lowe, Koch |
| | 2003 | Grand Teton, *Golden Pillar* | Collins, Johnstone |

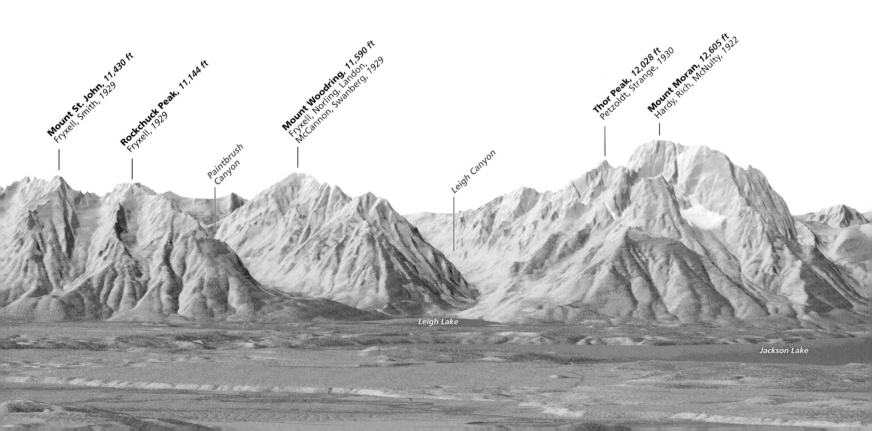

**Mount St. John, 11,430 ft**
Fryxell, Smith, 1929

**Rockchuck Peak, 11,144 ft**
Fryxell, 1929

Paintbrush
Canyon

**Mount Woodring, 11,590 ft**
Fryxell, Norling, Landon,
McCannon, Swanberg, 1929

Leigh Canyon

**Thor Peak, 12,028 ft**
Petzoldt, Strange, 1930

**Mount Moran, 12,605 ft**
Hardy, Rich, McNulty, 1922

Leigh Lake

Jackson Lake

# The Economy

As they have in much of the West, resource development, agriculture, and tourism have played important roles in the economic history of the Greater Yellowstone Area. For gateway communities such as Cody, Wyoming, and West Yellowstone, Montana, tourism has long been the mainstay of the economy. For communities more distant from the park, mining, oil and gas development, timber harvesting, farming, and ranching have been prime economic drivers. This began to change in the 1970s, 1980s, and particularly in the 1990s. In 1970 close to 30 percent of labor earnings in the region were from farming, ranching, mining, oil and gas development, and manufacturing (which includes the wood products industry). By 2000 these sectors accounted for only 12 percent of total labor earnings. By 2008—a period that included a resurgence in oil and gas development, particularly in Wyoming—the resource development sector accounted for 11 percent of total labor earnings.

The sources of fastest economic growth have been the service and professional industries and nonlabor sources of income, such as retirement and investment income. The service industries include sectors often associated with tourism and include components of retail trade, transportation (for example, tour buses and airlines), and amusement and recreation businesses (for example, fishing guides and private parks and zoos). Service sectors also include relatively high-wage sectors, such as architecture, engineering, software development, and legal and medical services. Because of advances in telecommunications (the Internet, cell phones), efficient and ubiquitous delivery services such as FedEx and UPS, and the growth of regional airports, many skilled professionals formerly tied to jobs located in larger metropolitan areas are now able to live in relatively remote, rural locations offering an attractive quality of life. In the 1980s and 1990s, the Greater Yellowstone Area was drawing increasing numbers of these new "amenity migrants." It became possible to work, for example, as an engineer in Bozeman, Montana, while the main factory or office was elsewhere and clients, suppliers, and colleagues were dispersed across the country or around the world. Fast-growing areas such as Gallatin County, Montana, Bonneville County, Idaho, and Teton County, Wyoming, offered a combination of recreational, environmental, and cultural amenities; an educated workforce; and ready access to the outside world via transportation networks, particularly daily commercial air service.

Nonlabor income sources accounted for half the net growth in real personal income from 1970 to 2000 and were 39 percent of total personal income by 2000 (42 percent by 2008). Nonlabor income consists of income from investments (dividends, interest, and rent) and government transfer payments such as retirement payments and Medicare. With significant growth in the stock market and the aging "baby boom" generation, these sources of income became a major driver for

many counties. Because this measure does not include private pensions and savings (for example, 401K retirement plans), it underestimates the true size of investment and retirement payments in the region.

As service sector earnings increased, and as more nonlabor sources of income flowed into the area, other sectors of the economy such as construction, government, and manufacturing (although not timber-related manufacturing) also grew.

## Decoupling

Historically, farming, ranching, and the resource industries were more important forces in the region than they are today because the area's economy was more specialized and less diverse. When farm and ranch income fluctuated, other sectors of the economy would fluctuate with them; the same was true for the timber industry and mining. These industries were, in a sense, the "horse pulling the cart." During the 1970s, 1980s, and 1990s, this image became less apt, with new sectors entering the economy that had little or nothing to do with agriculture and natural resource development. An example of this change is the growth of service industries. In the past, service sector employment consisted primarily of grocery store clerks, gas station attendants, and other occupations typically associated with the production of a service. If the lumber mill laid off people, mill workers would have less money to spend, and the service sectors would decline

along with the timber industry. By the late 1980s and early 1990s, the service sectors began to include new migrants working as financial consultants, software developers, and in other occupations still closely tied to goods production (for example, architects are part of the construction process), but "decoupled" from the traditional economic drivers in the region. Rather than being tied to the fluctuations of the local timber, mining, and agricultural sectors, these new service occupations are tied to economic forces outside the region.

## Total Personal Income by County, 1970–2000
### Values in millions of dollars

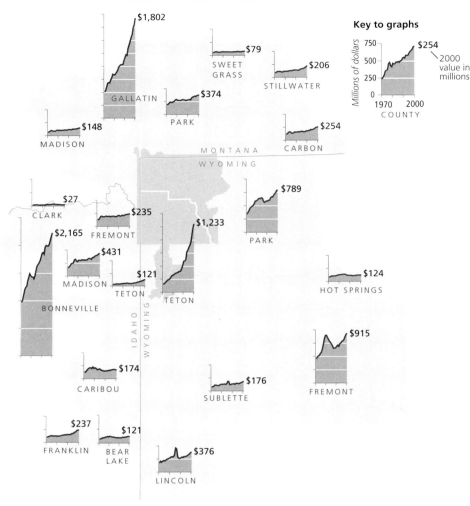

## Decline of Mining Income in Caribou County

## Growth of Service Industries in the GYA

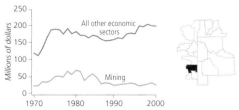

*includes wood products

# Total Personal Income from All Economic Sectors, 1970–2000

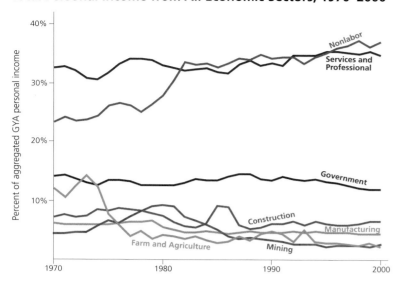

# Personal Income by Sector

Percentage of GYA sector total

| | | |
|---|---|---|
| ☐ <1% | | ▨ 6-10% |
| ☐ 1% | | ▩ 11-20% |
| ▨ 2-5 % | | ■ > 20% |

## Government

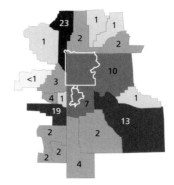

### Total Personal Income from Government, 2000
*Value in millions of dollars**

| | | | |
|---|---|---|---|
| Bear Lake (ID) | $20.0 | Lincoln (WY) | $52.2 |
| Bonneville (ID) | $246.6 | Madison (ID) | $55.0 |
| Carbon (MT) | $19.8 | Madison (MT) | $17.9 |
| Caribou (ID) | $23.3 | Park (MT) | $30.5 |
| Clark (ID) | $6.5 | Park (WY) | $132.4 |
| Franklin (ID) | $24.6 | Stillwater (MT) | $16.0 |
| Fremont (ID) | $41.8 | Sweet Grass (MT) | $9.6 |
| Fremont (WY) | $168.0 | Sublette (WY) | $24.8 |
| Gallatin (MT) | $302.3 | Teton (ID) | $14.5 |
| Hot Springs (WY) | $18.9 | Teton (WY) | $92.6 |

Aggregated GYA County Income — **$1,317.2** *12.0%*

## Construction

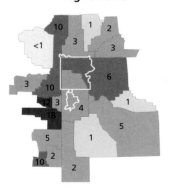

### Total Personal Income from Construction, 2000
*Value in millions of dollars**

| | | | |
|---|---|---|---|
| Bear Lake (ID) | $1.9 | Lincoln (WY) | $30.9 |
| Bonneville (ID) | $145.0 | Madison (ID) | $21.5 |
| Carbon (MT) | $7.2 | Madison (MT) | $11.6 |
| Caribou (ID) | $10.7 | Park (MT) | $16.3 |
| Clark (ID) | $0.1 | Park (WY) | $53.1 |
| Franklin (ID) | $4.3 | Stillwater (MT) | $4.7 |
| Fremont (ID) | $7.3 | Sweet Grass (MT) | $6.0 |
| Fremont (WY) | $62.8 | Sublette (WY) | $15.4 |
| Gallatin (MT) | $166.0 | Teton (ID) | $8.8 |
| Hot Springs (WY) | $3.8 | Teton (WY) | $142.4 |

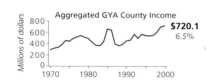

Aggregated GYA County Income — **$720.1** *6.5%*

## Manufacturing

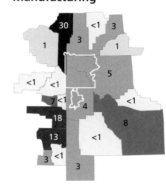

### Total Personal Income from Manufacturing, 2000
*Value in millions of dollars**

| | | | |
|---|---|---|---|
| Bear Lake (ID) | $2.4 | Lincoln (WY) | $16.2 |
| Bonneville (ID) | $90.1 | Madison (ID) | $33.8 |
| Carbon (MT) | $3.2 | Madison (MT) | $5.5 |
| Caribou (ID) | $63.4 | Park (MT) | $13.3 |
| Clark (ID) | $0.8 | Park (WY) | $25.9 |
| Franklin (ID) | $12.9 | Stillwater (MT) | $13.0 |
| Fremont (ID) | $2.4 | Sweet Grass (MT) | $2.1 |
| Fremont (WY) | $37.4 | Sublette (WY) | $1.4 |
| Gallatin (MT) | $146.3 | Teton (ID) | $2.1 |
| Hot Springs (WY) | $1.4 | Teton (WY) | $21.8 |

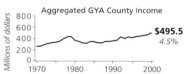

Aggregated GYA County Income — **$495.5** *4.5%*

## Farm and Agriculture

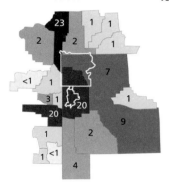

### Total Personal Income from Farm and Agricultural Services, 2000
*Value in millions of dollars**

| | | | |
|---|---|---|---|
| Bear Lake (ID) | $3.8 | Lincoln (WY) | $5.1 |
| Bonneville (ID) | $42.8 | Madison (ID) | $29.9 |
| Carbon (MT) | $6.4 | Madison (MT) | $0.2 |
| Caribou (ID) | $10.8 | Park (MT) | $6.9 |
| Clark (ID) | $7.3 | Park (WY) | $15.1 |
| Franklin (ID) | $23.7 | Stillwater (MT) | $4.1 |
| Fremont (ID) | $25.2 | Sweet Grass (MT) | $1.2 |
| Fremont (WY) | $11.2 | Sublette (WY) | $3.6 |
| Gallatin (MT) | $24.7 | Teton (ID) | $8.1 |
| Hot Springs (WY) | $1.3 | Teton (WY) | $9.3 |

Aggregated GYA County Income — **$240.7** *2.2%*

## Mining

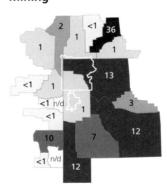

### Total Personal Income from Mining, 2000
*Value in millions of dollars**

| | | | |
|---|---|---|---|
| Bear Lake (ID) | n/d | Lincoln (WY) | $0.3 |
| Bonneville (ID) | $1.2 | Madison (ID) | $1.8 |
| Carbon (MT) | $2.3 | Madison (MT) | $2.4 |
| Caribou (ID) | $28.4 | Park (MT) | $1.5 |
| Clark (ID) | $1.2 | Park (WY) | $38.2 |
| Franklin (ID) | $0.7 | Stillwater (MT) | $104.5 |
| Fremont (ID) | $1.6 | Sweet Grass (MT) | $0.3 |
| Fremont (WY) | $34.9 | Sublette (WY) | $19.8 |
| Gallatin (MT) | $4.9 | Teton (ID) | n/d |
| Hot Springs (WY) | $9.5 | Teton (WY) | $4.2 |

Aggregated GYA County Income — **$289.7** *2.6%*

## Services and Professional

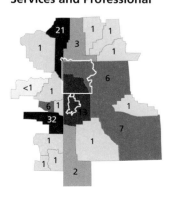

### Total Personal Income from Services and Professional, 2000
*Value in millions of dollars**

| | | | |
|---|---|---|---|
| Bear Lake (ID) | $26.0 | Lincoln (WY) | $87.8 |
| Bonneville (ID) | $1,239.0 | Madison (ID) | $222.1 |
| Carbon (MT) | $47.9 | Madison (MT) | $38.2 |
| Caribou (ID) | $33.6 | Park (MT) | $126.3 |
| Clark (ID) | $8.4 | Park (WY) | $226.4 |
| Franklin (ID) | $39.2 | Stillwater (MT) | $32.6 |
| Fremont (ID) | $41.8 | Sweet Grass (MT) | $20.0 |
| Fremont (WY) | $261.2 | Sublette (WY) | $39.1 |
| Gallatin (MT) | $817.2 | Teton (ID) | $22.4 |
| Hot Springs (WY) | $36.0 | Teton (WY) | $490.5 |

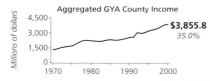

Aggregated GYA County Income — **$3,855.8** *35.0%*

## Nonlabor Sources

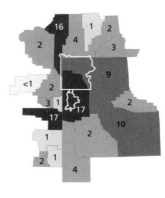

### Total Personal Income from Nonlabor Sources, 2000
*Value in millions of dollars**

| | | | |
|---|---|---|---|
| Bear Lake (ID) | $49.1 | Lincoln (WY) | $155.2 |
| Bonneville (ID) | $683.4 | Madison (ID) | $134.6 |
| Carbon (MT) | $123.4 | Madison (MT) | $80.6 |
| Caribou (ID) | $58.0 | Park (MT) | $169.8 |
| Clark (ID) | $6.6 | Park (WY) | $360.2 |
| Franklin (ID) | $72.2 | Stillwater (MT) | $85.4 |
| Fremont (ID) | $92.2 | Sweet Grass (MT) | $47.5 |
| Fremont (WY) | $412.2 | Sublette (WY) | $85.6 |
| Gallatin (MT) | $648.7 | Teton (ID) | $45.3 |
| Hot Springs (WY) | $61.2 | Teton (WY) | $711.1 |

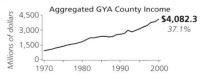

Aggregated GYA County Income — **$4,082.3** *37.1%*

*Adjusted to 2000 US Dollar Values*

51

# Labor and Employment

The labor force includes all a region's population currently employed or seeking employment. Most of the individuals who are not in the labor force are younger than sixteen years, past retirement age, or not actively seeking paid employment. In the Yellowstone region, the size of the labor force has increased over time as the size of the total population has grown and as more women have entered the labor force.

Across the three-state region, labor force participation rates vary. Idaho and Montana, on average, have participation rates similar to the national average, but many of the more isolated rural counties have lower participation rates. As a state, Wyoming has a higher participation rate than the national average, a long-standing trend. This is partially explained by the high levels of agricultural self-employment in the state.

The unemployment rate measures the portion of the labor force that is currently unemployed and looking for work. Across the Yellowstone region, unemployment rates have roughly followed national trends. Unemployment was high throughout the region in the mid-1980s as the national economy experienced a deep economic downturn. Unemployment rates declined throughout the 1990s and into the 2000s as the national economy experienced a period a relative prosperity but then rose sharply as the U.S. economy entered the recession of 2007 to 2009.

The three states have all had unemployment rates lower than the national average since 2001. Within the Yellowstone region, individual counties have weathered the recession of 2007 to 2009 very differently. Some of the more isolated rural counties, such as Sweet Grass in Montana and Sublette in Wyoming have had very low unemployment in the latest recession. These counties make up a small portion of total employment in the region, however, and the counties with larger total employment have a greater influence on average unemployment rates. The counties with the largest total employment—Bonneville in Idaho and Gallatin in Montana—have a larger

## Labor Force Participation, 2009

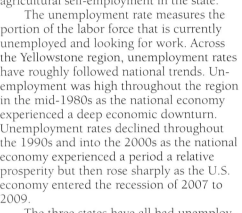

40  50  60  70  80  90%

## Tristate Labor Force by Gender, 1900–2000

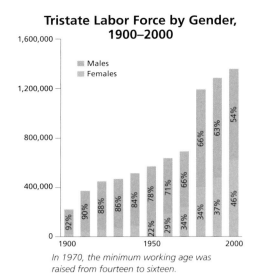

In 1970, the minimum working age was raised from fourteen to sixteen.

## Prevalent Industry, 1990

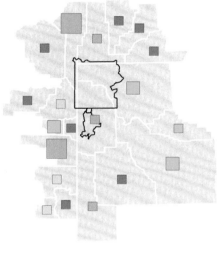

## Prevalent Industry, 2009

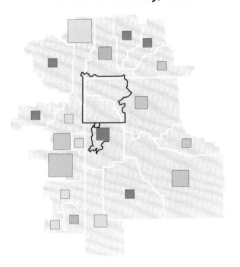

Total Employed
- ☐ <8,000
- ☐ 8,001–16,000
- ☐ 16,001–24,000
- ☐ 24,001–32,000
- ☐ >32,000

Industry
- Natural resources and mining
- Construction and manufacturing
- Trade, transportation, and utilities
- Education and health services
- Leisure and hospitality services

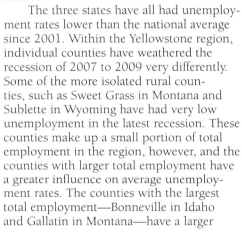

## Employees by Industry

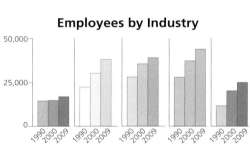

## Government Employment

### Total Government
Percent of total employment, 2009

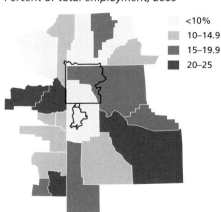

- ☐ <10%
- 10–14.9
- 15–19.9
- 20–25

### Federal Government
Percent of total employment, 2009

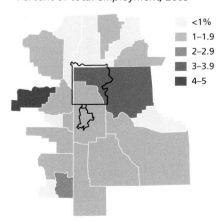

- ☐ <1%
- 1–1.9
- 2–2.9
- 3–3.9
- 4–5

### Local and State Government
Percent of total employment, 2009

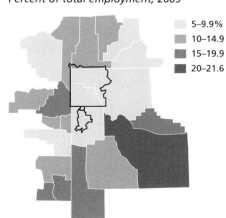

- ☐ 5–9.9%
- 10–14.9
- 15–19.9
- 20–21.6

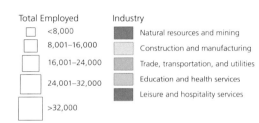

## Unemployment by County, 1985–2010

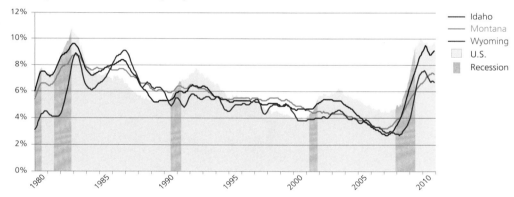

## Montana, Wyoming, Idaho, and U.S. Unemployment Rates
*Monthly, 1980–2010, Seasonally Adjusted*

- Idaho
- Montana
- Wyoming
- U.S.
- Recession

## Average Unemployment Rate

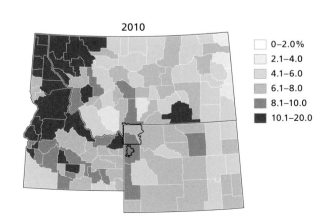

1990     2000     2010

- 0–2.0%
- 2.1–4.0
- 4.1–6.0
- 6.1–8.0
- 8.1–10.0
- 10.1–20.0

effect on the regional economy. Relative to the national average, these counties and the entire Yellowstone region have experienced relatively low unemployment rates during the most recent recession.

Government employment in the region includes federal employees working with public lands, state employees at universities and correctional institutions, and local government employees working as fire fighters, police, and teachers. Although a large portion of the region is federal public land, only a small percentage of total government jobs are federal jobs. Most of the government employment in the region is at the local and state level. Some smaller counties have high portions of employment in the government sector because a single public institution, such as a university or a prison, can dominate a small local economy.

Total employment grew across the greater Yellowstone region from 1990 to 2009, but it increased unevenly. The metro commuter counties grew more, in terms of absolute growth and percentage growth. The greatest growth, measured by the change in the number of jobs, was in the counties that were already large: Bonneville in Idaho and Gallatin in Montana. Those counties have some of the larger cities in the region—Idaho Falls and Bozeman—and those communities grew rapidly. Measuring growth by percent change, employment tripled in Teton County, Idaho. In Gallatin, Montana, and Teton, Wyoming, total employment doubled over the twenty-year period.

Since 1990, the regional economy has continued to shift away from extractive industries. The portion of employment in natural resources and mining dropped from 10 percent of total employment in the greater Yellowstone area to 8 percent. Conversely, leisure and hospitality services grew from 8 percent to 11 percent.

# Income

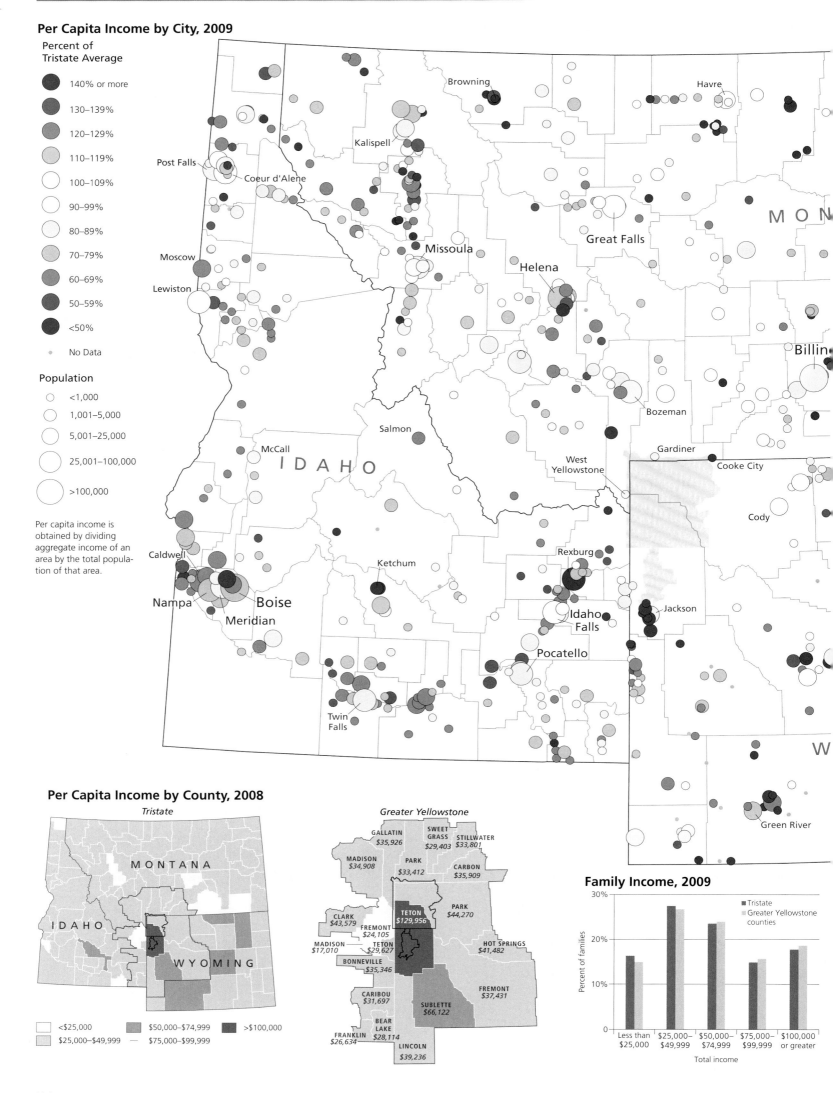

## Per Capita Income by City, 2009

Percent of
Tristate Average

- 140% or more
- 130–139%
- 120–129%
- 110–119%
- 100–109%
- 90–99%
- 80–89%
- 70–79%
- 60–69%
- 50–59%
- <50%
- No Data

### Population

- <1,000
- 1,001–5,000
- 5,001–25,000
- 25,001–100,000
- >100,000

Per capita income is
obtained by dividing
aggregate income of an
area by the total popula-
tion of that area.

Browning
Havre
Kalispell
Post Falls
Coeur d'Alene
MONTANA
Moscow
Great Falls
Missoula
Lewiston
Helena
Billin
Salmon
McCall
IDAHO
West Yellowstone
Gardiner
Cooke City
Bozeman
Cody
Rexburg
Caldwell
Ketchum
Nampa
Boise
Idaho Falls
Jackson
Meridian
Pocatello
W
Twin Falls
Green River

## Per Capita Income by County, 2008

### Tristate

MONTANA

IDAHO

WYOMING

- <$25,000
- $25,000–$49,999
- $50,000–$74,999
- $75,000–$99,999
- >$100,000

### Greater Yellowstone

| County | Value |
|---|---|
| GALLATIN | $35,926 |
| SWEET GRASS | $29,403 |
| STILLWATER | $33,801 |
| MADISON | $34,908 |
| PARK | $33,412 |
| CARBON | $35,909 |
| PARK | $44,270 |
| CLARK | $43,579 |
| FREMONT | $24,105 |
| TETON | $129,956 |
| MADISON | $17,010 |
| TETON | $29,627 |
| HOT SPRINGS | $41,482 |
| BONNEVILLE | $35,346 |
| CARIBOU | $31,697 |
| FREMONT | $37,431 |
| SUBLETTE | $66,122 |
| BEAR LAKE | $28,114 |
| FRANKLIN | $26,634 |
| LINCOLN | $39,236 |

## Family Income, 2009

- Tristate
- Greater Yellowstone counties

Percent of families

30%
20%
10%
0

Less than $25,000 | $25,000–$49,999 | $50,000–$74,999 | $75,000–$99,999 | $100,000 or greater

Total income

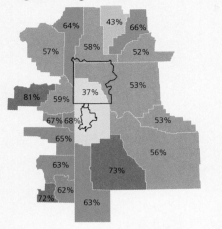

## Income by Type, 2008
Percentage of total income

<10% | 11–20% | 21–30% | 31–40% | 41–50% | 51–60% | 61–70% | >70%

**Wages, Earnings, and Salaries**

64%  43%  66%
57%  58%  52%
81%  59%  37%  53%
67% 68%  53%
65%
63%  73%  56%
62%
72%  63%

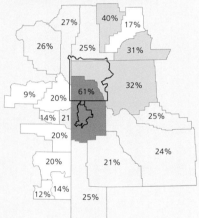

**Dividends, Interest, and Rent**

27%  40%  17%
26%  25%  31%
9%  20%  61%  32%
14%  21
20%  25%
20%  24%
20%  21%
12%  14%
25%

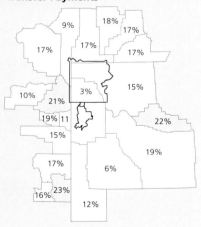

**Transfer Payments**

9%  18%
17%  17%  17%
17%  17%
10%  21%  3%  15%
19% 11
15%  22%
17%  6%  19%
16%  23%
12%

## Poverty by County, 2009

| County | % in poverty | County | % in poverty |
|---|---|---|---|
| Bear Lake (ID) | 13.12% | Lincoln (WY) | 6.89% |
| Bonneville (ID) | 10.32% | Madison (ID) | 34.51% |
| Carbon (MT) | 9.66% | Madison (MT) | 13.10% |
| Caribou, (ID) | 8.59% | Park (MT) | 13.03% |
| Clark (ID) | 15.75% | Park (WY) | 7.81% |
| Franklin (ID) | 14.34% | Stillwater (MT) | 8.86% |
| Fremont (ID) | 10.34% | Sweet Grass (MT) | 14.47% |
| Fremont (WY) | 14.08% | Sublette (WY) | 3.39% |
| Gallatin (MT) | 14.08% | Teton (ID) | 8.22% |
| Hot Springs (WY) | 8.60% | Teton (WY) | 7.70% |

## Per Capita Income by Type
*All Greater Yellowstone Area counties*

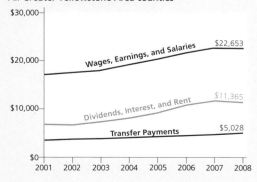

Per capita income is one measure used to compare average income across areas. It is the total of income from all sources in an area divided by the area's total population. Within the Yellowstone and the tristate region, average incomes vary widely, reflecting the mix of income sources.

Fifteen of the twenty counties within the Yellowstone region have average incomes below the national average of $40,166. This is not surprising since this is a largely non-urban region and such areas tend to have lower wages than urban areas. Countering this trend in the tristate area are small towns in generally rural settings with a significant government presence or with easy access to environmental amenities (for example, national parks or ski areas), as well as those located in areas where mining and energy extraction occur.

Personal income comes from three general sources: employment, investment income, and transfer payments. Income from employment—earnings, salary, and wages—makes up a smaller portion of total income in the Yellowstone region than in the United States. Nationally, employment income accounts for 67 percent of all personal income; in the Yellowstone region, only 58 percent. Households in Idaho tend to generate a higher portion of personal income from earnings than households in the rest of the region. Wyoming's Sublette County is one of the few counties in Wyoming that generates a large portion of its personal income from earnings (73 percent). The county has a very high per capita income ($66,122), the second highest average income in the region. Its high earnings and overall income stem from a boom in the natural gas industry.

Investment income—dividends, interest, and rent—includes the income received from investments, rent from real estate property, patents, copyrights, and natural resource rights. Nationwide, this income source accounts for 18 percent of all personal income; in the Yellowstone region, 29 percent. Teton County, Wyoming, generated 60 percent of its total income from dividends, interest, and rent and has an average per capita income more than three times the national average. Local resource amenities and lack of a state income tax attract wealthy households to this and similar counties. Such households typically have higher proportions of outside investment income. The region's boom in natural gas also has increased income from mineral rights.

Transfer payments are the payments made by governments to individuals, such as retirement and disability insurance benefits, Medicare, Medicaid, unemployment insurance compensation, and veteran's benefits. Nationwide, transfer payments make up 15 percent of personal income, compared to 13 percent in the Yellowstone region. Here, the high levels of investment income from the retired population dominate overall income levels, so only a relatively small portion of the overall income is derived from transfer payments.

55

# Agriculture

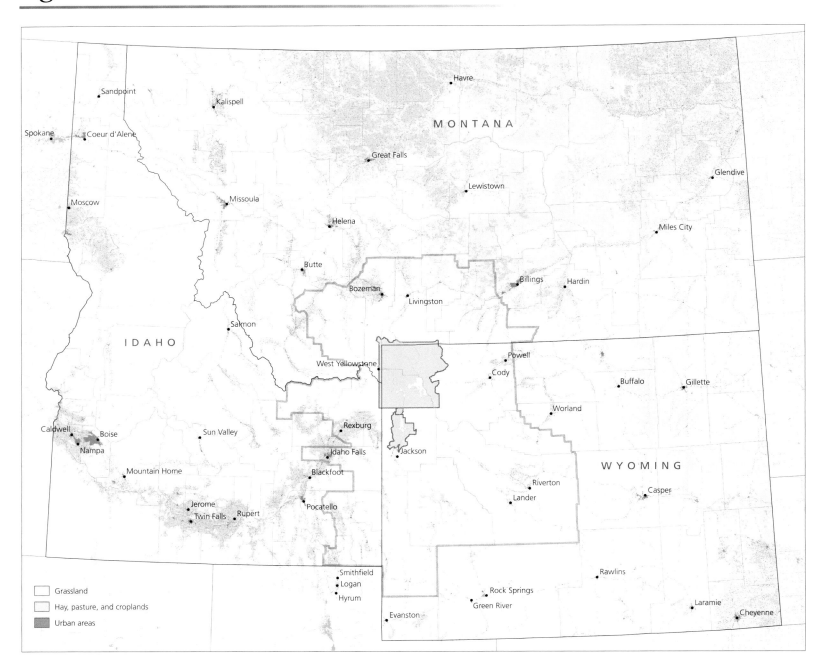

Grassland

Hay, pasture, and croplands

Urban areas

## Cropland Irrigation

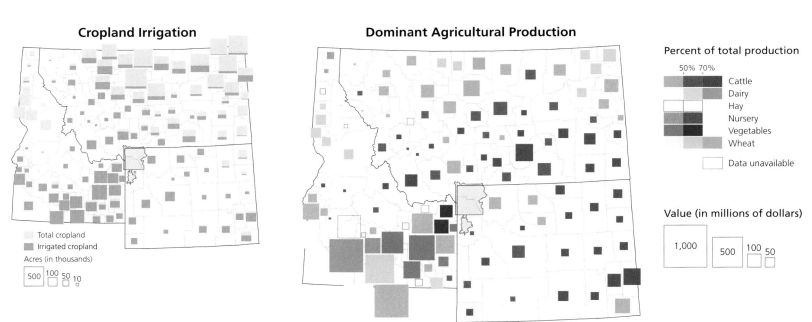

Total cropland
Irrigated cropland
Acres (in thousands)

500  100  50  10

## Dominant Agricultural Production

Percent of total production

50%  70%

Cattle
Dairy
Hay
Nursery
Vegetables
Wheat

Data unavailable

Value (in millions of dollars)

1,000  500  100  50

Agriculture in the three states surrounding Yellowstone is far from uniform. Wyoming is dominated by cattle ranching, particularly in the eastern portion of the state. The industry's critical hay-raising component is mostly confined to relatively narrow strips of bottom-land—nearly all of this acreage is irrigated. Dry farming takes place only in the state's northeast corner, an extension of the pattern seen in Montana and the Dakotas.

Idaho, in contrast, is predominantly a dairy and potato-growing state, though beef cattle and hay remain important nearly everywhere and are the primary crops in many counties. Irrigated dairy and farming operations in the Snake River Plain employ labor- and capital-intensive farming methods similar to those used in California. Northern Idaho's wheat belt is a continuation of the Columbia Basin's Palouse region while Montana's wheat country is the western end of the Great Plains grain belt. Precipitation in both these regions is too low to support continual dry farming, so grains are grown in annually alternating bands of cropland and cultivated fallow, conserving moisture for the subsequent year's crop.

Much of the Greater Yellowstone Area at higher elevation is set aside as national park, national forest, or wilderness; these federal lands are generally too cold or too dry for extensive farming. They are, however, surrounded by farming and ranching economies. This proximity affects wildlife management particularly near the national parks, where wildlife such as bison and wolves are managed to control competition for resources, predation, and spread of disease at park boundaries.

The dominant role of cattle in the landscape outside the parks is evident in the maps on these pages. Sheep, a much smaller part of the region's livestock industry, are concentrated mainly in east-central Wyoming, though significant numbers are found in many other areas as well.

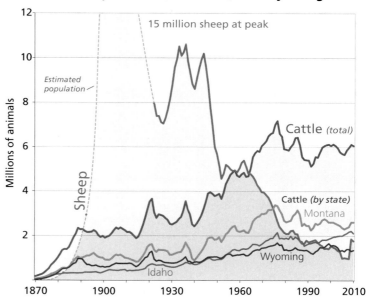

**Cattle and Sheep in Idaho, Montana, and Wyoming**

During the late nineteenth century the picture was very different. The near elimination of the buffalo by the mid-1880s opened the area to a rapid influx of sheep operations. The associated intensive and unsustainable grazing practices resulted in sheep numbers plummeting in the late-1920s. A rebound in the 1930s and 1940s has been followed by a nearly continuous decline since then. Cattle numbers peaked in the 1970s and 1980s but have remained within a fairly narrow range, around six million, for the past few decades.

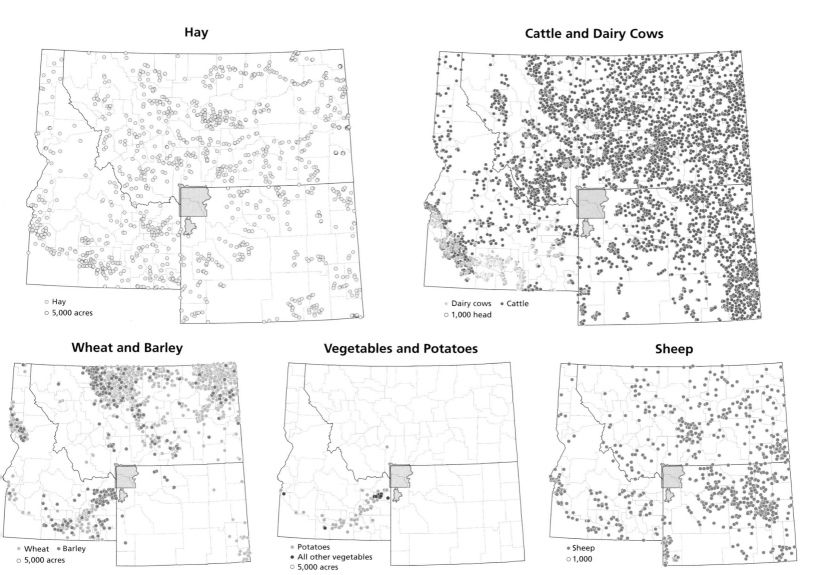

**Hay**

**Cattle and Dairy Cows**

**Wheat and Barley**

**Vegetables and Potatoes**

**Sheep**

# Market Access

## Market Access and Economic Structure

The greater Yellowstone region's economy has grown and diversified dramatically in the past forty years. The trend has been away from a heavy dependence on resource extraction and agriculture toward a largely service-based economy, with significant growth also coming from retirement and investment income. These changes, however, have not been uniform across all counties; some have grown, adding service and construction industries plus retirement and investments income, while others have lagged behind. A number of factors influence economic growth, but paramount among them is the ability for people to fly to larger markets and population centers; counties with airports have grown quickly while those without them have not.

The vast distances between population centers in the American West were for a long time a detriment to business; the isolation made it difficult to bring products to market and to unite suppliers, producers, and customers. More recently, the open spaces of the West—with national parks, wilderness areas, wild rivers, spectacular scenery, and abundant opportunities to ski, hike, fish, and hunt—have attracted many people and their businesses. Some occupations such as law, finance, insurance, real estate, business, health, and engineering are generally not location bound—workers in these fields often have greater options as to where they can find employment. Such workers are attracted to a place in large part because of its amenities; natural amenities are especially important in drawing knowledge-based workers. This employment group, which began to emerge at the end of the past century, includes analysts, information brokers, and technology workers—highly educated and highly paid individuals able to acquire and apply theoretical and analytical knowledge.

But environmental amenities alone are not enough to ensure an area's economic vigor. Another key is transportation access to larger markets and population centers. This is the case with regard to satisfying the needs of the higher-wage components of the service industries. Technology workers, for example, have been shown to travel by air 60 to 400 percent more frequently than those in the general workforce.

It wasn't until the mid-1980s, 1990s, and 2000s, with the advent of the Internet and the growth in numbers and capacity of regional airports, that the vast distances of greater Yellowstone stopped being a detriment, most notably for those communities with airports. The recreational and environmental amenities became an economic asset and major driver of the economy that has attracted "footloose" entrepreneurs and "amenity migrants," including retirees and people with investment income. The ideal combination for economic growth is a mix of world-class amenities and ready access via air travel. This explains why the fastest growth in the region is associated with the most "connected" places. The largest airports in the region (Billings and Bozeman, Montana; Jackson, Wyoming; and Idaho Falls, Idaho) are also cities where the economy has grown the fastest.

**Driving Time to Airports**

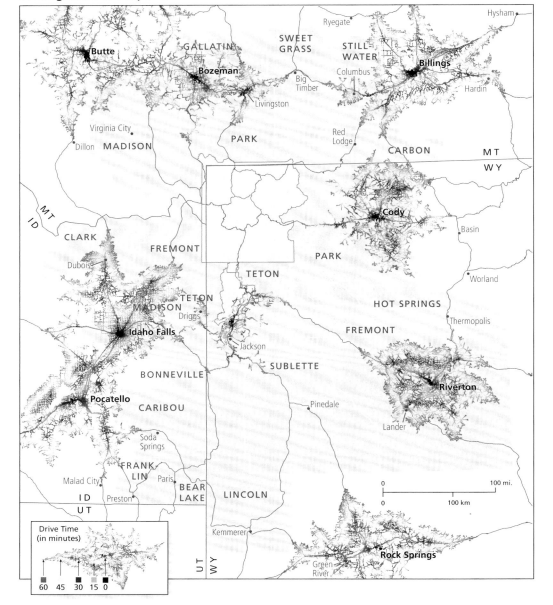

**Airport Enplanements, 2009**

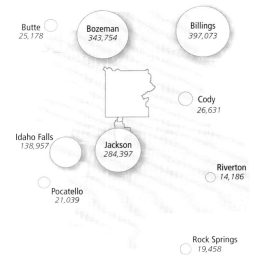

**Airport Enplanements, 2000–2009**

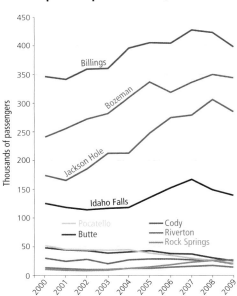

## The Three Wests

The counties of the West can be classified into three types. The "Metro Commuter" counties (accounting for more than 80 percent of the West's population) consist of a metropolitan statistical area surrounded by counties within easy commuting distance by road. "Rural Connected" counties have a large commercial airport with daily access to major population centers. Inhabitants in "Isolated Rural" counties cannot readily commute to work in a big city and are more than an hour's drive from an airport with daily flights to major cities.

The underlying logic behind these categories is that opportunities for economic development and individual opportunities for employment are limited by distance and access to markets. Those living in Metro Commuter areas have access to large and diversified labor markets. They either live in those markets or are within commuting distance. Others live in counties that are seemingly rural, but are connected to metro counties via air travel. In these Rural Connected counties, entrepreneurs, for example, may live in a remote setting but travel occasionally to the city to visit colleagues, clients,

and suppliers. Those living in truly remote areas (the Isolated Rural counties) have limited opportunities, normally constrained to natural resources, such as farming, ranching, and resource extraction.

Rural Connected counties show significant profile variation from one another, with their differences largely associated with distances to airports. Those closest to an airport have opportunities more like metropolitan areas, with high growth rates, a relatively more educated workforce, and employment in manufacturing, service, and professional industries. Those farther from an airport tend to follow the pattern of isolated rural areas, with slow growth rates, a higher relative dependence on investment and retirement income, and employment that is more likely to be in energy development or agriculture.

Compared to the West as a whole, the counties of greater Yellowstone with the fastest growth in real personal income, population, per capita income, and employment are the Rural Connected, underscoring the importance of airports to the regional economy.

### The Three Wests of the Western U.S.

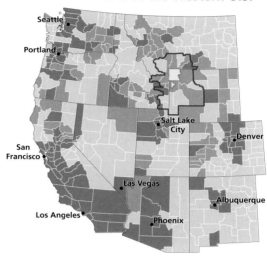

■ **Metro Commuter**
Counties that contain a metropolitan area or are within an hour's drive of one.

■ **Rural Connected**
Counties that are non-metro but have airports with daily commercial flights to major hubs and metro areas.

□ **Isolated Rural**
Counties that do not have a metropolitan area or airport within an hour's drive.

### Western U.S. Economic Change, 1970–2005

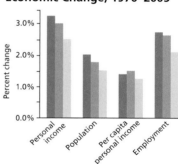

### Greater Yellowstone Economic Change, 1970–2005

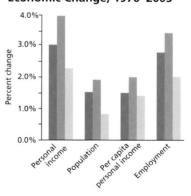

*Personal income is the total income by person from all sources. Per capita personal income is the sum of all personal income divided by population.*

### The Three Wests of Greater Yellowstone

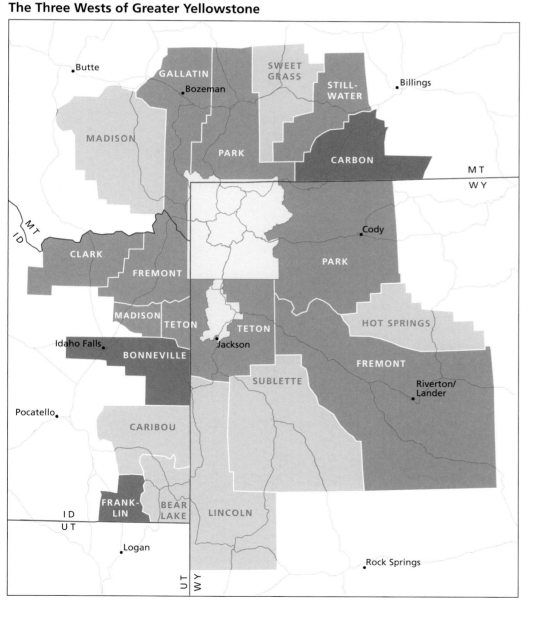

# Wildland Economies

In the past, the economic function of public lands was, to a large extent, to serve as a repository of raw materials to be harvested or extracted; in more recent years, this function has expanded to include providing opportunities for recreation and tourism. While resource development and tourism remain important, there is now yet another role for public lands: as a setting that makes adjacent communities attractive places to live and do business. This new phenomenon is not as easy to measure as board feet of lumber harvested or dollars collected from bed taxes, but it is undeniably a major driver in rural economic development in the West. Today, the wide open spaces administered by the Forest Service, Bureau of Land Management (BLM), and National Park Service help attract entrepreneurs who can locate anywhere, retirees seeking towns with a desirable quality of life, and "amenity migrants," people who choose where to live first and then either find work or create employment opportunities for themselves.

To put the changing economic role of public lands in perspective, those activities normally associated with uses of public lands and often perceived as the staples of the rural West have grown slowly and account for a smaller and smaller part of the overall economy. Cumulatively, farming, ranching, forestry, lumber and wood products manufacturing, hard rock mining, and fossil fuel development contributed less than 3 percent of total new jobs to the economy of the West from 1990 to 2008. In 2008, these sectors combined constituted roughly 7 percent of all jobs in the nonmetropolitan (rural) West, and 3 percent in the West as a whole. The value of public lands today extends far beyond those associated with extractive uses of the land to now include "nonuse values" such as scenery, solitude, a sense of place.

The body of literature documenting this new phenomenon is large and growing. Studies detailing the relationship between wilderness and other forms of protected public lands and economic development have found that counties in the West with wilderness, national parks, national monuments, and other protected public lands set aside for their natural characteristics stimulate economic growth. Lands with greater protection have stronger positive effects on growth.

Economic development is a complicated matter, and not all communities benefit equally from protected lands. Access to metropolitan areas via road and air travel is also important. The education of the workforce, the arrival of newcomers, and a number of other factors allow some areas to flourish and take advantage of protected public lands as part of an economic development strategy. In other words, the amenities of public lands are a necessary, but not sufficient, condition for growth.

One of the reasons greater Yellowstone is growing rapidly is because it is a large region, a wildland complex that extends far beyond the boundaries of Yellowstone National Park. There are a number of these wildland complexes in the West, areas with a national park at their core and surrounded by Forest Service, BLM, or other federally managed public lands.

## Rural Counties Adjacent to Large Wildland Complexes

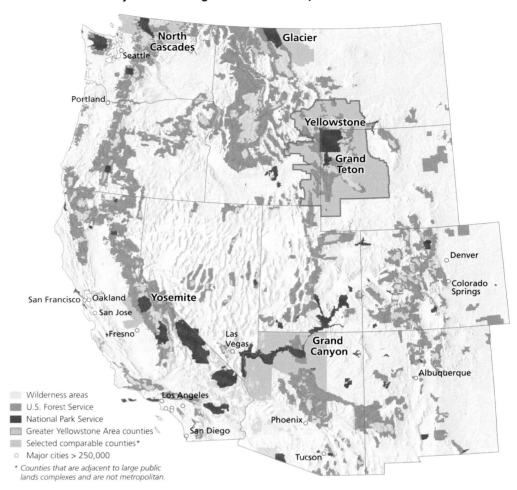

- Wilderness areas
- U.S. Forest Service
- National Park Service
- Greater Yellowstone Area counties
- Selected comparable counties*
- Major cities > 250,000

*Counties that are adjacent to large public lands complexes and are not metropolitan.*

## Economic Growth, 1990–2005
*Adjusted to 2005 U.S. dollar values*

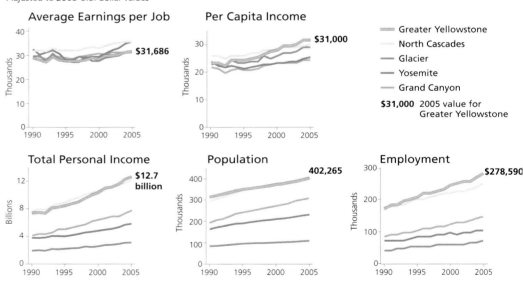

- Greater Yellowstone
- North Cascades
- Glacier
- Yosemite
- Grand Canyon

**$31,000** 2005 value for Greater Yellowstone

## Wildlands Comparison

In recent decades, the wildland regions shown here became magnets for economic growth. Greater Yellowstone has led the way in terms of growth of per capita income, is close to the top in terms of employment growth, and has comparatively much higher rates of education.

The fast growth in per capita income is due to the growth in jobs and labor earnings, but also due to the vigorous growth in non-labor sources of income, such as retirement and investment income. Greater Yellowstone also ranks highest in terms of the percentage of the workforce with a college degree, an important indicator of future development potential. In 2006 (before the recession of the late-2000s), greater Yellowstone also had the lowest rate of unemployment.

Even though average earnings per job are lower in Greater Yellowstone than the national average, the earnings buy more, as indicated by the housing affordability index. A higher index score reflects a smaller portion of income being used for housing. Relative affordability of housing is an important competitive advantage for the region. While wages are relatively lower in greater Yellowstone, residents of the region spend a smaller portion of income on housing than many wildland complexes that are closer to fast-growing metropolitan areas.

## Metropolitan Comparison

In the 1990s and early 2000s, boom years throughout the country, greater Yellowstone stood out as a remarkable economic performer. Even when compared to the fastest-growing metropolitan areas of the West—Silicon Valley, the Denver and Seattle metropolitan areas—greater Yellowstone grew faster in terms of employment and population and came in second in terms of growth of per capita income. It also had the lowest rate of unemployment. While average earnings per job were the lowest, it was also the most affordable place to live in terms of housing.

The very rapid growth, especially when compared to large, diverse, and fast-growing metropolitan regions of the West, is a clear indication that the economy of greater Yellowstone is successful, despite its relatively remote rural setting. While in the past the long distances between cities were a detriment to business, today these vast open spaces, and the recreational and environmental amenities they contain, are important economic drivers that attract people and business. Greater Yellowstone offers a high quality of life, a number of its communities are well connected to the rest of the world via airports, and housing is relatively affordable. It is a place where workers can ride their bicycles to work in the morning and fly fish in the afternoon. As the recruitment page of one high-tech employer stated, "Why live in Silicon Valley when you can live in Paradise Valley?"

## Economic Indicators

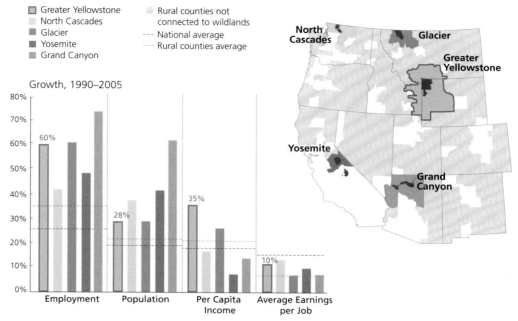

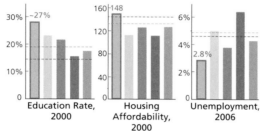

Education rate = % of population 25 and over who have a college degree, 2000 Census data

The housing affordability index indicates how well a median family can afford the median house; a higher score reflects more affordable housing. The housing affordability figures assume a 20% down payment and that no more than 25% of a family's income goes to paying the mortgage. It is based on an interest rate of 10.01% in 1990 and 8.03% in 2000. Use this statistic as a comparative, rather than absolute, measure.

## Economic Indicators

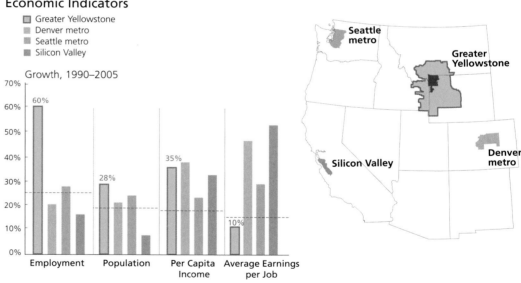

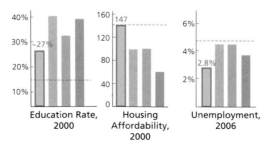

# Protected Areas

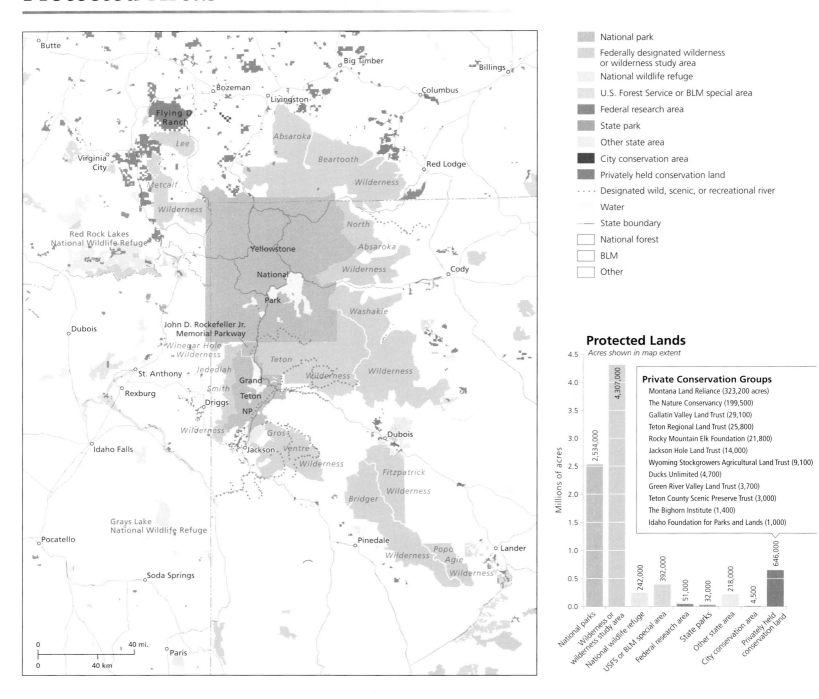

### Legend

- National park
- Federally designated wilderness or wilderness study area
- National wildlife refuge
- U.S. Forest Service or BLM special area
- Federal research area
- State park
- Other state area
- City conservation area
- Privately held conservation land
- Designated wild, scenic, or recreational river
- Water
- State boundary
- National forest
- BLM
- Other

## Protected Lands
*Acres shown in map extent*

Millions of acres

- National parks — 2,534,000
- Wilderness or wilderness study area — 4,307,000
- National wildlife refuge — 242,000
- USFS or BLM special area — 392,000
- Federal research area — 51,000
- State parks — 32,000
- Other state area — 218,000
- City conservation area — 4,500
- Privately held conservation land — 646,000

### Private Conservation Groups

Montana Land Reliance (323,200 acres)
The Nature Conservancy (199,500)
Gallatin Valley Land Trust (29,100)
Teton Regional Land Trust (25,800)
Rocky Mountain Elk Foundation (21,800)
Jackson Hole Land Trust (14,000)
Wyoming Stockgrowers Agricultural Land Trust (9,100)
Ducks Unlimited (4,700)
Green River Valley Land Trust (3,700)
Teton County Scenic Preserve Trust (3,000)
The Bighorn Institute (1,400)
Idaho Foundation for Parks and Lands (1,000)

## West Yellowstone and Centennial Valley

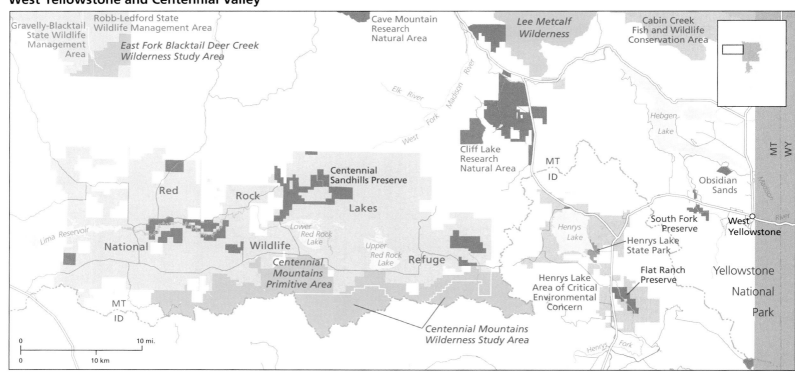

Public land ownership and management has played and continues to play a key role in protecting natural resources in the Greater Yellowstone Area. Yet amid this vast acreage of public lands are small communities and many parcels of privately owned land. These private lands are increasingly important to the ecological, scenic, and cultural vitality of the region because of their locations. Early settlers sought out the lower-elevation valleys of the region with plenty of water, tree cover, and milder temperatures. As a result, though the total acreage of the privately owned lands is small compared to the vast national forests, parks, and refuges, their ecological and economic importance is disproportionately high. These areas are essential winter habitats for wildlife. Migrating populations of elk, deer, antelope, and moose move along the river corridors. Many species of birds summer here or rest in these locations on their seasonal migrations.

In recent decades, the strong tourist economy, migration of people to amenity-rich areas, and booming regional energy development have rapidly expanded once-small communities, threatening the very resources that make these lands so attractive. The highest population growth, concentrated along the borders of the public lands, has made the remaining open ranchland even more important.

As private land-use planning is limited, local, regional, and national conservation organizations are stepping in, working to protect the remaining undeveloped lands—particularly the most vulnerable and irreplaceable parcels—by offering landowners flexible, incentive-based options. These landowner-conservation partnerships are identified as "privately held conservation land" on these maps.

## Cody, Wyoming

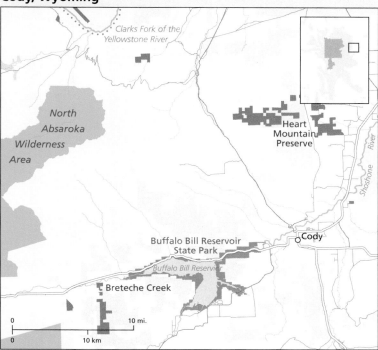

## Bozeman and Paradise Valley, Montana

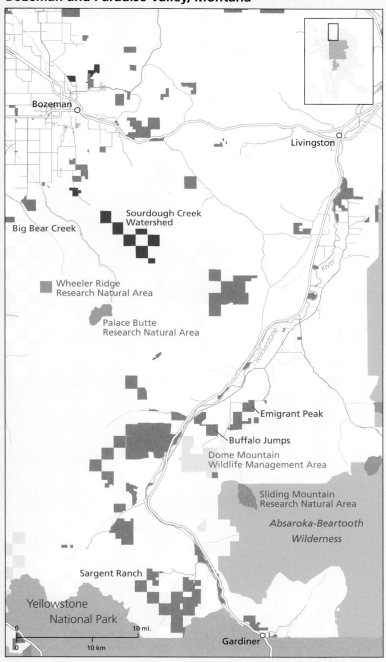

## Jackson, Wyoming

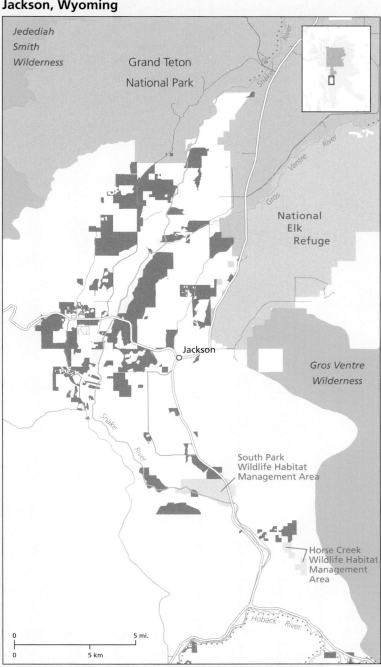

# Land Ownership

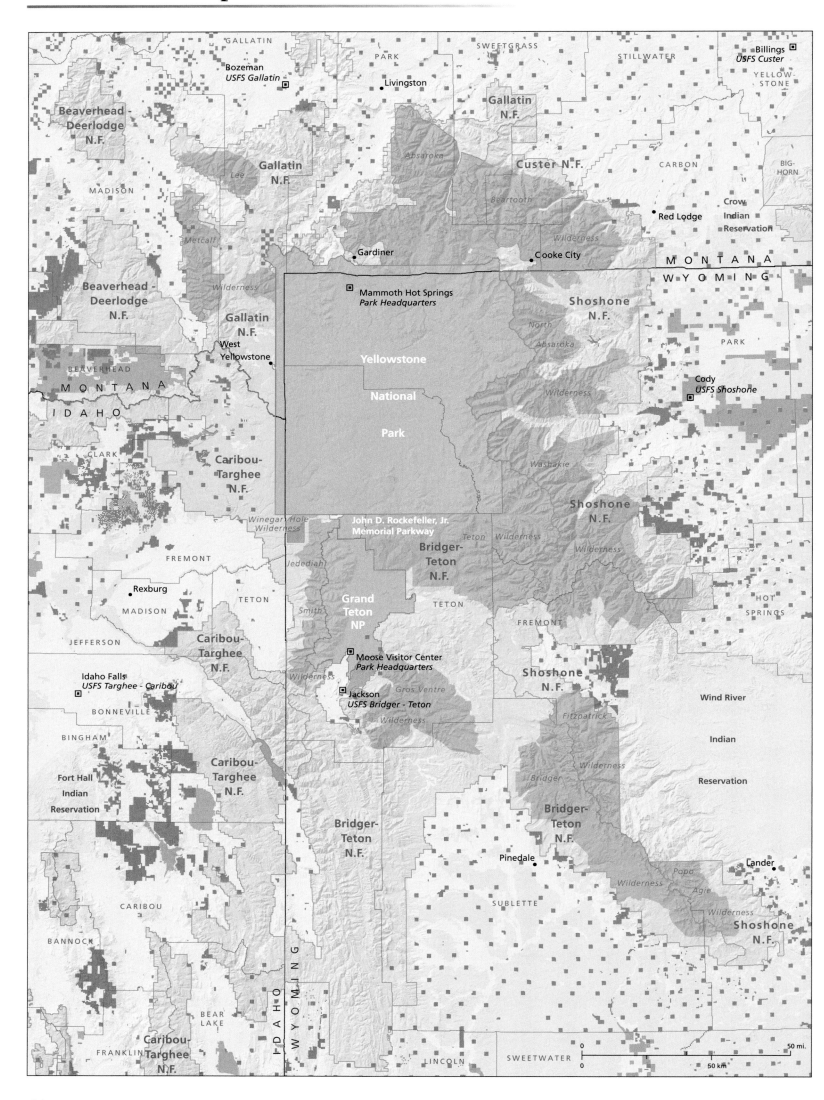

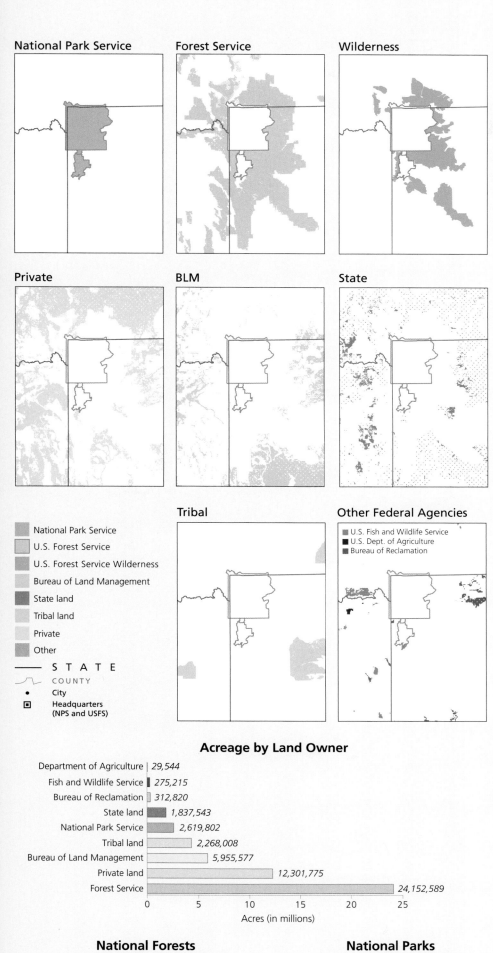

**National Park Service**

**Forest Service**

**Wilderness**

**Private**

**BLM**

**State**

**Tribal**

**Other Federal Agencies**
- ■ U.S. Fish and Wildlife Service
- ■ U.S. Dept. of Agriculture
- ■ Bureau of Reclamation

- ■ National Park Service
- ■ U.S. Forest Service
- ■ U.S. Forest Service Wilderness
- ■ Bureau of Land Management
- ■ State land
- ■ Tribal land
- ■ Private
- ■ Other

— **S T A T E**
⌐⌐ COUNTY
• City
▣ Headquarters (NPS and USFS)

### Acreage by Land Owner

| Land Owner | Acres |
|---|---|
| Department of Agriculture | 29,544 |
| Fish and Wildlife Service | 275,215 |
| Bureau of Reclamation | 312,820 |
| State land | 1,837,543 |
| National Park Service | 2,619,802 |
| Tribal land | 2,268,008 |
| Bureau of Land Management | 5,955,577 |
| Private land | 12,301,775 |
| Forest Service | 24,152,589 |

Acres (in millions) 0 5 10 15 20 25

### National Forests

|  | |
|---|---|
| Caribou-Targhee | |
| Bridger-Teton | |
| Shoshone | |
| Gallatin | |
| Custer | |
| Beaverhead-Deerlodge | |

Acres (in millions) 0 1 2 3 4

### National Parks

| Park | Acres |
|---|---|
| Yellowstone | 2,219,789 |
| Grand Teton | 307,745 |

Acres (in millions) 0 1 2 3

Many people know our public lands as the public domain, that vast network of property held "in common" for the people and managed for us by federal, state, and local governments—from national forests to city parks.

The public domain has a colorful and troubled past. Most of it was originally acquired through war, treaty, or purchase from European powers, Mexico, and Indian nations. Americans have rarely agreed on the purpose or fate of this sprawling government-controlled estate. Following the American Revolution, the fledgling federal government set out to transfer most of the public domain into private ownership. The Federal Land Act of 1796 authorized public auctions of lands, and the General Land Office was established in 1812 to implement land disposal. During the next 100 years, laws granted lands to homesteaders, miners, railroads, and states. By the early twentieth century, more than one billion acres—70 percent of the continental United States—had been granted or sold. Of those lands, 52 percent was sold in homestead or cash sales, 24 percent was granted to states, and 12 percent was conditionally granted to railroads. This transfer represents one of the greatest redistributions of wealth by any government ever.

From early in the nation's history, certain public lands had always been retained under federal control, most famously for military and Indian reservations. But by the second half of the 1800s, the demise of the frontier, decreasing resource abundance, and the widespread abuse of public-land laws prompted the first set-asides of the western public domain for purposes of conservation.

It was in that spirit that Congress created Yellowstone National Park in 1872. In 1891, the Yellowstone Timberland Reserve was established, and by 1905 it became the Shoshone National Forest, America's first national forest. Other national forests and wildlife refuges followed so that by 1935, with the creation of Red Rock Lakes National Wildlife Refuge, the basic administrative structure of what we now call the Greater Yellowstone Area was essentially complete.

The process of transferring federal lands to private hands has left an especially complex and controversial legacy in the American West. Additional complexity has arisen from the bewildering array of government entities charged with land and resource management. A single elk wandering in the Yellowstone ecosystem, for example, may, in the course of a month, fall under the administrative control of the federal government, Montana, Wyoming, and Idaho—and be subject to different hunting regulations in each. Counties, municipalities, other land-managing agencies, and private landowners also control hunting with varying rules. Similarly complicated oversight jurisdictions apply to all other species, as well as all other resources such as air, water, timber, oil, gas, and minerals.

# Population

## Population Density, 2010

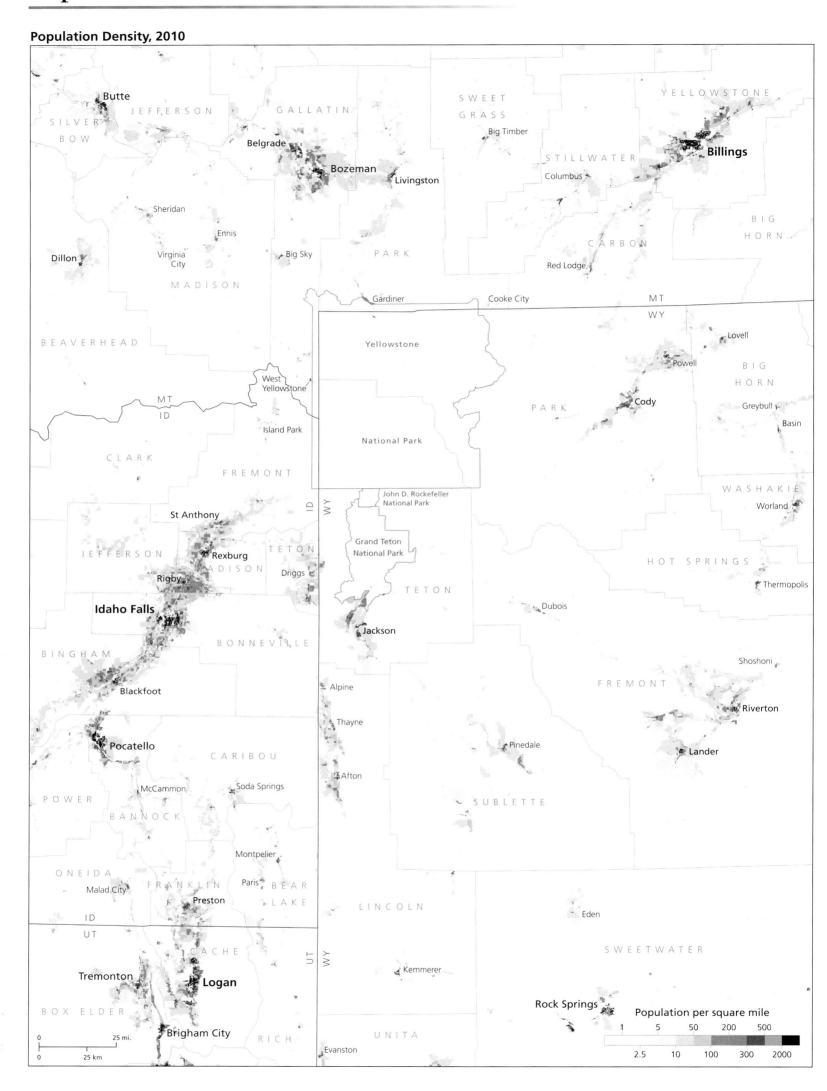

Population per square mile

1   5   50   200   500

2.5   10   100   300   2000

Relative to many parts of the country, the Yellowstone region is sparsely populated. This is due to the area's extreme terrain, severe weather, inaccessibility, limited infrastructure, and public land ownership. For the most part, population concentrates in urban places.

The age distribution in 1950 fit a classic "population pyramid," with the largest percentage of people under ten years old and ever smaller numbers in successively older age classes. Aging of baby boomers, out-migration of young working-age adults, and the amenity-driven in-migration of middle-aged and older adults increased the proportion of older people over the years. As a result, the

region has about the same number of people in each age class up to age sixty. Local variations occur, such as Madison County, Idaho, where Brigham Young University–Idaho adds young people to the population.

The proportion of males and females is approximately equal throughout the region. There is more variation at the county scale, where factors such as universities, ski areas, or the economic base sometimes alter gender ratios. In some cases, such as Clark County, Idaho, the numbers in each age class are so small that ratios may reflect the presence of only a few people rather than indicating larger trends.

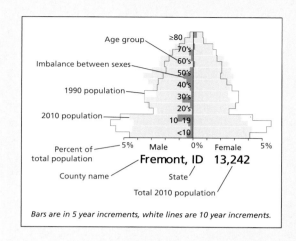

**Age and Sex Distribution**

**Twenty County Total**

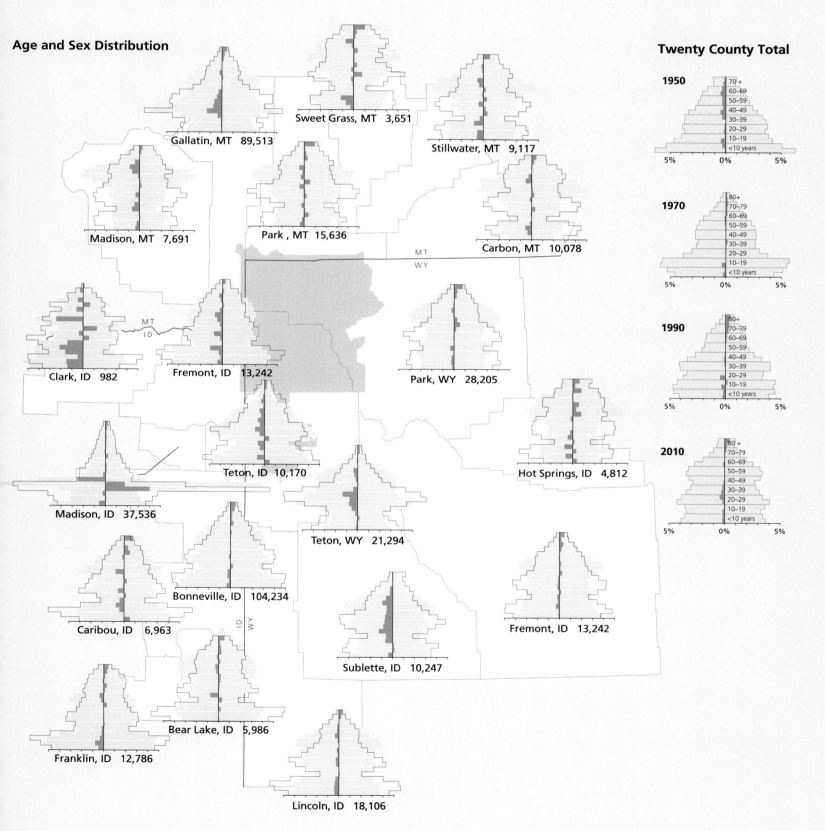

# County Population

In the early twenty-first century, the Greater Yellowstone Area was home to more than 400,000 people. Many of those residents arrived after 1970; the region's rapid growth has outpaced that of the overall West.

The vast areas remained sparsely populated during the late nineteenth century. By 1900, two clusters of denser population were evident. Southeastern Idaho, with its irrigated agriculture, proved attractive to farmers and ranchers and also was home to growing numbers of Mormons. Southwest Montana's river valleys (including the Gallatin and Yellowstone drainages) attracted people to settings such as Gallatin, Park, and Carbon counties.

Agricultural opportunities encouraged further growth between 1900 and 1920—land was cheap, farm prices were rising, better railroad connections improved accessibility. Wyoming's Fremont County saw its population double in twenty years as irrigation expanded and as former Indian reservation lands opened to settlers. Similar growth occurred in Idaho's Snake River Valley and across southwest Montana.

After 1920, however, the regional pattern became more complex.

## County Populations 1870–2000

*Note: Historical county information is located on the Political Boundaries page.*

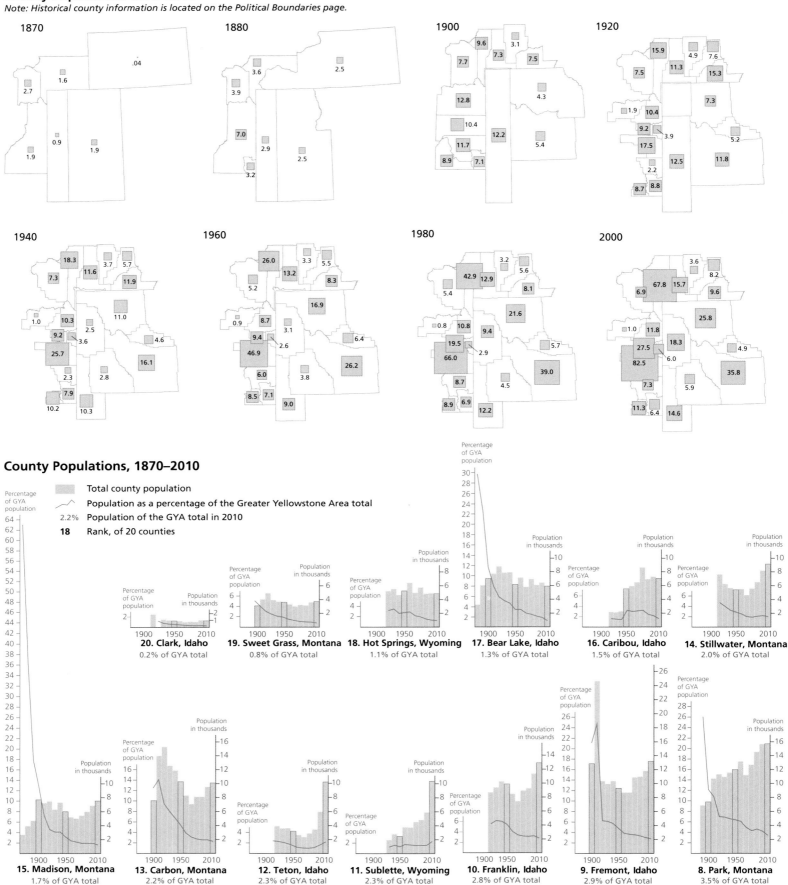

## County Populations, 1870–2010

Many farming and ranching counties saw their populations stagnate and even decline. Both Montana's Carbon County and Wyoming's Hot Springs County witnessed significant population losses between 1920 and 1940 as the area's agricultural sector and rural economy struggled. Other settings fared better. Bonneville County, Idaho, and Gallatin County, Montana, continued to grow thanks to their major cities (Idaho Falls, Bozeman) and more diverse economies.

Following World War II, the more populated counties fared best, though some sparsely settled areas also experienced renewed growth. Eastern Idaho—particularly Bonneville and Madison counties—benefited from excellent highway and rail connections, growing Mormon populations, industrial expansion (especially in Idaho Falls and Rexburg), and new employment in the health-care and education sectors. Montana's Gallatin County became that state's fastest growing area between 1980 and 2010, with the Gallatin Valley attracting tourists, retirees, and high-tech companies.

Smaller counties, especially those with amenity-oriented attractions, have also seen impressive gains since 1970. Wyoming's Teton County (Jackson Hole) illustrates the pattern. Its population of just over 3,000 people in 1960 grew to more than 18,000 in just forty years and has continued to draw migrants in the early twenty-first century. On a more modest scale, Idaho's Teton County (including Driggs) has been similarly attractive, reversing a long, slow decline in its largely agricultural population. Comparable trends are in play elsewhere, often bolstered by periodical increases in natural resource–based industries (coal, oil, metals). Montana's Stillwater County has benefited from its location near the Beartooth Range as well as by being home to one of North America's only platinum mines. Likewise, while most of Wyoming's coal boom lies farther east, developments in the energy sector have boosted populations in Lincoln, Sublette, and Park counties.

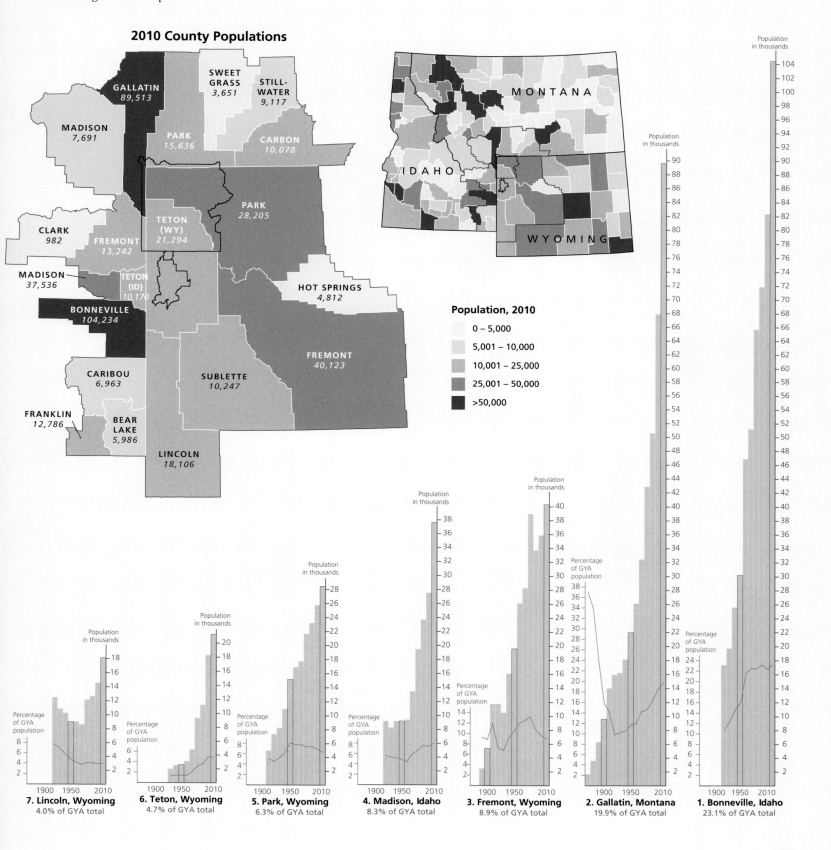

## 2010 County Populations

Population, 2010
- 0 – 5,000
- 5,001 – 10,000
- 10,001 – 25,000
- 25,001 – 50,000
- >50,000

**7. Lincoln, Wyoming**
4.0% of GYA total

**6. Teton, Wyoming**
4.7% of GYA total

**5. Park, Wyoming**
6.3% of GYA total

**4. Madison, Idaho**
8.3% of GYA total

**3. Fremont, Wyoming**
8.9% of GYA total

**2. Gallatin, Montana**
19.9% of GYA total

**1. Bonneville, Idaho**
23.1% of GYA total

# City Population

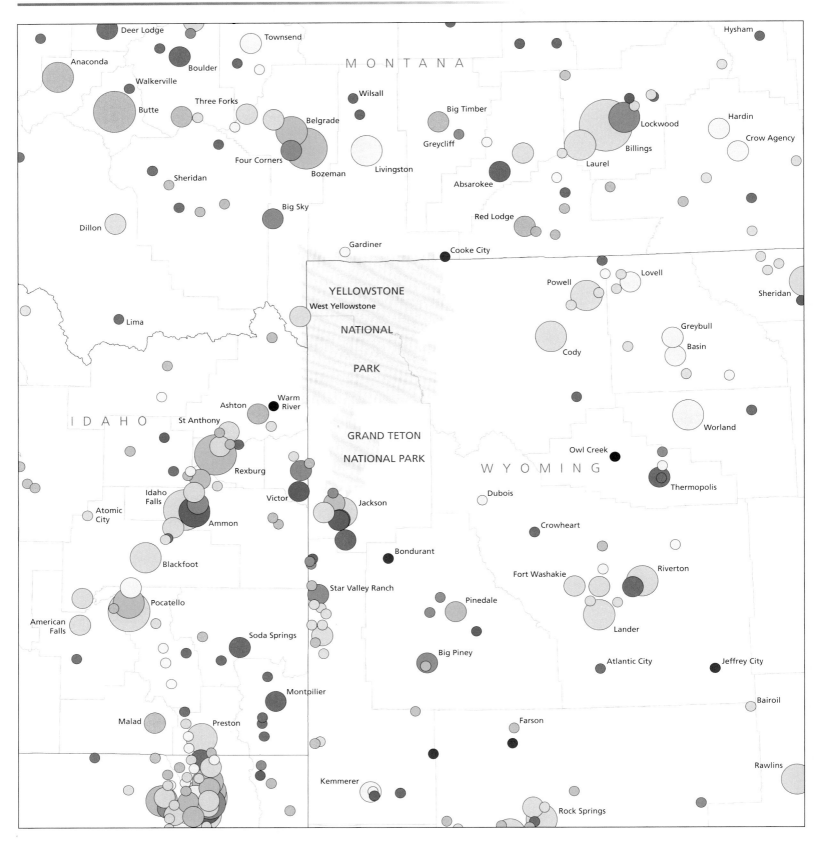

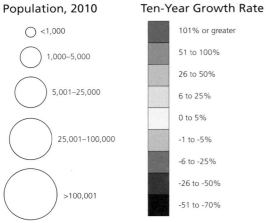
Urban populations within the Yellowstone region remained small during the nineteenth and early twentieth centuries. Most towns were agricultural centers that served farmers and ranchers in eastern Idaho, southwest Montana, and central Wyoming. Railroad lines promoted growth, fostering development in places such as Livingston, Montana, and Riverton, Wyoming.

After World War II, urban growth accelerated. A constellation of settlements in Idaho's Snake River Valley benefited from excellent transportation connections, growing Mormon populations, and an economy diversifying beyond agriculture. In Montana, both Bozeman and Belgrade experienced

rapid urbanization between 1980 and 2010 as the Gallatin Valley attracted amenity-oriented migrants and as the expanding health- care, educational, services, and industrial sectors offered new jobs. Slower growth characterized Wyoming's urban centers with the exception of amenity-rich Jackson, which has seen its population more than quintuple since 1960.

Residents of the region's blossoming cities make significant year-round use of local recreational resources, including Yellowstone National Park. At the same time, open spaces around these centers are rapidly disappearing as suburbs replace farm and ranch lands with subdivisions and shopping centers.

# City Population, 1880–2010

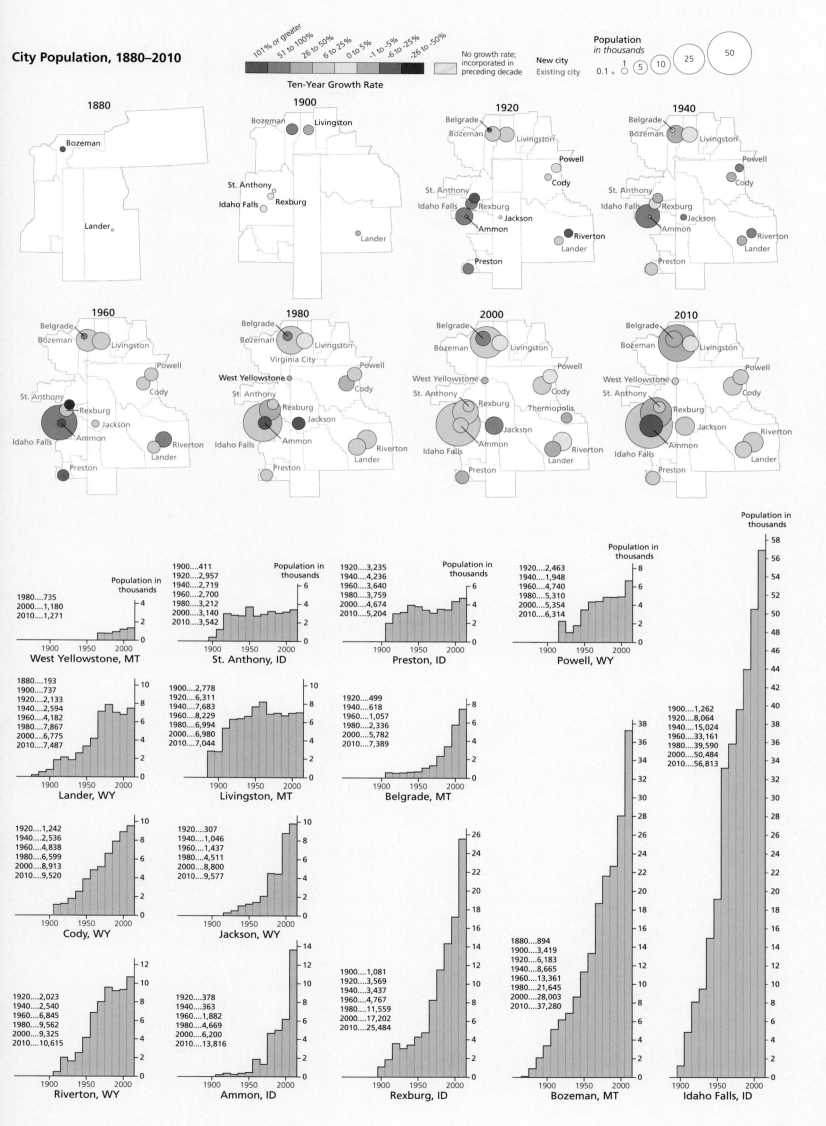

# Education

There is a long tradition of strong public support for education in the Yellowstone region. Religious groups interested in Christianizing local Indians founded the first schools; later, "subscription schools" were funded in early mining camps. Present school districts tend to be large in area but have small enrollments, reflecting relatively low population densities and the expanses between towns. These extremes of distance can pose challenges for student participation in athletics and other extracurricular activities.

All three states in the region had a higher high school completion rate in 2008 than the national average of 85 percent. Despite significant challenges—isolation, topographic barriers, a sparse population—the region's average level of completion of bachelor's degrees is only slightly lower than the national average (23 versus 24 percent).

Opportunities for higher education abound in this part of the American West with twenty-one private and public colleges and universities in Montana, nineteen in Idaho, and eight in Wyoming. Training in specialized fields such as medicine is provided through an innovative consortium that guarantees an admissions quota for students from all areas. This helps ensure the distribution of an educated and highly skilled population across the three-state region.

## School Districts and High Schools, 2010

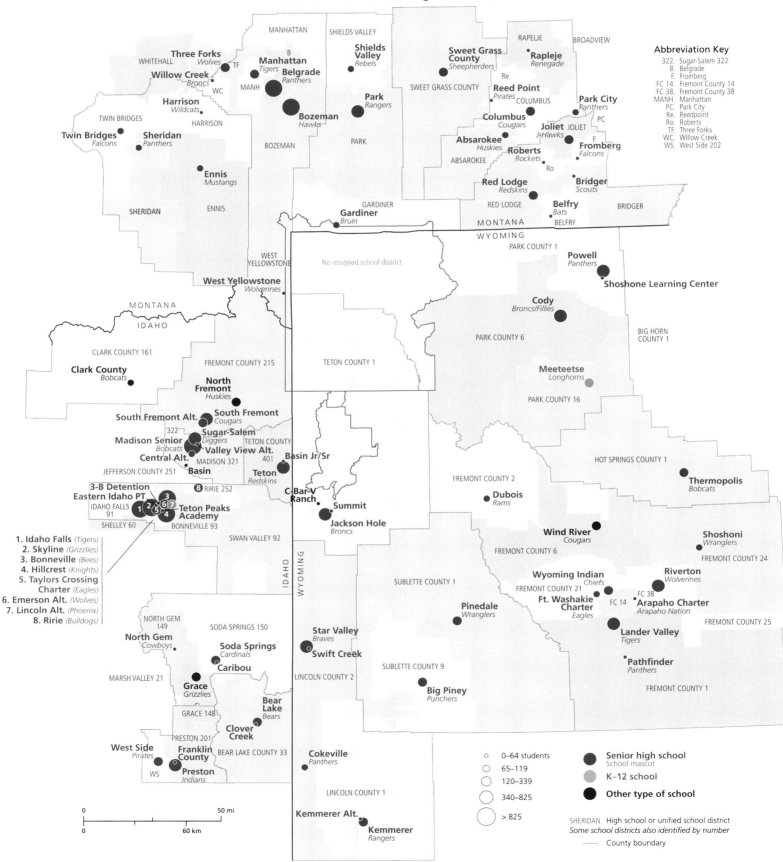

Abbreviation Key
322. Sugar-Salem 322
B. Belgrade
F. Fromberg
FC 14. Fremont County 14
FC 38. Fremont County 38
MANH. Manhattan
PC. Park City
Re. Reedpoint
Ro. Roberts
TF. Three Forks
WC. Willow Creek
WS. West Side 202

1. Idaho Falls (Tigers)
2. Skyline (Grizzlies)
3. Bonneville (Bees)
4. Hillcrest (Knights)
5. Taylors Crossing Charter (Eagles)
6. Emerson Alt. (Wolves)
7. Lincoln Alt. (Phoenix)
8. Ririe (Bulldogs)

○ 0–64 students
○ 65–119
○ 120–339
○ 340–825
○ > 825

● Senior high school
School mascot
● K–12 school
● Other type of school

SHERIDAN High school or unified school district
Some school districts also identified by number
— County boundary

# Educational Attainment, 2000

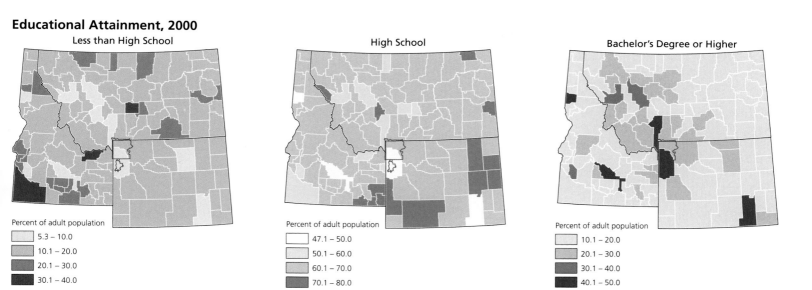

### Less than High School

**Percent of adult population**
- 5.3 – 10.0
- 10.1 – 20.0
- 20.1 – 30.0
- 30.1 – 40.0

### High School

**Percent of adult population**
- 47.1 – 50.0
- 50.1 – 60.0
- 60.1 – 70.0
- 70.1 – 80.0

### Bachelor's Degree or Higher

**Percent of adult population**
- 10.1 – 20.0
- 20.1 – 30.0
- 30.1 – 40.0
- 40.1 – 50.0

## Higher Education
### Degree-granting Institutions

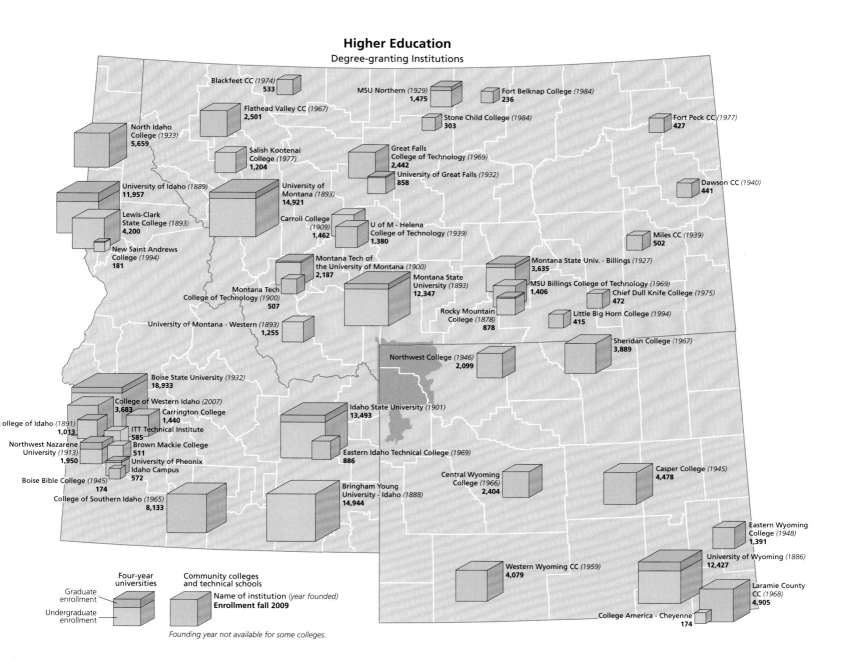

Blackfeet CC (1974) **533**

Flathead Valley CC (1967) **2,501**

MSU Northern (1929) **1,475**

Fort Belknap College (1984) **236**

Stone Child College (1984) **303**

Fort Peck CC (1977) **427**

North Idaho College (1933) **5,659**

Salish Kootenai College (1977) **1,204**

Great Falls College of Technology (1969) **2,442**

University of Great Falls (1932) **858**

Dawson CC (1940) **441**

University of Idaho (1889) **11,957**

University of Montana (1893) **14,921**

Lewis-Clark State College (1893) **4,200**

Carroll College (1909) **1,462**

U of M - Helena College of Technology (1939) **1,380**

Miles CC (1939) **502**

New Saint Andrews College (1994) **181**

Montana Tech of the University of Montana (1900) **2,187**

Montana State Univ. - Billings (1927) **3,635**

Montana Tech College of Technology (1900) **507**

Montana State University (1893) **12,347**

MSU Billings College of Technology (1969) **1,406**

Chief Dull Knife College (1975) **472**

University of Montana - Western (1893) **1,255**

Rocky Mountain College (1878) **878**

Little Big Horn College (1994) **415**

Sheridan College (1967) **3,889**

Northwest College (1946) **2,099**

Boise State University (1932) **18,933**

College of Western Idaho (2007) **3,683**

Carrington College **1,440**

Idaho State University (1901) **13,493**

College of Idaho (1891) **1,013**

ITT Technical Institute **585**

Northwest Nazarene University (1913) **1,950**

Brown Mackie College **511**

University of Pheonix Idaho Campus **572**

Eastern Idaho Technical College (1969) **886**

Casper College (1945) **4,478**

Boise Bible College (1945) **174**

College of Southern Idaho (1965) **8,133**

Central Wyoming College (1966) **2,404**

Bringham Young University - Idaho (1888) **14,944**

Eastern Wyoming College (1948) **1,391**

University of Wyoming (1886) **12,427**

Western Wyoming CC (1959) **4,079**

Laramie County CC (1968) **4,905**

College America - Cheyenne **174**

**Four-year universities**

Graduate enrollment
Undergraduate enrollment

**Community colleges and technical schools**

Name of institution (year founded)
**Enrollment fall 2009**

*Founding year not available for some colleges.*

73

# Race and Ethnicity

As in other parts of the interior West, white populations dominate maps of race and ethnicity in the states of Idaho, Montana, and Wyoming. Native Americans are the second largest group, with many concentrated on reservation land. African Americans and other minority groups have long been comparatively small in number, although thousands of Chinese miners and railroad workers once resided here.

Latino migration into the Yellowstone region has mirrored national growth trends since the 1980s, with Idaho emerging as one of the most popular destinations for Latinos in the twenty-first century. The local settlement of Mexicans has a long history beginning with mule packers, *vaqueros* (cowboys), and miners in the early 1850s. Other Mexicans came to work alongside Chinese immigrants on the railroads. Later in the nineteenth and early twentieth centuries, sugar companies aggressively recruited laborers from Mexico and the American Southwest to help cultivate beet fields in Idaho and Montana. The 1942 Bracero Program agreement between the United States and Mexico greatly expanded the number of Latino agricultural workers in places such as Treasure Valley in southwestern Idaho. Recently, more and more Mexicans have joined service staffs at area ski resorts; in Jackson Hole, Mexicans now account for nearly 20 percent of the population. The vitality of the growing Mexican American community is easily seen in the region's numerous *taquerias* (fast food eateries), bakeries, and small grocery stores, where Spanish and English are both commonly heard.

Evidence of early immigrant groups remains in the region's distinctive cultural landscape. Historic houses, ethnic churches, and other unique "landscape signatures" reflect the late nineteenth- and early twentieth-century settlement of the area's largest three ancestry groups—Germans, Finns, and Norwegians. Many of the European migrants were attracted by brochures in their homelands touting the opportunities of life in the area. Others came to join family members or to work as miners, farmers, or entrepreneurs. Germans helped establish the cattle industry in Wyoming as Basques played a vital role in the sheep industry. Finns settled in the Long Valley of Idaho and elsewhere in the area.

These groups, along with the Mormons and others, contributed to local culture and economic development. Interestingly, Scandinavians and Germans who converted to the Mormon faith in eastern Idaho show up as "English" or "American" on period census maps.

English is the region's dominant language followed by Spanish and Native American languages. Far more surprising are the languages spoken in remote Liberty and Toole counties in northern Montana, adjacent to the Canadian border. Here German-speakers make up more than one-quarter of the population, reflecting both the lingering effects of the large German immigrant population and the important role Canada played in early German migration to the region.

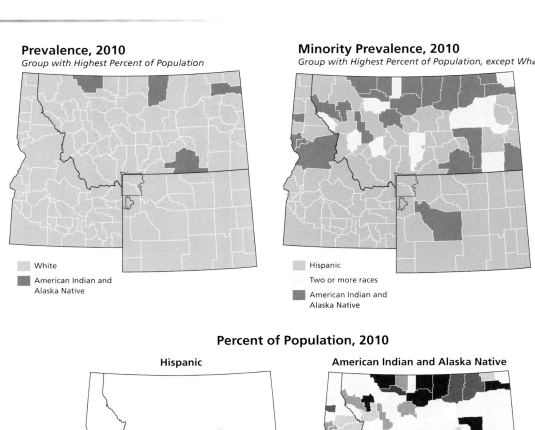

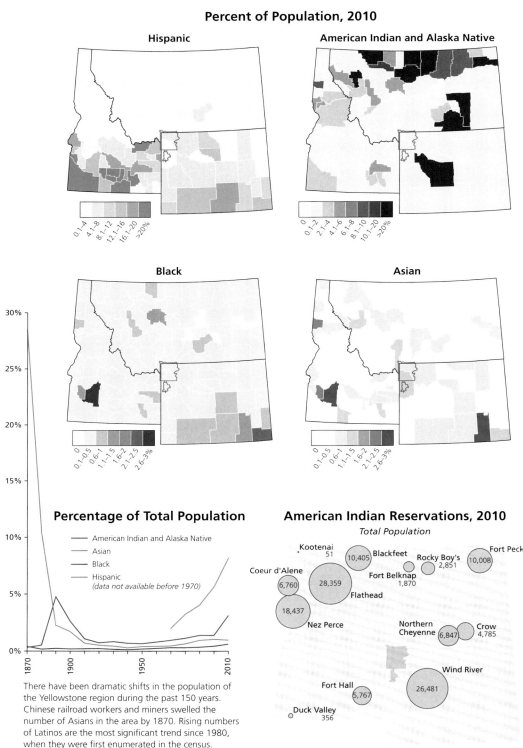

There have been dramatic shifts in the population of the Yellowstone region during the past 150 years. Chinese railroad workers and miners swelled the number of Asians in the area by 1870. Rising numbers of Latinos are the most significant trend since 1980, when they were first enumerated in the census.

## Dominant Ancestry, 2009

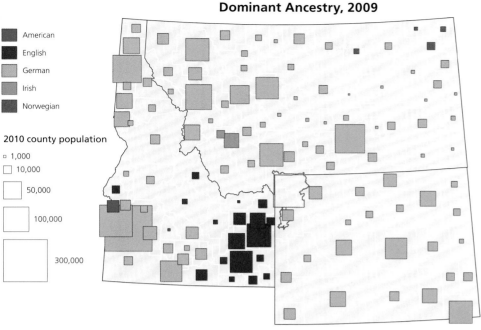

American
English
German
Irish
Norwegian

2010 county population
1,000
10,000
50,000
100,000
300,000

## Language Spoken at Home, 2009

### English

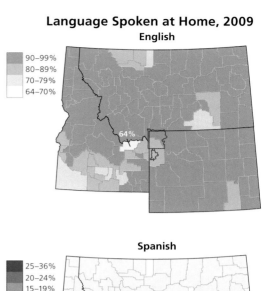

90–99%
80–89%
70–79%
64–70%

64%

### Spanish

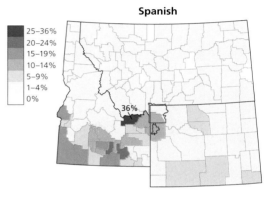

25–36%
20–24%
15–19%
10–14%
5–9%
1–4%
0%

36%

### German

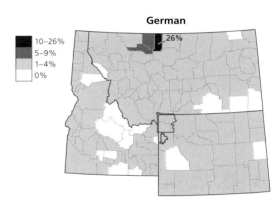

10–26%
5–9%
1–4%
0%

26%

### American Indian Languages

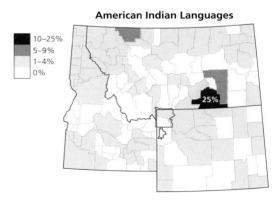

10–25%
5–9%
1–4%
0%

25%

### Other Languages

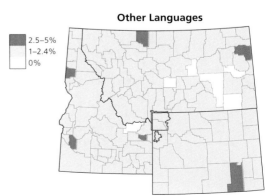

2.5–5%
1–2.4%
0%

## Race and Ethnicity in High Schools, 2008–2009

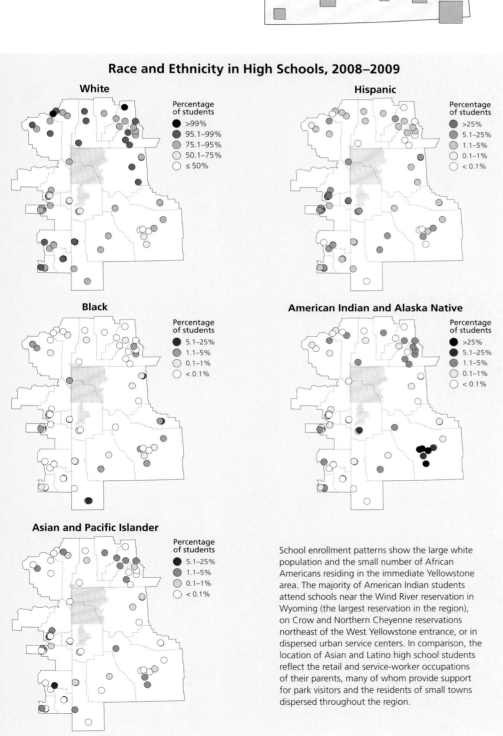

### White

Percentage of students
>99%
95.1–99%
75.1–95%
50.1–75%
≤ 50%

### Hispanic

Percentage of students
>25%
5.1–25%
1.1–5%
0.1–1%
< 0.1%

### Black

Percentage of students
5.1–25%
1.1–5%
0.1–1%
< 0.1%

### American Indian and Alaska Native

Percentage of students
>25%
5.1–25%
1.1–5%
0.1–1%
< 0.1%

### Asian and Pacific Islander

Percentage of students
5.1–25%
1.1–5%
0.1–1%
< 0.1%

School enrollment patterns show the large white population and the small number of African Americans residing in the immediate Yellowstone area. The majority of American Indian students attend schools near the Wind River reservation in Wyoming (the largest reservation in the region), on Crow and Northern Cheyenne reservations northeast of the West Yellowstone entrance, or in dispersed urban service centers. In comparison, the location of Asian and Latino high school students reflect the retail and service-worker occupations of their parents, many of whom provide support for park visitors and the residents of small towns dispersed throughout the region.

# Religion and Politics

## Religion

The religious makeup of the tristate region surrounding Yellowstone is varied. Catholicism is strong nearly everywhere, being either the first or second largest religious denomination throughout most of the region. In Montana the primary and secondary positions alternate between Catholic and Lutheran, resulting from the region's nineteenth-century German and Scandinavian immigrant base. Idaho is a center of Mormonism second only to Utah. Southern and especially southeast Idaho may be described as North Deseret, and Mormon influence extends well into Wyoming and Montana. Although not as prevalent as Mormons, large numbers of Catholics in Southern Idaho reflect a significant Latino farm worker population.

Baptists and Methodists are the second most numerous denominations in some central and eastern Wyoming counties, resulting from migration along the Platte River into the area from the South and lower Midwest where adherents of these religious groups are much more dominant. The higher numbers of Pentecostals in Northern Idaho represents the eastern fringe of their regional importance in Washington and Oregon. None of these groups has a strong presence in the immediate Yellowstone neighborhood.

### Primary Religious Denomination

20% — 40% — 60%

- Catholic
- Latter-Day Saints
- Lutheran
- United Methodist
- Baptist
- Pentecostal
- Other

Religions

2005 County Population
*in thousands*

500 | 250 | 100 | 50 | 25 | 10 5 1

For all large maps

### Religious Adherence

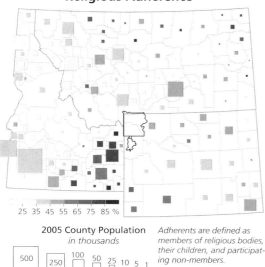

25 35 45 55 65 75 85 %

2005 County Population
*in thousands*

500 | 250 | 100 | 50 | 25 | 10 5 1

*Adherents are defined as members of religious bodies, their children, and participating non-members.*

### Denominational Strength, 2000

|  | Idaho | Mont. | Wyo. | Tristate | U.S. |
|---|---|---|---|---|---|
| Catholic | 10.1% | 18.8% | 16.3% | 14.1% | 22.0% |
| Baptist | 2.1% | 2.6% | 4.9% | 2.8% | 8.4% |
| Methodist | 1.6% | 2.2% | 3.0% | 2.0% | 3.9% |
| Lutheran | 2.1% | 7.4% | 4.4% | 4.3% | 2.9% |
| Jewish | 0.1% | 0.1% | 0.1% | 0.1% | 2.2% |
| Pentecostal | 2.3% | 3.1% | 1.5% | 2.4% | 1.9% |
| LDS | 24.1% | 3.7% | 9.6% | 14.6% | 1.5% |
| Christian | 1.2% | 1.0% | 1.3% | 1.2% | 1.5% |
| Presbyterian | 0.8% | 1.1% | 1.4% | 1.0% | 1.3% |
| Anglican | 0.6% | 0.7% | 1.8% | 0.8% | 0.8% |
| Other | 3.5% | 4.0% | 2.6% | 3.5% | 3.8% |
| All Denominations | 48.5% | 44.7% | 46.7% | 46.9% | 50.2% |
| Population, 2000 | 1,293,953 | 902,195 | 493,782 | 2,689,930 | 281,421,906 |

### Secondary Religious Denomination

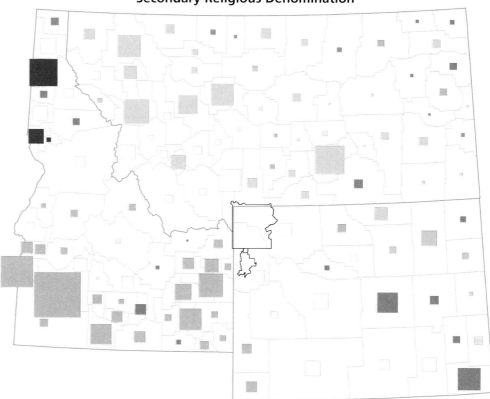

## 2008 U.S. Presidential Election

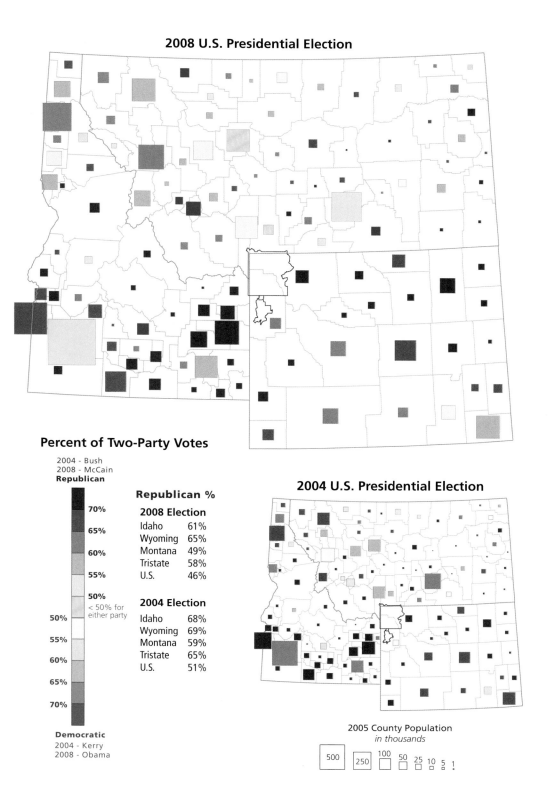

### Percent of Two-Party Votes

2004 - Bush
2008 - McCain
**Republican**

| | Republican % | |
|---|---|---|
| 70% | **2008 Election** | |
| 65% | Idaho | 61% |
| 60% | Wyoming | 65% |
| | Montana | 49% |
| 55% | Tristate | 58% |
| 50% | U.S. | 46% |
| < 50% for either party | | |
| 50% | **2004 Election** | |
| 55% | Idaho | 68% |
| | Wyoming | 69% |
| 60% | Montana | 59% |
| 65% | Tristate | 65% |
| 70% | U.S. | 51% |

**Democratic**
2004 - Kerry
2008 - Obama

## 2004 U.S. Presidential Election

**2005 County Population**
*in thousands*

500  250  100  50  25  10  5  1

## Politics

Idaho, Montana, and Wyoming have consistently voted conservatively since the mid-1960s. This tendency was a shift from the region's Progressive political character of earlier eras when, for example, Wyoming pioneered women's suffrage (1869), and La Follett took more than 35 percent of the three-state vote in the 1924 presidential election. The region voted for Franklin Roosevelt and the New Deal in about the same proportion as the country as a whole, and Idaho went for Truman by a slight margin in 1948. Since the late 1960s, however, the three states have voted more conservatively than the nation. Ross Perot took more than 25 percent of the vote in 1992, compared to less than 20 percent nationwide. The bar chart at the bottom of the page traces the long-term trend.

The Yellowstone states fit comfortably with the wider pattern of Intermountain and Great Plains region political conservatism in the later twentieth century. Local exceptions to the pattern are numerous, however. Counties that voted more liberally in the 2004 and 2008 presidential election are characterized by four quite distinct voting blocks. The Butte and Anaconda mining counties of Montana reflect a local labor tradition going back more than a century. Towns where state colleges and universities are located, such as Bozeman (Montana State University), Laramie (University of Wyoming), and Missoula (University of Montana), usually vote more liberally, although Pocatello (Idaho State University) bucks this trend. Indian reservation counties traditionally vote for more rather than less government. This trend also shows up in state capitals, although the effect is typically muted by the varied political perspectives of people in these locations. Finally, counties and towns that have outdoor amenity-based economies like Sun Valley and Jackson Hole are strongly inclined to environmental protection.

## Idaho, Montana, and Wyoming Historic Presidential Election Results

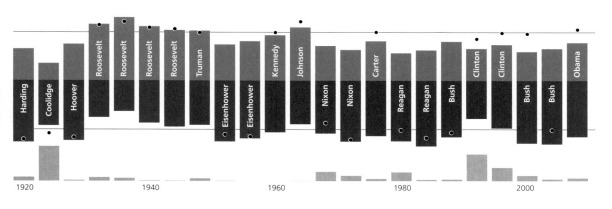

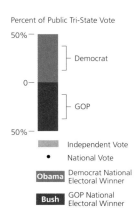

Percent of Public Tri-State Vote

50% — 
— Democrat
0 —
— GOP
50% —

Independent Vote
• National Vote

**Obama** Democrat National Electoral Winner
**Bush** GOP National Electoral Winner

# Elevation

Big Timber
el 4089

Columbus
el 3600

RIVER

Bozeman
el 4810

Livingston
el 4503

Jefferson River

RIVER

Boulder River

Stillwater

Yellowstone

BEARTOOTH MOUNTAINS

Red Lodge
el 5553

Virginia
City
el 5822

Lone Mtn
11166

Madison River

GALLATIN RANGE

YELLOWSTONE

Emigrant Peak
10921

ABSAROKA

Granite Peak
12799

Pilot Peak
11708

Clarks

Fork

Ruby River

MADISON RANGE

RUBY RANGE

Hebgen
Lake
6528

Electric Peak
10969

Mt Washburn
10243

Yellowstone National Park

ABSAROKA RANGE

Lamar River

Yellowstone

Red Rock River

CENTENNIAL MOUNTAINS

Island Park
Reservoir

7732

Shoshone
Lake
7791

YELLOWSTONE
LAKE

Eagle Peak
11367

Cody
el 5088

Buffalo Bill
Reservoir
5369

Shoshone River

Greybull River

Yellowstone River

Henrys

Fork

River

Warm

Snake

River

Dubois
el 5145

SNAKE RIVER PLAIN

St. Anthony
el 4972

Fork

Falls

River

Teton

John D. Rockefeller, Jr.
Memorial Parkway

TETON RANGE

JACKSON
LAKE
6772

Mt Moran
12605

Grand Teton
National Park

OWL CREEK MOUNTAINS

Driggs
el 6116

Grand Teton
13770

Rexburg
el 4865

River

Henrys

RIVER

Idaho Falls
el 4700

SNAKE

River

TETON RANGE

Gros

Ventre

River

Jackson
el 6208

GROS VENTRE RANGE

Wind

River

River

Wind

Gannett Peak
13804

Palisades
Reservoir
5620

Hoback

River

Fremont Peak
13745

CARIBOU RANGE

SNAKE

Salt

River

GREYS RIVER

WYOMING RANGE

New

Fork

Pinedale
el 7176

WIND RIVER RANGE

Lander
el 535

Blackfoot

River

Blackfoot
Reservoir
6111

River

SALT RIVER RANGE

River

Green

River

Popo

Agie

River

Portneuf

River

Soda Springs
el 5760

Bear

River

Wyoming Peak
11378

River

Paris
el 5968

Bear

Preston
el 4716

Bear
Lake
5923

River

Green

River

Big Sandy

River

0                    25 mi.

0                    25 km

3,000    4,000    5,000    6,000    7,000    8,000    9,000    10,000    12,000 Feet

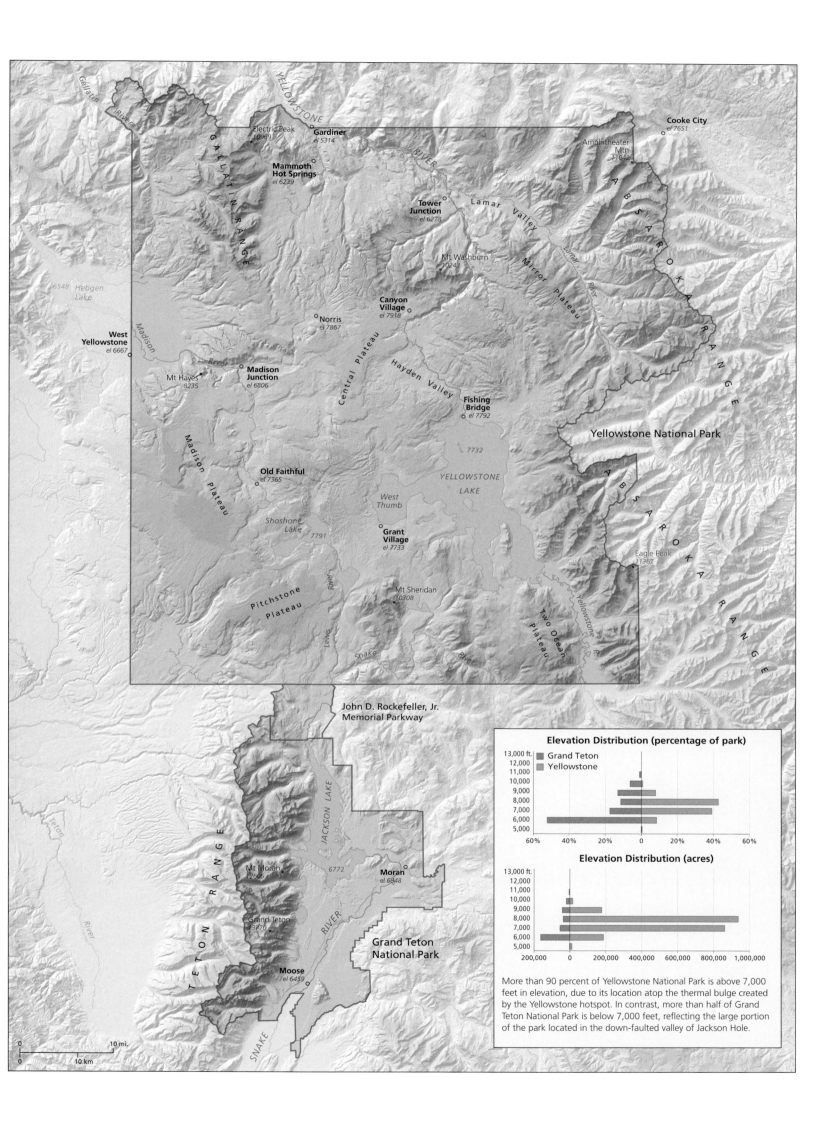

**Elevation Distribution (percentage of park)**

Grand Teton
Yellowstone

**Elevation Distribution (acres)**

More than 90 percent of Yellowstone National Park is above 7,000 feet in elevation, due to its location atop the thermal bulge created by the Yellowstone hotspot. In contrast, more than half of Grand Teton National Park is below 7,000 feet, reflecting the large portion of the park located in the down-faulted valley of Jackson Hole.

# Cross Sections

## Teton Range South to North

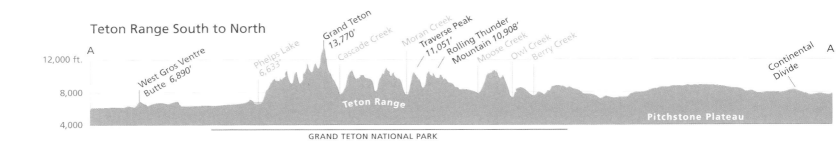

12,000 ft.

A · West Gros Ventre Butte 6,890' · Phelps Lake 6,633' · Grand Teton 13,770' · Cascade Creek · Moran Creek · Traverse Peak 11,051' · Rolling Thunder Mountain 10,908' · Moose Creek · Owl Creek · Berry Creek · Continental Divide · A

8,000

4,000

Teton Range — Pitchstone Plateau

GRAND TETON NATIONAL PARK

## Teton Range West to East

12,000 ft.

B · Teton Creek · Grand Teton 13,770' · Cottonwood Creek · Snake River · B

8,000

4,000

Big Hole Mountains — Teton Basin — Teton Range — Jackson Hole

IDAHO | WYOMING — GRAND TETON NATIONAL PARK

## Yellowstone Lake

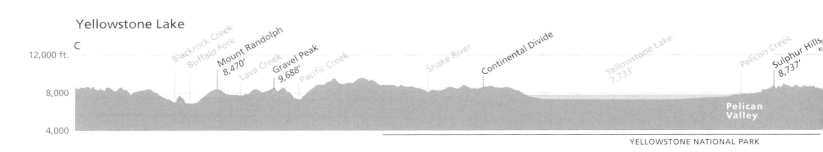

12,000 ft.

C · Blackrock Creek · Buffalo Fork · Mount Randolph 8,470' · Lava Creek · Gravel Peak 9,688' · Pacific Creek · Snake River · Continental Divide · Yellowstone Lake 7,733' · Pelican Creek · Sulphur Hills 8,737' · C

8,000

4,000

Pelican Valley

YELLOWSTONE NATIONAL PARK

## Snake River Plain to Bighorn Basin

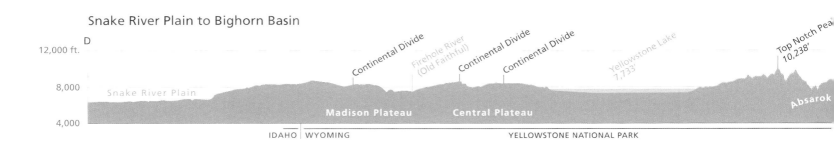

12,000 ft.

D · Continental Divide · Firehole River (Old Faithful) · Continental Divide · Continental Divide · Yellowstone Lake 7,733' · Top Notch Peak 10,238'

8,000

Snake River Plain

4,000

Madison Plateau — Central Plateau — Absarok

IDAHO | WYOMING — YELLOWSTONE NATIONAL PARK

## Snake River Plain to Beartooth Mountains

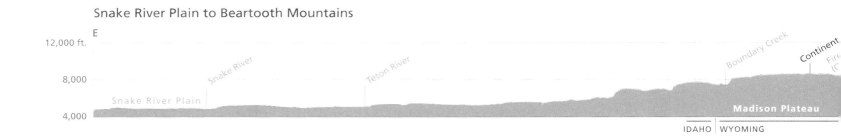

12,000 ft.

E · Snake River · Teton River · Boundary Creek · Continent · Fir (C

8,000

Snake River Plain

4,000

Madison Plateau

IDAHO | WYOMING

## Snake River Plain to Wind River Range

12,000 ft.

F · Snake River · Blackfoot River · Willow Creek · Clear Creek · McCoy Creek · Palisades Reservoir · Snake R · B

8,000

Snake River Plain · Blackfoot Mountains · Willow Creek Lava Field · Caribou Range · Sna

4,000

IDAHO | WYOMIN

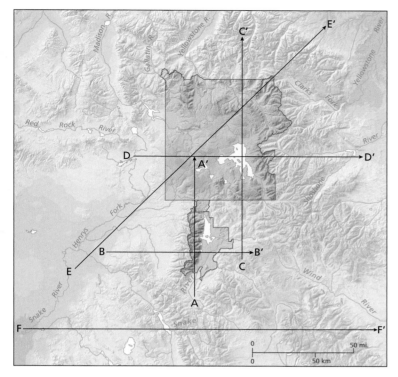

These cross sections illustrate the remarkable variety of topography across the Greater Yellowstone Area. They feature great vertical exaggeration, required at this scale to make the region's relief visible. If the vertical and horizontal scales were the same, the peak of Grand Teton would appear only a small fraction of an inch higher than Jackson Hole.

The transects through Grand Teton National Park (A and B) show the striking elevation changes that make this landscape so dramatic. The south-to-north transect runs along the crest of the Teton Range, highlighting the sharp peaks and deep gorges created by glacial incision at the range's core. The west-to-east transect demonstrates how regional uplift and faulting have pushed up the Tetons relative to the low-lying Snake River Plain to the west and the down-faulted valley of Jackson Hole to the east. Glaciers eroded deep valleys in the Tetons, leaving behind sharp peaks; related moraine deposits on the valley floor formed dams impounding Jenny Lake and numerous other bodies of water. Glaciers from the Yellowstone icecap extended to five miles south of Jackson Lake. Runoff from these glaciers formed the thick and relatively flat gravel deposits covering much of Jackson Hole's floor.

The three cross sections through Yellowstone National Park (C–E) portray the volcanic plateau of Yellowstone, much of which is between 8,000 and 9,000 feet. These high elevations result from uplift created by the Yellowstone hotspot, an area of long-lived volcanism. Past eruptions of the Yellowstone volcano above the hotspot generated vast deposits of ash and lava flows that form the rolling landscapes of the park's central plateau. The lofty elevations promote heavy snowfalls, with resulting snowmelt flowing into Yellowstone Lake.

To the west and southwest of the Yellowstone Plateau is the low-lying Snake River Plain (E). This plain once was above the hotspot, but the continent moved to the southwest and Yellowstone now sits over the hotspot. To the north and northeast of the plateau is the Beartooth Range, an area of ancient erosion-resistant rocks and home to Granite Peak, the highest point in Montana. Glaciers from the Beartooth icecap spilled into northern Yellowstone, helping to shape the broad valleys that became the drainages of the Lamar River, Slough Creek, and the lower Yellowstone River. Areas of lower elevation also occur where the process of headward erosion by rivers has cut canyons, as along the Grand Canyon of the Yellowstone. To the northeast and east of the central plateau is the Absoraka Range, a landscape of poorly consolidated lavas and volcanic mudflows, which glaciers eroded to form deep valleys.

The Greater Yellowstone Area includes large expanses to the south of the national parks (F). Moving from west to east, one encounters a series of smaller mountain ranges before cresting over the Snake River Range into the high basin of the Green River, which has its headwaters in the vast granitic core of the Wind River Mountains. These glacier-carved mountains include Gannett Peak, the highest point in Wyoming.

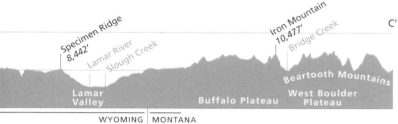

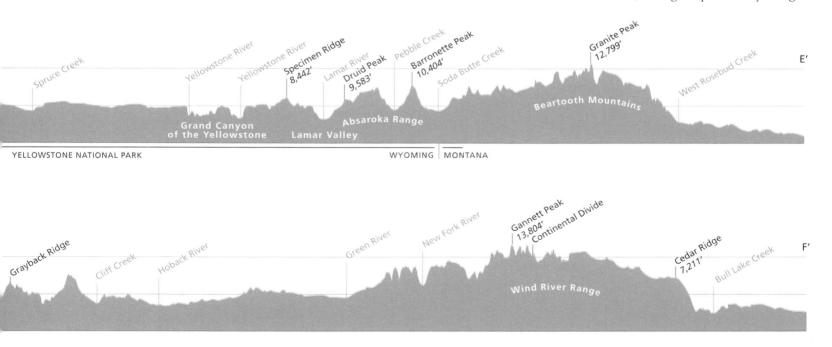

The Yellowstone hotspot explains many of the large-scale landforms found in Yellowstone and Grand Teton national parks. The hotspot left a track of northeastward-progressing volcanism as the North American tectonic plate moved southwest across a relatively fixed thermal mantle plume. Over the past few million years, this hotspot has driven Yellowstone volcanism, Grand Teton faulting, and the large area of uplift centering on Yellowstone. Related to the uplift, the Continental Divide crosses Yellowstone from southeast to northwest. Twenty-five major streams flow outward in a radial pattern from the Greater Yellowstone Area, which is thousands of feet higher than adjacent terrain.

The centerpiece of Yellowstone is a rhyolitic volcanic plateau between 8,000 and 9,000 feet in elevation. The rhyolite (as much as 77 percent silica) forms sandy to silty soils poor in clay and nutrients. Two different kinds of volcanic activity produced the rhyolite. Huge caldera-forming eruptions sent hundreds to thousands of cubic kilometers of red-hot ash and related materials blasting upward and spreading outward, coming to rest in deposits exposed today mostly outside the perimeter of the Yellowstone caldera. Large eruptions of viscous lava—tens of cubic kilometers of material—also spread rhyolite in the interior and near the margins of the Yellowstone Caldera. These eruptions created broad domes, with gently sloping hummocky tops and much steeper sides. The younger rhyolite flows cover much of the central and southwest part of Yellowstone. The streams draining the Yellowstone volcanic plateau begin with gentle gradients that steepen near the area's margins. Larger streams flowing off the plateau have eroded deep rocky gullies and created scenic canyons such as the Lewis River Canyon in the south, the Madison Canyon in the west, and the Grand Canyon of the Yellowstone in the north.

Yellowstone National Park's high central plateau is flanked on three sides by mountain areas. To the east is the Absaroka Range of

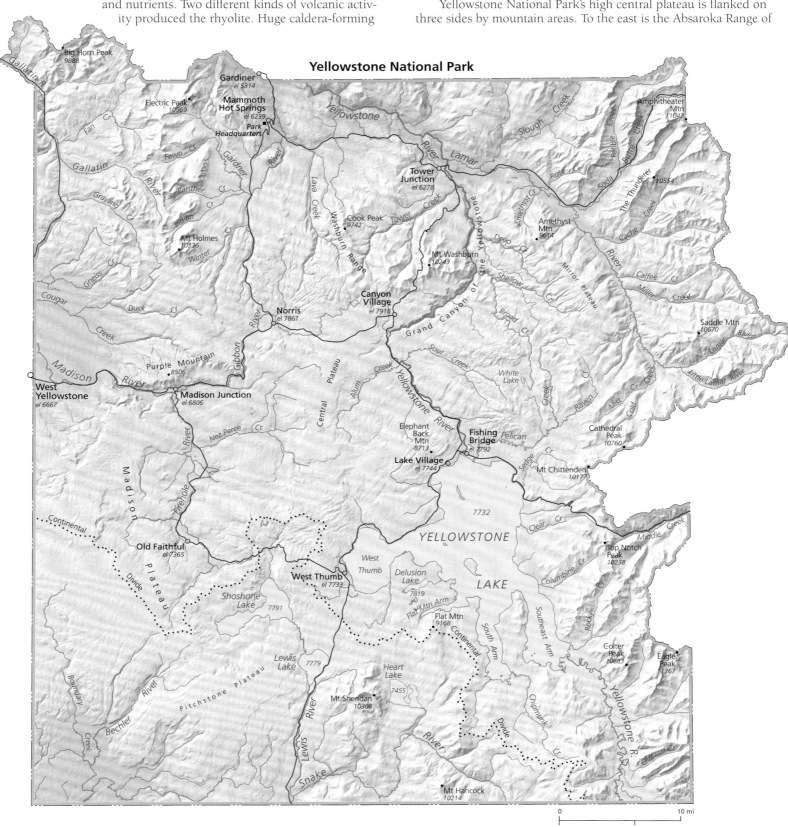

**Yellowstone National Park**

forty- to fifty-million-year-old andesitic lava flows and mud flows. The Absaroka volcanic units are poorly consolidated and readily eroded to form steep mountainous terrain; glaciated valleys are as deep as 3,000 feet. From northern Yellowstone and further to the northeast is the high Beartooth Uplift, formed of Pre-Cambrian erosion-resistant rocks (typically granitic). This uplift includes the highest mountain in Montana, Granite Peak. The Beartooth Uplift was the source of large Pleistocene glaciers more than half a mile thick that flowed south into Yellowstone. The Gallatin Range, to the northwest, rises above the central plateau and is formed of erosion-resistant rocks ranging from Pre-Cambrian granites in the south, through cliffs of Paleozoic limestone and sandstone in the center, to more readily eroded Mesozoic sandstones and shales in the north. Exceptions to this pattern, such as Electric Peak, exist where these Mesozoic sedimentary rocks were hardened by intrusions of hot magma.

High mountains of readily erodible rocks are best explained as having been formed by geologically recent uplift associated with the Yellowstone hotspot. One example lies southeast of the Yellowstone Plateau and northeast of Jackson Hole, where a high upland of Mesozoic shales and sandstones has peaks above 9,500 feet in elevation. Another example is the high uplands formed of Absaroka volcanic units that extend east to and beyond the park boundary and feature peaks above 11,000 feet.

Grand Teton National Park is divided into two contrasting structural blocks on either side of the Teton fault—the Teton Range to the west and Jackson Hole to the east. The fault results from a stretching (extension) of the Earth's crust. A ten-foot offset along the Teton fault would result in the Tetons rising approximately one foot and Jackson Hole lowering about nine feet. As Jackson Hole recedes, space is created for the accumulation of gravels and other sediment. Such deposits have a flat surface, somewhat like soup in a bowl—this explains the generally smooth floor of Jackson Hole. The "soup" here is largely quartzite-rich gravel carried by glaciers and glacial streams from the northeast.

On the west side of the Teton fault is the imposing front of the Teton Range, more than a mile high. This steep vertical relief creates powerful gravitational energy for erosion of the resistant rocks of the Teton Range; this force has been augmented by the scouring action of multiple glaciations during the Pleistocene epoch. Glacial erosion creates the spectacular sharp peaks and shear valley walls of the Teton Range. Young and continuing offset of the Teton fault creates the abrupt, awe-inspiring front of the Teton Range.

The cold, snowy climate of the Yellowstone and Grand Teton area is created not only from regional uplift, but also from the large lowland track to the southwest left by the Yellowstone hotspot. It provides

a 300-mile-long conduit for moist winter air masses to traverse the Snake River Plain northeastward until they reach the high mountains and plateaus at the plain's upper east end. Here the air masses rise as much as 4,000 feet and produce deep snows and water for Yellowstone's hydrothermal system. Hotspot-related topography thus has created environmental conditions including high terrain, cold temperatures, heavy winter snows, late spring snowmelt with significant runoff, and deep soil wetting. During glacial times, these cold, snowy conditions fostered the buildup of large glaciers that covered nearly all of Yellowstone and much of Grand Teton.

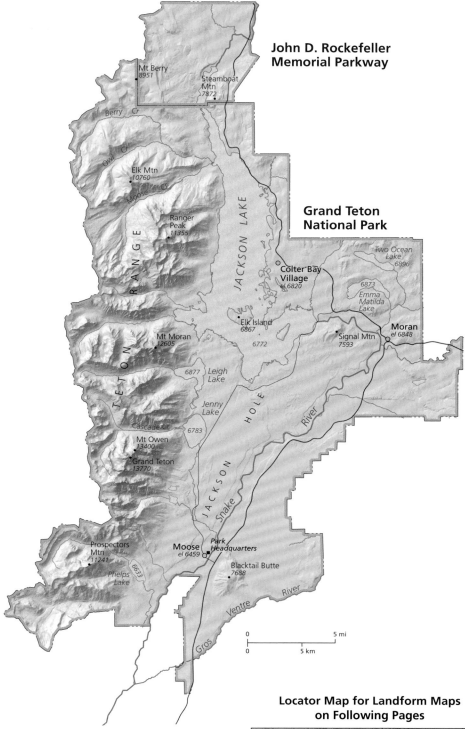

**John D. Rockefeller Memorial Parkway**

**Grand Teton National Park**

**Locator Map for Landform Maps on Following Pages**

# Landforms: Park Headquarters

Yellowstone National Park headquarters is located on a bedrock ridge capped by travertine above the Gardner River. This site was selected because of the spectacular travertine deposits at Mammoth Hot Springs and because the location was accessible via railroad in the early years of the park. The landscape surrounding Mammoth may be subdivided into four areas: Sepulcher Mountain (not shown on the map below), the Gardner River Canyon, Mount Everts, and the Yellowstone River Canyon. Sepulcher Mountain is underlain by volcanic rocks about 40 million years old that are highly faulted and fractured and rest upon much older Paleozoic marine limestones. Hot geothermal waters dissolve calcium carbonate ($CaCO_3$) from the limestone. When these waters rise to the surface through fractures, cooling precipitates the $CaCO_3$ to form the beautiful travertine terraces and thermal pools at Mammoth. The Gardner River Canyon cuts deeply into ancient seafloor shales deposited some 90 million years ago. These gray shales are easily seen to the east of Mammoth Hot Springs—they are also the source of debris flows that come off Mount Everts. Mount Everts is the highest point on a bedrock highland of Cretaceous sedimentary rocks, locally capped by a cliff of Huckleberry

Ridge tuff from the eruption of the Yellowstone caldera 2 million years ago. The Yellowstone River near the top of the map has cut its deep canyon into metamorphic rocks (gneiss and schist more than 2.5 billion years old) belonging to the ancient craton of North America.

Grand Teton National Park headquarters at Moose, Wyoming, lies on the floor of Jackson Hole. This area has been down-dropped thousands of feet along the Teton fault as a result of stretching of the Earth's crust. The modern channel of the Snake River has cut downward through glacial outwash deposits from the most recent (Pinedale) glaciation, which spread southward from the terminal moraine that dammed the south shore of Jackson Lake. Dried-up outwash channels can be seen south of Spalding Bay and elsewhere on the high bench of Baseline Flat. Timbered Island is a mound of glacial debris from the Bull Lake glaciation (about 140,000 year ago). Blacktail Butte is an outcrop of old, southwest-dipping sedimentary rocks (Paleozoic-Mesozoic), locally covered by much younger Miocene sediments. Glacial erosion sculpted the dramatic Teton Range topography, as is discussed on other pages in this section.

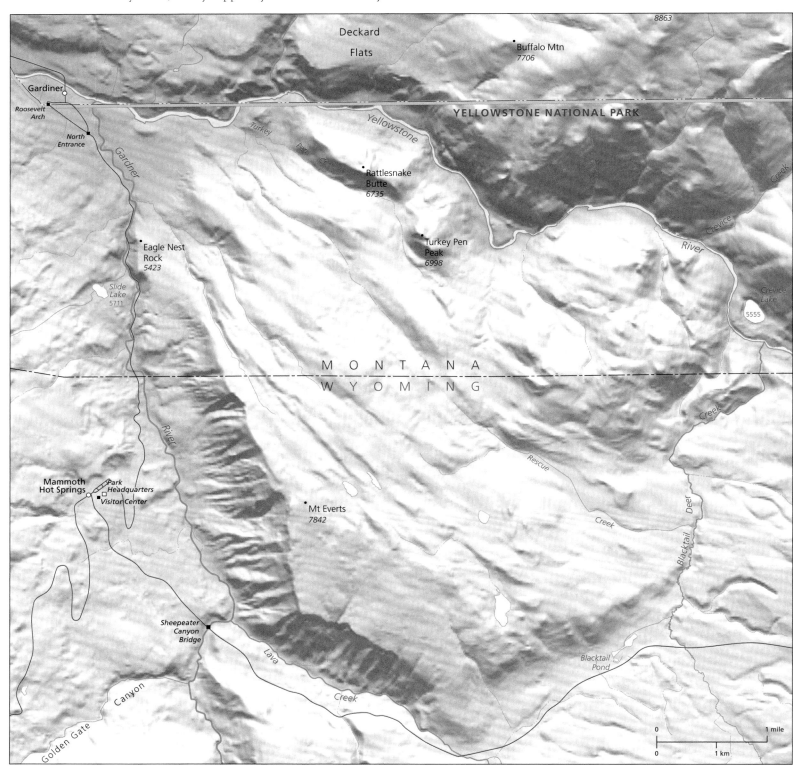

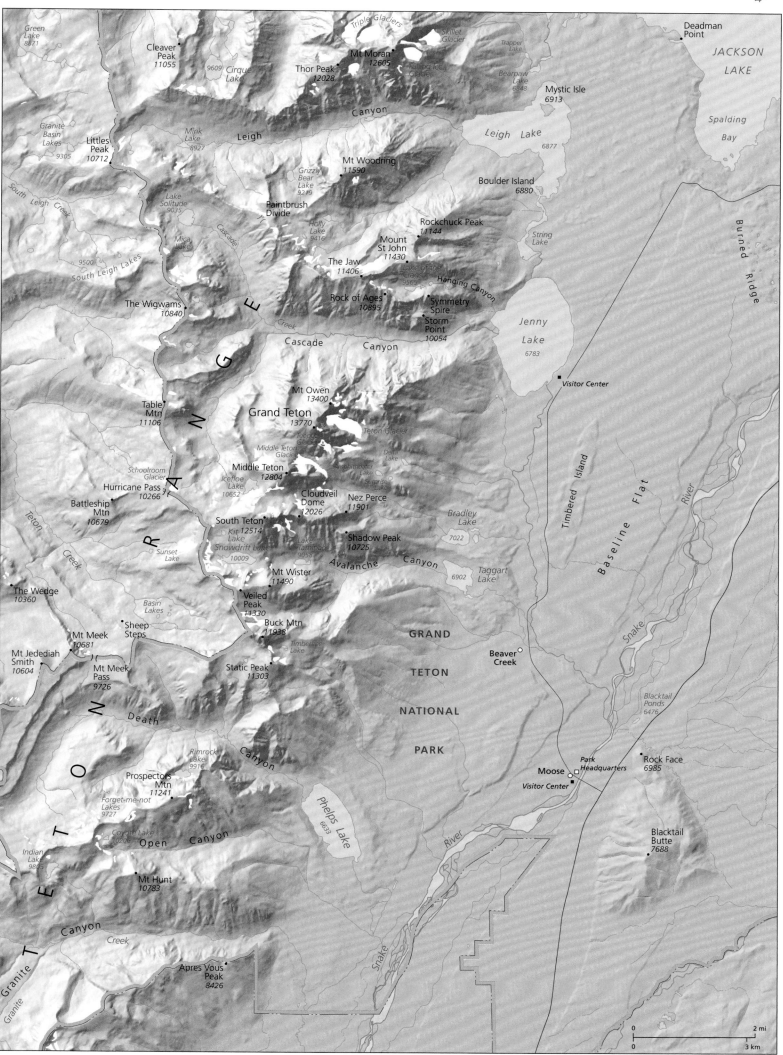

Green
Lake
*8871*

Cleaver
Peak
*11055*

Cirque
Lake
*9609*

Triple Glaciers

Skillet
Glacier

Mt Moran
*12605*

Trapper
Lake

Deadman
Point

*JACKSON
LAKE*

Thor Peak
*12028*

Falling Ice
Glacier

Bearpaw
Lake
*6848*

Granite
Basin
Lakes
*9305*

Littles
Peak
*10712*

Mink
Lake
*8927*

Leigh            Canyon

Mystic Isle
*6913*

Spalding
Bay

South Leigh Creek

Lake
Solitude
*9035*

Grizzly
Bear
Lake
*9219*

Mt Woodring
*11590*

Leigh  Lake
*6877*

Boulder Island
*6880*

Mica
Lake

Paintbrush
Divide

Holly
Lake
*9416*

Rockchuck Peak
*11144*

String
Lake

*9500*

South Leigh Lakes

Mount
St John
*11430*

The Jaw
*11406*

Lake of the
Crags
*9565*

Hanging Canyon

The Wigwams
*10840*

Cascade

Rock of Ages
*10895*

Symmetry
Spire

Storm
Point
*10054*

Jenny
Lake
*6783*

R  A  N  G  E

Creek

Cascade          Canyon

Table
Mtn
*11106*

Mt Owen
*13400*

Grand Teton
*13770*

Teton Glacier

Delta
Lake

Visitor Center

Schoolroom
Glacier

Middle Teton
Glacier

Icefloe
Lake
*10652*

Middle Teton
*12804*

Amphitheater
Lake

Surprise
Lake

Hurricane Pass
*10266*

Cloudveil
Dome
*12026*

Nez Perce
*11901*

Battleship
Mtn
*10679*

Bradley
Lake

*7022*

Sunset
Lake

South Teton
*12514*

Kit
Lake

Snowdrift Lake
*10009*

Lake
Taminah
*9058*

Shadow Peak
*10725*

Avalanche        Canyon

The Wedge
*10360*

Basin
Lakes

Mt Wister
*11490*

*6902*

Taggart
Lake

Veiled
Peak
*11330*

Sheep
Steps

Buck Mtn
*11938*

Mt Meek
*10681*

Timberline
Lake

*GRAND*

Mt Jedediah
Smith
*10604*

Mt Meek
Pass
*9726*

Static Peak
*11303*

Beaver
Creek

*TETON*

T  E  T  O  N

Death       Canyon

*NATIONAL*

Rimrock
Lake
*9916*

Prospectors
Mtn
*11241*

*PARK*

Rock Face
*6985*

Forget-me-not
Lakes
*9727*

Coyote Lake
*10206*

Open     Canyon

Phelps Lake
*6633*

Moose

Park
Headquarters

Visitor Center

Indian
Lake
*9885*

Mt Hunt
*10783*

Blacktail
Butte
*7688*

Canyon

Creek

Granite

Apres Vous
Peak
*8426*

Snake

River

Timbered  Island

Baseline  Flat

Snake

River

Blacktail
Ponds
*6476*

Burned  Ridge

0                    2 mi

0                    3 km

Garnet
Hill
7047

Crescent
Hill
7894

Floating
Island
Lake

The Cut
7571

Elk

Creek

Slough

Creek

Junction
Butte
6588

Lamar

River

Tower
Junction

NORTHEAST

ENTRANCE

ROAD

Lost
Lake

The
Narrows
6132

Prospect
Peak
9525

Tower
Fall

Lost

Creek

ROAD

Creek

Agate

Creek

YELLOWSTONE

Tower

Creek

Creek

GRAND

LOOP

Deep

Creek

Camellan

Creek

Antelope

Mt Washburn
10243

THE

Dunraven
Peak
9904

Dunraven
Pass
8859

RANGE

OF

Broad

Hedges
Peak
9685

WASHBURN

River

Creek

Sulpher

CANYON

Moss

Creek

GRAND

Canyon
Village

Yellowstone

Inspiration
Point

Ribbon
Lake

Artist
Point

Crystal
Falls

Lower
Falls

Clear
Lake

Upper
Falls

0     2 mi

0     3 km

The Grand Canyon of the Yellowstone formed as the Yellowstone River cut through the uplifted north flank of the Yellowstone caldera after its eruption 640,000 years ago. Following the master stream, tributaries such as Carnelian Creek also cut deep, narrow gorges through the caldera's flank. Dunraven Peak and other Washburn Range summits line the caldera rim here, with gentler topography of the caldera floor extending to the south. Mount Washburn is a much older, deeply eroded remnant of a large cone-shaped stratovolcano. It was active about 50 million years ago, part of a chain of volcanoes that ejected the andesitic Absaroka volcanics of northern and eastern Yellowstone.

Glacial activity is evident in small lakes surrounding Tower Junction, in depressions scoured by glacial erosion, and in kettles formed when ice blocks buried in glacial sediments melted away—as recently as 14,000 years ago. During maximum icecap buildup several thousand years earlier, glacial ice more than 3,000 feet thick in the Yellowstone Lake basin flowed out over the Washburn Range, leaving telltale striations scratched on the highest summits.

Shortly after the cataclysmic caldera-forming eruption, magma resurgence uplifted the caldera floor north of Yellowstone Lake, raising the football-shaped Sour Creek Dome, with Stonetop Mountain at its southern crest. The expanding dome

was split by numerous faults and fractures, evident in linear valleys that cut the dome from southeast to northwest. Evidence of younger uplift and faulting lies in similar northeast-trending valleys cutting Elephant Back Mountain, itself a rhyolite lava flow approximately 150,000 years old. Yellowstone Lake occupies the only part of the caldera depression not raised in a resurgent dome or filled in by thick rhyolite lava flows. Indian Pond and Turbid Lake are craters formed in postglacial time when shallow hydrothermal systems exploded, showering the surrounding landscape with steaming hot mud and rocks (note the obvious rim around Turbid Lake). Mary Bay also contains several large hydrothermal explosion craters.

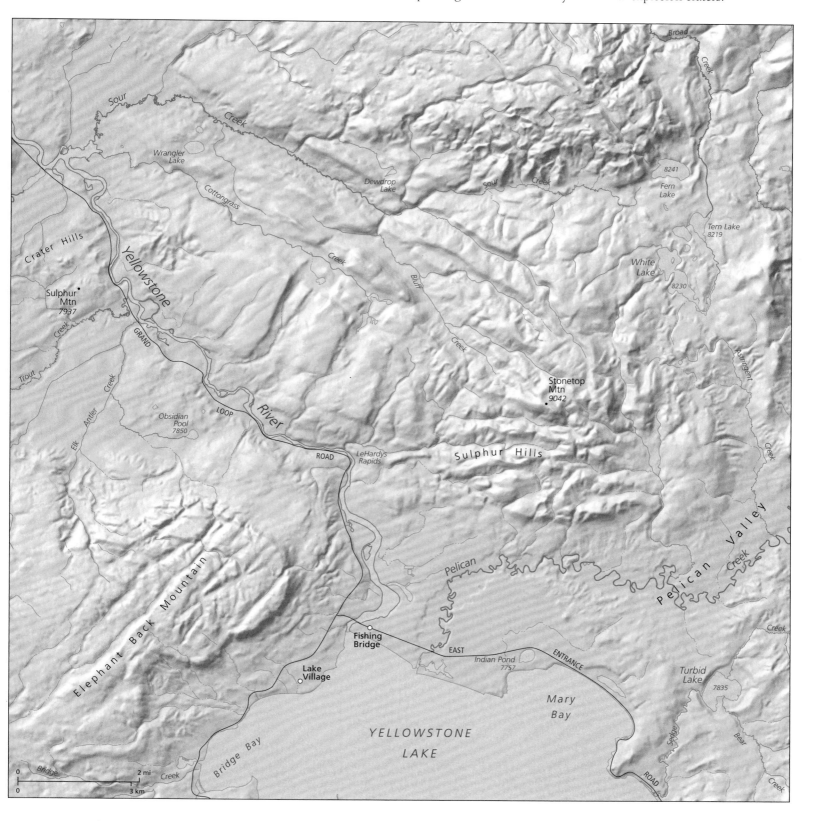

# Landforms: Lava Flows and Glacial Erosion

In the southwestern part of Yellowstone known as Cascade Corner, streams descend the precipitous margins of huge rhyolite lava flows. These flows are some of the largest in the world and include the paw-like lobe of the Pitchstone Plateau flow east of Bechler Meadows. This flow, which erupted 70,000 years ago, represents the youngest rhyolitic volcanism in the park and is little modified by glacial and stream erosion. The steep fronts and curving pressure ridges on this and other flows attest to the highly viscous character of rhyolite lava. Storms funneled along the Snake River Plain ramp up on the Yellowstone Plateau here, producing the heaviest snowfall in the park and abundant runoff over numerous waterfalls.

Yellowstone's northeast corner is cut by large glacial trough valleys, so deeply filled by ice during the maximum phase of the last glaciation that only the summits of the highest mountains like Barronette Peak protruded above the ice. The last remnants of ice existed in glacier-carved bowls (cirques), as on the north side of Abiathar Peak. Local bedrock is dominated by andesitic mudflow deposits of the Absaroka volcanics, overlying Paleozoic limestone, shale, and other sedimentary rocks. Valley walls up to 3,000 feet high are oversteepened by glacial erosion; in areas with highly erodible or unstable rock types this can result in rock falls, debris flows, and landslides. A massive landslide produced the rumpled lobe of debris that contains Trout Lake. The deep, round depression holding the lake suggests the landslide buried retreating glacial ice; when the ice melted, the unsupported debris collapsed to form the basin. Only a few thousand years ago, the steep toe of the Trout Lake landslide slumped downward, temporarily damming Soda Butte Creek and creating the broad sediment-filled meadow called Round Prairie to the northeast. Other large landslides are evident southwest of Trout Lake and west of Silver Gate. Storms often transform the numerous small streams draining the erodible valley sides into muddy torrents and debris flows. Such flows have deposited large fans and aprons of sediment that blanket much of the valley sides.

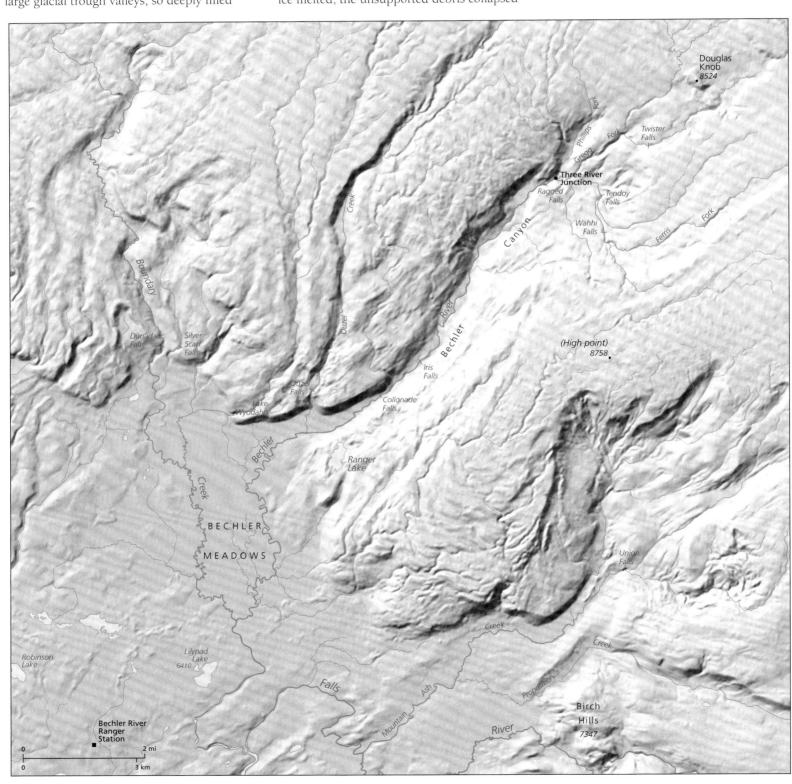

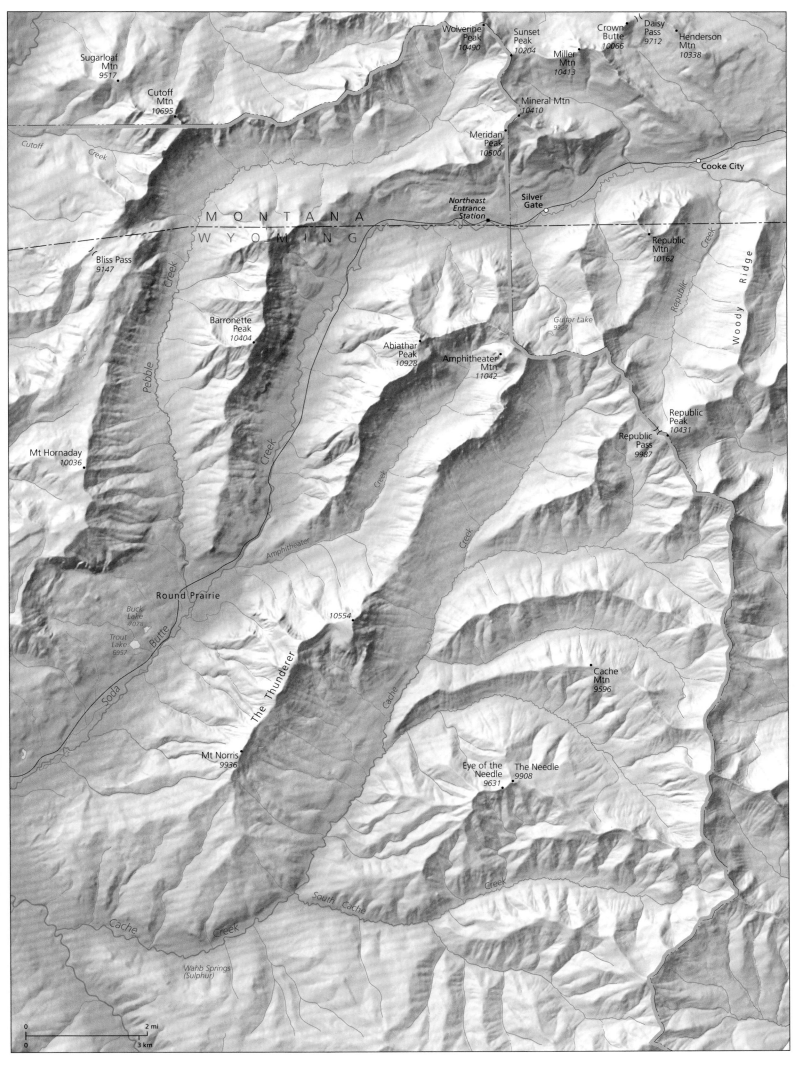

Sugarloaf
Mtn
*9517*

Cutoff
Mtn
*10695*

Wolverine
Peak
*10490*

Sunset
Peak
*10204*

Miller
Mtn
*10413*

Crown
Butte
*10066*

Daisy
Pass
*9712*

Henderson
Mtn
*10338*

Mineral Mtn
*10410*

Meridan
Peak
*10500*

Cooke City

*Northeast
Entrance
Station*

Silver
Gate

M O N T A N A

W Y O M I N G

Republic
Mtn
*10162*

Bliss Pass
*9147*

*Pebble*

*Cutoff*          *Creek*

*Creek*

Barronette
Peak
*10404*

*Guitar Lake*
*9321*

Abiathar
Peak
*10928*

Amphitheater
Mtn
*11042*

*Woody     Ridge*

*Republic     Creek*

Mt Hornaday
*10036*

*Creek*

Republic
Peak
*10431*

Republic
Pass
*9987*

*Amphitheater*

*Creek*

Round Prairie

Buck
Lake
*7078*

*10554*

*Creek*

Trout
Lake
*6957*

*Butte*

*Soda*

*The Thunderer*

Cache
Mtn
*9596*

*Cache*

Mt Norris
*9936*

Eye of the
Needle
*9631*

The Needle
*9908*

*South Cache*

*Creek*

*Cache*

*Creek*

*Wahb Springs
(Sulphur)*

0                    2 mi

0                    3 km

91

# Landforms: Overthrust Belt and Glacial Features

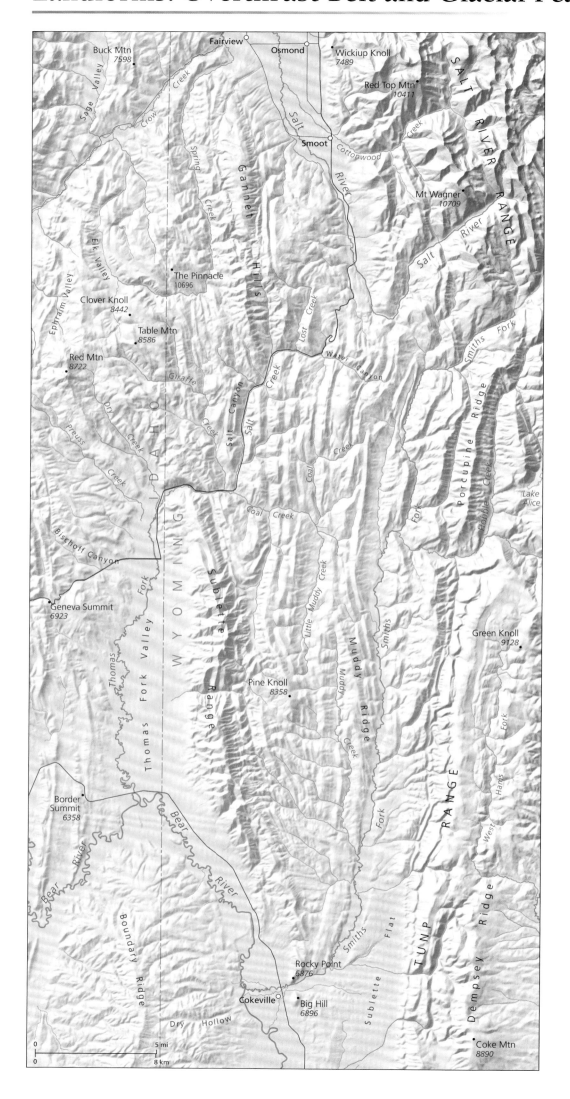

The "overthrust belt" of western Wyoming and southeast Idaho is part of an enormous linear zone of rock deformation that spans western North America from the Brooks Range in Alaska to the Sierra Madre Oriental of eastern Mexico. The overthrust belt formed during a major mountain-building event during the Cretaceous and early Paleogene periods, from about 140 to 55 million years ago. It is called the overthrust belt because great slabs of sedimentary rock, called thrust sheets, were carried up and over younger rock layers to the east by underlying thrust faults. The end result resembles overlapping roof shingles. Successive mountain ranges were formed in this manner from west to east across the width of the overthrust belt, each representing a major thrust sheet, a large anticline (upward fold) within the sedimentary succession, or both. For example, the Salt River and Sublette ranges in western Wyoming are underlain by the Absaroka and Crawford-Cokeville thrust faults, respectively; these thrusts, along with others in the region, dip westward and eventually merge with a low-angle, master thrust fault (décollement) near the base of the sedimentary section. This topography created by north-trending thrust sheets and related faults results in major rivers generally flowing in north-south oriented valleys with short east-west connecting segments. The cause of this deformation was deep-seated compression of western North America, generated by the collision and accretion of massive oceanic plateaus and island arcs along the West Coast, as well as convergence and subduction of an oceanic plate into the Earth's mantle beneath western North America.

As illustrated on the map to the right, the glacial geology along the southwest flank of the northern Wind River Range is a prime example of alpine-glacial geomorphology. Glaciers formed in cirques (bowl-shaped basins on mountain flanks) and coalesced as an icecap atop the Wind River Plateau. Valley glaciers flowed off the plateau down preexisting valleys, with some glaciers extending to the floor of the Wind River Basin. Lateral and terminal moraines (poorly sorted, hummocky deposits of sand, gravel, and boulders) mark the former extent of these glaciers, such as Fremont Ridge and the ridgeline that surrounds Willow Lake. As the ice melted, receding back into the high Wind River Range, water became dammed by the moraines and formed magnificent lakes, some of which are segmented by recessional moraines. U-shaped valleys bear testimony to the erosive power of former valley glaciers along the flanks of the range.

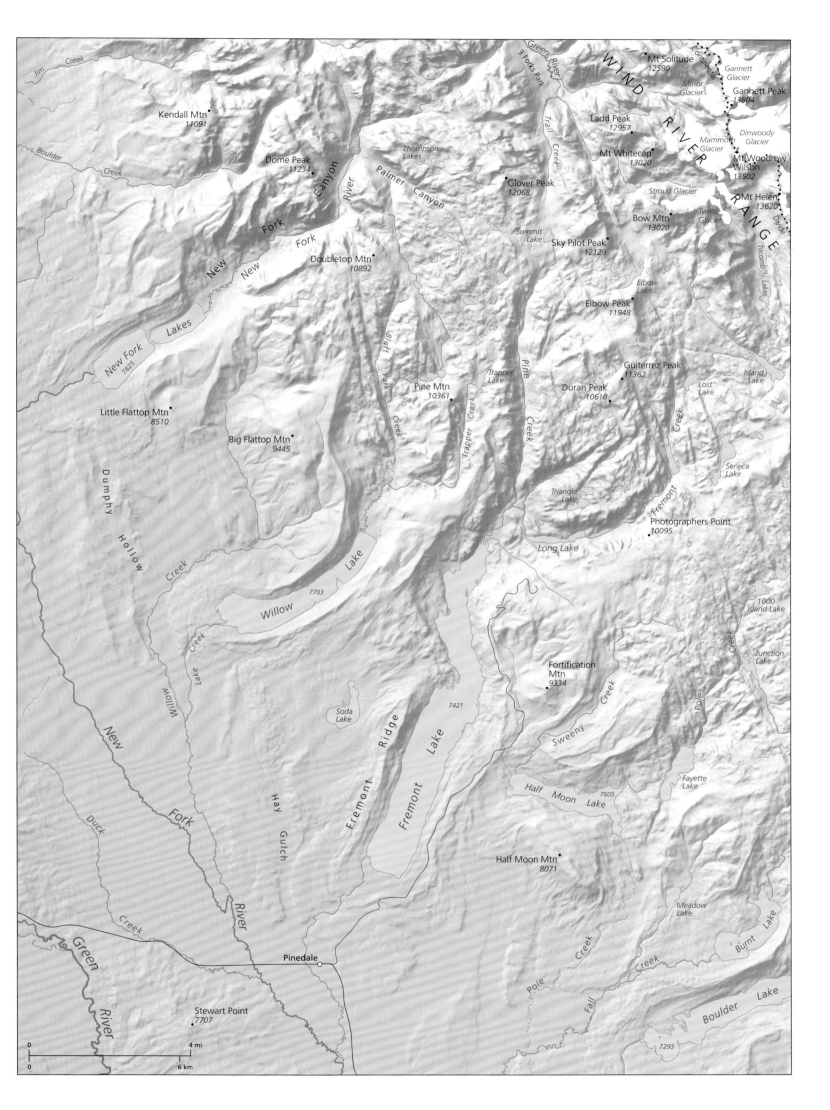

Jim Creek

Boulder Creek

Green River

3 Forks Park

Continental

WIND

Mt Solitude
12590

Gannett Glacier

Gannett Peak
13804

Kendall Mtn
11091

Ladd Peak
12957

Minor Glacier

RIVER

Dome Peak
11234

Thompson Lakes

Trail Creek

Mt Whitecap
13020

Mammoth Glacier

Dinwoody Glacier

New Fork Canyon

Palmer Canyon

Glover Peak
12068

Mt Woodrow Wilson
13502

Summit Lake

Stroud Glacier

RANGE

New Fork

New Fork

Doubletop Mtn
10892

Sky Pilot Peak
12129

Bow Mtn
13020

Twins Glacier

Mt Helen
13620

Lakes

Bluff Park

Elbow Lake

Elbow Peak
11948

Tricomb Lakes

Divide

New Fork
7825

Pine Creek

Trapper Lake

Island Lake

Little Flattop Mtn
8510

Pine Mtn
10361

Trapper Creek

Duran Peak
10618

Guiterrez Peak
11362

Lost Lake

Dumphy Hollow

Big Flattop Mtn
9445

Trapper Creek

Seneca Lake

Triangle Lake

Fremont Creek

Photographers Point
10095

Willow Lake
7703

Long Lake

1000 Island Lake

Willow

Creek

Junction Lake

Lake Creek

Soda Lake

Fremont Ridge

7421

Fortification Mtn
9334

Sweeny Creek

Pole Creek

New Fork

Duck Creek

Hay Gulch

Fremont Lake

Half Moon Lake

Fayette Lake

Half Moon Lake
7605

River

Half Moon Mtn
8071

Meadow Lake

Green River

Pinedale

Pole Creek

Fall Creek

Creek

Burnt Lake

Boulder Lake

Stewart Point
7707

7293

0        4 mi

0        6 km

93

# Grand Teton Geology

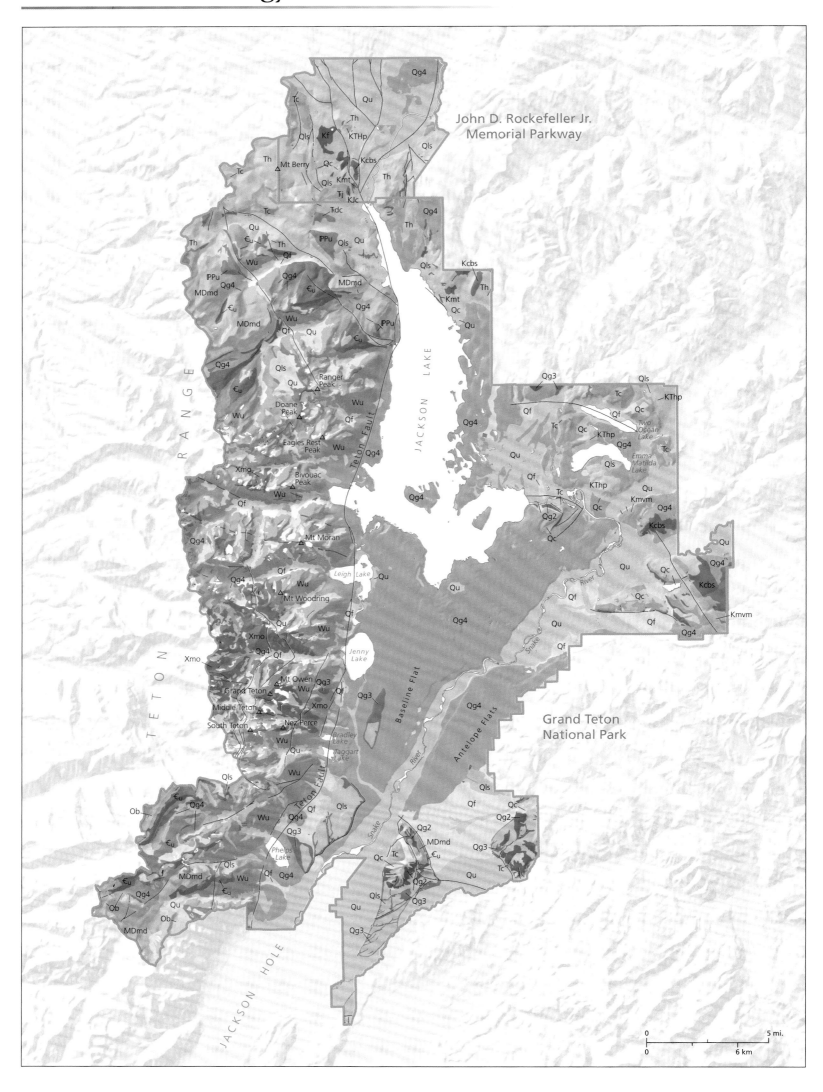

The landscapes of Grand Teton National Park are the result of young, ongoing geological processes. The park encompasses two major landforms: the jagged peaks of the Teton Range and the expansive floor of Jackson Hole.

The impressive east face of the Teton Range shoots up nearly 7,000 feet from the floor of Jackson Hole to the summit of Grand Teton at 13,770 feet. The Teton Range's tremendously steep east front abruptly ends in the flat floor of Jackson Hole along the north-trending, geologically young Teton normal fault. A normal fault occurs when the crust is extended (pulled apart) and ruptures, so that one side drops down while the other side tilts up. Jackson Hole comprises the down-dropped "hanging wall block" of the east-dipping Teton normal fault; the Teton Range comprises the uplifted "footwall block." The extension along this fault causes the valley floor to drop downward at approximately nine feet for every one foot that the Tetons tilt upward.

The reasons for this tectonic activity relate to regional forces. The Tetons sit at the eastern margin of the geologically active Basin and Range Province. This region makes up a large portion of the western United States between the Rocky Mountains and the Sierra Nevada and is slowly extending in the same fashion as the Tetons. The underlying tectonic reasons for this extension are complicated but are fundamentally related to high heat flow from the Earth's mantle to the surface. In addition to warming the crust, this anomalously high heat flow has resulted in regional uplift of the western United States, crustal thinning, distributed extension among countless normal faults that separate individual basins and ranges, and widespread bimodal volcanic activity (basalt and rhyolite) over approximately the past twenty million years. The origin of all this young geologic activity ultimately lies in the tectonic interactions of the North American and Pacific plates as they attempt to move past one another along the San Andreas fault and the inland "soft plate boundary" of the Basin and Range Province. The Tetons are the easternmost result of these processes.

As the Teton Range slowly rises, competing forces erode and wear down the range. In addition to the erosive power of modern streams that drain the spectacular canyons of the range, glaciers have periodically carved and sculpted the Teton Range for almost two million years. The effects of the most recent Pinedale glaciation are most easily seen: the terminal glacial moraines of poorly sorted boulders, cobbles, and sand that dam the lakes along the base of the range (for example, Jenny, Bradley, Taggart, and Phelps lakes) and the hummocky moraine at the south end of Jackson Lake. Outwash from the melting glaciers produced broad flat deposits in the valley bottom, which the Snake River eroded, creating the broad terraced valley bottom of Jackson Hole.

**Qu** **Quaternary 2**
Alluvium, colluvium, talus, lacustrine, swamp, landslide, glacial, loess, and travertine deposits, undivided

**Qc** **Quaternary 1**
Surficial detritus ranging in size from silt to boulders, derived from underlying and adjacent rock units

**Qf** **Alluvial fans**
Crudely stratified deposits of boulders, gravel, sand, and silt, commonly found at the base of steep slopes where streams exit narrow canyons and flow onto valley floors

**Qls** **Landslides**
Chaotically mixed boulders and finer rock debris; includes older (stabilized) and still-active landslide deposits

**Qg4** **Glaciation 4**
Drift and outwash deposits related to most recent glaciation: characterized by steep-sided moraine ridges with little or no soil development (often covered with coniferous trees) and small closed depressions (kettles); may include older, reworked glacial till

**Qg3** **Glaciation 3**
Drift and outwash deposits related to glaciation 3: characterized by moderately weathered erratics of granitic rock capped by loess in many places; subdued topography

**Qg2** **Glaciation 1 and 2**
Drift and outwash deposits related to glaciations 1 and 2: characterized by deep weathering of erratics and till, extensive soil development, little or no topographic expression (often expressed as lag deposits), generally does not support coniferous trees

**Th** **Huckleberry Ridge Tuff**
Compound cooling unit from Yellowstone's first volcanic cycle, composed of gray-brown rhyolitic ash-flow tuff, densely welded and devitrified but locally glassy (e.g., at the base); abundant phenocrysts of quartz, sanadine, and sodic plagioclase

**Tc** **Eocene Volcanics, undivided**
Includes several formations that are variously composed of volcanic mudflow breccia, conglomerate, sandstone, tuff, limestone, claystone, and mafic volcanic rocks

**KThp** **Harebell and Pinyon Conglomerates**
Quartzite conglomerate, sandstone, claystone, and tuff

**Kmvm** **Mesa Verde and Meeteetse Formations**
White porous sandstone, gray shale, coal, slabby tuff, and bentonite beds

**Kcbs** **Cody, Bacon Ridge, and Sohare Formations**
Thick dull-gray shale, thick-bedded sandstone, and lenticular fine-grained sandstone interbedded with dark-gray shale

**Kf** **Frontier Formation**
Gray sandstone interbedded with black shale and thin coal beds; bentonite beds in lower part

**Kmt** **Thermopolis, Muddy, and Mowry Formations**
Black shale, gray sandstone and siltstone

**KJc** **Morrison and Cloverly Formations**
Buff sandstone with red-green-gray siltstone and claystone; rusty-colored sandstone and variegated claystone with cream-colored limestone

**Trj** **Nugget, Gypsum Springs, and Sundance Formations**
Salmon-pink, fine-grained, cliff-forming sandstone; dark-red soft shale and gypsum; glauconitic, calcareous buff sandstone

**Trdc** **Dinwoody and Chugwater Formations**
Brown, slabby dolomitic siltstone; red shale and siltstone with purple limestone

**PPu** **Amsden, Tensleep, and Phosphoria Formations**
Limestone interbedded with red-green shale; yellowish-brown sandstone; black phosphatic shale with beds of chert, carbonate, and mudstone

**MDmd** **Darby and Madison Formations**
Brown, vuggy, fetid dolomite; light- to dark-gray, massive- to thin-bedded, cherty, fossiliferous limestone

**Ob** **Bighorn dolomite**
Light-gray, mottled, siliceous dolomite

**€u** **Cambrian strata, undivided**
Dark-gray limestone mottled with tan-yellow dolomite, flat-pebble conglomerate, apple-green shale; basal unit is maroon, cross-bedded sandstone (Flathead sandstone)

**p€d** **Precambrian (Proterozoic) dikes**
Dark-gray or green-gray, diabase dikes with tabular plagioclase phenocrysts

**Xmo** **Mount Owen quartz monzonite**
Light-colored biotite-quartz monzonite with muscovite-biotite pegmatite

**Wu** **Late Archean rocks, undivided**
Biotite gneiss, migmatite, amphibolite, metagabbro, iron formation and associated impure marble, and pods of ultramafic rocks

☐ Streams, rivers, and lakes
☐ Glaciers
—— Faults

Qs
TMₔs
Ti
Tv
Tyh
TMₔs
Qs
Tv
TMₔs
Ti
Tv
pCm
Tyh
Qtn
Tj
Tv
Qyl
Qpb
pCm
TMₔs
Qs
Pzs
TMₔs
pCm
Ti
TMₔs
Tyh
Tj
Tv
Qhi
Qtn
Ti
Qs
Tv
TMₔs
Tv
Tyh
Qh
Ti
Qyl
Qpb
Qpo
Tj
Qs
pCm
Ti
TMₔs
Qyl
Qpb
Tj
Qyl
Qpb
Qyl
Tyh
Qpo
*
TMₔs
Qpo
Qpr
Qhe
Tyh
Qyl
Qyl
Tv
Ti
Qm
Qhe
Qyl'
Qyl
Qpb
Tv
Qhe
Qyl
**MADISON VALLEY**
Tv
Tyh
Qyl
TMₔs
Qpb
Qpr
**Norris Geyser Basin**
TMₔs
Qh
Qyl
Qhe
Qpb
Qhi
Qyl
Qh
Qpb
Qhi
Qm
*Qm
Qs
Qpo
**CENTRAL PLATEAU**
Qh
Qpu
**SOUR CREEK DOME**
Qhe
Tyh
Qh
*
Qs
Qpr
Qhi
Tyh
Qhe
Qh
Qm
Qyl
Qh
Qpc
Qs
*
Qyl
Qhi
**MADISON PLATEAU**
Qm
Qh
*Qs
Qhi
Qhe
Qpu
**Lower Geyser Basin**
*
Qs
Qhi
Qs
Hayden Valley
Qs
Qm
Qyl
Qm
*
Qhi
Qs
Qpc
Qpc
Qhe
Pelican Valley
Qyl
Qpm
Qhe
Qh
Qpu
**MALLARD LAKE DOME**
Qhi
Qs
**Midway Geyser Basin**
*
Qh
Qs
Ti
Qpu
**Upper Geyser Basin**
Qs
**YELLOWSTONE LAKE**
Qhi
Qh
Qyl
Qs
*
Qh
**West Thumb Caldera**
Qs
Qhe
Shoshone Lake
Qs
Qs
*
Qhi
Qh
Qs
Tv
Qhi
Qs
Qs
Qpc
Lewis Lake
Lava Creek Caldera
Qh
Qm
Qyl
**South Arm**
**Southeast Arm**
Ti
Qs
Qh
Qhi
Qs
Tyh
Heart Lake
Qs
Ti
Qs
Qpc
Qhi
**PITCHSTONE PLATEAU**
Tyh
Qyl
Qs
Qpb
Qs
Qpb
Qyl
Qpb
Qh
Qlc
Qyl
TMₔs
Qh
TMₔs
Tv
Qpb
**Bechler Meadows**
Tv
Ti
Tyh
Qyl
Qlc
Qpc
TMₔs
Qm
**TWO OCEAN PLATEAU**
TMₔs
Tv

0                    10

0          10 km

Yellowstone National Park has a rich geologic history. The park sits on the ancient Archean crust of the Wyoming Province, which dates to roughly 2.5 to 3.5 billion years ago and forms part of the old, stable core of the North American plate. Though the park is exposed as dry land today, this was not always the case; tectonic forces and erosion caused alternating periods of uplift and subsidence, resulting in periodic flooding by ancient Paleozoic and Mesozoic seas. Dynamic episodes of mountain building, volcanism, glaciation, and crustal extension repeatedly reshaped the landscape during Cenozoic times. Yellowstone as we know it today, an active volcano with vigorous hydrothermal activity, is only the most recent face of an ever-changing landscape.

The modern Yellowstone volcanic system is superimposed on older (Eocene) volcanic rocks of the Absaroka Volcanic Supergroup. These andesite-dacite volcanic rocks crop out extensively in the north and east sides of the park. Subsequent volcanic eruptions, some of the largest known in Earth history, produced explosive ash plumes and thick and slow-moving lava flows that covered vast areas. Three major volcanic episodes occurred at 2.1 million, 1.3 million, and 640 thousand years ago, when the extremely explosive Lava Creek Tuff eruption partially emptied a large subterranean magma reservoir, causing collapse of the overlying crust to form the Yellowstone caldera. Caldera collapse occurred along two concentric ring fracture zones around the Mallard Lake and Sour Creek resurgent domes. Many more eruptions occurred between these three events and after the Lava Creek eruption, including the large eruption that formed the West Thumb of Yellowstone Lake approximately 173,000 years ago. After caldera collapse, additional magma oozed out, covering much of present-day Yellowstone National Park with relatively young lava flows.

Yellowstone boasts one of the world's highest concentrations of hydrothermal features, including hot springs, geysers, fumaroles, mud pots, and hydrothermal explosion craters. Most are within or near the Yellowstone caldera; however, significant hydrothermal activity also occurs along a linear band extending from Norris Geyser Basin to Mammoth Hot Springs. The distribution of hydrothermal features is controlled by the locations of faults: surface waters percolate into the crust through faults, becoming heated by magma at depth. Extensive faulting along the ring fracture zone of the Yellowstone caldera, and regional faulting in the Norris Geyser Basin area, provided the necessary passageways to establish Yellowstone's large hydrothermal system.

**Qs** **Detrital Deposits**
Unconsolidated gravel, sand, silt, and clay sediments that cover much of the ground surface and fill stream and river valleys.

**Qh** **Hot Spring Deposits**
Siliceous and calcareous sedimentary deposits that precipitated from hot spring waters.

**Qhi** **Ice Contact Deposits Cemented by Hot Springs**
Zeolite and opal cemented mounds that are relict features from glacial times, when sediment accumulated in depressions of retreating glaciers.

**Qhe** **Hydrothermal Explosions**
Craters formed by the explosive ejection of steam, water, and rock with no associated volcanism.

**Qpc** **Central Plateau Member**
Exceptionally large and thick rhyolite lava flows that partially fill the Yellowstone caldera basin, representing the youngest episode of volcanism after caldera collapse.

**Qpr** **Roaring Mountain Member**
Small rhyolite lava flows that erupted to the north of the Yellowstone caldera basin.

**Qpm** **Mallard Lake Member**
A single rhyolite lava flow uplifted on the Mallard Lake Dome in the western portion of the Yellowstone caldera.

**Qpo** **Obsidian Creek Member**
Small rhyolite lava flows that occur in a linear corridor that extends from the northern caldera margin in the vicinity of Norris Geyser Basin northward toward Mammoth Hot Springs.

**Qpu** **Upper Basin Member**
Rhyolite lava flows that erupted near the Mallard Lake and Sour Creek resurgent domes of the Yellowstone caldera, representing the earliest episode of volcanism after caldera collapse.

**Qyl** **Lava Creek Tuff**
Pyroclastic ash flows that erupted 640,000 years ago, partially emptying a large subsurface magma reservoir that caused collapse of the overlying crust to form the Yellowstone caldera.

**Qm** **Mount Jackson Rhyolite**
Several large rhyolite lava flows that erupted from the ring fracture zone of the Yellowstone caldera prior to the climactic Lava Creek Tuff eruption.

**Qlc** **Lewis Canyon Rhyolite**
Rhyolite lava flows that erupted south of the ring fracture zone of the Yellowstone caldera prior to the climactic Lava Creek Tuff eruption.

**Qtn** **The Narrows**
Interlayered gravels and basalts, with minor proportions of fine-grained sediments, glacial deposits, and ash beds from explosive volcanic eruptions.

**Qpb** **Pleistocene Basalts**
Basalt lava flows that erupted around the Yellowstone caldera; no basalt flows are known to have erupted within the Yellowstone caldera basin.

**Tyh** **Huckleberry Ridge Tuff**
Pyroclastic ash flows that erupted during the major Yellowstone volcano explosive episode 2.1 million years ago.

**Jb** **Junction Butte Basalt**
Thick basalt lava flows exposed in two areas north of the Yellowstone caldera, representing some of the oldest products of Yellowstone volcanism.

**Tv** **Tertiary Volcanic**
Predominantly andesite lava flows of the Absaroka Volcanic Supergroup, formed by numerous volcanic eruptions related to subduction zone volcanism about 50 million years ago.

**Ti** **Tertiary Intrusive**
Eroded plutons from magma intrusions into the crust, most related to Absaroka volcanism about 50 million years ago.

**TMzs** **Sedimentary (Paleozoic and Mesozoic)**
Limestones, sandstones, and shales that were deposited as deep seafloor sediments, shallow, muddy sea sediments, and as sand dunes and tidal flats in intervening dry periods.

**p€m** **Precambrian Metamorphic**
Granitic gneisses and amphibolites that form the cores of ancient mountain ranges, uplifted by tectonic forces and eroded by glaciers.

☐ Water
\* Hydrothermal vents
— Faults
— Caldera boundaries

# Yellowstone Volcano

## Path of the Yellowstone Hotspot

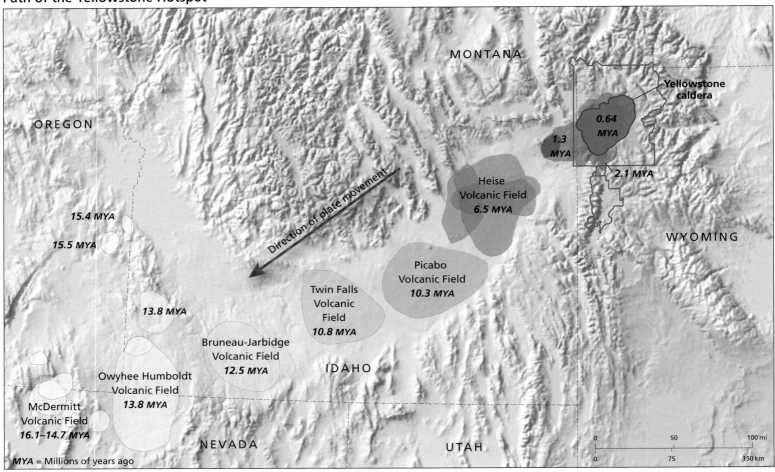

MONTANA

Yellowstone caldera

0.64 MYA

1.3 MYA

2.1 MYA

OREGON

WYOMING

Direction of plate movement

15.4 MYA

15.5 MYA

Heise Volcanic Field
6.5 MYA

Picabo Volcanic Field
10.3 MYA

Twin Falls Volcanic Field
10.8 MYA

13.8 MYA

Bruneau-Jarbidge Volcanic Field
12.5 MYA

IDAHO

Owyhee Humboldt Volcanic Field
13.8 MYA

McDermitt Volcanic Field
16.1–14.7 MYA

NEVADA

UTAH

MYA = Millions of years ago

0   50   100 mi
0   75   150 km

Far beneath Yellowstone volcano is a fixed heat source. From it, heat from the Earth's interior rises upward and fuels melting of the overriding crust. This process generates magmas that erupt episodically as well as catastrophically. Three such catastrophic eruptions have occurred in the past 2.1 million years, leaving behind huge calderas in eastern Idaho and western Wyoming. Yellowstone's eruptions were some of the largest in Earth's history.

Yellowstone volcano is not alone in having produced such enormous volcanic eruptions. Over the past 16 million years, caldera-forming volcanic fields have propagated in a northeasterly direction. The track of volcanism corresponds to southwest movement of the North American plate over a fixed heat source, which exposed new materials for melting in conveyor belt fashion. Northwestern Wyoming's current position over the plume of heat is the cause of Yellowstone's volcanic activity. Older volcanic fields to the southwest are now extinct. Volcanism located at Yellowstone National Park, too,

will diminish over time as the relative location of the hot spot moves farther to the northeast. Eventually—far, far in the future—the entire volcanic hot spot track will die.

The last giant caldera-forming eruption occurred 640,000 years ago. Following this eruption, large rhyolite lava flows and lesser amounts of basalt reached the surface, forming the large, relatively flat plateaus of Yellowstone National Park. Other major eruptions also have occurred episodically since that time. Approximately 173,000 years ago, for example, the West Thumb eruption yielded a caldera-forming eruptive volume much larger than that of Mount St. Helens.

Amassing large quantities of magma requires time, typically hundreds of thousands of years for the huge volumes erupted at Yellowstone. The 640,000 years that have elapsed since the last major caldera-forming eruption at Yellowstone may be sufficiently long that a new phase of magma production is under way. However, even if this is so, it is extremely unlikely that another major caldera-forming eruption will happen in our lifetimes.

Geophysicists are actively studying Yellowstone. Seismic imaging reveals a velocity contrast that is interpreted to be a plume of heat beneath the park. This low-velocity anomaly means that seismic waves travel more slowly through its relatively hot and less dense rock. The imaging reveals a three-dimensional structure about 60 miles wide that plunges northwest from Yellowstone to a depth of more than 300 miles beneath Dillon, Montana. Scientists are still debating numerous ideas to explain the subtle, deep structure surrounding Yellowstone.

Like anything hot, magma rises through the mantle. As iron-rich pods of magma ascend, they melt the overlying silica-rich crust. The crustal melts are viscous and can generate large caldera-forming eruptions when trapped gases reach a critical pressure and release catastrophically. Following the forceful venting of magma, silica-rich lavas erupt slowly to form the rhyolite lava plateaus of the region. Eventually, basaltic magma can reach the surface and cover previously erupted, silica-rich lavas. This is why basalt covers large areas, such as the Snake River Plain, or occurs outside the 640,000-year-old Yellowstone caldera.

## Yellowstone's Mantle Plume

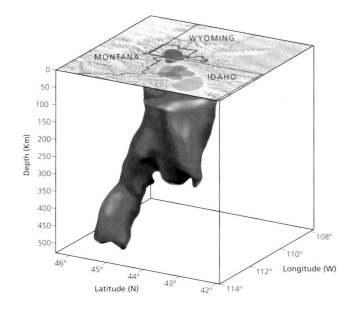

WYOMING

MONTANA

IDAHO

Depth (Km)
0
50
100
150
200
250
300
350
400
450
500

108°
110°
112°
114°
Longitude (W)

46°   45°   44°   43°   42°
Latitude (N)

# Yellowstone Caldera

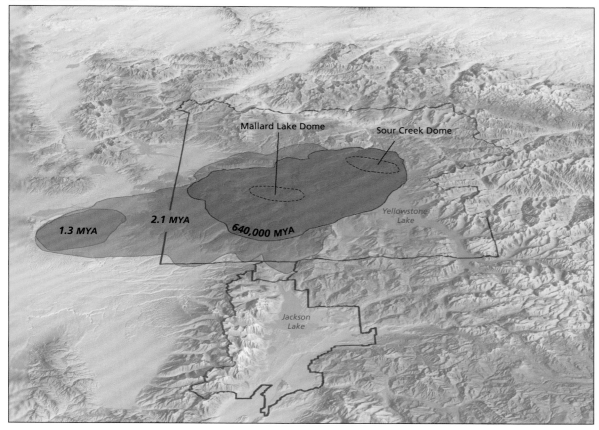

Mallard Lake Dome

Sour Creek Dome

2.1 MYA

1.3 MYA

640,000 MYA

*Yellowstone Lake*

*Jackson Lake*

During its 2 million-year lifespan, the Yellowstone volcano has generated three major caldera-forming eruptions. Eruptions of hot ash and gas partially emptied subterranean magma chambers, causing the overlying crust to collapse. The three calderas are so large that they cannot be seen in their entirety from ground level. Together, they span more than 50 miles and cover a large portion of Yellowstone National Park. The youngest of these major calderas is the Yellowstone caldera, produced after a cataclysmic eruption 640,000 years ago. Since that event, there have been eighty lava flows or eruptions, approximately thirty within the caldera. Two prominent resurgent domes, the Mallard Lake Dome and Sour Creek Dome, mark the locations of two separate ring-fracture zones where eruption vents formed. Rather than erupting from a single vent, the climactic eruption that produced the Yellowstone caldera was sourced from numerous vents within these two concentric fault zones. Following the caldera-forming eruption 640,000 years ago, thick, slow-moving lava flows partially filled the Yellowstone caldera basin and spilled over the western caldera rim. Combined, these postcaldera lavas have an extremely large volume, roughly half that of the climactic caldera-forming eruption.

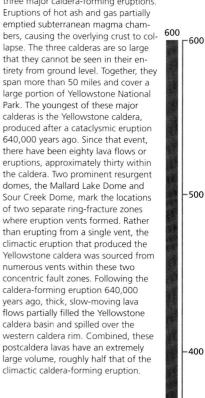

## Precaldera Lava Flows
### (0.6 MYA or older)

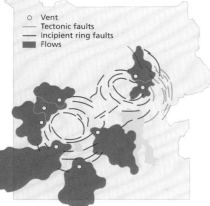

o  Vent
— Tectonic faults
— Incipient ring faults
▪ Flows

Episodic            lava flows occurred for approximately a million years before the Yellowstone caldera formed. Tectonic faults and incipient ring-faults define two concentric eruptive zones.

## Postcaldera Lava Flows
### (0.5–0.3 MYA)

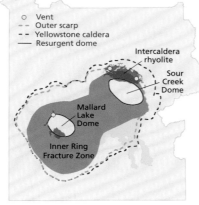

o  Vent
-- Outer scarp
- - Yellowstone caldera
— Resurgent dome

Intercaldera rhyolite

Sour Creek Dome

Mallard Lake Dome

Inner Ring Fracture Zone

The Yellowstone caldera subsided along an inner ring fracture zone (shaded) defined by incipient ring-faults. Caldera-filling eruptions occurred near the current Mallard Lake and Sour Creek resurgent domes.

## Postcaldera Lava Flows
### (0.2–0.07 MYA)

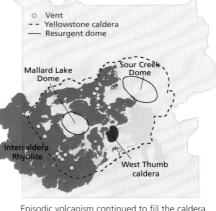

o  Vent
- - Yellowstone caldera
— Resurgent dome

Mallard Lake Dome

Sour Creek Dome

Intercaldera Rhyolite

West Thumb caldera

Episodic volcanism continued to fill the caldera and spill outside its boundaries along two linear vent zones that bisect the Yellowstone caldera. An explosive eruption 173,000 years ago during this postcaldera phase formed the smaller West Thumb caldera.

Major caldera-forming eruptions fragment huge quantities of magma into small ash particles that are blown skyward by gas pressure. Near the eruptive center, the column of hot rock and ash collapse to form pyroclastic flows that avalanche across the landscape at high speed for many tens of miles. At distances of hundreds to even a thousand miles away, ash falls, blanketing the ground. Yellowstone produced two of the largest caldera-forming eruptions in Earth's history, 2.1 million years ago and 640,000 years ago. Ash from these eruptions covered most of the western United States, reaching to the Pacific Ocean and Gulf of Mexico—far greater than the ashfall resulting from the 1980 eruption of Mount St. Helens. The volumes of ash generated by Yellowstone's caldera-forming eruptions are staggering, totaling about 600 cubic miles for the eruption 2.1 million years ago and 240 cubic miles for the eruption 640,000 years ago. There is no historic precedent for eruptions of this magnitude. However, far smaller eruptions from Mount St. Helens and Mount Pinatubo underscore the potential danger and disruption to human civilization.

## Ashfall Extent of Major Historic Eruptions

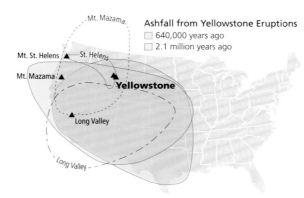

Mt. Mazama

Ashfall from Yellowstone Eruptions
☐ 640,000 years ago
☐ 2.1 million years ago

Mt. St. Helens ▲ St. Helens

Mt. Mazama ▲

**Yellowstone**

Long Valley ▲

Long Valley

## Cubic Miles of Ash Ejected by Major Volcanic Eruptions

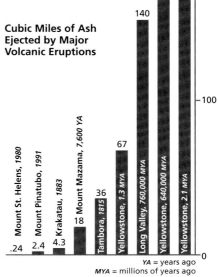

Mount St. Helens, 1980  .24
Mount Pinatubo, 1991  2.4
Krakatau, 1883  4.3
Mount Mazama, 7,600 YA  18
Tambora, 1815  36
Yellowstone, 1.3 MYA  67
Long Valley, 760,000 MYA  140
Yellowstone, 640,000 MYA  240
Yellowstone, 2.1 MYA  600

600
500
400
300
240
200
140
100

YA = years ago
MYA = millions of years ago

# Geothermal Activity

There are more than 10,000 geothermal features in Yellowstone National Park. Some geothermal waters near Mammoth Hot Springs are rich in calcium carbonate that precipitates to form travertine terraces. Geothermal waters throughout the rest of the park contain dissolved silica, which forms silica-rich sediments. Siliceous features include fumaroles, mudpots, pools, hot springs, and geysers. If water flow in a conduit is small and the heat high, pure steam roars from the ground to form fumaroles, as occurs at Roaring Mountain. Mudpots form at places such as Mud Volcano, where small amounts of hot, geochemically active water change volcanic rock to clay. Hot springs

have more water; silica coats the conduit walls so mud does not form near the surface. Geysers are hot springs that explode episodically with steam. Sometimes the steam explodes so violently that a crater is formed. Such craters range in diameter from twelve feet at Porkchop in Norris Geyser Basin to two miles at Mary Bay and range in age from 22 to 13,000 years. The floor of Yellowstone Lake has explosion craters, hydrothermal vents, and siliceous spires but no geysers because water pressure prevents steam formation.

Remote sensing measurements reveal varying ground temperatures. In some areas, the highest temperatures form linear patterns along fractures. More random patterns occur

### Number of Thermal Features

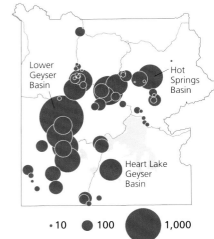

### Thermal Areas of Yellowstone

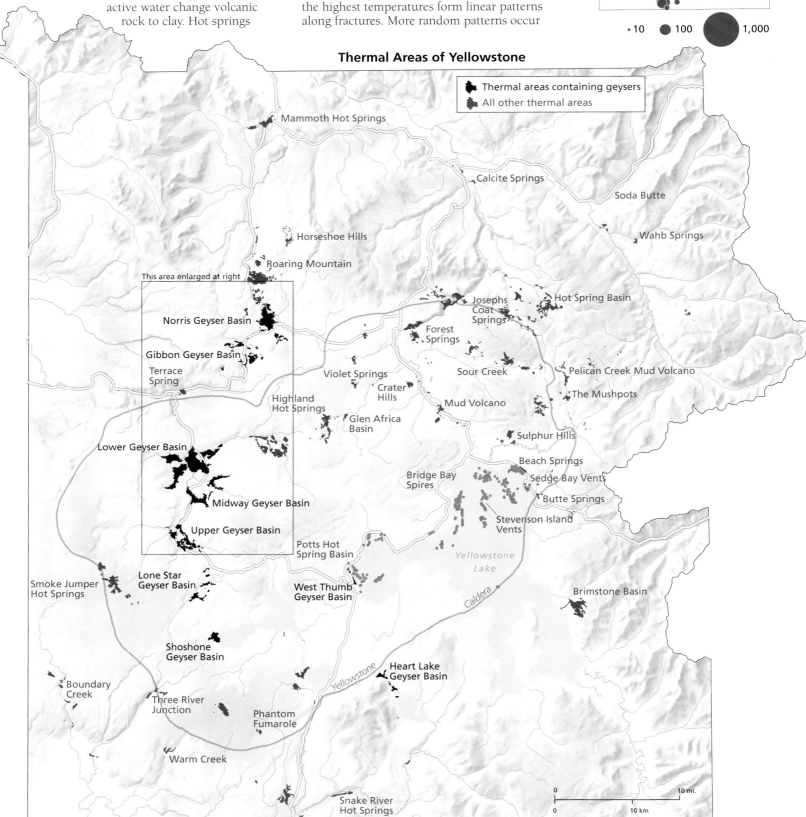

100

where geothermal waters rise along intersections of fractures or fracture swarms. A circle of high temperatures southwest of Lower Geyser Basin reflects the Twin Buttes hydrothermal explosion crater.

Between the Norris and Upper geyser basins water acidity ranges from pH values of one (the acidity of stomach acid) to ten (the pH of ammonia). Seven is a neutral pH. Many factors influence pH, including gases in the water, materials dissolved from surrounding rock, mixing of water from different sources, and surface interactions with the atmosphere and microbes. Given this complexity, the breadth of pH found in the geyser basins is not surprising.

Norris Geyser Basin, the hottest geothermal area in the park, has 550 geothermal features. The surface water temperatures of these features varies from 57 to 199°F; at a depth of 1,100 feet temperatures are almost 400°. Geochemical models suggest the temperatures may be as high as 644° at greater depths. The locations of the geothermal features and high ground temperatures parallel fractures in the rock.

Approximately 94 percent of the chloride leaving the park's four major rivers has a geothermal origin. Monitoring the chloride content of rivers, therefore, is a tool for recording changes in the flux of geothermal waters in Yellowstone park. The total mass of geothermal chloride in major rivers leaving the park declined approximately 10 percent from 1982 to 2002. One explanation is that this decline is related to deflation of the Yellowstone caldera, but more recent data suggest possible reinflation. The scientific jury is still out on the cause of this decline.

## Norris to Upper Geyser Basins

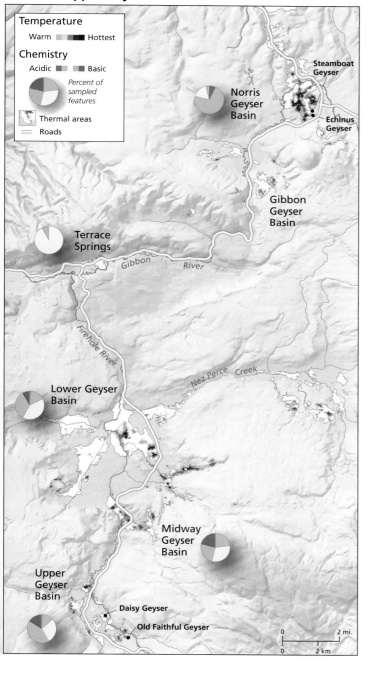

## Norris Geyser Basin

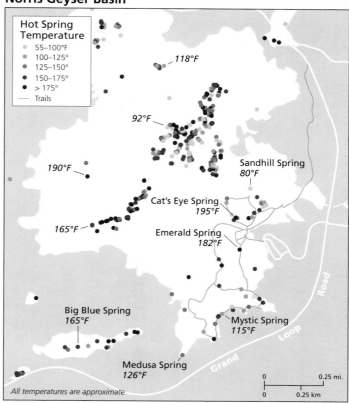

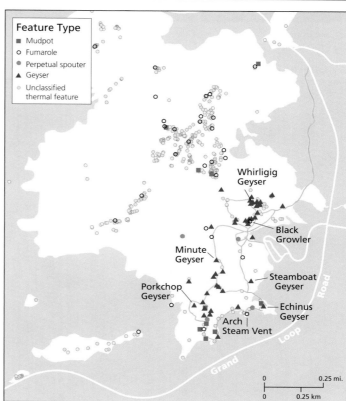

## Chloride Flux from Yellowstone Rivers, 1982–2002

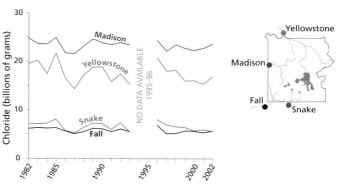

# Hydrothermal Areas

## Fountain Group, Lower Geyser Basin

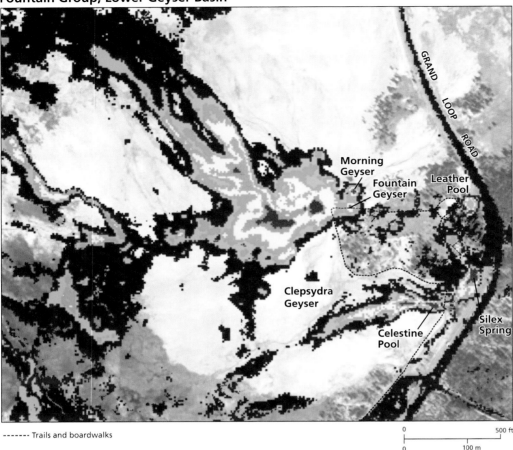

------ Trails and boardwalks

```
0                    500 ft
0              100 m
```

## Midway Geyser Basin

```
0                    1000 ft
0              200 m
```

Yellowstone's hydrothermal features were a primary reason for the creation by Congress of the world's first national park. Hydrothermal features include geysers, hot springs, mudpots, and fumaroles—the most visible elements of complex hydrothermal systems. One tool researchers use to map, monitor, and understand these systems is airborne thermal infrared imagery, which records emitted thermal energy. Acquired at night to avoid measuring heating from the sun, thermal infrared imagery provides an estimate of the temperatures of hydrothermal and other features. In these images, the temperature maps are overlaid on high resolution topographic imagery derived from laser-based radar, or lidar.

The Fountain Group sits on a thick deposit of sediments that cover the underlying volcanic rocks. The vivid colors indicate where heat and water flow through cracks and reach the surface. Rich in silicon and oxygen, these hydrothermal waters are the source of extensive hot-spring deposits shown in pink. The areas that display gray to white tones have temperatures less than 5°C (41°F) and are not assigned a temperature-related color.

Midway Geyser Basin contains the famous Grand Prismatic Spring. This and other hot springs provide thermal input to the Firehole River, making its temperature relatively warm for a high-elevation stream. Rhyolitic lava flows formed the hills to the east and west; terraces and outwash plains from melting glacier ice more than 10,000 year ago form the landscape between the volcanic hillsides.

### Hydrothermal Areas

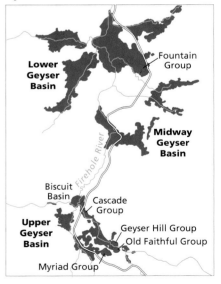

### Temperature C° (F°)

- 40–70° (104–158°)
- 30–40° (86–104°)
- 20–30° (68–86°)
- 15–20° (59–68°)
- 10–15° (50–59°)
- 5–10° (21–50°)

## Biscuit Basin and Cascade Group, Upper Geyser Basin

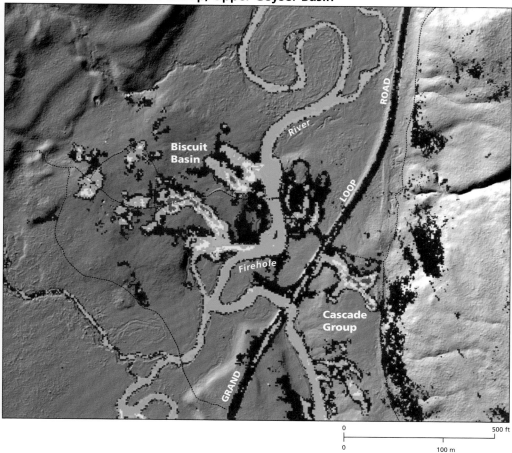

Biscuit Basin sits within Yellowstone's extensive Upper Geyser Basin. Rhyolitic lava flows, eroded and smoothed by glacial ice, crop out on the hillsides. The area is noted for forceful geyser eruptions and small hydrothermal explosions.

Visited by millions of people each year, the Old Faithful area is perhaps the most famous geothermal site in the world. The extraordinary concentration of thermal features in the Upper Geyser Basin result from a unique combination of heat, fluids, geology, and a natural plumbing system. Water heated at depth flows through fractures to the surface. Earthquakes keep them open. The pattern of low temperature (blue) around Old Faithful geyser shows the effect of human structures such as buildings, roads, water lines, and sewers; in contrast, the temperature patterns at Geyser Hill reflect a natural hydrothermal system. This kind of information helps resource managers in their efforts to balance protection of the park's world-class hydrothermal system with providing visitor access.

## Old Faithful, Myriad, and Geyser Hill Groups, Upper Geyser Basin

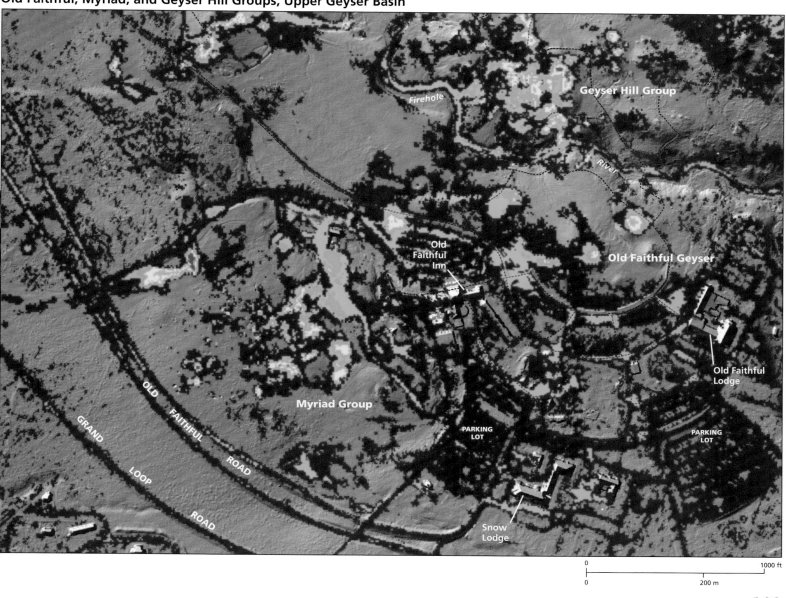

103

# Geysers

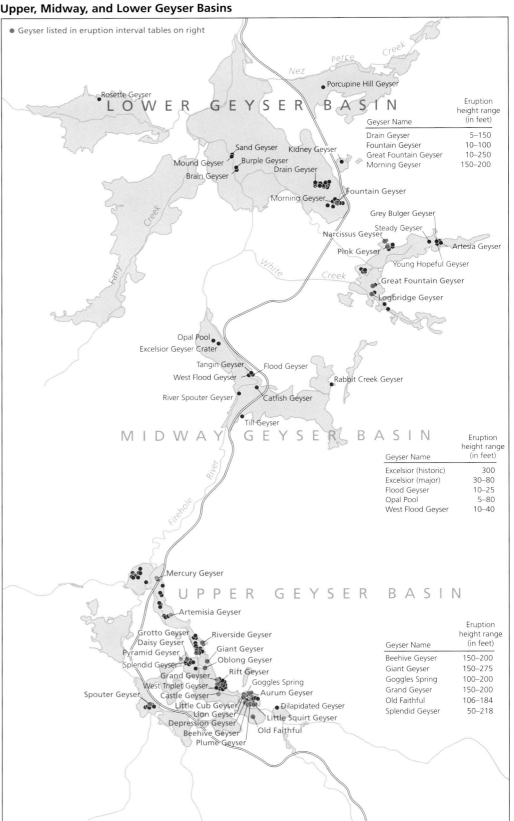

## Yellowstone's Geyser Basins

*(Map labels)*
Norris Geyser Basin
Gibbon Geyser Basin
Lower Geyser Basin
Upper Geyser Basin
Midway Geyser Basin
Lone Star Geyser Basin
West Thumb Geyser Basin
Shoshone Geyser Basin
Heart Lake Geyser Basin

## Named Geysers

### LOWER GEYSER BASIN

Angle Geyser
Artesia Geyser
Bear Claw Geyser
Bellfontaine Geyser
Blowout Geyser
Burple Geyser
Brain Geyser
Clepsydra Geyser
Collapsed Geyser
Deep Blue Geyser
Drain Geyser
Earthquake Geyser
Ferric Geyser
Firehosue Geyser
Fountain Geyser
Great Fountain Geyser
Grey Bulger Geyser
Honeycomb Geyser
Impatient Miser Geyser
Jet Geyser
Kaleidescope Geyser
Kidney Geyser
Little Crack Geyser

Logbridge Geyser
Mask Geyser
Morning Geyser
Morning Geyser's Thief
Mound Geyser
Narcissus Geyser
New Bellfontaine Geyser
Pair Geyser
Pink Geyser
Porcupine Hill Geyser
Rosette Geyser
Sand Geyser
Spasm Geyser
Steady Geyser
Sub Geyser
Super Frying Pan Geyser
Three Vent Geyser
Twig Geyser
Vertical Geyser
Volcanic Tableland Geyser
West Sprinkler Geyser
Young Hopeful Geyser

### MIDWAY GEYSER BASIN

Catfish Geyser
Excelsior Geyser
Flood Geyser
Opal Pool
Rabbit Creek Geyser

River Spouter Geyser
Tangin Geyser
Till Geyser
West Flood Geyser

### UPPER GEYSER BASIN

Artemisia Geyser
Atomizer Geyser
Aurum Geyser
Baby Daisy Geyser
Ball Cap Geyser
Bank Geyser
Beehive Geyser
Big Cub Geyser
Biscuit Basin Geyser
Black Pearl Geyser
Boardwalk Geyser
Bulger Geyser
Cascade Geyser
Cauliflower Geyser
Churn Geyser
Cliff Geyser
Comet Geyser
Coral Geyser
Coronet Geyser
Culvert Geyser
Daisy Geyser
Daisy's Thief Geyser
Depression Geyser
Dilapidated Geyser
Dome Geyser
Dusty Geyser
East Triplet Geyser
Fan Geyser
Giant Geyser
Giantess Geyser
Grand Geyser
Grotto Fountain Geyser
Grotto Geyser
Hankerchief Geyser
Hillside Geyser
Infant Geyser
Inverted Geyser
Island Geyser
Jagged Geyser
Jewel Geyser
Link Geyser
Lion Geyser
Lioness Geyser
Little Cub Geyser
Little Squirt Geyser
Marmot Cave Geyser
Model Geyser
Mortar Geyser

North Geyser
North Goggles Geyser
North Triplet Geyser
Oblong Geyser
Old Faithful
Old Tardy Geyser
Park Place Geyser
Pear Geyser
Penta Geyser
Percolator Geyser
Plate Geyser
Plume Geyser
Pump Geyser
Radiator Geyser
Restless Geyser
Rift Geyser
Riverside Geyser
Rocket Geyser
Roof Geyser
Rusty Geyser
Sawmill Geyser
Seismic Satellite Geyser
Seismic Geyser
Silver Globe Silt Geyser
Silver Globe Geyser A
Silver Globe Geyser B
Silver Globe Geyser D
Silver Globe Geyser E
Slide Geyser
Slot Geyser
South Anemone Geyser
South Grotto Fountain Geyser
Spa Geyser
Spasmodic Geyser
Spiteful Geyser
Splendid Geyser
Sponge Geyser
Spouter Geyser
Startle Geyser
Surge Geyser
Tardy Geyser
Turban Geyser
Vault Geyser
Vent Geyser
West Geyser
West Triplet Geyser
Whistle Geyser

Yellowstone has approximately 500 geysers—more than are found in all the Earth's other geyser regions combined. The world's greatest concentration of geysers is in the Lower, Midway, and Upper Geyser basins, where there are approximately 150 geysers; the precise number is hard to define because geysers sometimes disappear and reappear over time. The remarkable abundance of geysers and geothermal wonders were a major reason that Yellowstone was set aside as a national park.

Geysers occur where the water and subterranean plumbing system combine with heat and pressure to produce intermittent ejections of steam and water. In Yellowstone, faults and fractures associated with the caldera and ongoing geologic activity provide pathways for water to move through the silica-rich, volcanic rhyolite. The circulating water is heated at depth and dissolves silica from the surrounding rock. As the water rises, it begins to cool and silica precipitates, creating a mineral that coats, smoothes, and seals the conduit walls. The seal isolates the water from the rock so that the water remains free of mud. If a geyser eruption is sufficiently violent, the seal can be broken,

## Upper, Midway, and Lower Geyser Basins

● Geyser listed in eruption interval tables on right

*(Map labels — Lower Geyser Basin)*
LOWER GEYSER BASIN
Rosette Geyser
Porcupine Hill Geyser
Sand Geyser
Kidney Geyser
Mound Geyser
Burple Geyser
Brain Geyser
Drain Geyser
Morning Geyser
Fountain Geyser
Grey Bulger Geyser
Steady Geyser
Narcissus Geyser
Pink Geyser
Artesia Geyser
Young Hopeful Geyser
Great Fountain Geyser
Logbridge Geyser

| Geyser Name | Eruption height range (in feet) |
|---|---|
| Drain Geyser | 5–150 |
| Fountain Geyser | 10–100 |
| Great Fountain Geyser | 10–250 |
| Morning Geyser | 150–200 |

*(Map labels — Midway Geyser Basin)*
MIDWAY GEYSER BASIN
Opal Pool
Excelsior Geyser Crater
Tangin Geyser
Flood Geyser
West Flood Geyser
Rabbit Creek Geyser
River Spouter Geyser
Catfish Geyser
Till Geyser

| Geyser Name | Eruption height range (in feet) |
|---|---|
| Excelsior (historic) | 300 |
| Excelsior (major) | 30–80 |
| Flood Geyser | 10–25 |
| Opal Pool | 5–80 |
| West Flood Geyser | 10–40 |

*(Map labels — Upper Geyser Basin)*
UPPER GEYSER BASIN
Mercury Geyser
Artemisia Geyser
Grotto Geyser
Daisy Geyser
Riverside Geyser
Pyramid Geyser
Giant Geyser
Splendid Geyser
Oblong Geyser
Grand Geyser
Rift Geyser
West Triplet Geyser
Goggles Spring
Spouter Geyser
Castle Geyser
Aurum Geyser
Little Cub Geyser
Dilapidated Geyser
Lion Geyser
Little Squirt Geyser
Depression Geyser
Beehive Geyser
Old Faithful
Plume Geyser

| Geyser Name | Eruption height range (in feet) |
|---|---|
| Beehive Geyser | 150–200 |
| Giant Geyser | 150–275 |
| Goggles Spring | 100–200 |
| Grand Geyser | 150–200 |
| Old Faithful | 106–184 |
| Splendid Geyser | 50–218 |

*(Scale)* 0 — 2 mi. / 0 — 2 km

mud and rock can clog the pathway, and geyser activity may cease.

The eruptive behavior of geysers is due to several factors. The conduit of an active geyser is believed to contain a chamber with a constriction near the top and a narrow vertical pathway to the surface. At Old Faithful, there is a chamber about fifty feet below the surface (which is as far as cameras have penetrated). As heated water rises from below and fills the chamber, the pressure from the overlying water prevents boiling. The water can actually get hotter than boiling due to this pressure, but eventually steam bubbles form and push upward through the constricted chamber roof into the narrow vertical conduit. As the superheated, steam-charged water emerges from the geyser throat, the pressure from the overlying water drops and steam bubbles form rapidly. The steam-water mixture begins to shoot up into the air, the hot water in the chamber is evacuated, the eruption subsides, and the process begins again. Thus, the factors needed to form geysers in Yellowstone are heat, water, rhyolite, vertical fractures, and one or more silica-coated chambers with small openings near the top. Geysers are rare because the conditions needed to form them are rare.

Yellowstone National Park has nine geyser basins as well as several isolated geysers. Six of the basins are readily accessible by boardwalks near roads (West Thumb, Norris, Gibbon, Lower, Midway, and Upper geyser basins); the other three (Shoshone, Lone Star, and Heart Lake) require backcountry travel. Norris Geyser Basin has the earliest dated geyser activity, with mineral deposits dating to 12,000 years ago and possibly as far back as 150,000 years. Norris is also home to the tallest active geyser, Steamboat

Geyser, although it can go decades between eruptions.

The protection of Yellowstone National Park does not necessarily guarantee the preservation of geysers within its confines. Geothermal exploration outside the park can damage geysers (see sidebar). Prior to the modern understanding of how fragile geysers are, roads and walkways built on or near thermal features altered surface runoff, reduced groundwater percolation, and loosened sediment, all of which may have altered geyser behavior. In the past, tourists have damaged geysers by throwing items into vents and clogging the conduits, which reduced the size and frequency of eruptions.

## Steamboat Geyser Eruptions, 1878–2005

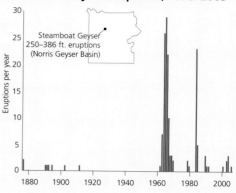

## Geyser Eruption Intervals (hours)

*Data on geyser eruption intervals collected in 2010*

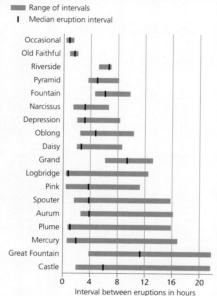

## Geyser Eruption Intervals (days)

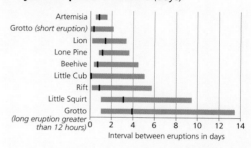

Geyser eruption intervals range from minutes to decades. Steamboat Geyser may go more than fifty years between eruptions. Old Faithful ranges from 30 to 127 minutes between eruptions. A variety of factors affect eruptions. Precipitation of silica can reduce conduit size and stop eruptions, as can trash thrown in geyser throats. The frequency of geysers' eruptions increases in Norris Geyser Basin in early autumn as ground water levels drop. In Upper Geyser Basin, eruption intervals are related to long-term trends in snow accumulation and melt. Large earthquakes in and near Yellowstone cause eruption patterns to change—even an Alaskan earthquake has been associated with changes in eruptive behaviors. Earthquake waves break silica seals and change the number, length, and depth of fractures.

## Geothermal Energy Development

Geothermal development has destroyed many geysers: half of Iceland's geysers, three quarters of New Zealand's geysers, and all geysers outside Yellowstone in the lower forty-eight states. Yellowstone's national park status has helped protect its geothermal features. However, in 1970, Congress identified the Island Park Known Geothermal Resource Area (KGRA) and the Corwin Springs KGRA as areas for geothermal energy development. Research has shown heat flow declines, land subsides, and pressure drops at distances up to twenty-two miles from geothermal power plants and resorts; Yellowstone geyser basins are within eleven miles of Island Park KGRA and twenty-one miles of the Corwin Springs KGRA. Concerns increased in 1986 when flow at La Duke Hot Springs six miles outside the park declined 92 percent after testing at an exploratory geothermal well across the Yellowstone River. In 1994 a compact between Montana and the federal government established the Yellowstone Controlled Groundwater Area to monitor and control ground water development outside the park; a proposed seventy-five mile geyser protection area centered at Old Faithful would include Idaho and Wyoming as well.

### Geothermal Resources Areas near Yellowstone

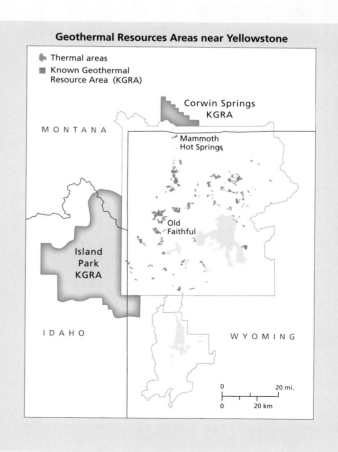

### Geyser Basins and Geothermal Energy Development

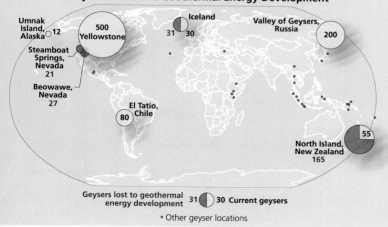

# Earthquakes

**Snake River Parabola, 1973–2010**

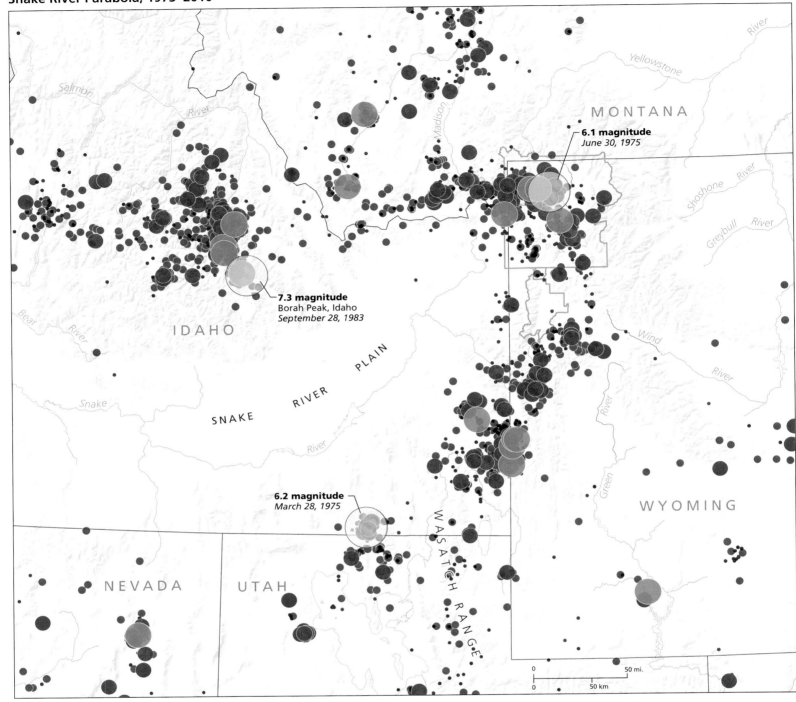

**6.1 magnitude**
*June 30, 1975*

**7.3 magnitude**
Borah Peak, Idaho
*September 28, 1983*

**6.2 magnitude**
*March 28, 1975*

MONTANA

IDAHO

SNAKE RIVER PLAIN

NEVADA

UTAH

WYOMING

WASATCH RANGE

Yellowstone is earthquake country. Earthquakes with magnitudes less than 3.0 occur frequently. More rarely, earthquakes with magnitudes greater than 7 occur, as with the 7.3 Borah Peak, Idaho, event (1983) and the 7.5 Hebgen Lake, Montana, temblor (1959).

Three interrelated factors drive modern-day earthquakes in the region. First, Yellowstone National Park lies at the apex of an arc of seismicity that extends from central Idaho to Yellowstone before curving southward through Grand Teton National Park and into Utah. This seismic arc wraps around the track of the ancestral Yellowstone hotspot along the eastern Snake River Plain and matches a region of uplift and rugged mountains adjacent to both sides of this plain. Geologists interpret this arc as a thermally uplifted "crustal shoulder" adjacent to the path of the hotspot. In contrast, the plain itself has cooled and subsided in the wake of its passage over the hotspot. The stresses within the crust and upper mantle resulting from the Yellowstone hotspot produce this arc of seismicity.

Second, Yellowstone National Park lies along the Intermountain Seismic Belt, a zone of active seismicity extending from eastern California to the Wasatch Front in central Utah, then northward into southeast Idaho, Yellowstone, and western Montana to the Flathead Valley. This belt marks the eastern limit of active crustal extension

(pulling apart) associated with the Basin and Range Province. The Basin and Range spans the west from the eastern slopes of the Sierra Nevada to the Wasatch Mountains and has been extending for approximately twenty million years, dramatically thinning the crust. The Basin and Range Province has been described as a "soft plate boundary," accommodating some of the oblique motion between the North American and Pacific plates through a complex array of normal faults, strike-slip faults, and magmatic lineaments, including the Yellowstone hotspot.

Third, earthquakes in Yellowstone National Park occur due to the injection of gas and magma into fracture systems of the caldera, particularly along its northwest margin near West Yellowstone. Although present-day Yellowstone is between volcanic eruptions, it nevertheless burps, shifts, and adjusts to small changes taking place in the equilibrium of rock, gas, magma, and fluids in the subsurface. This "living caldera" is the reason why Yellowstone has so many small- to medium-sized earthquakes on a regular basis.

Many of the earthquakes within Yellowstone park are restricted to the perimeter of the 640,000-year-old Yellowstone caldera and a zone of faults and small igneous features trending northward from the north flank of the caldera to Mammoth Hot Springs, the so-called

## Earthquake Magnitude, 1973–2010

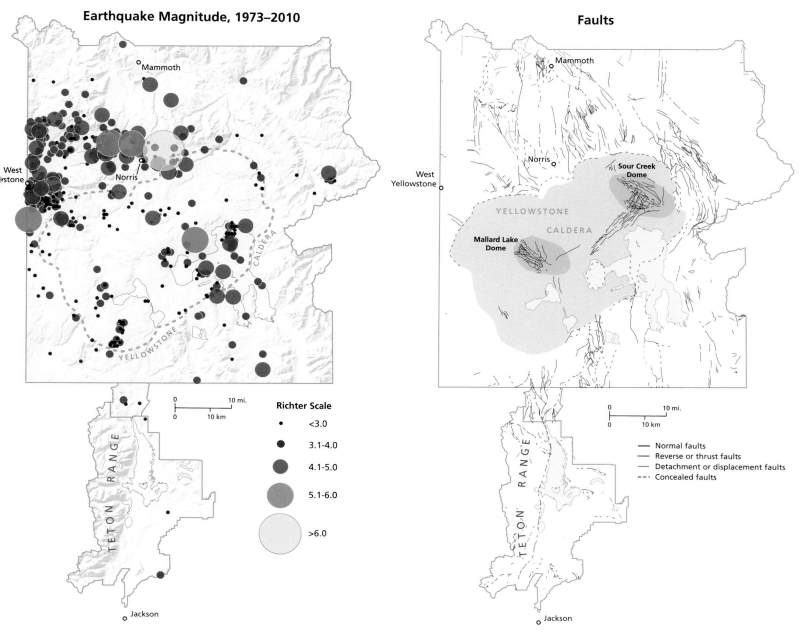

**Richter Scale**

- •    <3.0
- ●    3.1–4.0
- ●    4.1–5.0
- ●    5.1–6.0
- ○    >6.0

## Faults

- —— Normal faults
- —— Reverse or thrust faults
- —— Detachment or displacement faults
- - - - Concealed faults

## Yellowstone Earthquakes

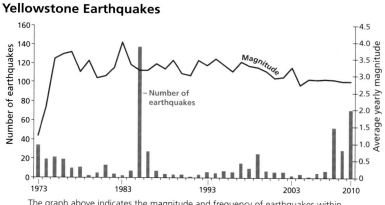

The graph above indicates the magnitude and frequency of earthquakes within the Yellowstone National Park boundary. During the same time period, Grand Teton National Park only experienced six earthquakes ranging from 2.5 to 3.1 on the Richter scale.

"Norris-Mammoth corridor." The spatial distribution of active earthquakes and thermal features are similar, aligning with concentric and radial surface faults that surround the caldera. In comparison, seismicity in Grand Teton National Park is notably lacking, despite the enormous normal fault bounding the eastern front of the Teton Range. This fault has been active in the past 10,000 years. At present, the Teton fault may be locked and building up elastic strain energy, or there may be other processes accommodating regional extension in western Wyoming.

## Seismic Monitoring Stations

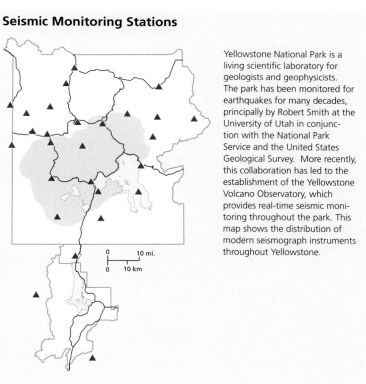

Yellowstone National Park is a living scientific laboratory for geologists and geophysicists. The park has been monitored for earthquakes for many decades, principally by Robert Smith at the University of Utah in conjunction with the National Park Service and the United States Geological Survey. More recently, this collaboration has led to the establishment of the Yellowstone Volcano Observatory, which provides real-time seismic monitoring throughout the park. This map shows the distribution of modern seismograph instruments throughout Yellowstone.

# Glaciers

## Glacial Landscapes of the Teton Range

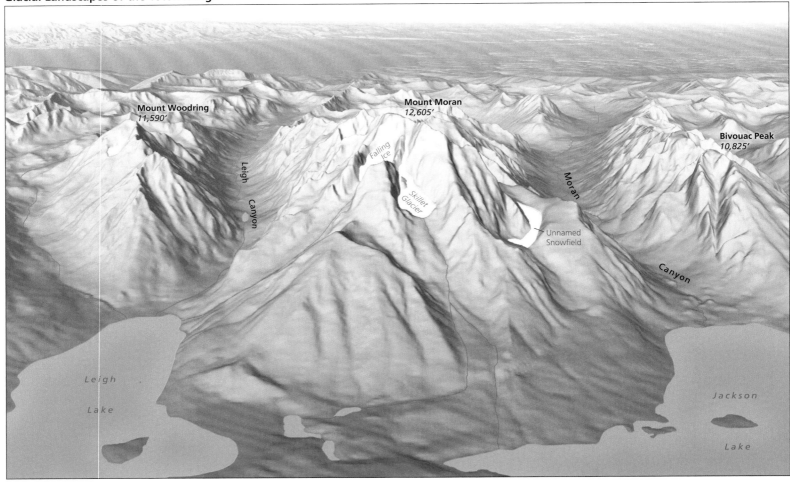

Ice and glaciers have sculpted much of the terrain in Yellowstone and Grand Teton national parks—shaping mountains, carving valley walls, altering valley floors. Glaciers form where more snow falls in the winter than melts in the summer. Driven by gravity, a glacier in motion is basically a slow-moving river of ice, its vast energy dynamically altering the landscape through which it passes. Glacial moraines are deposits of debris incorporated in a flowing glacier and deposited at the down-valley glacier terminus. A glacial moraine consists of ridges of generally unsorted clay, silt, sand, and gravel that are commonly studded with large boulders. Glacial meltwater streams carry and deposit outwash gravel. Sliding at the base of a glacier molds unconsolidated deposits into streamlined forms such as those found in the Gardiners Hole and Coulter Bay areas. This force also erodes resistant bedrock into rounded and polished knobs.

Deposits from the two most recent Pleistocene glaciations are well represented in the parks. The Bull Lake glaciation (which filled all of Jackson Hole and much of the West Yellowstone basin with ice) culminated 150,000 years ago; the Pinedale glaciation peaked 17,000 to 20,000 years ago.

Pinedale ice sheets first formed in the high mountains to the southeast (Absaroka Range), northeast (Beartooth Uplift), and west (Gallatin Range) of the Yellowstone Plateau. Over time, a glacial system built

upward and outward from these mountains and onto the central Yellowstone Plateau, where deep snows raised the glacial surface to elevations of 11,000 feet. More than 95 percent of Yellowstone National Park was covered by glacial ice. Glaciers extended well beyond park boundaries; for example, the glacier flowing north down the Yellowstone River valley extended forty miles outside the present-day park, terminating in Paradise Valley. Oddly, on the Yellowstone Plateau, which was the center of glacial buildup and outflow, only subtle evidence of glacial erosion and deposition can be recognized today, apparently because the rhyolite bedrock crumbles so readily to sand and silt.

In the Grand Teton National Park area, Pinedale glaciers along the southern margin of the greater Yellowstone glacial system advanced into northern Jackson Hole and terminated about five miles south of Jackson Lake. Glacial outwash streams deposited thick gravel terraces, eventually supporting sagebrush flats. Meanwhile, valley glaciers formed in cirques in the Teton Range and flowed to the floor of Jackson Hole, depositing the moraines that now encompass Jenny, Bradley, Taggart, and Phelps lakes. These Pleistocene Teton glaciers also helped shape the Tetons by sharpening peaks, carving sharp divides, and sculpting U-shaped valleys.

Sagebrush grasslands that form Yellowstone's Northern Range are founded on Pine-

dale glacial deposits and demonstrate the profound influence of geology upon ecology. Most of the glacial deposits in the Northern Range were laid down during Pinedale glacial recession about 15,000 years ago. These deposits retain available soil moisture at shallow depths and thus favor grassland vegetation rather than forests. Other prominent glacier-formed landscape features of the Northern Range are deep channels eroded into bedrock. Phantom Lake and the Mammoth-Tower highway occupy one such low-gradient river valley.

## Global Ice Volume

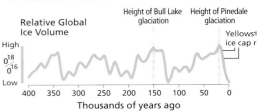

$\frac{O^{18}}{O^{16}}$ Higher oxygen isotope ratios of $O^{18}$ to $O^{16}$ equate to greater ice volumes.

## Greater Yellowstone 17,000 Years Ago

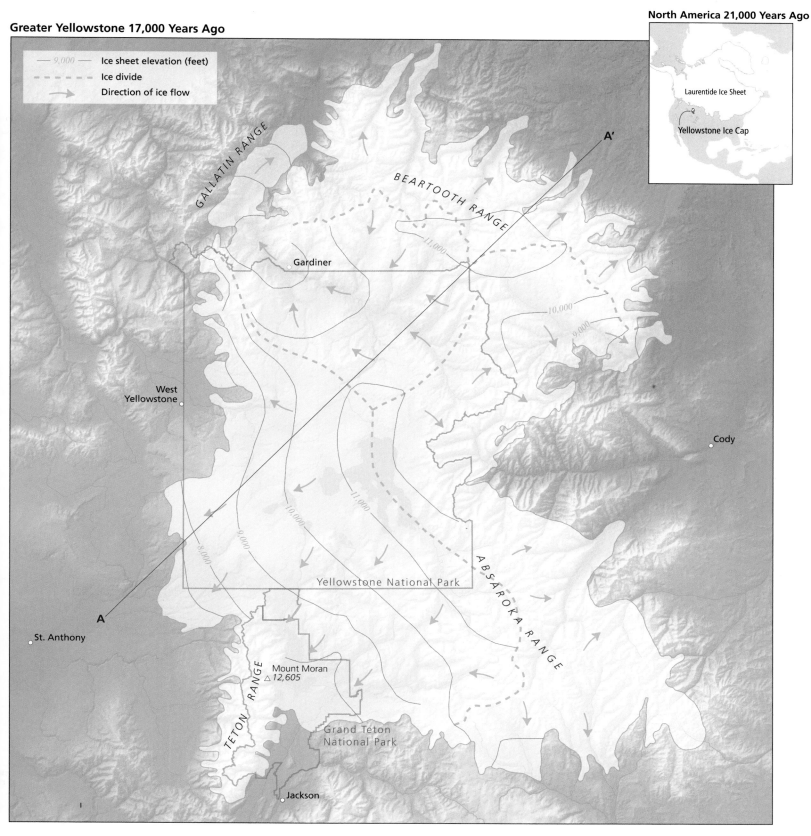

### North America 21,000 Years Ago

Laurentide Ice Sheet

Yellowstone Ice Cap

**Legend:**
- — 9,000 — Ice sheet elevation (feet)
- --- Ice divide
- → Direction of ice flow

GALLATIN RANGE

BEARTOOTH RANGE

11,000

10,000

9,000

Gardiner

West Yellowstone

Cody

11,000

10,000

9,000

8,000

Yellowstone National Park

ABSAROKA RANGE

St. Anthony

TETON RANGE

Mount Moran
△ 12,605

Grand Teton National Park

Jackson

A

A'

## Winter Weather Systems over Yellowstone 17,000 Years Ago

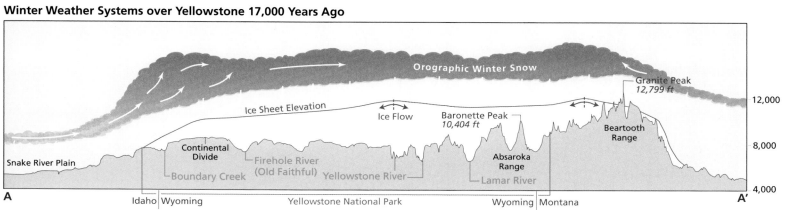

Orographic Winter Snow

Granite Peak
12,799 ft

Ice Sheet Elevation

Ice Flow

Baronette Peak
10,404 ft

Beartooth Range

Snake River Plain

Continental Divide

Firehole River
(Old Faithful)

Yellowstone River

Absaroka Range

Lamar River

Boundary Creek

A

Idaho | Wyoming

Yellowstone National Park

Wyoming | Montana

A'

12,000

8,000

4,000

# Yellowstone Lake

Covering 286 square miles, Yellowstone Lake is the largest high-altitude lake in North America. With an average depth of 138 feet, it contains just over twelve million acre-feet of water and is covered by ice from mid-December through May. Entering the lake are 141 tributaries, while only one river, the Yellowstone, flows both into and out of this body.

Yellowstone Lake straddles the southeast edge of the 640,000-year-old Yellowstone Caldera, one of the most geologically dynamic places on Earth. Powerful geologic forces—including volcanism, faulting, glaciation, and hydrothermal activity—have shaped Yellowstone Lake, which sits directly above the Yellowstone magma chamber. In a typical year 1,000 to 3,000 generally small earthquakes are recorded

## Elevation Profile through Yellowstone Lake

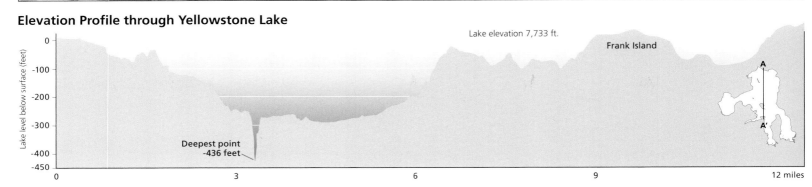

on the Yellowstone Plateau, many occurring beneath Yellowstone Lake. Uplift and subsidence events regularly take place in the Yellowstone Caldera. Such active deformation affects the lake's ever-changing shoreline and is believed related to geologic activity such as magma recharge and crystallization and the development and migration of hydrothermal fluids.

Yellowstone Lake has been scientifically studied since the 1871 Hayden expedition. Henry Elliott, a member of the survey party, produced the first bathymetric map of Yellowstone Lake. Successive bathymetric maps have been produced with ever-improving accuracy. The most recent mapping effort employed differential GPS and high-resolution multibeam swath sonar coupled with sub-bottom seismic reflection. Hydrothermal vent fluids and mineral deposits on the lake bottom were sampled with a remotely operated submersible vehicle (ROV).

Yellowstone Lake can be divided into two different geologic domains. The northern two-thirds of the lake were shaped primarily by volcanic and hydrothermal processes, while glacial and alluvial processes formed the southern third. Faulting, uplift, and subsidence of the Yellowstone Caldera have affected the entire lake area. A number of large-scale advances and retreats of glacial ice over the lake occurred in the past 200,000 years. About 3,300 feet of ice accumulated over the central basin of Yellowstone Lake during the most recent glacial advance beginning approximately 25,000 years ago.

Yellowstone Lake emerged from under its icecap about 15,000 years ago and is young with respect to its flora and fauna. The lake is home to the largest population of Yellowstone cutthroat trout in North America. These fish are a critical food source for grizzly and black bears, bald eagles, white pelicans, osprey, otters, and possibly thirty-five other wildlife species. The Molly Islands in southern Yellowstone Lake contain one of the only continuously surviving white pelican breeding colonies in North America. They are joined there by a noisy collection of nesting cormorants and gulls. All are sustained by a diet rich in fish, which thrive in invertebrate-laden waters. Dozens of actively spewing hydrothermal vents influence the lake's water chemistry; the degree to which these hydrothermal fluids affect the lake biota is under investigation.

## Northern Basin

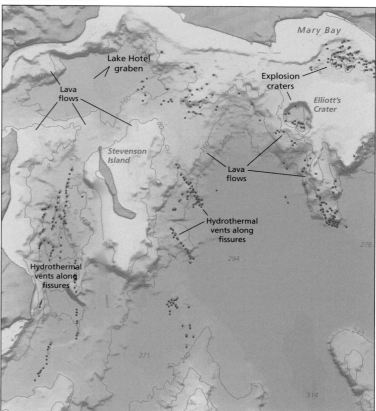

Recent high-resolution maps of the lake reveal many previously unknown or partially known features. The lake's floor is marked by large post-caldera rhyolitic lava flows, at least 650 hydrothermal vents, more than five large hydrothermal explosion craters (greater than 1,650 feet in diameter), spires and domed sediments related to hydrothermal activity, fissures and faults, landslide deposits, former shorelines, glacial meltwater deposits, and a recently active graben. Modern understanding of the lake suggests a number of potential hazards exist: hydrothermal explosions, earthquakes, earthquake-generated waves, and the collapse of caprock structures over active hydrothermal systems.

## Geology

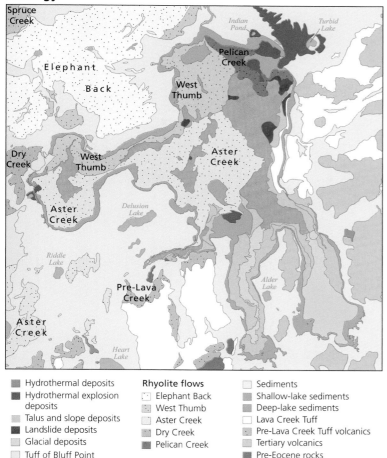

- ■ Hydrothermal deposits
- ■ Hydrothermal explosion deposits
- ■ Talus and slope deposits
- ■ Landslide deposits
- □ Glacial deposits
- □ Tuff of Bluff Point

**Rhyolite flows**
- ▩ Elephant Back
- ▨ West Thumb
- ▨ Aster Creek
- ▨ Dry Creek
- ▩ Pelican Creek

- □ Sediments
- ■ Shallow-lake sediments
- ■ Deep-lake sediments
- □ Lava Creek Tuff
- ▨ Pre-Lava Creek Tuff volcanics
- □ Tertiary volcanics
- ■ Pre-Eocene rocks

## Structural Geology

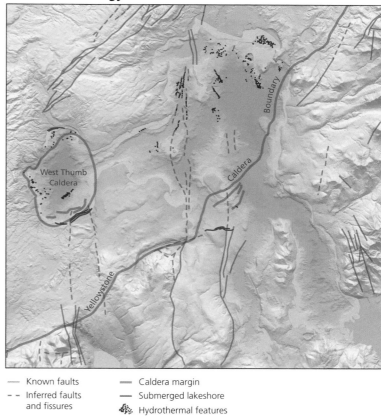

- — Known faults
- - - Inferred faults and fissures
- — Caldera margin
- — Submerged lakeshore
- ⚶ Hydrothermal features

# Drainage Basins

Drainage basins represent the land surface area that contributes runoff to a river. Basins (or watersheds) are defined by drainage divides, which are ridges or elevated areas that determine the direction water will flow. Basins of the Greater Yellowstone Area drain to the Pacific Ocean, the Gulf of Mexico, and the Gulf of California (the Great Divide basin to the southeast is a closed catchment with no outlet).

Any precipitation falling within a watershed's boundaries will ultimately flow downstream to the basin mouth—assuming the water is not lost to evaporation, long-term storage (groundwater or ice), or diversion out of the basin by humans. Disturbances in an upstream basin therefore affect water quality and supply in all the downstream basins, making watersheds especially important in planning and management efforts. Most of the basins in Yellowstone and Grand Teton national parks are headwater basins or drain pristine areas; this helps to preserve the aquatic ecosystems of these parks.

## Headwaters of the Nation

**•·•˙• Continental Divide**

**ᴧ∿⌒ Basin boundary**

## Greater Yellowstone Area Drainage Basins

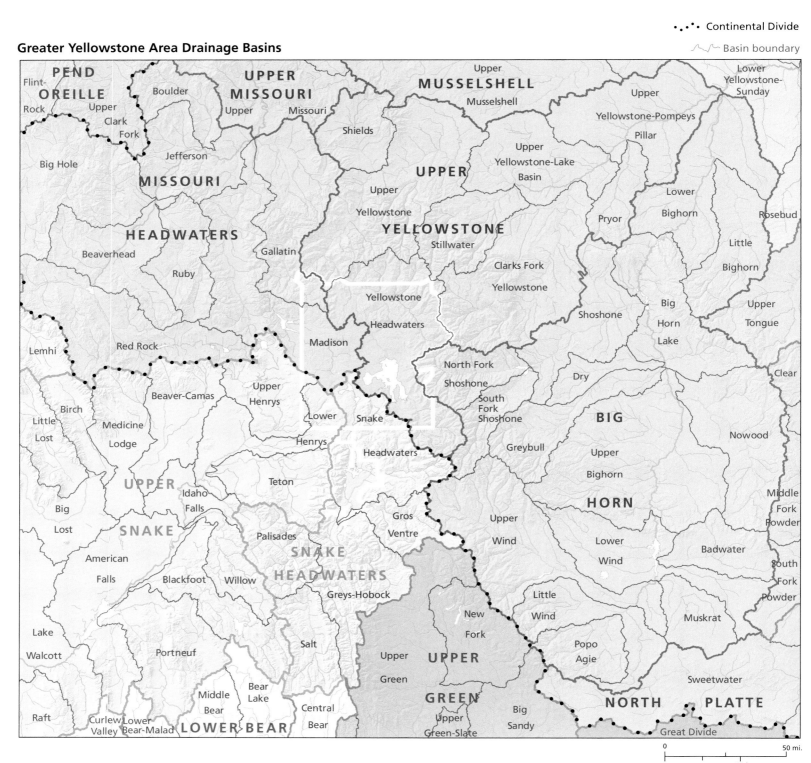

# Yellowstone and Grand Teton Drainage Basins

*Only subbasins intersecting Grand Teton and Yellowstone national parks are shown*

〰️ Basin boundary

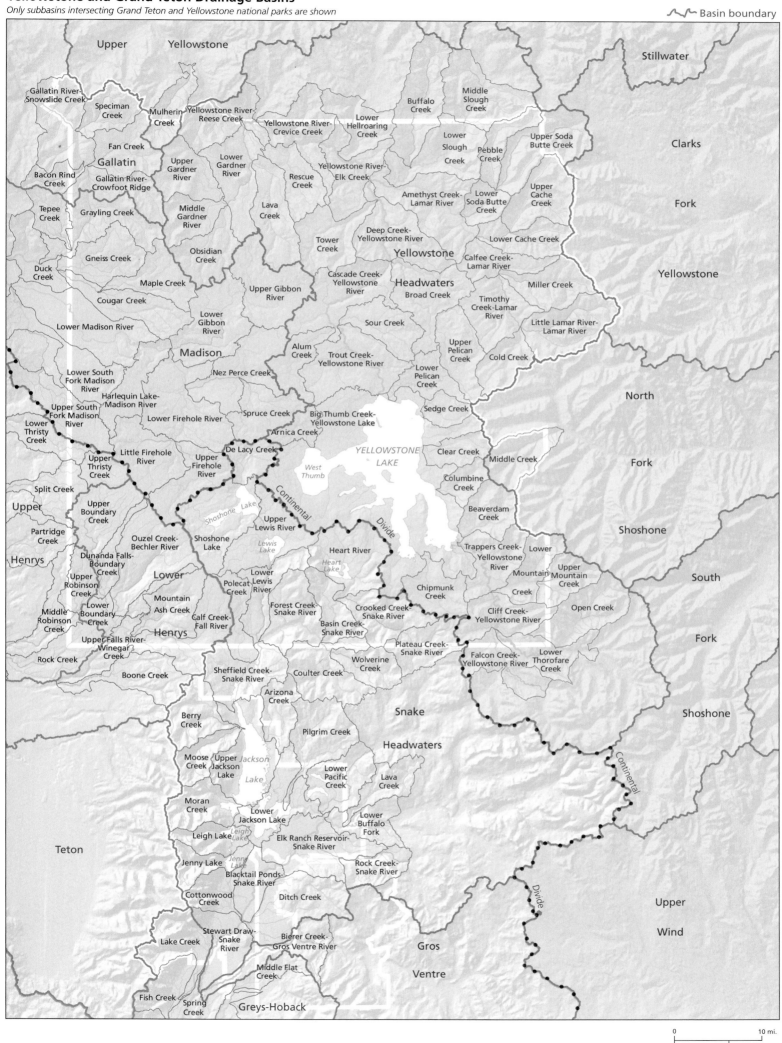

Upper Yellowstone

Stillwater

Gallatin River-Snowslide Creek

Speciman Creek

Mulherin Creek

Yellowstone River-Reese Creek

Yellowstone River-Crevice Creek

Lower Hellroaring Creek

Buffalo Creek

Middle Slough Creek

Clarks

Fan Creek

Gallatin

Upper Gardner River

Lower Gardner River

Rescue Creek

Yellowstone River-Elk Creek

Lower Slough Creek

Pebble Creek

Upper Soda Butte Creek

Fork

Bacon Rind Creek

Gallatin River-Crowfoot Ridge

Amethyst Creek-Lamar River

Lower Soda Butte Creek

Upper Cache Creek

Tepee Creek

Grayling Creek

Middle Gardner River

Lava Creek

Yellowstone

Gneiss Creek

Obsidian Creek

Tower Creek

Deep Creek-Yellowstone River

Headwaters

Calfee Creek-Lamar River

Lower Cache Creek

Miller Creek

Yellowstone

Duck Creek

Maple Creek

Upper Gibbon River

Broad Creek

Timothy Creek-Lamar River

Cougar Creek

Cascade Creek-Yellowstone River

Sour Creek

Little Lamar River-Lamar River

Lower Madison River

Lower Gibbon River

Madison

Alum Creek

Trout Creek-Yellowstone River

Upper Pelican Creek

Cold Creek

North

Lower South Fork Madison River

Nez Perce Creek

Lower Pelican Creek

Harlequin Lake-Madison River

Lower Firehole River

Spruce Creek

Big Thumb Creek-Yellowstone Lake

Sedge Creek

Fork

Upper South Fork Madison River

Arnica Creek

De Lacy Creek

YELLOWSTONE LAKE

Clear Creek

Lower Thristy Creek

Upper Thristy Creek

Little Firehole River

Upper Firehole River

West Thumb

Shoshone Lake

Middle Creek

Split Creek

Columbine Creek

Upper

Upper Boundary Creek

Ouzel Creek-Bechler River

Shoshone River

Shoshone Lake

Lewis Lake

Upper Lewis River

Continental Divide

Heart River

Beaverdam Creek

Shoshone

Partridge Creek

Dunanda Falls-Boundary Creek

Lower

Heart Lake

Trappers Creek-Yellowstone River

Lower Mountain Creek

Upper Mountain Creek

South

Henrys

Upper Robinson Creek

Mountain Ash Creek

Polecat Creek

Lower Lewis River

Forest Creek-Snake River

Chipmunk Creek

Fork

Middle Robinson Creek

Lower Boundary Creek

Calf Creek-Fall River

Basin Creek-Snake River

Crooked Creek-Snake River

Cliff Creek-Yellowstone River

Open Creek

Upper Falls River-Winegar Creek

Henrys

Plateau Creek-Snake River

Falcon Creek-Yellowstone River

Lower Thorofare Creek

Shoshone

Rock Creek

Boone Creek

Sheffield Creek-Snake River

Coulter Creek

Wolverine Creek

Fork

Arizona Creek

Berry Creek

Snake

Pilgrim Creek

Headwaters

Continental Divide

Moose Creek

Upper Jackson Lake

Jackson Lake

Lower Pacific Creek

Lava Creek

Moran Creek

Lower Jackson Lake

Leigh Lake

Leigh Lake

Elk Ranch Reservoir-Snake River

Lower Buffalo Fork

Teton

Jenny Lake

Jenny Lake

Blacktail Ponds-Snake River

Rock Creek-Snake River

Cottonwood Creek

Ditch Creek

Upper

Stewart Draw-Snake River

Bierer Creek-Gros Ventre River

Gros

Wind

Lake Creek

Middle Flat Creek

Ventre

Fish Creek

Spring Creek

Greys-Hoback

Continental Divide

0 ——— 10 mi.

0 ——— 10 km

# Rivers

In the 1760s mapmaker Jonathan Carver, aiding Robert Rogers in his search for the Northwest Passage, speculated that a height of land spawned all the great rivers of the continent. Carver's musings were ultimately shown to be erroneous in many details, but accurate in concept. Greater Yellowstone is such a place. At Three Waters Mountain, south of Yellowstone park, a single gust of wind can deposit snow that will eventually end up in the Gulf of Mexico, the Gulf of California and the Pacific Ocean. The waters that flow from the remote mountains and high plateaus of the Greater Yellowstone Area are the lifeblood of much of the arid west.

Fully two-thirds of the region's water reaches the Missouri River via two routes. The Madison and Gallatin rivers build along the west side of Yellowstone park, combine with the Jefferson, and form the Three Forks of the Missouri. The Madison is unique in that it drains Yellowstone's major geyser basins, making its waters warmer and chemically richer than other rivers. Fabled as sport fisheries, the Madison and Gallatin are among the most popular rivers in the ecosystem.

The Yellowstone River supplies a large fraction of the remaining water flowing to the Missouri. Draining the central portion of the park, the river and its tributaries are integral with many iconic images of Yellowstone: the wildlife vistas of the Hayden and Lamar valleys, the spectacular Lower Falls, and the jewel of Yellowstone Lake. After the confluence with the Bighorn River, the Yellowstone enters the Missouri near Montana's border with North Dakota. It remains the longest undammed river in the lower forty-eight states, although some of its tributaries well downstream of Yellowstone National Park are extensively dammed.

The geologic history of the region is on display as the Snake River passes beneath the spires of the Tetons in Jackson Hole. The distinctive river terraces on the valley floor formed as the river cut into glacial deposits left from Pleistocene times.

Meandering through a wide sagebrush-covered valley between the Wind River and Wyoming mountain ranges, the Green is the third of greater Yellowstone's major rivers. The main tributary of the Colorado, it supplies much of the dry southwest with water.

Because approximately 75 percent of greater Yellowstone is in public ownership—much held as national park and wilderness areas—the region's rivers are generally healthy, though problems exist. Oil and gas development in the Green River Basin threatens many of the Green's tributaries, while dams on the Madison, Shoshone, Henrys Fork and the Snake have already altered downstream riparian ecosystems. Soda Butte Creek near Yellowstone park's northeast entrance once received waste from gold mining operations near Cooke City, Montana. Recent efforts have stopped the mine waste from entering the park, but to this day the floodplains of Soda Butte are a repository for contaminated mine sediments.

**Headwaters of the Nation**

**Greater Yellowstone Area Rivers**

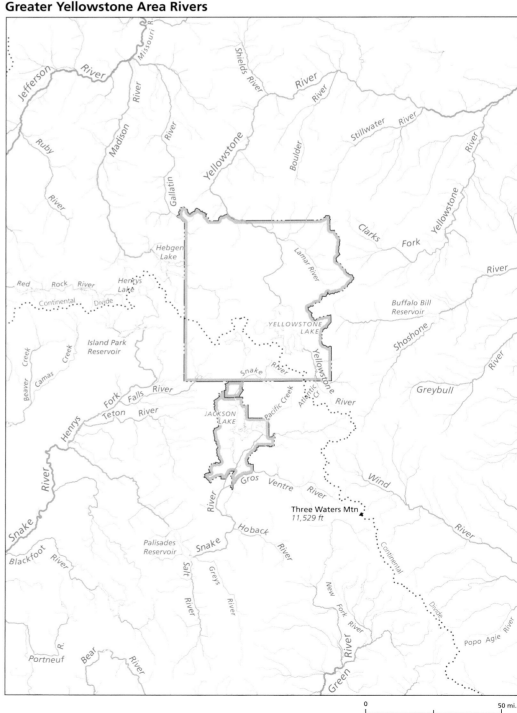

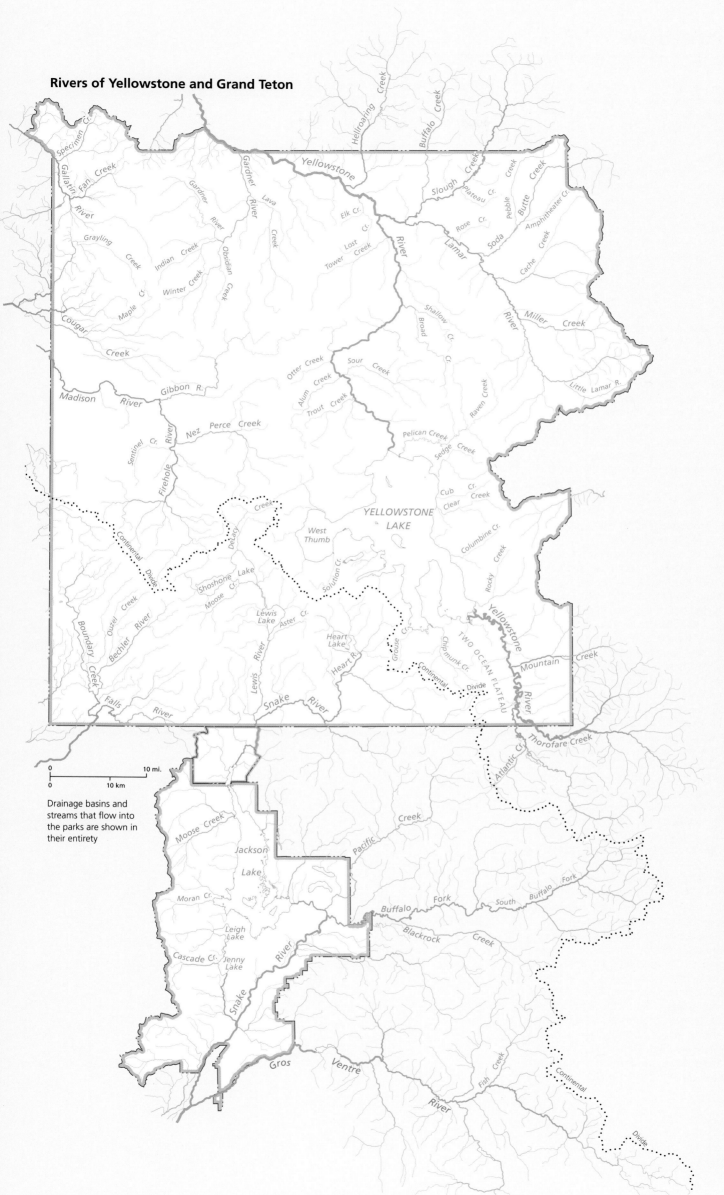

**Rivers of Yellowstone and Grand Teton**

Specimen Cr.
Gallatin
Fan Creek
River
Grayling
Creek
Indian Creek
Winter Creek
Maple Cr.
Gardner
River
Obsidian Creek
Gardner River
Lava
Creek
Cougar
Creek
Madison
River
Gibbon R.
Nez Perce Creek
Sentinel Cr.
Firehole
River
Continental
Divide
Creek
DeLacy
Shoshone Lake
Moose Cr.
Lewis
Lake
Aster Cr.
Ouzel Creek
Bechler River
Boundary
Creek
Lewis
River
Falls
River
Snake
River

Yellowstone
Helloaring Creek
Buffalo Creek
River
Slough Creek
Plateau Cr.
Pebble Creek
Butte Creek
Amphitheater Cr.
Rose Cr.
Soda
Cache Creek
Lamar
River
Miller
Creek
Little Lamar R.
Elk Cr.
Lost Creek
Tower Creek
Broad Cr.
Shallow Cr.
Raven Creek
Otter Creek
Alum Creek
Trout Creek
Sour Creek
Pelican Creek
Sedge Creek
Columbine Cr.
Cub Cr.
Clear Creek
Rocky Creek

YELLOWSTONE
LAKE
West
Thumb
Solution Cr.
Heart
Lake
Heart R.
Grouse Cr.
Chipmunk Cr.
Continental
Divide
TWO OCEAN PLATEAU
Yellowstone
River
Mountain Creek
Thorofare Creek
Atlantic Cr.

0        10 mi.
0        10 km

Drainage basins and
streams that flow into
the parks are shown in
their entirety

Moose Creek
Jackson
Lake
Moran Cr.
Leigh
Lake
Cascade Cr.
Jenny
Lake
Snake
River
Gros
Ventre
River

Pacific
Creek
Buffalo
Fork
South Buffalo Fork
Blackrock Creek
Fish Creek
Continental
Divide

# Streamflow

Water is life in the arid American West. Riparian vegetation, fish, and a host of other organisms evolved to take advantage of water availability or to survive water scarcity at specific times of year. Farmers, anglers, municipalities, and rafters all have reasons for closely observing fluctuations in their local rivers. In the Greater Yellowstone Area, tracking and assessing variations in rivers is essential to understanding ecological processes and managing water resources. Flow gauges help in this effort, gathering critical information for downstream water allocation and flood warning and also providing data for resource managers monitoring aquatic species and other aspects of biological health in specific watersheds.

The seasonal cycle of stream flow in Yellowstone rivers is largely controlled by snowpack. A single peak, usually sometime in June, stands in contrast to the lower flows during the remainder of the year. Flows are lowest in January or February, when most surface water is locked up in snow and ice; as a result, near-surface groundwater that can contribute to stream flow is not being replenished. Some rivers, such as the Lamar, derive the majority of their flow from snowmelt and have a dramatic spike in spring runoff. Other rivers, such as the Madison, receive a larger proportion of runoff from groundwater and have less variation between peak and low flows.

Reservoirs impounded behind dams and natural lakes also modulate river levels by disrupting unchecked runoff flow. The Yellowstone River (naturally controlled at Lake Outlet) and the Snake River (dam-controlled near Moran) both have a lower proportion of their flows during the peak runoff period compared to less-regulated streams. All the rivers within Yellowstone National Park retain their natural flow regimes without modification by dams. In the United States, it is rare to find such a large area where this is the case.

As with seasonal variability, year-to-year changes in total annual flow are also affected by the proportion of runoff from groundwater relative to snowmelt and rainfall. Rivers with major groundwater inputs such as the Madison have smaller annual variations in discharge compared to snowmelt-driven rivers like the Lamar, where total flows can vary by a factor of two from one year to the next.

Floods (peak annual flows) in the region's major rivers are also controlled by both snowfall and snowmelt. Greatly varying amounts of snow—between 100 and 350 inches—fall each winter in the high country. If warm temperatures begin in April and continue through June, the pace at which the snow melts is moderated and peak flows are constrained. But if warm temperatures are delayed until June, much of the melt is compressed into a few short weeks. When such delayed warming is coupled with high snowpack, rivers flood, as occurred throughout much of the region with the record flows of 1996.

## Mean Daily Streamflow

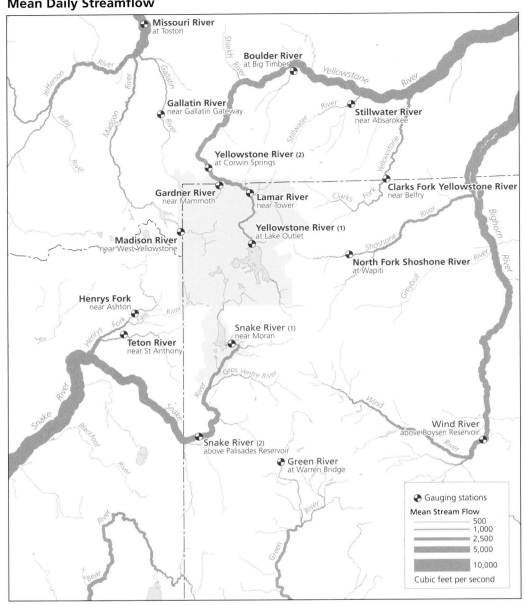

## Annual Mean Discharge

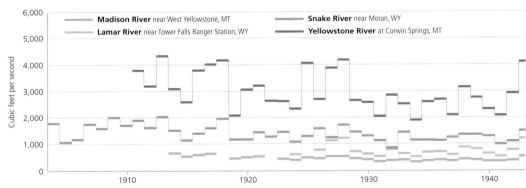

## Annual Peak Streamflow

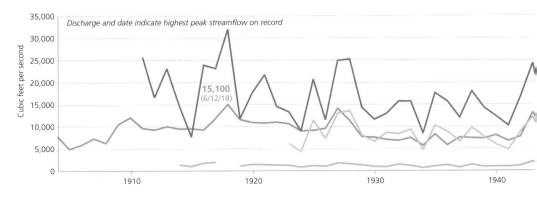

116

## Average Flow Seasonality

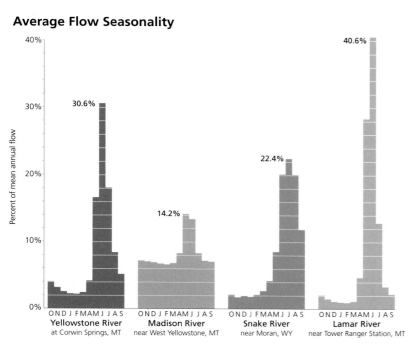

40%

30%

20%

10%

0%

Percent of mean annual flow

30.6%

14.2%

22.4%

40.6%

O N D J F M A M J J A S
**Yellowstone River**
at Corwin Springs, MT

O N D J F M A M J J A S
**Madison River**
near West Yellowstone, MT

O N D J F M A M J J A S
**Snake River**
near Moran, WY

O N D J F M A M J J A S
**Lamar River**
near Tower Ranger Station, MT

## Average Annual Total Runoff

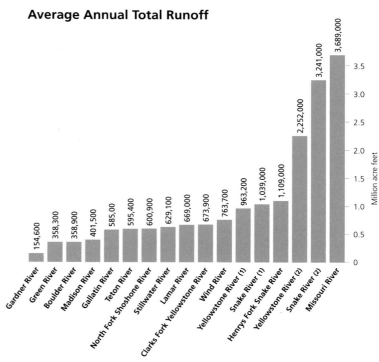

Million acre feet

3.5

3.0

2.5

2.0

1.5

1.0

0.5

0

154,600 — Gardner River
358,300 — Green River
358,900 — Boulder River
401,500 — Madison River
585,00 — Gallatin River
595,400 — Teton River
600,900 — North Fork Shoshone River
629,100 — Stillwater River
669,000 — Lamar River
673,900 — Clarks Fork Yellowstone River
763,700 — Wind River
963,200 — Yellowstone River (1)
1,039,000 — Snake River (1)
1,109,000 — Henrys Fork Snake River
2,252,000 — Yellowstone River (2)
3,241,000 — Snake River (2)
3,689,000 — Missouri River

## Historical Flows

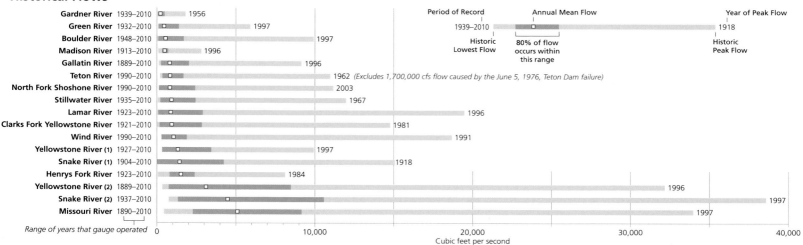

| River | Years | Low Flow Year | Peak Flow Year |
|---|---|---|---|
| Gardner River | 1939–2010 | 1956 | |
| Green River | 1932–2010 | 1997 | |
| Boulder River | 1948–2010 | 1997 | |
| Madison River | 1913–2010 | 1996 | |
| Gallatin River | 1889–2010 | 1996 | |
| Teton River | 1990–2010 | 1962 *(Excludes 1,700,000 cfs flow caused by the June 5, 1976, Teton Dam failure)* | |
| North Fork Shoshone River | 1990–2010 | 2003 | |
| Stillwater River | 1935–2010 | 1967 | |
| Lamar River | 1923–2010 | 1996 | |
| Clarks Fork Yellowstone River | 1921–2010 | 1981 | |
| Wind River | 1990–2010 | 1991 | |
| Yellowstone River (1) | 1927–2010 | 1997 | |
| Snake River (1) | 1904–2010 | 1918 | |
| Henrys Fork River | 1923–2010 | 1984 | |
| Yellowstone River (2) | 1889–2010 | 1996 | |
| Snake River (2) | 1937–2010 | 1997 | |
| Missouri River | 1890–2010 | 1997 | |

Period of Record
1939–2010

Annual Mean Flow

Year of Peak Flow
1918

Historic Lowest Flow

80% of flow occurs within this range

Historic Peak Flow

*Range of years that gauge operated*

0    10,000    20,000    30,000    40,000
Cubic feet per second

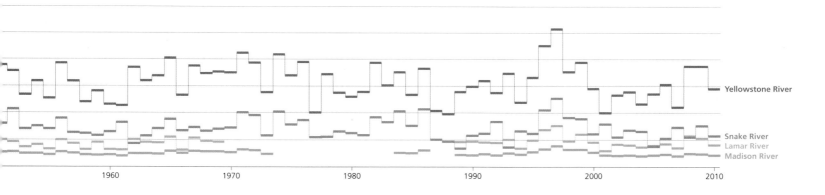

Yellowstone River

Snake River
Lamar River
Madison River

1960    1970    1980    1990    2000    2010

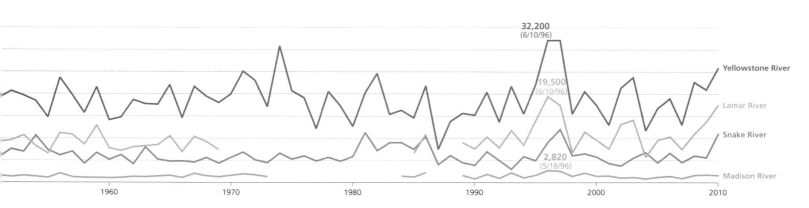

**32,200**
(6/10/96)

19,500
(6/10/96)

2,820
(5/18/96)

Yellowstone River

Lamar River

Snake River

Madison River

1960    1970    1980    1990    2000    2010

**117**

# Flow Regimes

The main stem of the Yellowstone River is the longest undammed river in the lower forty-eight states. Climate is the primary control on the timing and quantity of its streamflow. Each year, peak runoff occurs in late May through June when snow melts in the high country. Minimum annual flows are in midwinter, when surface water is frozen and streamflow is maintained by groundwater or outflow from Yellowstone Lake. Variations between years reflect droughts and wet periods. Maximum and minimum annual flows were notably smaller, for example, during the Dust Bowl years of the mid-1930s, while the high snowpack years of 1995 to 1997 generated spring floods, greater groundwater recharge, and higher sustained flows throughout the year. Although the runoff amount varies between years, the annual timing of peak and low streamflows remains steady. Native species that evolved in response to this hydrologic regime thus can carry on their life cycle behaviors (for example, spawning) even as flow volumes fluctuate.

In contrast, flow in the Snake River near Moran, Wyoming, has been profoundly modified by Jackson Lake Dam. Prior to 1957, Jackson Lake Dam was the primary storage reservoir for farmers' irrigation water in the Snake River Plain. Peak flows were delayed by one to two months to accommodate late summer irrigation needs and flows were released abruptly, rather than climbing gradually as occurs with naturally melting snow. The size of peak flows was generally unchanged by the dam, although winter low flows were significantly lower. In 1957, Palisades Reservoir was completed farther downstream on the Snake River and became the primary irrigation storage facility. This meant that water from Jackson Lake Dam could be released more uniformly throughout the summer and peak flows were reduced.

Before 1957, channel shifting and floodplain disturbance prevented expansion of some vegetation communities, most notably blue spruce forests. Subsequent expansion of blue spruce forests created more habitat for some species, such as bald eagles, that use the forest for nesting. This came at a cost, however, of decreasing willow and alder habitat, important in providing winter food for species such as moose.

The more constant river flow and reduced channel shifting also meant that side channels were no longer inundated every year by floods. Beaver colonized these less flood-prone side channels, built dams, and disconnected the smaller streams from the river. The river therefore increasingly follows a single channel rather than its previous course along interconnected braids. This improves boating safety because canoeists and rafters avoid being stranded in blocked side channels. Snake River cutthroat trout, however, prefer smaller side streams for spawning habitat, so the reduced floods may promote increased reliance on hatchery fish. These tradeoffs highlight the difficult management decisions faced by dam regulators.

## Natural Flow Regime
**Mean Daily Flow**
**Yellowstone River at Corwin Springs, MT (Water Year, 1911–2003)**
Mean daily flow ranges from 380 to 32,000 cubic feet per second.

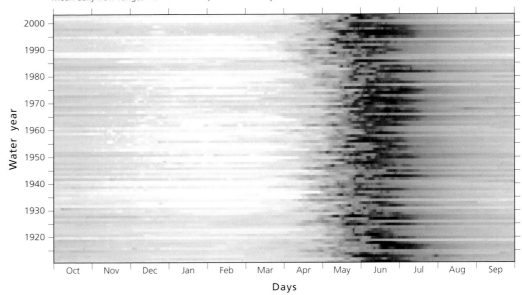

## Regulated Flow Regime
**Mean Daily Flow**
**Snake River near Moran, WY (Water Year, 1904–2003)**
Mean daily flow ranges from 0.3 to 14,700 cubic feet per second.

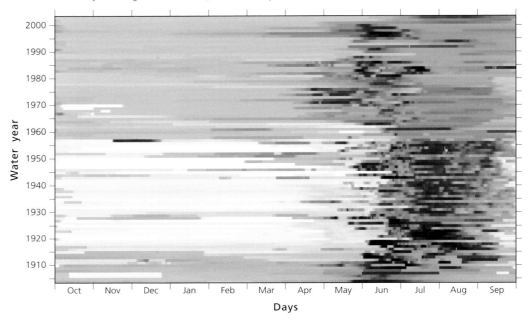

### Gauging Stations

Yellowstone River at
■ Corwin Springs, MT

Snake River
near Moran, WY

The depictions of flow regimes on these pages are "raster hydrographs" portraying variations in discharge over time. The advantage of a raster hydrograph is that it displays every single day of data for the entire period of record, enabling the viewer to see both short-term and long-term trends. The hydrographs on this page display 33,603 daily observations at Corwin Springs from 1911 to 2003 and 36,529 observations at Moran from 1904 to 2003. Reading the raster hydrograph is like reading a calendar. Rows represent the progression of days in a single water year (which by convention starts on October 1), with the amount of flow indicated by color (darker blue is more water). Columns represent the same day of the year in different years. Displaying the raster hydrographs in three-dimensional form (right page) helps to visualize the variations in discharge within and between years, but at the cost of obscuring some data points behind higher values of discharge.

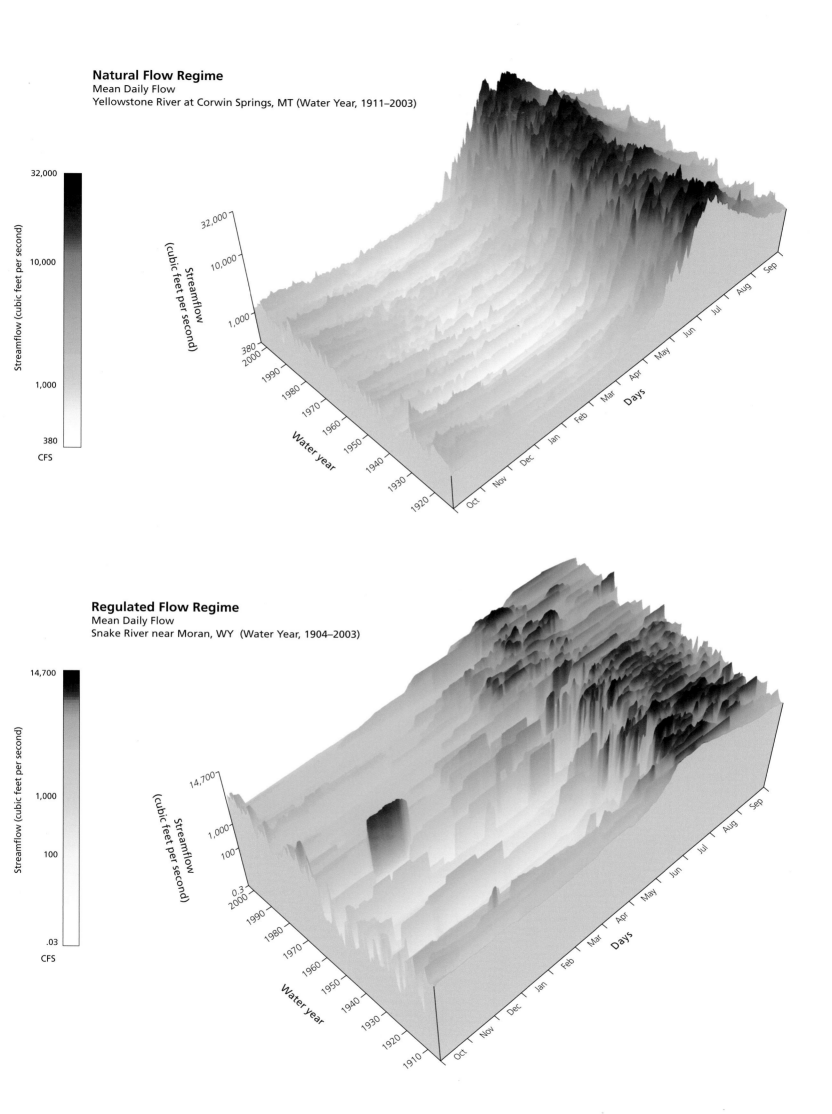

**Natural Flow Regime**
Mean Daily Flow
Yellowstone River at Corwin Springs, MT (Water Year, 1911–2003)

Streamflow (cubic feet per second)

32,000

10,000

1,000

380

CFS

Streamflow
(cubic feet per second)

32,000

10,000

1,000

380

2000
1990
1980
1970
1960
1950
1940
1930
1920

Water year

Days

Oct Nov Dec Jan Feb Mar Apr May Jun Jul Aug Sep

**Regulated Flow Regime**
Mean Daily Flow
Snake River near Moran, WY  (Water Year, 1904–2003)

Streamflow (cubic feet per second)

14,700

1,000

100

.03

CFS

Streamflow
(cubic feet per second)

14,700

1,000

100

0.3

2000
1990
1980
1970
1960
1950
1940
1930
1920
1910

Water year

Days

Oct Nov Dec Jan Feb Mar Apr May Jun Jul Aug Sep

# Waterfalls

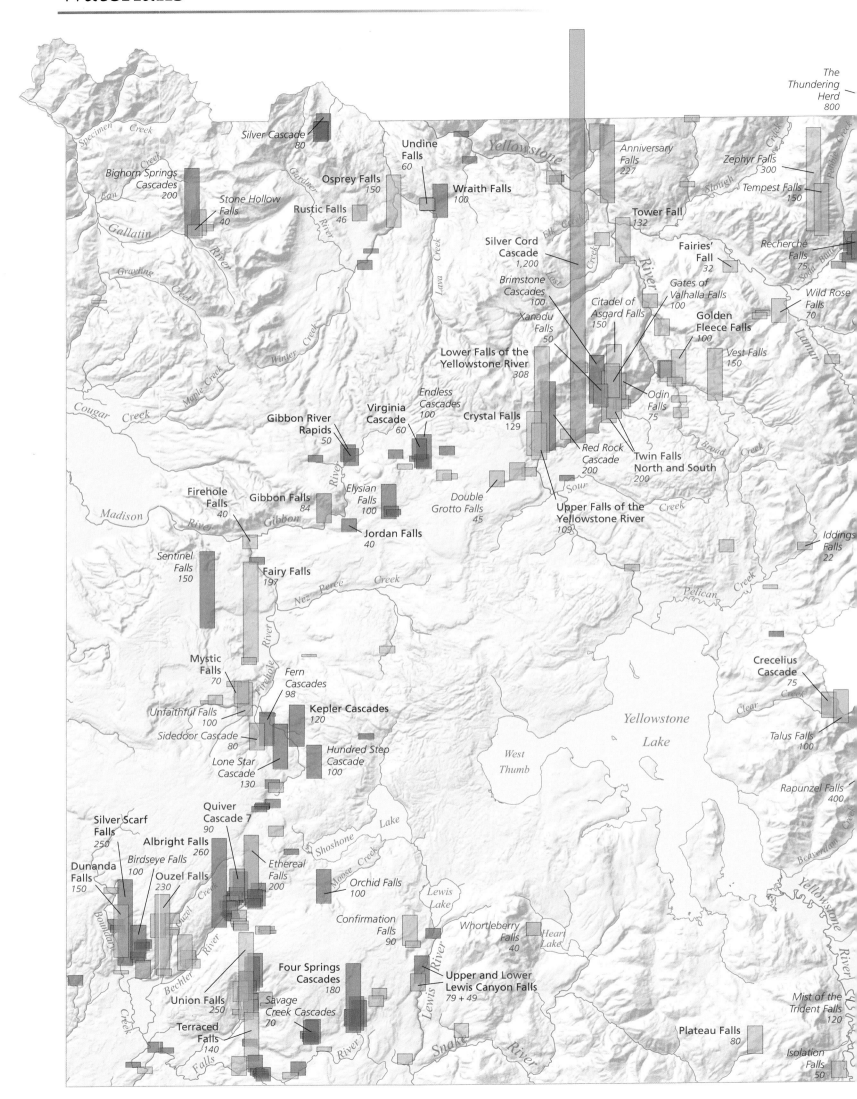

Silver Cascade
80

Undine
Falls
60

Osprey Falls
150

Wraith Falls
100

Bighorn Springs
Cascades
200

Stone Hollow
Falls
40

Rustic Falls
46

Anniversary
Falls
227

The
Thundering
Herd
800

Zephyr Falls
300

Tempest Falls
150

Tower Fall
132

Silver Cord
Cascade
1,200

Fairies'
Fall

Recherché
Falls
75

Brimstone
Cascades
100

Gates of
Valhalla Falls
100

Golden
Fleece Falls
100

Wild Rose
Falls
70

Xanadu
Falls
50

Citadel of
Asgard Falls
150

Lower Falls of the
Yellowstone River
308

Vest Falls
150

Gibbon River
Rapids
50

Virginia
Cascade
60

Endless
Cascades
100

Crystal Falls
129

Odin
Falls
75

Red Rock
Cascade
200

Twin Falls
North and South
200

Firehole
Falls
40

Gibbon Falls
84

Elysian
Falls
100

Double
Grotto Falls
45

Upper Falls of the
Yellowstone River
109

Jordan Falls
40

Iddings
Falls
22

Sentinel
Falls
150

Fairy Falls
197

Mystic
Falls
70

Crecelius
Cascade
75

Fern
Cascades
98

Unfaithful Falls
100

Kepler Cascades
120

Talus Falls
100

Sidedoor Cascade
80

Hundred Step
Cascade
100

Lone Star
Cascade
130

Yellowstone
Lake

West
Thumb

Rapunzel Falls
400

Quiver
Cascade 7
90

Silver Scarf
Falls
250

Albright Falls
260

Ethereal
Falls
200

Birdseye Falls
100

Dunanda
Falls
150

Ouzel Falls
230

Orchid Falls
100

Lewis
Lake

Confirmation
Falls
90

Whortleberry
Falls
40

Heart
Lake

Union Falls
250

Four Springs
Cascades
180

Savage
Creek Cascades
70

Upper and Lower
Lewis Canyon Falls
79 + 49

Mist of the
Trident Falls
120

Terraced
Falls
140

Plateau Falls
80

Isolation
Falls
50

*Enchantress Falls 90*

Some of Yellowstone's most plentiful and spectacular features are its waterfalls. About 350 waterfalls of more than fifteen feet are currently identified. Yellowstone may be the world's oldest national park, but geologically speaking it is a young landscape. Ongoing volcanic activity involving its rhyolite, basalt, and granite rocks has formed a mountainous, multileveled landmass. This topography, sculpted by glaciation and blessed with plentiful year-round precipitation, makes Yellowstone an ideal setting for waterfalls.

The waterfalls are concentrated in the northeast and southwest corners of the park as well as in the side canyons of the Yellowstone River. Eighty percent of Yellowstone's waterfalls are tucked away in the deepest recesses of its wilderness and not visible from roads or trails.

Discovery of virtually all of the park's waterfalls occurred during four periods. The expeditions of Washburn, Hayden, Barlow, and Hague in the 1870s and 1880s documented about fifty well-known waterfalls. During the 1920s, explorers William C. Gregg, C.H. Birdseye, and Jack Haynes found dozens of falls in the park's southwest corner, called the Bechler region. U.S. Fish and Wildlife Service stream surveys of the 1970s and 1980s located several dozen previously unknown falls, but these findings were confined to obscure archival reports. Finally, a large of number waterfalls have been formally documented in recent years: while conducting extensive backcountry explorations during the 1990s, park employees Paul Rubinstein, Mike Stevens, and Lee Whittlesey identified several hundred previously undocumented waterfalls. Since 2001, Stevens and Whittlesey have documented an additional sixty or more waterfalls.

The map displays many of Yellowstone's waterfalls by height in bar-graph form. The chart to the right lists Yellowstone's tallest waterfalls by height and type, including a height-range where applicable that represents the waterfall's plunge and its associated cascades. The Citadel of Asgard Falls, for example, has a 150-foot drop with 400 feet of additional cascades.

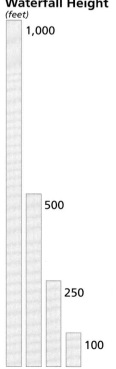

*Pollux Peak Falls 100*

## Waterfall Types

Yellowstone's waterfalls are of three basic kinds (plunge, horsetail, cascade), each of which can be divided into subcategories: fan, segmented, tiered, and serial. Here, waterfalls are typed by their dominant feature.

**◻ Plunge**

Water is free falling for some or all of its descent without coming into contact with the underlying rock.

**◼ Cascade**

Water flows at an angle over a series of rocks or down a broad rock face with many small leaps or segments. A cascade is often the result of a waterfall either eroding upstream or breaking down into smaller components through erosion.

**▨ Horsetail**

Water descends rapidly down a near-vertical wall, continually maintaining some contact with the underlying rock.

**▨ Mixed/Other**

A waterfall combining two or more of the basic types or a waterfall primarily described by subcategory characteristics

*Note: Italicized waterfall names indicate unofficially named waterfalls that are identified by their proposed local names.*

## Waterfall Height
(feet)

1,000

500

250

100

## Yellowstone Waterfalls

| Name | Min height | Max height | Height Range |
|---|---|---|---|
| **Silver Cord Cascade** | | | 1,200 |
| *The Thundering Herd* | | | 800 |
| *Rapunzel Falls* | | | 400–500 |
| **Lower Falls of the Yellowstone River** | | | 308 |
| *Zephyr Falls (seasonal)* | | | 300 |
| **Albright Falls** | | | 260 |
| **Silver Scarf Falls** | | | 250 |
| **Union Falls** | | | 250 |
| **Ouzel Falls** | | | 230 |
| *Anniversary Falls (seasonal)* | | | 227 |
| *Red Rock Cascade* | | | 200–250 |
| *Ethereal Falls* | | | 200 |
| **Twin Falls (north)** | | | 200 |
| **Twin Falls (south)** | | | 200 |
| **Fairy Falls** | | | 197 |
| **Four Springs #4** | | | 180 |
| *Citadel of Asgard Falls (seasonal)* | | | 150–550 |
| **Dunanda Falls** | | | 150 |
| **Osprey Falls** | | | 150 |
| *Sentinel Falls* | | | 150 |
| *Tempest Falls* | | | 150 |
| *Vest Falls* | | | 150 |
| **Terraced Falls** | | | 140 |
| **Tower Fall** | | | 132 |
| *Lone Star Cascade* | | | 130 |
| **Crystal Falls** | | | 129 |
| **Kepler Cascades** | | | 120 |
| *Mist of the Trident Falls* | | | 120 |
| **Upper Falls** | | | 109 |
| *Birdseye Falls* | | | 100 |
| *Breathtaking Falls* | | | 100 |
| *Brimstone Cascades* | | | 100 |
| *Elysian Falls* | | | 100 |
| *Endless Cascades* | | | 100 |
| **Four Springs #3** | | | 100 |
| *Gates of Valhalla Falls* | | | 100 |
| **Golden Fleece Falls** | | | 100 |
| *Heavenly Staircase* | | | 100 |
| *Hundred Step Cascade (seasonal)* | | | 100 |
| *Orchid Falls* | | | 100 |
| *Pollux Peak Falls* | | | 100 |
| *Talus Falls (seasonal)* | | | 100 |
| *Unfaithful Falls (seasonal)* | | | 100 |
| **Wraith Falls** | | | 100 |
| *Fern Cascades* | | | 98 |
| *Enchantress Falls* | | | 90 |
| **Quiver Cascade #7** *(Hourglass Falls)* | | | 90 |
| *Sidedoor Cascade* | | | 80 |
| *Silver Cascade* | | | 80 |
| *Confirmation Falls (seasonal)* | | | 75–90 |
| *Recherché Falls* | | | 75 |

500 ft        1,000 ft

# Precipitation

■ Weather observation station and average annual precipitation

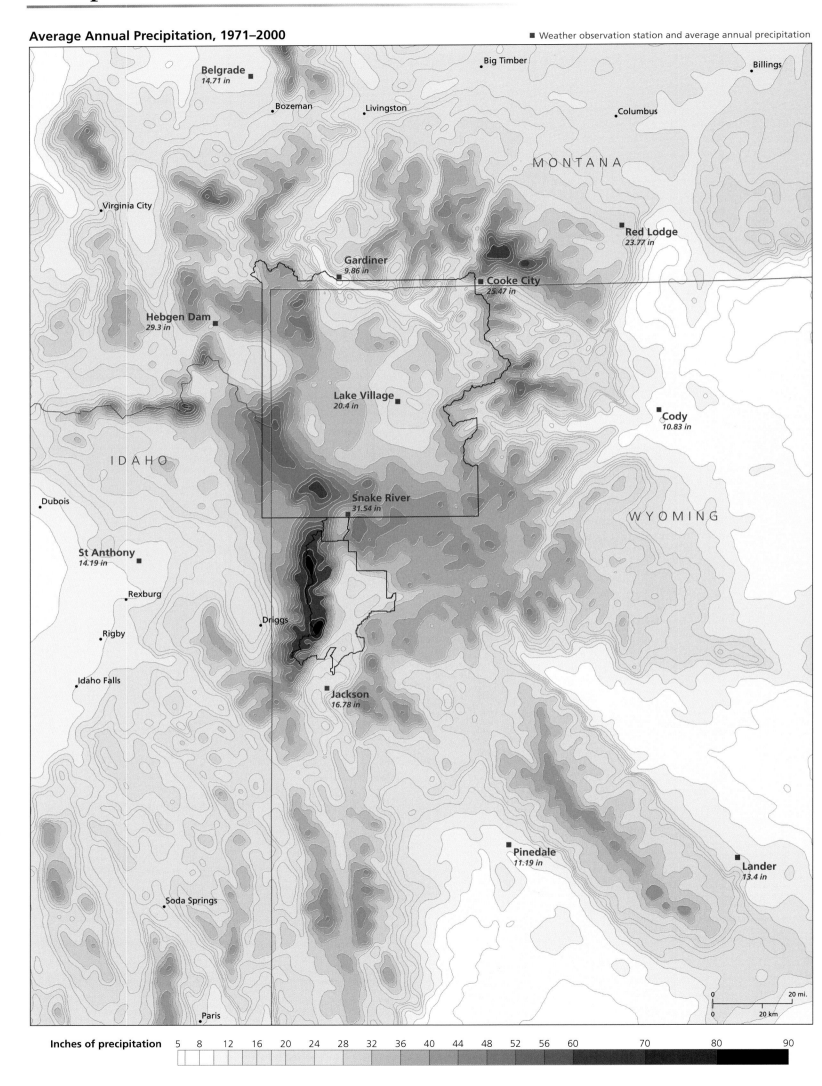

Belgrade
14.71 in

Big Timber

Billings

Bozeman

Livingston

Columbus

MONTANA

Virginia City

Red Lodge
23.77 in

Gardiner
9.86 in

Cooke City
25.47 in

Hebgen Dam
29.3 in

Lake Village
20.4 in

Cody
10.83 in

IDAHO

Dubois

WYOMING

Snake River
31.54 in

St Anthony
14.19 in

Rexburg

Rigby

Driggs

Idaho Falls

Jackson
16.78 in

Pinedale
11.19 in

Lander
13.4 in

Soda Springs

Paris

0                  20 mi.

0                  20 km

Inches of precipitation    5    8    12    16    20    24    28    32    36    40    44    48    52    56    60    70    80    90

Precipitation, especially winter snowfall, is crucial to water supply in the Yellowstone region and to downstream users. Yellowstone precipitation is controlled by large-scale atmospheric circulation, the regional influence of mountainous topography, and the local influence of summer convective (heat-driven) storms.

Winter precipitation dominates the western Yellowstone region, as exemplified by monthly precipitation at Snake River, Wyoming. The "winter wet" pattern occurs as the jet stream directs Pacific storms and moisture into the Yellowstone region. The Snake River Plain funnels this moisture into the Tetons and onto the Yellowstone Plateau, making these areas relatively wet in winter and spring compared to similar elevations to the east.

A "summer wet" period is typical of much of the eastern Yellowstone region (Cody, Wyoming, for example). Although westerly flows continue to dominate during this period, most Pacific storms have moved north and a "trough" of low pressure develops on the eastern slope of the Rockies. This trough pulls in moisture from the south, much of which is originally from the Gulf of Mexico, having precipitated and re-evaporated before reaching Yellowstone. Thunderstorms during this period are fueled by that moisture and triggered by local heating and interactions of winds with topography, which cause the air to rise and cool, forming clouds and precipitation.

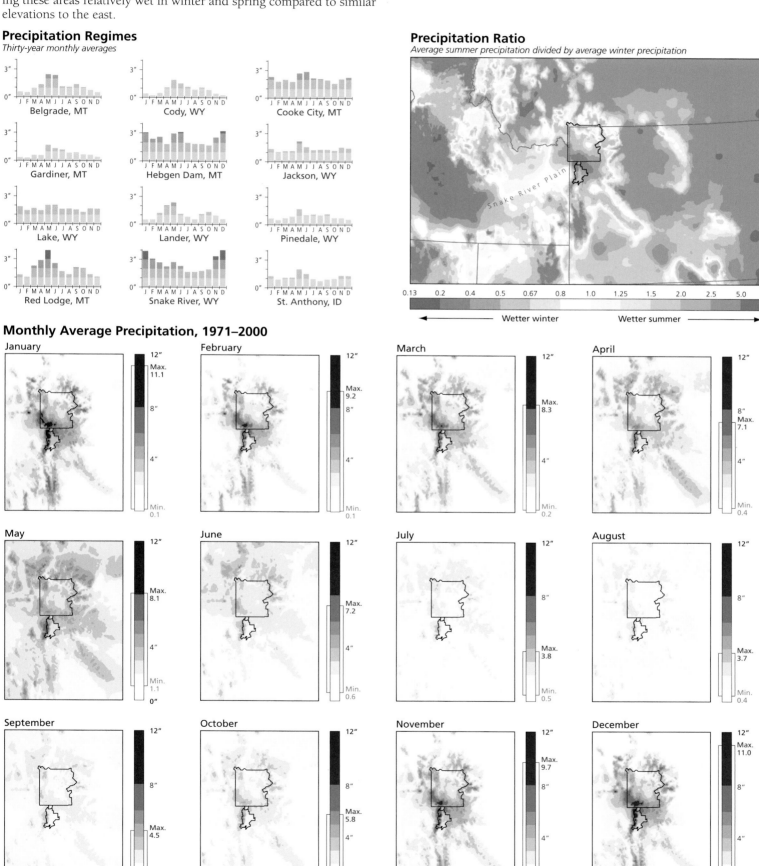

## Precipitation Regimes
*Thirty-year monthly averages*

Belgrade, MT

Cody, WY

Cooke City, MT

Gardiner, MT

Hebgen Dam, MT

Jackson, WY

Lake, WY

Lander, WY

Pinedale, WY

Red Lodge, MT

Snake River, WY

St. Anthony, ID

## Precipitation Ratio
*Average summer precipitation divided by average winter precipitation*

Snake River Plain

0.13   0.2   0.4   0.5   0.67   0.8   1.0   1.25   1.5   2.0   2.5   5.0   7.5

← Wetter winter          Wetter summer →

## Monthly Average Precipitation, 1971–2000

January — Max. 11.1, Min. 0.1

February — Max. 9.2, Min. 0.1

March — Max. 8.3, Min. 0.2

April — Max. 7.1, Min. 0.4

May — Max. 8.1, Min. 1.1

June — Max. 7.2, Min. 0.6

July — Max. 3.8, Min. 0.5

August — Max. 3.7, Min. 0.4

September — Max. 4.5, Min. 0.6

October — Max. 5.8, Min. 0.4

November — Max. 9.7, Min. 0.2

December — Max. 11.0, Min. 0.1

## Temperature Regimes, 1971–2000

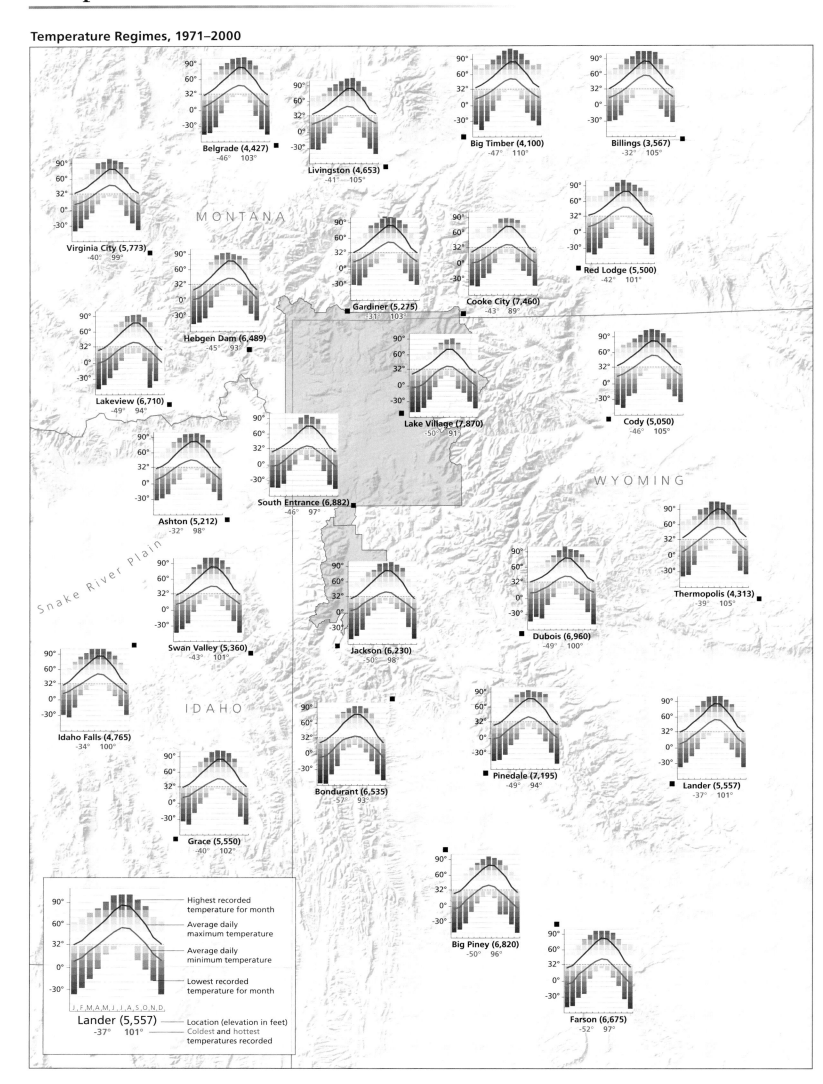

Belgrade (4,427)
-46° 103°

Livingston (4,653)
-41° 105°

Big Timber (4,100)
-47° 110°

Billings (3,567)
-32° 105°

Virginia City (5,773)
-40° 99°

MONTANA

Red Lodge (5,500)
-42° 101°

Hebgen Dam (6,489)
-45° 93°

Gardiner (5,275)
-31° 103°

Cooke City (7,460)
-43° 89°

Lakeview (6,710)
-49° 94°

Lake Village (7,870)
-50° 91°

Cody (5,050)
-46° 105°

Ashton (5,212)
-32° 98°

South Entrance (6,882)
-46° 97°

WYOMING

Snake River Plain

Swan Valley (5,360)
-43° 101°

Jackson (6,230)
-50° 98°

Dubois (6,960)
-49° 100°

Thermopolis (4,313)
-39° 105°

Idaho Falls (4,765)
-34° 100°

IDAHO

Bondurant (6,535)
-57° 93°

Pinedale (7,195)
-49° 94°

Lander (5,557)
-37° 101°

Grace (5,550)
-40° 102°

Big Piney (6,820)
-50° 96°

Farson (6,675)
-52° 97°

Highest recorded
temperature for month

Average daily
maximum temperature

Average daily
minimum temperature

Lowest recorded
temperature for month

J, F, M, A, M, J, J, A, S, O, N, D,

Lander (5,557) —— Location (elevation in feet)
-37° 101° —— Coldest and hottest
temperatures recorded

## Temperature Range
*Difference between average daily high and average daily low temperature*

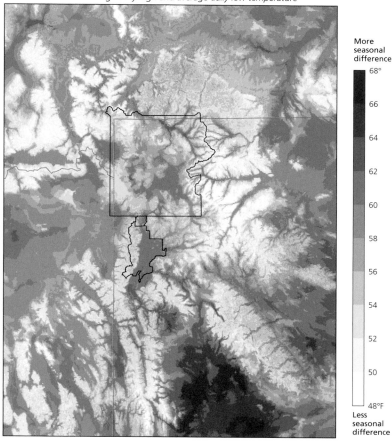

More seasonal difference

68°
66
64
62
60
58
56
54
52
50
48°F

Less seasonal difference

Three major factors drive the annual pattern of temperature in the greater Yellowstone region: the seasonal cycle of solar radiation, the influence of cold polar air delivered by the jet stream, and the diverse local topography that ranges from the high mountains of the Rockies to the low-lying Snake River Plain.

Temperatures in the greater Yellowstone region exhibit strong seasonal extremes, typical of interior continental and high elevation locations. The long and cold winters are a combination of short days with low sun angles and polar air. The coldest months are November through February, which receive the least amount of sunlight and have maximum temperatures near freezing. In contrast, the short and relatively warm summers feature long days with relatively high sun angles. The warmest months are July and August because it takes a while for the solar radiation maximum of June to heat the surface. Spring and fall are the shortest seasons with cold temperatures persisting in the high elevations of the surrounding mountain ranges.

In general, temperatures decrease at higher elevations. Variations in temperature across the region therefore largely mirror the topography. Overall, the annual range of temperatures is generally greater at lower elevations, which heat up much more than higher elevations during the summer.

During the winter, the low-lying Snake River Plain serves as a conduit allowing cold air to drain from the western portion of the Yellowstone Plateau. This same topographic feature helps funnel moisture from the Pacific Ocean during winter, leading to wetter winters in the Tetons and west Yellowstone region compared to the eastern portion of the park.

## Monthly Average Temperature, 1971–2000

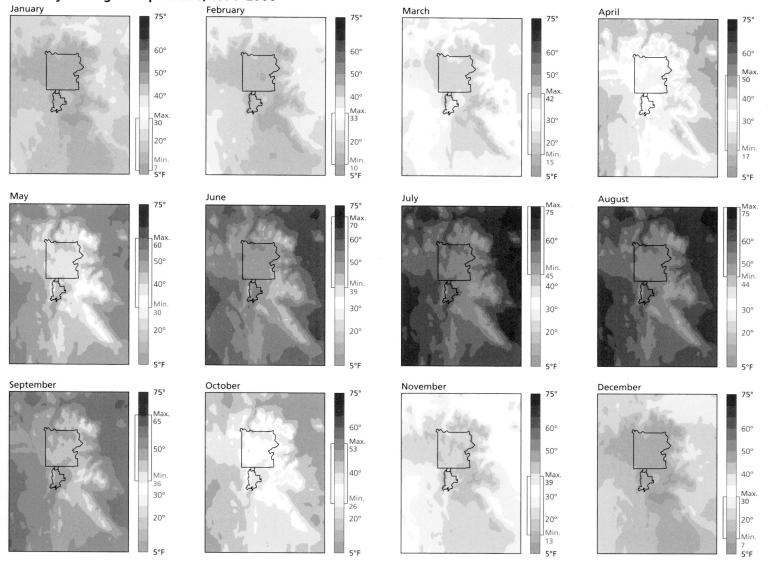

125

# Climate Change

The climate of any region changes continuously. On very long time scales, climate moves between cold ice ages and warm interglacial periods, while on shorter time scales, climate varies from decade to decade, and weather differs from one year or season to the next. The controls of climate variations include "external" changes in the amount of radiant energy received from the sun and the composition of the atmosphere (including greenhouse gases such as carbon dioxide and dust from volcanic eruptions) as well as "internal" changes in ocean temperatures, snow cover, and soil moisture, which can all influence atmospheric circulation and storm tracks.

In the past 250 years, the burning of fossil fuels and other activities have dramatically increased the amount of carbon dioxide in the atmosphere to levels greater than any experienced in the recent geo-

logical past, and the rate of increase is itself increasing. Carbon dioxide and other gases in the atmosphere, such as water vapor and methane, control the greenhouse effect—the trapping of heat that would otherwise be lost to space. As the greenhouse effect increases, Earth's atmosphere warms up and becomes moister in general. The changes in heating also have an effect on storm tracks, precipitation (and whether it falls as rain or as snow), evaporation, and the summer-winter temperature range.

The maps below show climate changes observed over the twentieth century and projected for the twenty-first century using a global climate model. Such computer models are related to those used to forecast weather on a day-to-day basis; the simulated climates include the effects of the past and projected future changes in greenhouse gases and other controls.

The overall trend and decade-to-decade variations in the climate of the Yellowstone area can be seen across these pages. The final decade of the twentieth century was the warmest but also had relatively high values of "snow water equivalent" (the amount of water that would be produced if the snowpack as it exists on April 1 were instantly melted). Greater snowfall is one of the consequences of a warmer climate in this relatively cold region.

The twenty-first century projections show a continuation and increase in the overall temperature trend. By the end of the century, the average year-round temperature is above freezing over nearly all of the mapped region, and July temperatures at some higher elevations exceed those observed for the Snake River Plain today. The trend in snowpack dramatically reverses, as higher temperatures prevail over any

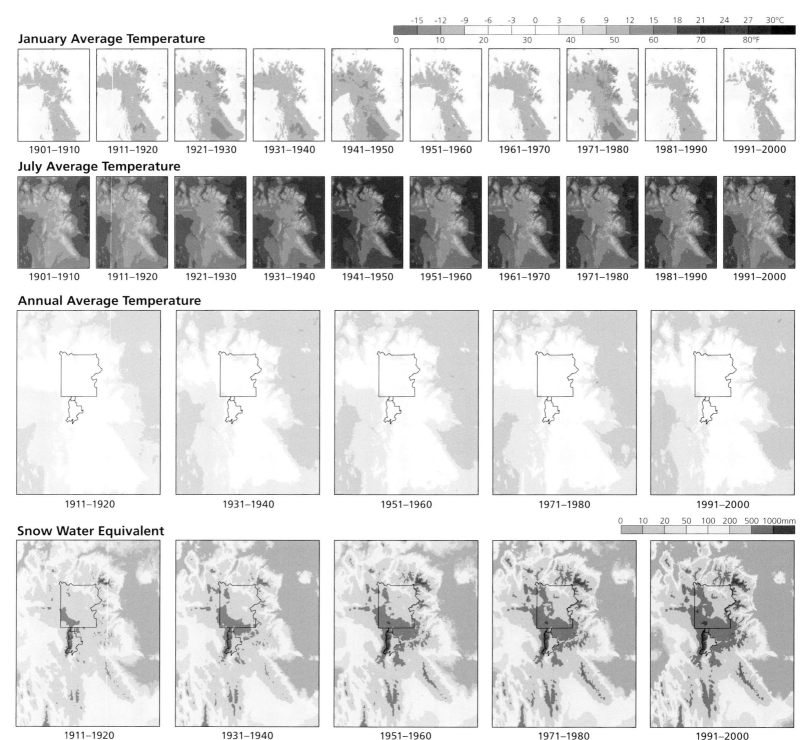

126

further increase in snowfall; at the end of the century, spring snowpack resembles that of mid-summer today.

The maps to the right illustrate projected climate change using Yellowstone Lake as a point of reference. Climate currently found at Yellowstone Lake may move northward into Canada and upward into higher elevations as temperatures increase (left-hand map). The projected future climate for Yellowstone Lake occurs today at lower elevations in the region—places that are warmer year-round and both wetter in the winter (with more rain than snow) and drier in the summer as compared to Yellowstone today (right-hand map).

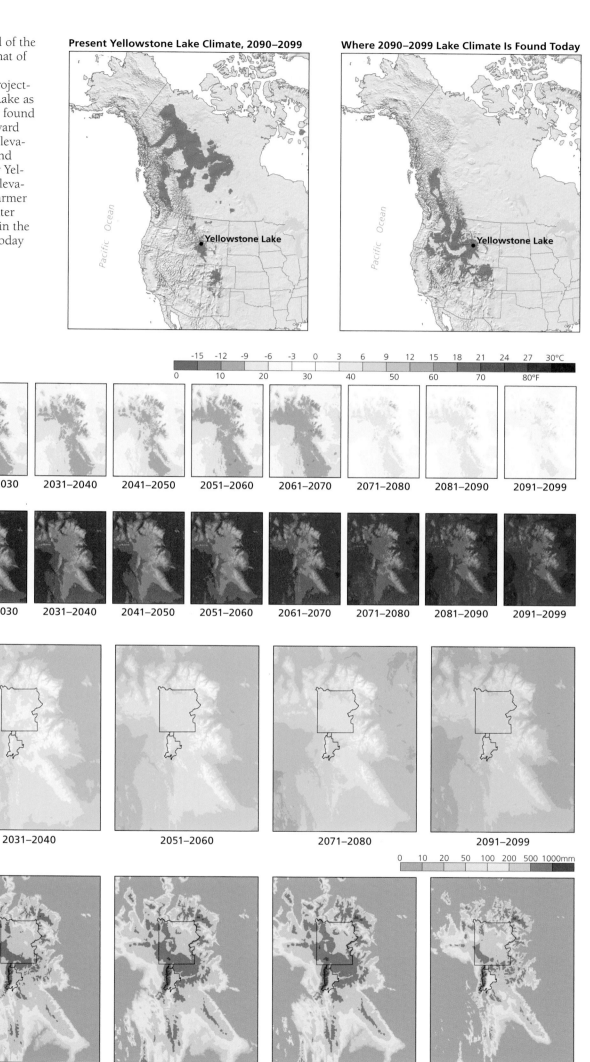

**Present Yellowstone Lake Climate, 2090–2099**

**Where 2090–2099 Lake Climate Is Found Today**

Yellowstone Lake

-15  -12  -9  -6  -3  0  3  6  9  12  15  18  21  24  27  30°C
0       10      20      30      40      50      60      70      80°F

2001–2010  2011–2020  2021–2030  2031–2040  2041–2050  2051–2060  2061–2070  2071–2080  2081–2090  2091–2099

2001–2010  2011–2020  2021–2030  2031–2040  2041–2050  2051–2060  2061–2070  2071–2080  2081–2090  2091–2099

2011–2020  2031–2040  2051–2060  2071–2080  2091–2099

0   10   20   50   100   200   500   1000mm

2011–2020  2031–2040  2051–2060  2071–2080  2091–2099

127

# Wetlands

Slough
Creek

Swan
Lake
Flat

Lamar
Valley

Norris
Geyser
Basin

Hayden
Valley

Pelican
Valley

Lower
Geyser
Basin

Upper
Geyser
Basin

*YELLOWSTONE*

*LAKE*

Bechler
Meadows

Two Ocean
Plateau

Wetlands—areas covered by standing water or where the water table is generally at or near the surface—have often been considered useless wastelands to be replaced and improved. Yet decades of research have revealed wetlands to be places of extraordinary biological diversity and productivity. While covering only 6.4 percent of the planet's surface, wetlands produce a quarter of the plant and animal matter on Earth. The 5 percent of the United States categorized as wetlands is home to nearly a third of the nation's known plant species.

More than 10 percent of Yellowstone National Park and 14 percent of Grand Teton are covered by a variety of wetlands. Thermal areas host hydrothermal microbes and rare plant species on surrounding land. Flooded meadows and seeps are home to amphibians and other wetland-dependent species. Streams and shores are essential to the biodiversity of fish and many mammals. The array of wetlands in Yellowstone and Grand Teton provide different environments key to the survival of thousands of unique plant, animal, and microbe species, some of which are found nowhere else on Earth.

Thirty-eight percent of Yellowstone's 1,200 plant species, including more than half of its ninety rare plant species, are associated with wetlands. These plants have developed specific adaptations that make life in waterlogged and oxygen-depleted soils possible.

Particularly significant are fens, where organic matter production exceeds decomposition due to waterlogging and peat accumulates. Peat environments reduce the availability of oxygen and nutrients, creating a challenging environment for plants. Accordingly, fens typically are characterized by high concentrations of rare species. On a state-wide basis, more than 10 percent of the Wyoming plant species of concern occur in peatlands. Within Yellowstone park, 61 percent of the Wyoming species of concern are associated with wetlands and 34 percent occur in peatlands.

Yellowstone's thermal areas support especially unusual and rare species growing in assemblages recorded nowhere else in the world. A single site can contain sources of both cool and warm water and provide heated air throughout the year, allowing plants that normally grow hundreds or thousands of miles apart to exist side by side.

## Wetland Locations

Riverine and lacustrine wetlands are associated with rivers and lakes, respectively. Palustrine wetlands include bogs, marshes, and other terrestrial sites with little flowing water.

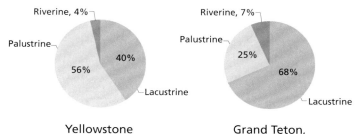

**Yellowstone**

Total acres: 257,109

Riverine, 4%
Palustrine 56%
40%
Lacustrine

**Grand Teton, John D. Rockefeller Jr.**

Total acres: 49,267

Riverine, 7%
Palustrine 25%
68%
Lacustrine

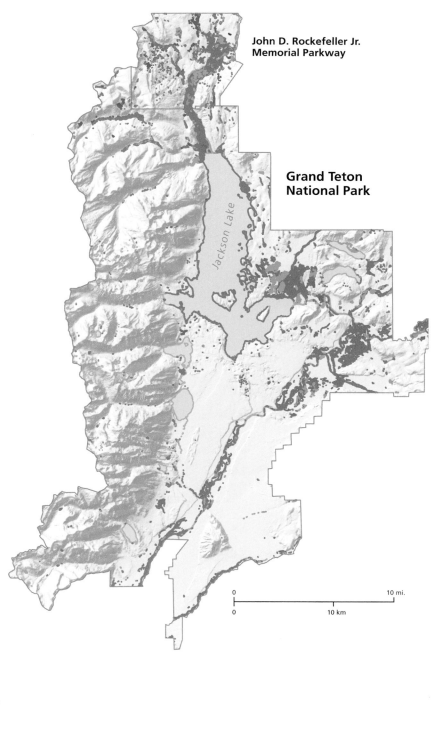

John D. Rockefeller Jr. Memorial Parkway

Grand Teton National Park

Jackson Lake

0        10 mi.
0     10 km

**Permanently flooded:** Water covers the land surface throughout the year in all years.

**Intermittently flooded:** The substrate is usually exposed, but surface water is present for variable periods without detectable seasonal periodicity.

**Seasonally flooded:** Surface water is present for extended periods especially early in the growing season but is absent by the end of the season in most years.

**Semipermanently flooded:** Surface water persists throughout the growing season in most years.

**Artificially flooded:** The amount and duration of flooding is controlled by means of pumps or siphons in combination with dikes or dams.

**Temporarily flooded:** Surface water is present for brief periods during the growing season.

**Intermittently exposed:** Surface water is present throughout the year except in years of extreme drought.

**Saturated:** The substrate is saturated to the surface for extended periods during the growing season, but surface water is seldom present.

## Wetland Acreage

Thousands of acres

250

102,670 acres

200

150
6,880
76,510

100

50
3,140
15,330
5,250
31,210
39,090
11,960
1,320
15,210
9,190
4,030

0
Yellowstone     Grand Teton and John D. Rockefeller Jr.

# Soils

More than eighty different soil types have been described within Yellowstone National Park. They occur in repeating patterns on the landscape, reflecting the combined influence of five soil-forming factors: type of original rock or other parent material; climate (temperature and precipitation); vegetation; landscape position; and the time that parent material is subject to soil formation. A hierarchical classification system divides every soil in the United States into twelve orders based on soil properties and soil-forming conditions. Because of the cold temperatures, low precipitation, and relatively short duration of soil development (ice covered the park up until 15,000 years ago), only six of the twelve soil orders occur within Yellowstone. Most of the park's eighty soil types fall into only three soil orders: Inceptisols, Mollisols, and Alfisols.

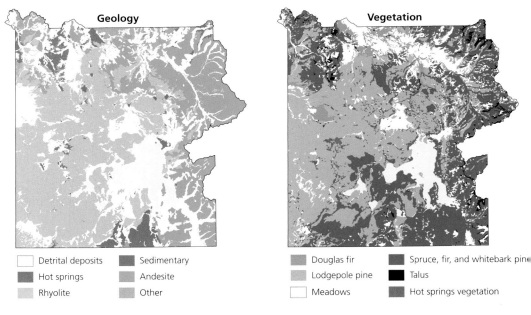

**Geology**

- Detrital deposits
- Hot springs
- Rhyolite
- Sedimentary
- Andesite
- Other

**Vegetation**

- Douglas fir
- Lodgepole pine
- Meadows
- Spruce, fir, and whitebark pine
- Talus
- Hot springs vegetation

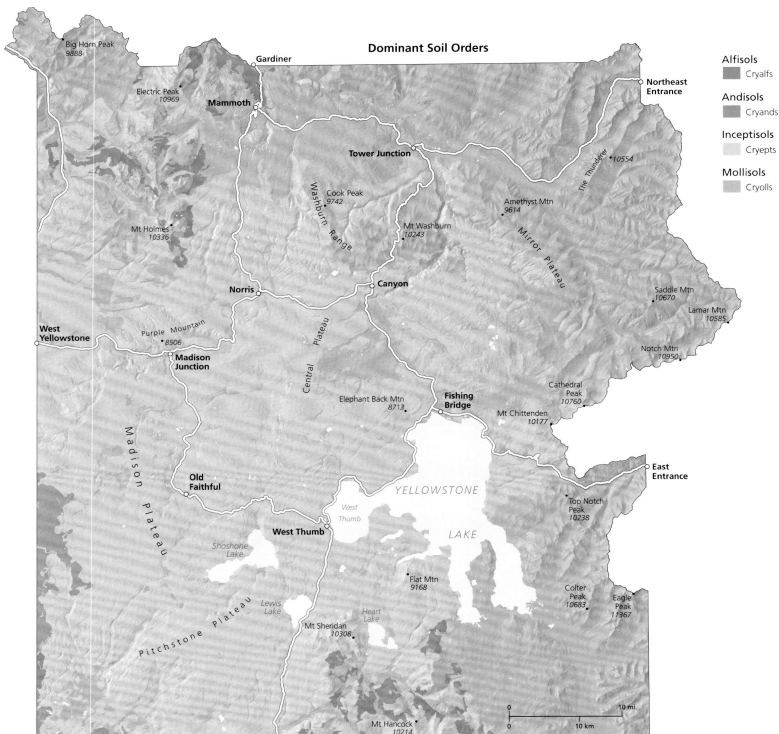

**Dominant Soil Orders**

**Alfisols**
- Cryalfs

**Andisols**
- Cryands

**Inceptisols**
- Cryepts

**Mollisols**
- Cryolls

Big Horn Peak 9888
Electric Peak 10969
Gardiner
Mammoth
Northeast Entrance
Tower Junction
Washburn Range
Cook Peak 9742
Mt Washburn 10243
The Thunderer 10554
Amethyst Mtn 9614
Mirror Plateau
Mt Holmes 10336
Norris
Canyon
Saddle Mtn 10670
Lamar Mtn 10585
Purple Mountain 8506
West Yellowstone
Central Plateau
Notch Mtn 10950
Madison Junction
Elephant Back Mtn 8713
Fishing Bridge
Cathedral Peak 10760
Mt Chittenden 10177
Madison Plateau
Old Faithful
East Entrance
YELLOWSTONE LAKE
Top Notch Peak 10238
West Thumb
West Thumb
Shoshone Lake
Flat Mtn 9168
Colter Peak 10683
Eagle Peak 11367
Pitchstone Plateau
Lewis Lake
Heart Lake
Mt Sheridan 10308
Mt Hancock 10214

0 10 mi.
0 10 km

## Alfisols

Alfisols have thin surface horizons and subsoil accumulations of clay. They occur throughout the forested north and east part of the park but dominate in areas weathering from sedimentary rocks. Cryalfs are the most common suborder, forming under very cold conditions.

Soil is present but not dominant *(for all maps)*

## Andisols

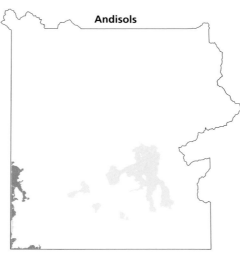

Andisols develop from volcanic ash. These soils occur along the southwestern boundary of the park where volcanic ash—from the eruption 7,700 years ago of Mount Mazama, now Crater Lake, Oregon—influences the upper portion of the soil.

## Inceptisols

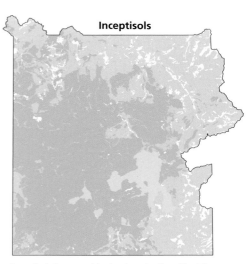

Inceptisols have weakly developed soil profiles. They are the most common soil order in Yellowstone and completely dominate the central and southwestern parts of the park. Cryepts are the most common suborder, forming under very cold conditions.

## Entisols

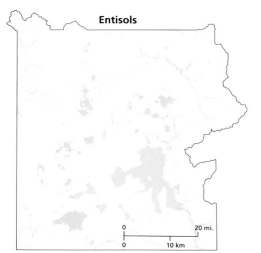

Entisols are typically too young to have developed soils horizons. They occur along streams where flooding adds new sediments; on the Pitchstone Plateau, where harsh winter conditions slow soil formation on recent lava flows; and in thermal areas. Orthents, the most common suborder, have medium textures and can be shallow.

## Histosols

Histosols are composed almost entirely of the decayed remains of plants in marshy areas. Histosols are not the dominant soil anywhere, but do occur throughout the park in wetlands. Three suborders occur within the park (Fibrist, Hemist, and Saprist) and are differentiated by the degree of decomposition of the decaying plant material.

## Mollisols

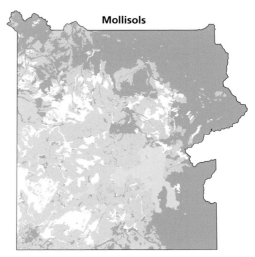

Mollisols have relatively thick, dark surface horizons and are rich in organic matter. They are usually associated with grassland in Yellowstone, but also occur in forests across the north and east sections of the park. Mollisols are the second most common soil order in the park; the suborder Cryoll is the most common type of Mollisol.

Most of the park's soils come from volcanic parent materials. The soils in the western and central plateau areas formed from parent material derived from rhyolite lava flows and ash-flow tuffs. Andesitic parent material from the Eocene Absaroka Volcanics is weathering into soil along the northwest, northeast, and eastern boundaries. Weathering products of these two rock types differ distinctly in mineral nutrient content and water-holding capacity—factors of primary importance to soil development and plant growth. The minerals (quartz and potassium feldspars) in rhyolite weather to form a nutrient-poor soil with a sandy texture and lots of rocks. The minerals (sodium plagioclase and hornblende) in andesite weather into clays more easily and produce loamy-textured soils that store more water and nutrients. This weathering process also releases more nutrients into the andesitic soils. Soils with more available water and nutrients grow more plants, resulting in more organic matter added to the soils when the plants die. The added organic matter creates a dark mollic surface horizon; its thickness is a measure of soil fertility. Organic matter further increases the soil's capacity to store water and slowly releases nutrients that increase the soil fertility. Over time these weathering and soil-forming cycles further differentiate the soils across the park, generating soils with dark surface horizons (Mollisols) across the north and east sections of the park and sandy, light-colored soils with thin surface layers (Inceptisols) in the center and southwest part of the park. This invisible, below-the-surface distribution of soil properties, inherited from the underlying geology, strongly influences the visible pattern of vegetation above the surface.

## Hydric Soils

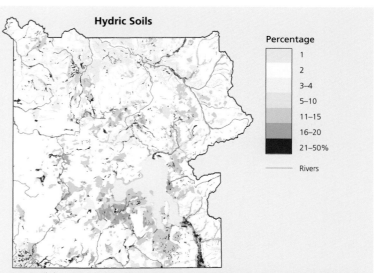

Most soils have pore spaces filled with enough oxygen to support common plants and soil life. If soils near the surface become saturated with water for long periods, new oxygen can't get into the soil, existing oxygen is depleted, and the soils become anaerobic. If this condition occurs annually during the growing season, this "hydric" soil develops identifiable properties. Most hydric soils in Yellowstone are mineral soils, but in some wetland conditions enough organic matter accumulates to form small areas of Histosols; these support unique communities of vegetation that can survive in low-oxygen environments. Wetlands are protected within the park, and the presence of hydric soils is a defining criterion of a wetland.

# Ecoregions

## Level III and IV Ecoregions

Level III Ecoregion Boundaries ——————    Level IV Ecoregion Boundaries (Labeled by number and letter code) ——————

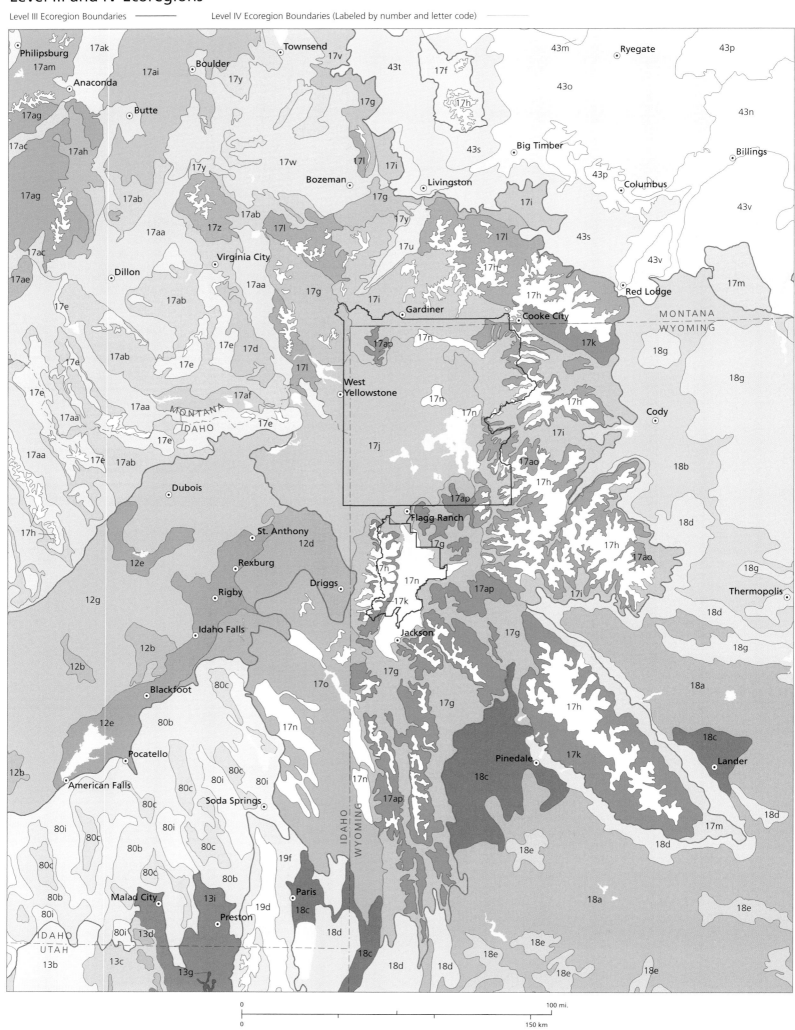

# Level III and IV Ecoregions

## 12. Snake River Plain
- 12b  Lava Fields
- 12d  Dissected Plateaus and Teton Basin
- 12e  Upper Snake RIver Plain
- 12g  Eastern Snake River Basalt Plains

## 13. Central Basin and Range
- 13b  Shadscale-Dominated Saline Basins
- 13c  Sagebrush Basins and Slopes
- 13d  Woodland and Shrub-Covered Low Mountains
- 13i  Malad and Cache Valleys
- 13g  Wetlands

## 17. Middle Rockies
- 17aa  Dry Intermontane Sagebrush Valleys
- 17ab  Dry Gneissic-Schistose-Volcanic Hills
- 17ac  Big Hole
- 17ae  Forested Beaverhead Mountains
- 17af  Centennial Basin
- 17ag  Pioneer-Anaconda Ranges
- 17ah  Eastern Pioneer Sedimentary Mountains
- 17ai  Elkhorn Mountains-Boulder Batholith
- 17ak  Deer Lodge-Philipsburg-Avon Grassy Intermontane Hills and Valleys
- 17am  Flint Creek-Anaconda Mountains
- 17ao  Absaroka Volcanic Subalpine Zone
- 17ap  Sedimentary Subalpine Zone
- 17d  Eastern Gravelly Mountains
- 17e  Barren Mountains
- 17f  Crazy Mountains
- 17g  Midelevation Sedimentary Mountains
- 17h  High Elevation Rockland Alpine Zone
- 17i  Absaroka-Gallatin Volcanic Mountains
- 17j  Yellowstone Plateau
- 17k  Granite Subalpine Zone
- 17l  Gneissic-Schistose Forested Mountains
- 17m  Dry Midelevation Sedimentary Mountains
- 17n  High Elevation Valleys
- 17o  Partly Forested Mountains
- 17u  Paradise Valley
- 17v  Big Belt Forested Highlands
- 17w  Townsend Basin
- 17y  Townsend-Horseshoe-London Sedimentary Hills
- 17z  Tobacco Root Mountains

## 18. Wyoming Basin
- 18a  Rolling Sagebrush Steppe
- 18b  Bighorn Basin
- 18c  Subirrigated High Valleys
- 18d  Foothill Shrublands and Low Mountains
- 18e  Salt Desert Shrub Basins and Slopes
- 18g  Bighorn Salt Desert Shrub Basin

## 19. Wasatch and Uinta Mountains
- 19d  Wasatch Montane Zone
- 19f  Semiarid Foothills

## 43. Northwestern Great Plains
- 43m  Judith Basin Grasslands
- 43n  Central Grassland
- 43o  Unglaciated Montana High Plains
- 43p  Ponderosa Pine Forest-Savanna Hills
- 43s  Noncalcareous Foothill Grasslands
- 43t  Shield-Smith Valley
- 43v  Pryor-Bighorn Foothills

## 80. Northern Basin and Range
- 80b  Semiarid Hills and Low Mountains
- 80c  High Elevation Forests and Shrublands
- 80h  Saltbush-Dominated Valley
- 80i  Sagebrush Steppe Valleys

- Water

Ecoregions denote areas of generally similar ecosystems with comparable types, qualities, and quantities of environmental resources. They serve as a spatial framework for research, assessment, management, and monitoring of ecosystems and ecosystem components. Ecoregions mapping is based on the premise that these regions can be identified through the analysis of patterns of biotic, abiotic, terrestrial, and aquatic phenomena, including geology, physiography, vegetation, climate, soils, land use, wildlife, and hydrology. The relative importance of each characteristic varies from one ecological region to another regardless of the hierarchical level. Level I is the coarsest level of ecoregions, dividing North America into fifteen regions, whereas at Level II the continent is subdivided into fifty classes. Refining further, levels III and IV are the subdivisions of ecological regions shown on this map of the Greater Yellowstone Area.

The ecoregions of Yellowstone and surrounding territories are organized by five major physiographic areas. In the west, the Snake River Plain and the Basin and Range ecoregions rise to the mountain ranges and plateaus of the Middle Rockies ecoregion. In the east, the arid intermontane Wyoming Basin ecoregion drains northward into the broad alluvial valleys of the Northwestern Great Plains ecoregion.

## 12. Snake River Plain
This portion of the xeric (dry) intermontane basin and range area of the western United States is considerably lower in elevation and more gently sloping than surrounding ecoregions. With abundant water for irrigation, a large percentage of the alluvial valleys bordering the Snake River are in agriculture, with sugar beets, potatoes, and vegetables being the principal crops. Cattle feedlots and dairy operations are also common in the river plain. Except for the scattered barren lava fields, the plains and low hills in the ecoregion have a sagebrush steppe potential natural vegetation and are now used for cattle grazing.

## 13. Central Basin and Range
This ecoregion is internally drained and characterized by a mosaic of xeric basins, scattered low and high mountains, and salt flats. It has a hotter and drier climate, more shrubland, and more mountain ranges than the Snake River Plain. Basins are covered by Great Basin sagebrush or saltbush-greasewood vegetation that grow in aridisols; cool season grasses are less common than in the mollisols of the Snake River Plain.

## 17. Middle Rockies
The Middle Rockies ecoregion is marked by high, steep-crested mountains and plateaus covered by open-canopy coniferous forests and alpine areas. Douglas fir, subalpine fir, and Engelmann spruce are common except where lodgepole pine dominates on plateaus with volcanic substrates. Pacific tree species are never dominant. Forests can be open. Foothills are partly wooded or shrub- and grass-covered. Intermontane valleys are covered in grasses, shrubs, or a mix of the two and contain a mosaic of terrestrial and aquatic fauna that is distinct from those found in the nearby mountains. Many mountain-fed, perennial streams occur and differentiate the intermontane valleys from the Northwestern Great Plains. Granitics and associated management problems are less extensive than in the Idaho Batholith. Recreation, logging, mining, and summer livestock grazing are common land uses.

## 18. Wyoming Basin
This ecoregion is a broad intermontane basin dominated by arid grasslands and shrublands and is interrupted by high hills and low mountains. Nearly surrounded by forest-covered mountains, the region is somewhat drier than the Northwestern Great Plains to the northeast. Much of the region is used for livestock grazing, although many areas lack sufficient vegetation to support this activity. The region contains major producing natural gas and petroleum fields.

## 19. Wasatch and Uinta Mountains
This ecoregion is composed of a core area of high, precipitous mountains with narrow crests and valleys flanked in some areas by dissected plateaus and open high mountains. The elevational banding pattern of vegetation is similar to that of the Southern Rockies except that aspen, chaparral, pinyon-juniper, and oak are more common at middle elevations. This characteristic, along with a far lesser extent of lodgepole pine and greater use of the region for grazing livestock in the summer months, distinguish the Wasatch and Uinta Mountains ecoregion from the more northerly Middle Rockies.

## 43. Northwestern Great Plains
The Northwestern Great Plains ecoregion encompasses the Missouri Plateau section of the Great Plains. It is a semiarid rolling plain of shale and sandstone punctuated by occasional buttes. Native grasslands, largely replaced on level ground by spring wheat and alfalfa, persist in rangeland areas on broken topography. Agriculture is restricted by erratic precipitation and limited opportunities for irrigation.

## 80. Northern Basin and Range
This ecoregion contains arid tablelands, intermontane basins, dissected lava plains, and scattered mountains. Nonmountain areas have sagebrush steppe vegetation. Cool season grasses and mollisols are more common than in the hotter and drier basins of the Central Basin and Range where aridisols are dominated by sagebrush, shadscale, and greasewood. Ranges are generally covered in mountain sagebrush, mountain brush, and Idaho fescue at lower and middle elevations; Douglas fir and aspen are common at higher elevations. The ecoregion is higher and cooler than the Snake River Plain. Rangeland is common. Dryland and irrigated agriculture occur in eastern basins.

# Vegetation

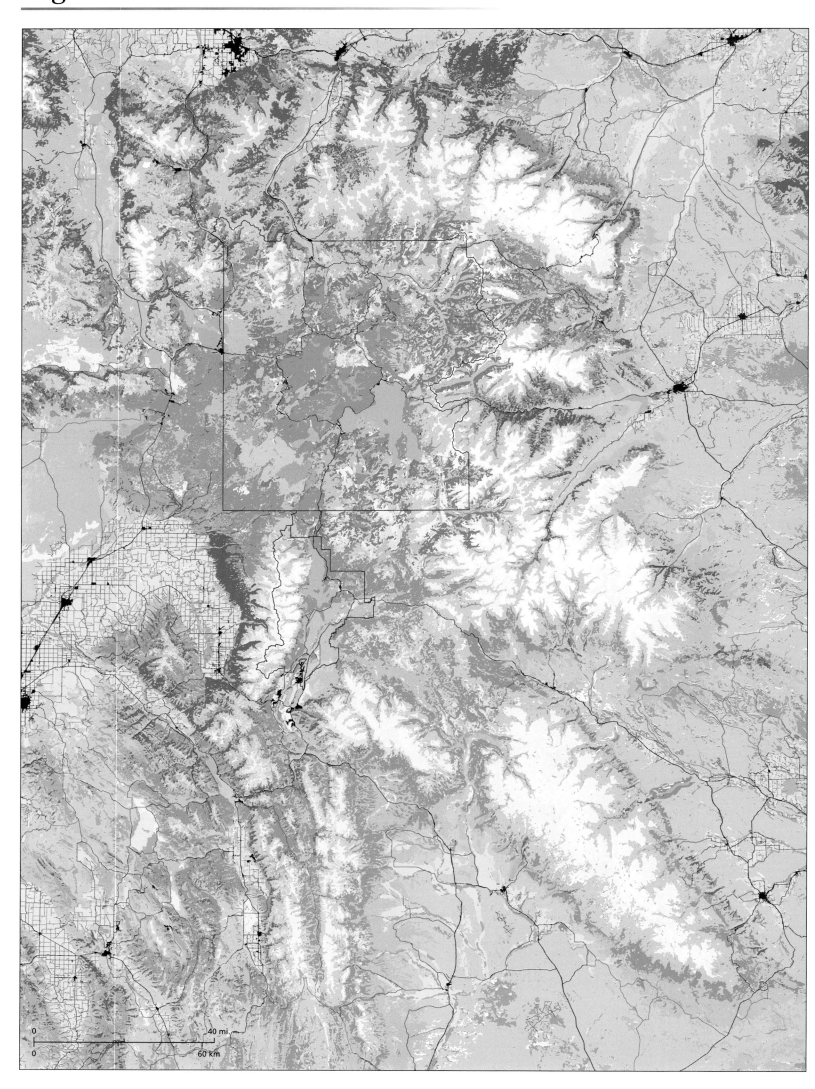

The vegetation of the Greater Yellowstone Area, ranging from dry sagebrush shrublands to verdant parklands and dense forests, reflects pronounced variations in elevation, rain and snow, and soil types. These environmental factors interact to affect the temperature and soil moisture during the growing season. They are the primary controls on vegetation over both the broad region (for example, from the Snake River Plain to the Bighorn Basin) and over short distances (such as from stream sides to dry hill slopes).

Forests occur between 6,000 and 10,000 feet above sea level. The western portion of the region receives more rain, supporting a Douglas-fir forest and parkland, especially at the upper reaches of the Snake River Plain. Douglas-fir forest is interrupted along streams, where wet soils and cold-air drainage favors subalpine fir. In the drier eastern portion, pinyon-juniper woodlands occur at lower treeline. Above about 7,000 feet, forests are dominated by subalpine fir with some Engelmann spruce in drier areas. Lodgepole pine dominates forests at elevations up to 9,000 feet on rhyolite soils and after fires. Aspen occurs in patchy stands on moist sites in the western and northern part of the region at elevations below 8,000 feet. At the highest elevations (above 9,000 feet), subalpine fir mixes with whitebark pine.

Nonforest vegetation occurs below 6,000 feet as grasslands and shrublands, above 10,000 feet as alpine meadows, and at specific microsites within the forested elevations. Several subspecies of sagebrush dominate at low elevations, though places receiving some summer rainfall have greater amounts of grasses. Montane meadows where trees are absent due to snowpack, persistent wet soils, or fire support dense grasslands and forbs.

These maps are derived from satellite images taken between 1999 and 2001, elevation data, and models that estimate vegetation type. The modeled data may be incorrect at individual sites (for example, mesic meadow in Pelican Valley is classified as alpine meadow). A number of thermal areas are missed because the scale of mapping and modeling is too coarse to pick up these relatively small features. Nevertheless, the maps accurately reflect general patterns of vegetation across the landscape.

*Detailed vegetation classes shown on following pages.*

## Forests and Woodlands

**rh** **Recently harvested**
Forests harvested in the recent past and currently regenerating with trees, shrubs, grasses, or a mix of these plants.

**rb** **Recently burned**
Burned areas, mainly dating from the 1988 fires.

**sf** **Spruce-fir**
Engelmann spruce and subalpine fir forests, marked by patches of lodgepole pine on dry sites, increased spruce on wet sites, and, occasionally, woodlands of whitebark pine at the highest elevations. The highest elevation forest type throughout the Rocky Mountains.

**lp** **Lodgepole pine**
Lodgepole pine forest typically is established after fire and is successional to spruce-fir forest; on gravelly, well-drained pumice soils lodgepole pine dominates due to conditions too extreme for other conifers.

**df** **Douglas fir**
Forests and woodlands dominated by Douglas fir. Common forest type of the lower treeline in the western half of the region.

**as** **Aspen**
Areas where aspens account for more than 75 percent of the trees. Established by large disturbances (fire, insects, etc.) and limited to montane areas that can provide sufficient moisture to this deciduous broadleaf tree.

**pj** **Limber pine-juniper**
Woodlands with an open tree canopy of limber pine, juniper, or a mix of the two. Restricted to thin soils and fractured bedrock near lower treeline or within sagebrush in the eastern half of the region.

**pp** **Ponderosa pine**
From sparse to continuous canopy cover of ponderosa pine with some Douglas fir. Found at lower treeline on warm sites along escarpments, buttes, and canyons in the northeastern corner of the region.

## Shrublands and Steppe

**ds** **Dwarf Wyoming big sagebrush**
Wyoming big sagebrush that is wind-dwarfed to less than one foot (thirty centimeters) tall. Found along the upper slopes of sagebrush basins predominantly in the southern half of the region.

**ms** **Mountain big sagebrush**
Mountain big sagebrush (occasionally with bitterbrush) with a diverse perennial forb layer. Found throughout the region at montane to subalpine elevations, generally in areas with deep soils and some summer moisture source.

**ws** **Wyoming big sagebrush**
Wyoming big sagebrush with significant amounts of perennial grasses (especially western wheatgrass). Widespread in basins throughout the eastern half of the region. Includes small areas of mountain mahogany on rocky sites.

**sg** **Saltbush and greasewood flats**
Dwarf shrublands of pure saltbush or saltbush codominated with greasewood. Located on arid, windswept basins and plains on saline deposits that may flood for a short part of the year.

## Grassland and Meadow

**am** **Alpine meadow**
Dense turf and wet meadows of grasses, sedges, forbs, and dwarf shrubs. Located above timberline and intermingled with alpine bedrock and scree.

**sg** **Subalpine grassland**
Lush perennial grasses and forbs on dry sites. Located near timberline typically on south-facing slopes.

**mm** **Montane and subalpine mesic meadow**
Seasonally moist dense grasslands and forb meadows. Located within and below timberline where snow or dry conditions limits tree establishment.

**mg** **Lower montane grassland**
A variety grassland and dwarf shrubland, usually transitional to big sagebrush shrubland and sometimes dominated by invasive annual grasses. Located on upper terraces with deep fine-to-sandy loam, with finer loam soils supporting prairie.

## Riparian and Wetlands

**sr** **Subalpine riparian**
Alder, willow, shrub birch, spruce, and fir. Located in seasonally flooded forest and shrublands that line mountain streams.

**lr** **Lower montane and plains riparian**
Various types of riparian vegetation extending from the foothills into the lowlands. Marked by cottonwood though often degraded by agriculture and other land use. Annual flooding results in wet soils most of the year.

**dw** **Depression wetlands**
Saline depressions and freshwater emergent marshes.

## Barren

**ab** **Alpine bedrock and scree**

**du** **Dunes**
Active migrating dunes and older stabilized former dunes. Restricted here to west-central and southern areas.

**ba** **Badlands**
Dryland shrubs and bare ground are common on these eroded unconsolidated clay soils. Forms rounded hills and plains.

## Developed

**hs** **Geysers and hot springs**

**pc** **Pasture and cultivated cropland**

**qu** **Quarries, mines, and gravel pits**

**dl** **Developed land and roads**

## Water

# Vegetation: North

### Forests and Woodlands

| | |
|---|---|
| rh | Recently harvested |
| rb | Recently burned |
| sf | Spruce-fir |
| lp | Lodgepole pine |
| df | Douglas fir |
| as | Aspen |
| pj | Limber pine-juniper |
| pp | Ponderosa pine |

### Shrublands and Steppe

| | |
|---|---|
| ds | Dwarf Wyoming big sagebrush |
| ms | Mountain big sagebrush |
| ws | Wyoming big sagebrush |
| sg | Saltbush and greasewood flats |
| se | Deciduous shrubland |

### Grassland and Meadow

| | |
|---|---|
| am | Alpine meadow |
| sg | Subalpine grassland |
| mm | Montane and subalpine mesic meadow |
| mg | Lower montane grassland |

### Riparian and Wetlands

| | |
|---|---|
| sr | Subalpine riparian |
| lr | Lower montane and plains riparian |
| dw | Depression wetlands |

### Barren

| | |
|---|---|
| ab | Alpine bedrock and scree |
| du | Dunes |
| ba | Badlands |
| hs | Geysers and hot springs |

### Developed

| | |
|---|---|
| pc | Pasture and cultivated cropland |
| qu | Quarries, mines, and gravel pits |
| dl | Developed land and roads |

### Water

0           20 mi.

0           20 km

# Vegetation: South

## Forests and Woodlands

- rh — Recently harvested
- rb — Recently burned
- sf — Spruce-fir
- lp — Lodgepole pine
- df — Douglas fir
- as — Aspen
- pj — Limber pine-juniper
- pp — Ponderosa pine

## Shrublands and Steppe

- ds — Dwarf Wyoming big sagebrush
- ms — Mountain big sagebrush
- ws — Wyoming big sagebrush
- sg — Saltbush and greasewood flats
- se — Deciduous shrubland

## Grassland and Meadow

- am — Alpine meadow
- sg — Subalpine grassland
- mm — Montane and subalpine mesic meadow
- mg — Lower montane grassland

## Riparian and Wetlands

- sr — Subalpine riparian
- lr — Lower montane and plains riparian
- dw — Depression wetlands

## Barren

- ab — Alpine bedrock and scree
- du — Dunes
- ba — Badlands
- hs — Geysers and hot springs

## Developed

- pc — Pasture and cultivated cropland
- qu — Quarries, mines, and gravel pits
- dl — Developed land and roads

## Water

0           20 mi.

0           20 km

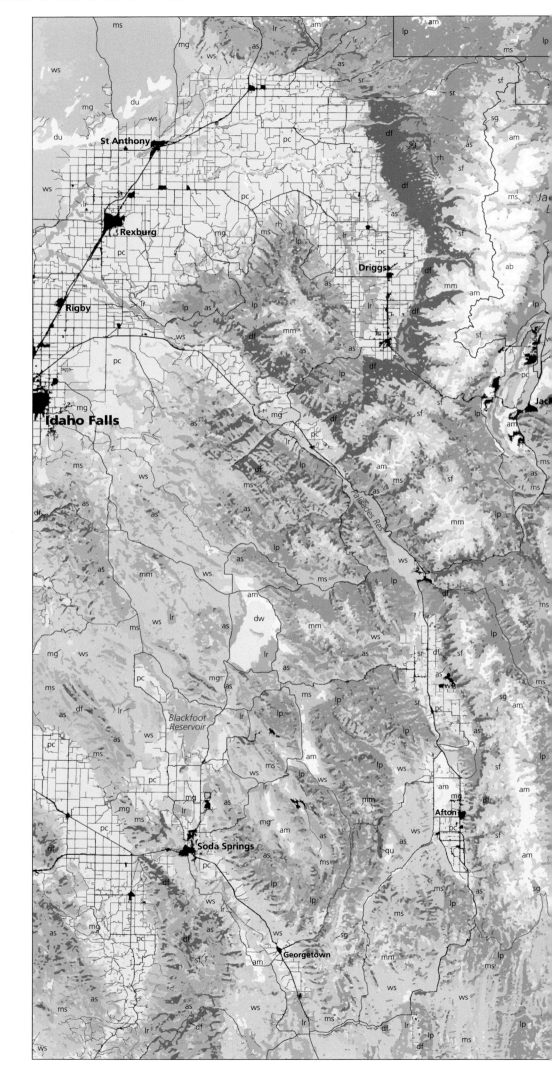

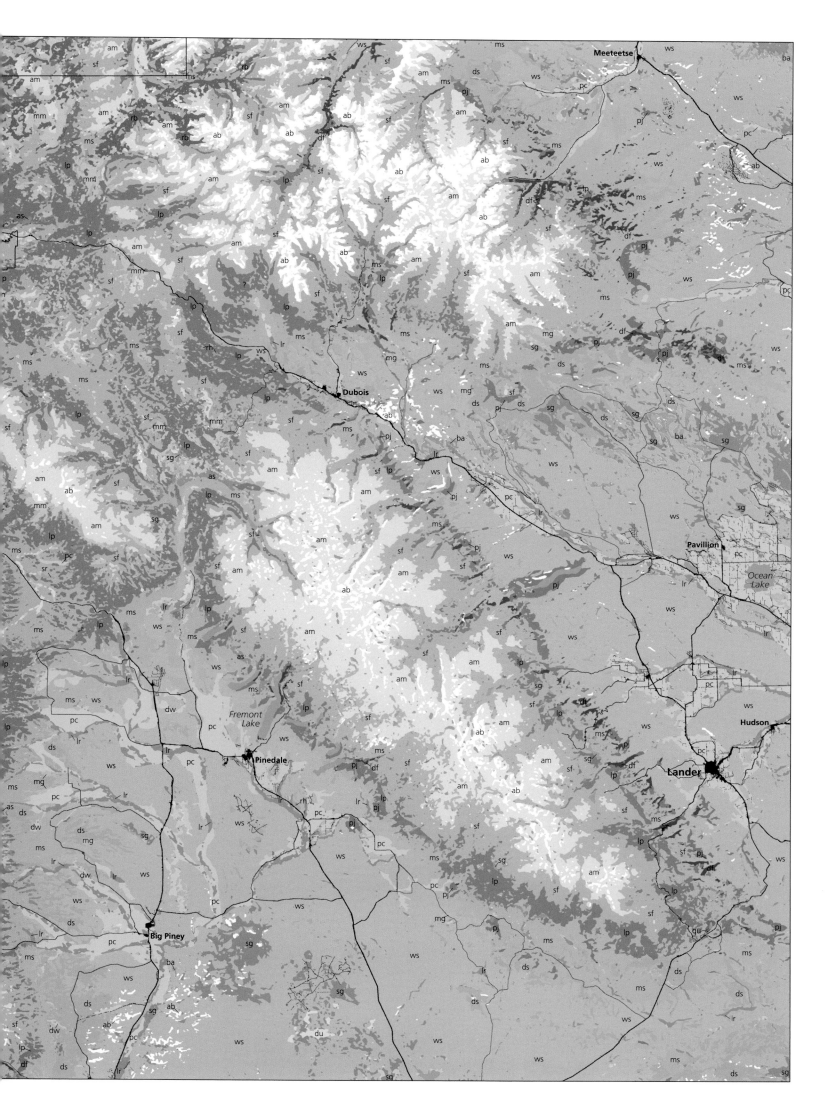

139

# Landscape Change

On this Landsat image mosaic, moist vegetation is bright green, conifers are dark green, open water is black, and snow is blue. Reddish tones indicate exposed surfaces, sagebrush, and recent burns.

140

The vast Greater Yellowstone Area has many remote wilderness areas; mapping the entire region from the ground is difficult. Yet landscape maps are crucial to monitoring, understanding, and managing resources in the region. Remotely sensed imagery from satellites and airplanes provides the information needed to create these maps.

Starting in the 1950s, the U.S. government acquired aerial photographs every few years throughout the Yellowstone region. The frequency of coverage improved dramatically in July 1972 with the launch of the first Landsat satellite. Its images, though coarser than aerial photos, covered the entire region on a monthly basis or more often, though clouds sometimes interfered. Satellites now acquire high-resolution imagery several times a month as well as less-detailed data multiple times a day.

The first systematic coverage in the 1950s used black-and-white aerial photos, followed by true color photos in the late 1960s and color infrared imagery in the late-1980s. In addition to recording visible light, color infrared imagery records shortwave light—invisible to the human eye—that highlights variations in vegetation, moisture, and other landscapes features. Landsat and many later satellites also collected data in the visible and infrared range. The region-wide extent of the satellite imagery enabled landscape mapping and monitoring of change across the entire Greater Yellowstone Area. Space-borne instruments now include thermal recorders that map surface temperature and radar to map topography.

The regional image mosaic to the left is developed from Landsat data. The imagery records light reflected from the Earth in the green, red, and near-infrared wavelengths. Subtle differences are captured, such as variations in crop cover in the agricultural fields of the Snake River Plain in the left center of the image. The image colors can be matched with ground data to generate regional maps (the technique used to create the land cover maps and vegetation maps in this atlas).

Repeat photography or satellite imagery is particularly valuable for documenting locations and rates of landscape change. Change detection images provide information about the forces driving landscape change and guide management efforts.

**Beetle Kill** 2002

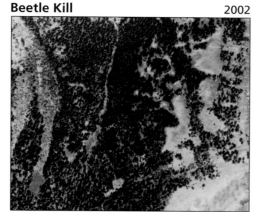

Combining visible and near-infrared light helps map forest health. The colors on this false color image are altered to show healthy trees in green.

2009

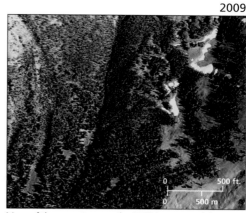

Many of the green trees on the 2002 image are pink or red in 2009, indicating trees that have been killed by pine beetles.

**Logging** 1979

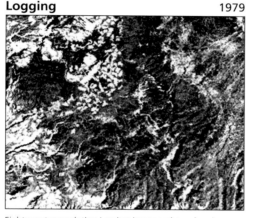

Eighty-meter resolution Landsat imagery shows logging (patchy white) to the northwest of the north-south Targhee National Forest–Yellowstone boundary.

1997

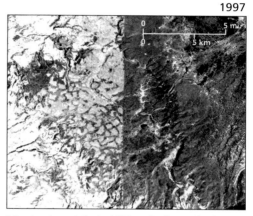

Extensive clear cuts in the 1980s on Targhee National Forest delineate the national park boundary on this thirty-meter, near-infrared Landsat imagery.

**Energy** 1994

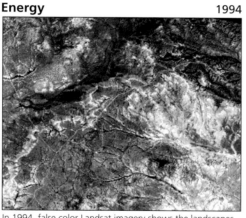

In 1994, false color Landsat imagery shows the landscapes of Sublette County, Wyoming, as largely undisturbed sagebrush flats with limited development.

2007

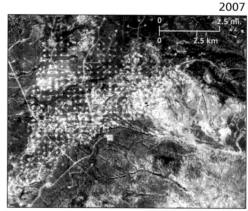

The natural gas boom of the early 2000s led to extensive road construction and drilling that fragmented the sagebrush-steppe habitat.

**Fire** Prefire, 1987

Landsat false color imagery highlights Shoshone Lake (black), Upper to Lower geyser basins (white), and extensive forest cover (reds and greens) in 1987.

During Fire, 1988

Satellite imagery clearly displays burned areas (green hues) and penetrates smoke (bluish-white) to show underlying land cover.

Postfire, 1989

Satellite imagery enabled mapping of areas burned in 1988, making it a valuable tool for understanding and responding to the effects of fire.

# Fire History

Fire routinely alters Yellowstone's landscape and is important to the area's ecology. While early park caretakers emphasized fire suppression, firefighting effectiveness was limited until after World War II, when better equipment and aircraft became available. In 1972, park administrators enacted a limited suppression policy to promote a more natural function of fire in Yellowstone. Some people incorrectly attributed the extensive 1988 fires to this policy; in fact, park managers attempted to suppress all fires after 21,000 acres had burned, but exceptionally dry and windy conditions caused fires to grow, consuming a total of 1.1 million acres. Firefighters have applied suppression techniques to most of the larger post-1988 fires, though dry conditions limited the effectiveness of these efforts. Climate, rather than suppression policy, is the greatest determinant of area burned.

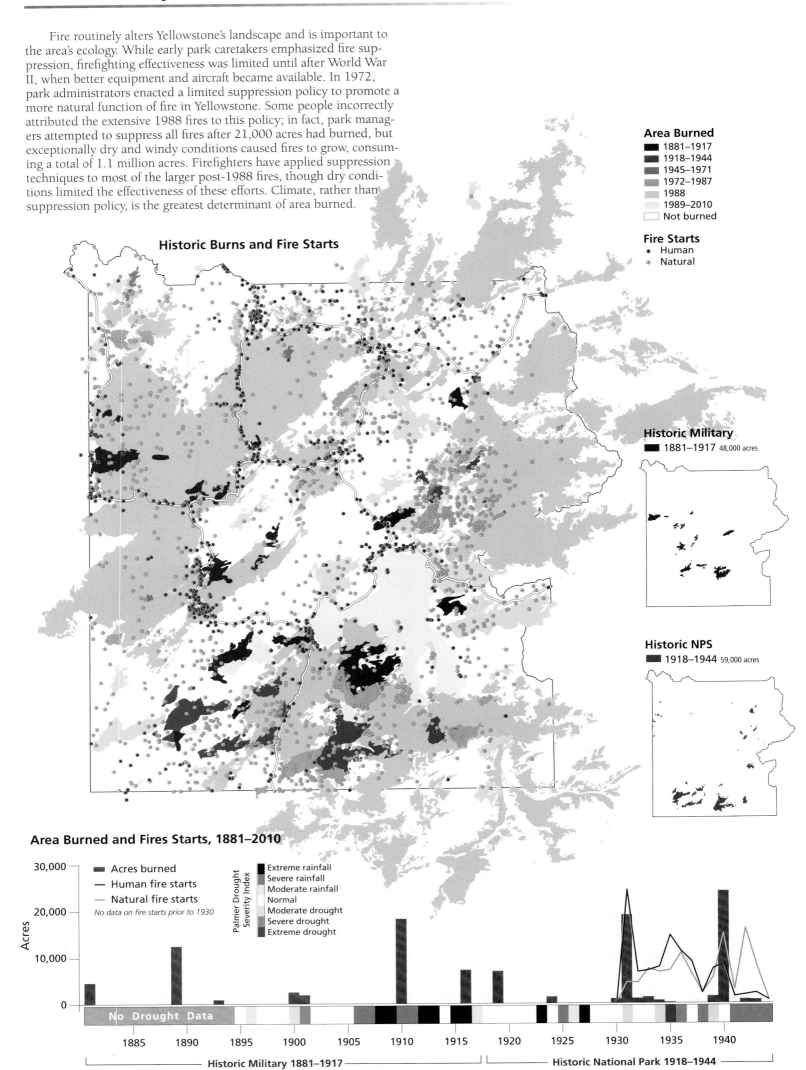

**Area Burned**
- 1881–1917
- 1918–1944
- 1945–1971
- 1972–1987
- 1988
- 1989–2010
- Not burned

**Fire Starts**
- Human
- Natural

**Historic Burns and Fire Starts**

**Historic Military**
1881–1917 48,000 acres

**Historic NPS**
1918–1944 59,000 acres

**Area Burned and Fires Starts, 1881–2010**

- Acres burned
- Human fire starts
- Natural fire starts
- *No data on fire starts prior to 1930*

**Palmer Drought Severity Index**
- Extreme rainfall
- Severe rainfall
- Moderate rainfall
- Normal
- Moderate drought
- Severe drought
- Extreme drought

Acres: 30,000 / 20,000 / 10,000 / 0

No Drought Data

1885  1890  1895  1900  1905  1910  1915  1920  1925  1930  1935  1940

Historic Military 1881–1917 — Historic National Park 1918–1944

Fires have long been part of the Yellowstone landscape. A petrified cone from an Eocene fossil forest shows that ancestors of the fire-adapted lodgepole pine were present more than 34 million years ago. Abundant pine pollen in lake sediments indicates the dominance of fire-adapted trees that flourished between glacial pulses in Pleistocene forests (1.8 million to 10,000 years ago) and over the majority of the Holocene epoch (the most recent 10,000 years).

The magnitude and frequency of wildfires are dependent on a complex interaction of ignition sources, fuels, topography, weather, and climate. A fire start requires dry fuels and an ignition source. Once ignited, a fire will spread only with sufficient dry fuel downwind. Finally, a fire will grow large only if there are strong, persistent winds and negligible rain or snow. Occasionally, the many factors necessary for fire expansion occur at the same time, but 80 percent of fires remain less than one acre in size, whether suppressed or allowed to burn as a natural fire.

Fires and fire starts in Yellowstone have distinct seasonal and spatial patterns. Spring snowmelt soaks dead fuels and saturates the soil. Lush new leaves, high in water content, begin to grow and, along with changing weather conditions, diminish soil moisture. By mid- to late-summer, the dead fuels have dried and the once-lush leaves are now dry enough to burn, creating the conditions for the onset of fire season. Lightning strikes and park visitation peak in the hot dry months of July and August, providing ample opportunity for ignitions. Lightning ignitions occur throughout the landscape, while human-caused ignitions usually start near roads. Lightning is often accompanied by rain, but coals from, for example, a lightning-struck tree can smolder for days or even months until drier conditions allow the fire to reignite. Cooling temperatures and precipitation throughout the region in mid- to late-September generally signal the end of fire season.

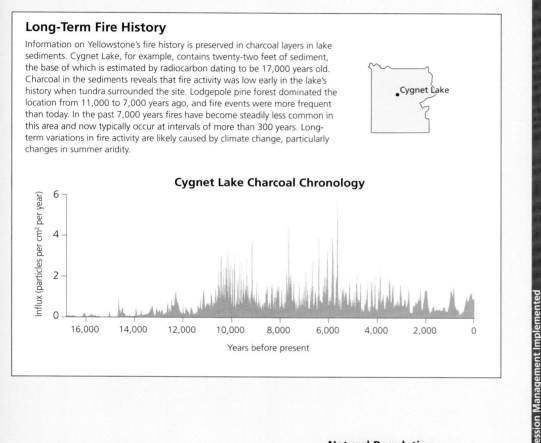

## Long-Term Fire History

Information on Yellowstone's fire history is preserved in charcoal layers in lake sediments. Cygnet Lake, for example, contains twenty-two feet of sediment, the base of which is estimated by radiocarbon dating to be 17,000 years old. Charcoal in the sediments reveals that fire activity was low early in the lake's history when tundra surrounded the site. Lodgepole pine forest dominated the location from 11,000 to 7,000 years ago, and fire events were more frequent than today. In the past 7,000 years fires have become steadily less common in this area and now typically occur at intervals of more than 300 years. Long-term variations in fire activity are likely caused by climate change, particularly changes in summer aridity.

Cygnet Lake

### Cygnet Lake Charcoal Chronology

Influx (particles per cm² per year)

Years before present

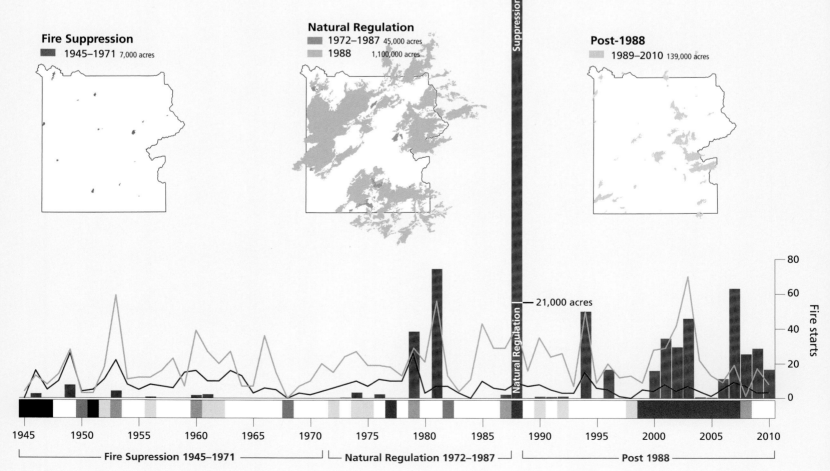

1,100,000
(includes acreage burned outside YNP boundary)

Suppression Management Implemented

**Fire Suppression**
■ 1945–1971  7,000 acres

**Natural Regulation**
■ 1972–1987  45,000 acres
■ 1988  1,100,000 acres

**Post-1988**
■ 1989–2010  139,000 acres

Natural Regulation

21,000 acres

Fire starts

1945  1950  1955  1960  1965  1970  1975  1980  1985  1990  1995  2000  2005  2010

├── Fire Supression 1945–1971 ──┤ └ Natural Regulation 1972–1987 ┘ └── Post 1988 ──┤

143

# 1988 Fires

Yellowstone's 1988 fires burned far more park area than had burned in all other years combined since the park was established in 1872. Weather was the primary factor contributing to the severe fire conditions. Above-average precipitation in April and May promoted vigorous growth of ground vegetation. Low precipitation and record high temperatures in June, July, and August dried out forests and ground cover. High winds in August and September fanned explosive growth of fires in the dry fuels. Despite extensive firefighting efforts, what brought the 1988 fire season to a close was low temperatures and precipitation in September.

## 1988 Fire Perimeters by Month

■ June 15–July 14    ■ July 15–Aug 15    ■ Aug 16–Sept 15
                                          Sept 16–Oct 1

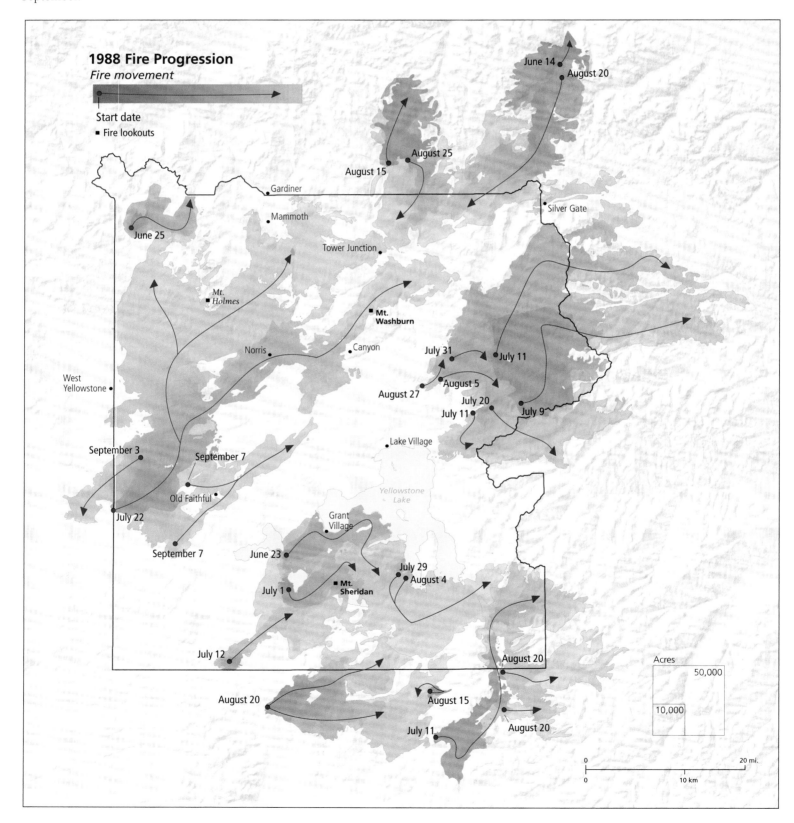

**1988 Fire Progression**
*Fire movement*

Start date
■ Fire lookouts

## Weather during the 1988 Fire Season
*Data from Mt. Holmes (~10,300 ft.)*

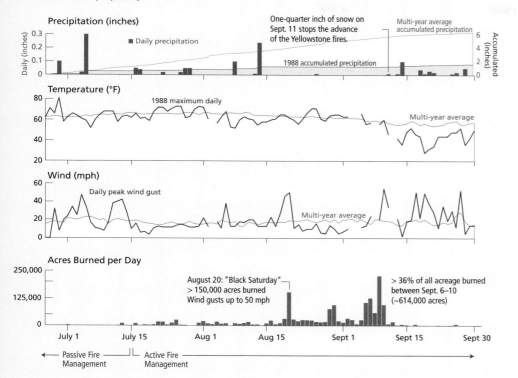

**Precipitation (inches)**

■ Daily precipitation

One-quarter inch of snow on Sept. 11 stops the advance of the Yellowstone fires.

Multi-year average accumulated precipitation

1988 accumulated precipitation

**Temperature (°F)**

1988 maximum daily

Multi-year average

**Wind (mph)**

Daily peak wind gust

Multi-year average

**Acres Burned per Day**

August 20: "Black Saturday"
> 150,000 acres burned
Wind gusts up to 50 mph

> 36% of all acreage burned between Sept. 6–10 (~614,000 acres)

July 1 — July 15 — Aug 1 — Aug 15 — Sept 1 — Sept 15 — Sept 30

←— Passive Fire Management —→ ←— Active Fire Management —→

The 1988 fire season began normally. As of June 23, only three acres had burned within the park, although the lightning-caused Storm Creek fire was burning eighteen miles northeast of the park boundary.

In the final ten days of June, however, record-breaking temperatures in the 90-degree range at low elevations and in the 80-degree range at high elevations created conditions for rapid fire growth. As of July 1, the Fan fire and three fires that would later merge to form the Snake River Complex had burned 1,100 acres within the park. By July 15, twenty-two fires had occurred within the park and eleven were actively burning a total of 8,600 acres. The active fires included the Falls fire, which would later merge into the Snake River Complex, and the Clover and Mist fires in the northeast. Three of the twenty-two park fires were human caused and suppressed; eight fires self-extinguished. Outside the park, the Mink fire was burning on 9,100 acres of the Bridger-Teton National Forest.

Until July 15, natural-ignition fires burned without human interference in accord with a management strategy called "wildland fire use for resource benefit." Such fires, however, were not ignored. Fire management personnel used active observation rather than active suppression; maps of daily fire growth, hourly weather observations, and fuel and forest data helped analysts continuously monitor and assess fire behavior. Persistent extreme fire conditions prompted park managers on July 15 to start suppressing new fires and attempting to control existing ones, with emphasis on protecting structures and human safety.

Despite these efforts, the fires grew. Long-duration, high-intensity crown fire was intermixed with lower-intensity ground burning. On July 22, the human-caused North Fork fire started outside the western park boundary and rapidly spread into the park, bringing the total park area burned to 18,000 acres. By August 1, thirty-six fires (four human-caused) had ignited within the park. Many smaller fires merged with larger fires, forming large complexes that covered 122,000 acres. Continued below-average precipitation through July and August, abnormally high winds, and large active fire fronts contributed to extreme fire behavior. On one day alone, August 20, known locally as "Black Saturday," high winds caused more than 150,000 acres to burn—more park area than had burned in total over the previous century. Two additional human-caused fires (Huck and Hell Roaring) originated outside Yellowstone and burned into the park. Conditions remained extreme in early September. Between September 6 and 10, 36 percent of all the acreage burned within the park that season went up in flames.

On September 11, after burning 1.1 million acres, the conflagration halted as plummeting temperatures and one-quarter inch of snow accomplished what nearly two months of vigorous firefighting could not.

## Major Fires

Hell Roaring · Storm Creek · Fan · North Fork · Clover Mist · Snake River Complex · Mink · Huck

## Fire Intensity

The large fires resulted in a pattern of unburned vegetation thoroughly intermixed with patches of canopy burn and mixed burn (a combination of canopy and ground burn). By one estimate, approximately 10 percent of the area within the fire perimeters was actually unburned.

■ **Canopy Burn:** Forest burned; bare, black branches remain; soil charred

■ **Mixed Burn:** Forest burned, some trees remain green and alive; some soil charring

■ **Nonforested Burn:** Grassland and sagebrush areas burned; some soil charring

□ **Unburned:** No burns detected at this mapping resolution

### Firefighting Resources

| | | |
|---|---|---|
| **Firefighters** | >25,000 | 11,700 military personnel and 9,600 firefighters at one time |
| **Fire retardant** | 1.4 million gallons | Dropped by fixed-wing aircraft |
| **Water** | >10 million gallons | Carried by helicopters in canvas buckets or slings |
| **Firebreaks** | 802 miles | 665 miles manually dug 137 miles bulldozed |
| **Transportation** | >100 fire engines >100 aircraft | Included 77 helicopters and 150 newly created helispots |
| **Logistical support** | >$120 million | $33 million were direct payments for services including gasoline, meals, lodging, rental items, and wages for nongovernment help. Most expenditures made in greater Yellowstone communities. |
| **Flight hours logged** | 18,000 | |

# Grizzly Bears

## Grizzly Bear Range

North American grizzly bears once roamed from northern Alaska to northern Mexico and from the Pacific Coast to western Missouri. Bear numbers and range decreased dramatically following European American exploration and settlement west of the Mississippi River. Ranches, farms, and cities encroached on grizzly habitat; people shot, trapped, and poisoned bears to protect cattle, sheep, and poultry from attack. Important bear foods such as salmon, bison, and elk grew ever harder to find as humans built dams, hunted, and introduced livestock. By 1975, grizzly bears had been extirpated from Mexico and all but 2 percent of their historic range in the lower forty-eight states. Bears remain abundant in Alaska and northern Canada, their populations stable except in a few areas of rapid human settlement and development. Farther south, however, habitat fragmentation, alteration, and destruction threaten the bears. The Greater Yellowstone Ecosystem is the southernmost grizzly range as of 2007 and plays a key role in maintaining grizzly bear populations and genetic diversity.

### Historic and Current Range

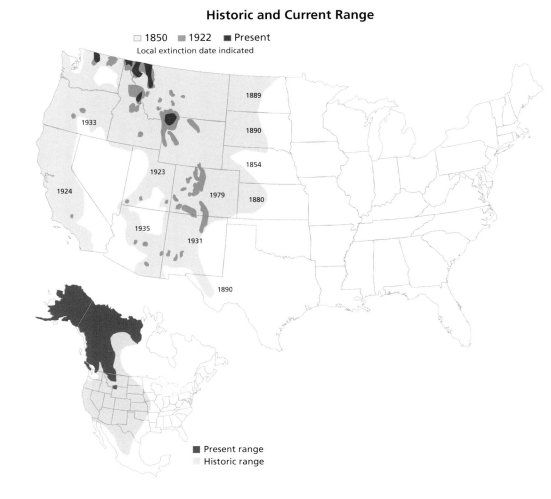

□ 1850   ■ 1922   ■ Present
Local extinction date indicated

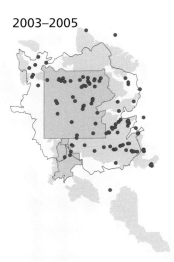

■ Present range
▢ Historic range

## Female Grizzly Bears with Cubs

▢ National Forest   • Sighting
▢ Wilderness area   — Bear recovery zone
▢ National Park

**1979–1981**

**1999–2001**

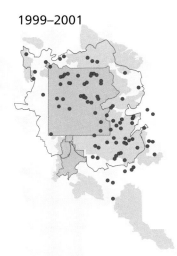

**2003–2005**

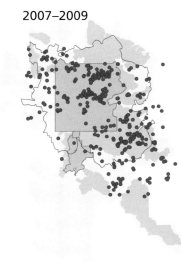

**2007–2009**

## Endangered Species Protection

Yellowstone area grizzly bears received federal protection as a threatened species in 1975. State and federal wildlife and habitat management agencies have worked to reduce both conflicts between bears and people and human-caused bear deaths as well as to increase cub production and survival. The area occupied by the bears has grown from approximately 15,000 square kilometers in the 1970s to 17,000 (1980s) to 34,000 (1990s) to 37,000 (2005). Annual counts of females with cubs are used to estimate grizzly bear population numbers and trends. Both the number of females producing cubs and the total number of cubs produced annually has increased since the mid-1980s. Due to the significant increase in bear numbers and range since then, the U.S. Fish and Wildlife Service removed the bears from threatened status in 2007. The matter is not settled, however; in response to a lawsuit filed by bear advocacy groups, a federal judge ordered grizzly bears returned to threatened status in 2009—a ruling currently under appeal by the U.S. Fish and Wildlife Service.

### Sow and Cub Population 1973–2009

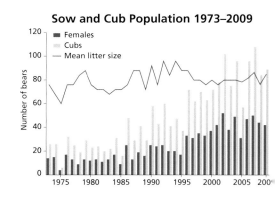

■ Females
▢ Cubs
— Mean litter size

148

## Seasonal Habitat and Food

Grizzly bears are omnivores, eating a wide variety of plants, animals, and insects. In the Greater Yellowstone Ecosystem, grizzly bears prefer concentrated high-energy foods such as the carcasses of large mammals, elk calves, spawning cutthroat trout, army cutworm moths, whitebark pine seeds, and clover. Spending approximately five months without food while hibernating in winter dens, the bears must ingest a year's food supply in just seven months. Preferred bear foods are seasonal in nature and fluctuate in abundance from year to year. As a consequence, grizzly bears require large home ranges to ensure they can meet their energy needs. Prime spring habitat includes elk and bison wintering areas where bears scavenge winterkilled animals, and areas of early plant growth where bears graze succulent vegetation. In summer, bears favor elk calving areas, streams with spawning cutthroat trout, and high elevation talus slopes where they eat large quantities of army cutworm moths. Preferred fall habitat contains cone-producing whitebark pine trees. Grizzly bears obtain the fat- and protein-rich whitebark pine seeds by raiding red squirrel caches. In years of low pine nut production, grizzlies forage more extensively on biscuit roots and yampa and eat more meat. Grizzly bears scavenge wolf-killed elk and bison during the spring, summer, and fall.

## Food Availability

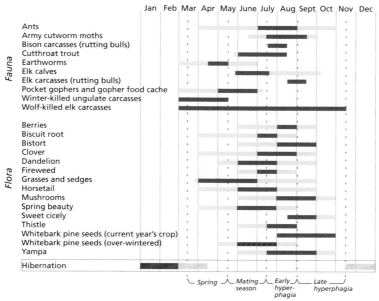

Peak food availability

Annual variability

Only available following a mast year

Yellowstone area grizzly bears eat a breadth of foods: insects, vertebrates, fungi, seeds. Unlike many other members of the order Carnivora, bears eat a significant amount and variety of vegetation. Bears have several adaptations for eating plants, including large chewing surfaces on molars and long claws for digging. Grizzlies are shrewd feeders, maximizing the quality of their food intake by foraging for many kinds of plants and seeking each plant when it is most nutritious and digestible.

Hyperphagia is the period of intensive foraging and eating preceeding hibernation.

## Seasonal Habitat Preference

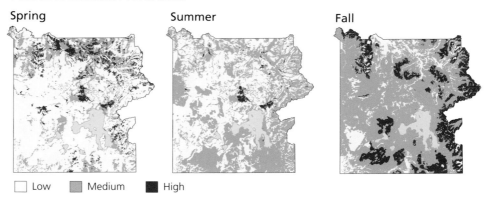

Spring          Summer          Fall

☐ Low    ▨ Medium    ■ High

## Mortality

Grizzly bears in the Greater Yellowstone Ecosystem die from many causes including old age, starvation, drowning, avalanche, and den collapse as well as when killed by other bears, wolves, or humans. A larger proportion of dependent young bears (cubs and yearlings) than adult bears die from natural causes. Adult grizzly bear deaths—85 percent of which are caused by humans—result mostly from management removal of bears involved in conflicts with people, defense of life or property by private citizens, legal hunting, mistaken identification by black bear hunters, poaching, vehicle strikes, and electrocution by downed power lines. Bears come into conflict with people more often during years with poor availability of their preferred foods, especially fall foods; conversely, fewer conflicts and human-caused bear deaths tend to occur in years when food is abundant. The proportion of bear deaths due to natural causes as compared to human causes is generally higher within national parks, whereas the opposite ratio is generally found outside park boundaries.

## Deaths, 1975–2005

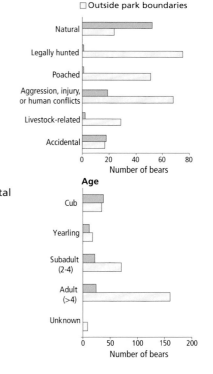

Natural Causes          Hunting

• Legally hunted
• Poached

Aggression or Injury          Livestock-Related and Accidental

• Livestock-related
• Accidental

## Causes of Mortality

▨ Inside park boundaries
☐ Outside park boundaries

Natural

Legally hunted

Poached

Aggression, injury, or human conflicts

Livestock-related

Accidental

Number of bears
0    20    40    60    80

**Age**

Cub

Yearling

Subadult (2-4)

Adult (>4)

Unknown

Number of bears
0    50    100    150    200

# Wolves

## Historic Range

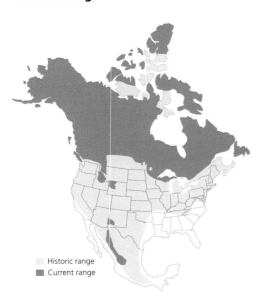

Historic range
Current range

Gray wolves (*Canis lupus*) roamed Yellowstone at the time the park was established in 1872. Following this time, humans systematically eliminated wolves from the park, killing the last remaining individual in 1926.

Through 1995 there were occasional reports of wolves—most not verified—moving through the Yellowstone area, but there was no established wolf population. This condition met a key criterion for the reintroduction of wolves under the Endangered Species Act. Reintroduction began in the mid-nineties with forty-one animals: thirty-one from Canada (Alberta in 1995 and British Columbia in 1996) and ten from northwest Montana (1997). Biologists acclimated the animals to local conditions in one-acre pens for ten weeks, then released them in selected locations.

The wolves thrived in the park, some of the best wolf habitat in the world. Abundant elk, deer, and bison provided food and supported population growth of roughly 17 percent per year from 1995 to 2005. Initially packs clashed as they established territorial boundaries, but within months the wolves settled into fairly well-defined ranges, mostly in the park's northern reaches. Wolves soon expanded their territories throughout the park, reaching relatively high densities for the species. Wolf packs now inhabit most areas where year-round prey is available. Fiercely territorial, wolves fight to protect or expand their range. Squabbles end up in boundary shifts, displacements, and often death, which may serve as an important population control mechanism in areas with a high density of wolves.

Yellowstone's wolf population peaked in 2003 at 174 wolves in fourteen packs. Half of these animals lived on the park's northern range (only 11 percent of the park's total area) making wolf density comparatively high in the north. The population crashed in 2005, declining 30 percent because of a disease outbreak. This followed a similar, though less severe, decline in 1999, also associated with disease. Researchers are still investigating the cause of these deaths, but canine distemper is a primary suspect (canine parvovirus and canine adenovirus may also have contributed). From 2005 to 2007, the population rebounded up to 171 wolves in eleven packs. In 2008, however, another disease outbreak caused a steep decline in wolf-pup survival rate.

Because wildlife laws shield Yellowstone National Park wolves from human hunting (the primary cause of wolf deaths outside of protected areas), packs inside the park tend to be larger and have wolves of higher average age. Wolves in the park die at an average age of four to five years, though many are living longer—a rare individual may live a decade. Average pack size in 2007 was fourteen wolves and many of the animals were relatively old. The effect these older wolves have on the pack is not yet known. It is possible they help defend territory and kill prey, especially bison, which can weigh twenty times the average 100-pound wolf.

The restoration process is drawing to a close with population stabilization. During the next phase, population growth is expected to slow or even turn negative. Once that occurs, wolf numbers should stabilize at a lower level as the animals find equilibrium with each other and their prey—unless, of course, disease or some major ecological disturbance interferes.

## Yellowstone Wolf Reintroduction, 1995

Rose Creek Pen (3 wolves)

Soda Butte Pen (5 wolves)

Crystal Creek Pen (6 wolves)

### Wolves Released in Yellowstone, 1995

| Pen | Sex | Color | Age | Wt. (lbs) |
|---|---|---|---|---|
| Crystal Creek Pen | | | | |
| | M | Black and silver | Pup | 77 |
| | M | Black | Pup | 80 |
| | M | Black | Adult | 98 |
| | F | Light gray | Adult | 98 |
| | M | Black | Pup | 75 |
| | M | Gray | Pup | 72 |
| Rose Creek Pen | | | | |
| | F | Red gray | Pup | 77 |
| | F | Black | Adult | 98 |
| | M | Light gray | Adult | 122 |
| Soda Butte Pen | | | | |
| | F | Gray | Adult | 92 |
| | M | Black | Adult | 122 |
| | M | Gray and black | Old adult | 113 |
| | F | Gray | Adult | 89 |
| | M | Black and silver | Pup | 75 |

○ Wolf Location

Wildlife biologists reintroduced wolves to Yellowstone from Canada in 1995 and 1996. They carefully selected release sites for a combination of accessibility (for easy feeding and protection by rangers) and remoteness (away from stressing road noises). Suitable locations also needed a water source as well as vegetative cover for comfort and shade. Biologists released most of the animals in Yellowstone's northern range, prime wolf habitat with year-round road access and an abundance of elk.

## Wolf Population, 1870–1970

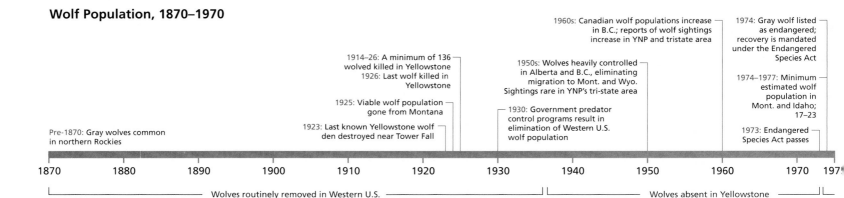

Pre-1870: Gray wolves common in northern Rockies

1914–26: A minimum of 136 wolved killed in Yellowstone
1926: Last wolf killed in Yellowstone

1925: Viable wolf population gone from Montana

1923: Last known Yellowstone wolf den destroyed near Tower Fall

1950s: Wolves heavily controlled in Alberta and B.C., eliminating migration to Mont. and Wyo. Sightings rare in YNP's tri-state area

1930: Government predator control programs result in elimination of Western U.S. wolf population

1960s: Canadian wolf populations increase in B.C.; reports of wolf sightings increase in YNP and tristate area

1974: Gray wolf listed as endangered; recovery is mandated under the Endangered Species Act

1974–1977: Minimum estimated wolf population in Mont. and Idaho; 17–23

1973: Endangered Species Act passes

1870   1880   1890   1900   1910   1920   1930   1940   1950   1960   1970   197

Wolves routinely removed in Western U.S.

Wolves absent in Yellowstone

# Yellowstone Wolf Population, 2006

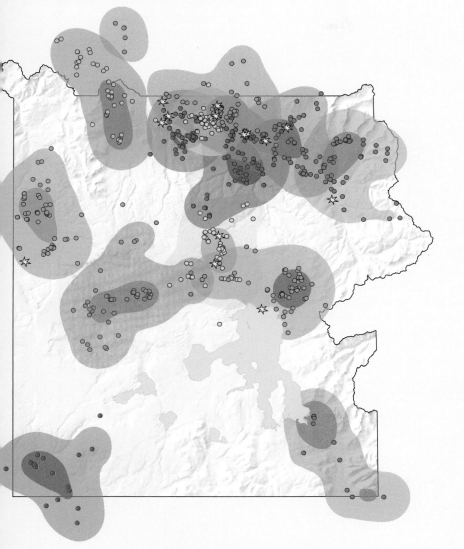

| Pack | Adults | Pups |
|---|---|---|
| Agate Creek | 7 | 6 |
| Druid Peak | 3 | 9 |
| Hellroaring | 5 | 1 |
| Leopold | 7 | 12 |
| Oxbow Creek | 4 | 8 |
| Slough Creek | 8 | 0 |
| Swan Lake | 2 | 3 |
| Bechler | 8 | 5 |
| Cougar Creek | 4 | 0 |
| Gibbon Meadows | 8 | 4 |
| Hayden Valley | 3 | 2 |
| Mollie's* | 6 | 5 |
| Yellowstone Delta | 11 | 5 |

**Total wolves in 2006** — 76 adults, 60 pups

136 wolves

Core range
Periphery range

☆ Wolf deaths as a result of territory disputes between wolves of different packs

○ Wolf observation

*Each point represent an alpha wolf location tracked by radio collar telemetry; points are not exclusive and may represent the same wolf more than once*
*\*Ages of Mollie's pack unconfirmed*

## Gray Wolf Population, 1995–2009

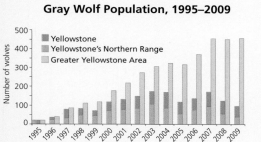

Number of wolves

- Yellowstone
- Yellowstone's Northern Range
- Greater Yellowstone Area

## Yellowstone Wolf Mortality, 1995–2009
*(collared wolves originating in Yellowstone)*

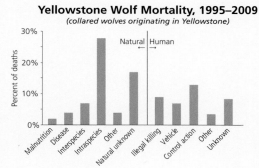

Percent of deaths

Natural ← | → Human

Malnutrition, Disease, Interspecies, Intraspecies, Other, Natural unknown, Illegal killing, Vehicle, Control action, Other, Unknown

## Yellowstone Pup Survival, 1995–2009
*(collared wolves originating in Yellowstone)*

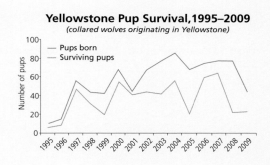

Number of pups

- Pups born
- Surviving pups

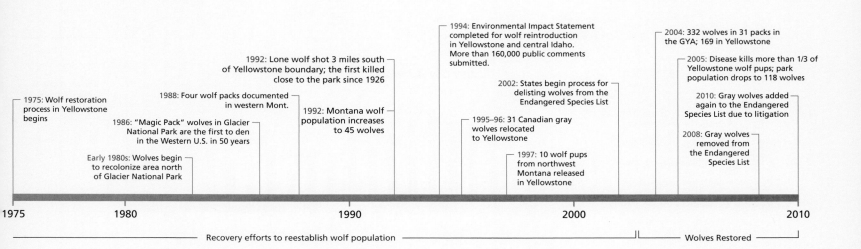

1975: Wolf restoration process in Yellowstone begins

1986: "Magic Pack" wolves in Glacier National Park are the first to den in the Western U.S. in 50 years

Early 1980s: Wolves begin to recolonize area north of Glacier National Park

1988: Four wolf packs documented in western Mont.

1992: Lone wolf shot 3 miles south of Yellowstone boundary; the first killed close to the park since 1926

1992: Montana wolf population increases to 45 wolves

1994: Environmental Impact Statement completed for wolf reintroduction in Yellowstone and central Idaho. More than 160,000 public comments submitted.

2002: States begin process for delisting wolves from the Endangered Species List

1995–96: 31 Canadian gray wolves relocated to Yellowstone

1997: 10 wolf pups from northwest Montana released in Yellowstone

2004: 332 wolves in 31 packs in the GYA; 169 in Yellowstone

2005: Disease kills more than 1/3 of Yellowstone wolf pups; park population drops to 118 wolves

2010: Gray wolves added again to the Endangered Species List due to litigation

2008: Gray wolves removed from the Endangered Species List

1975 — 1980 — 1990 — 2000 — 2010

Recovery efforts to reestablish wolf population | Wolves Restored

# Coyotes

## Coyote Distribution

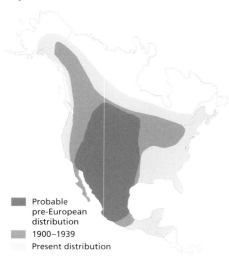

Probable
pre-European
distribution
1900–1939
Present distribution

The coyote (*Canis latrans*), a North American midsized carnivore, occupies an ultrageneralist niche with respect to resources, occurring in habitats as varied as deserts, plains, and mountains, as well as in urban and suburban settings. As humans have eradicated large carnivores throughout their historic ranges, coyotes have filled in—as predators and scavengers—and are now the apex predator in approximately 45 percent of the continental United States.

Predators play critical roles in stabilizing ecosystems, and throughout North America coyotes provide an essential component of system stability. It is coyotes' storied adaptability—switching from one food source to another—that equalizes predation pressures and dampens population oscillations in other species, from small mammals to ungulates. Ecologists are beginning to understand that ecosystems without predators are more vulnerable to destabilization and degradation, and that both generalist and specialist predators are important in stabilizing these systems.

Because coyotes have been historically treated as "varmints" (and hunted, trapped, snared, poisoned, and bountied with unregulated take across North America), their ecology in this vast region has been generally understood through the lens of a disrupted set of adaptations to intense human exploitation. Yellowstone, in contrast, represents a fully protected, intact, and natural ecosystem, providing rare and valuable insights into coyote social ecology. With a clear understanding of integrated ecological processes—for example, predator-prey dynamics—in an essentially undisturbed landscape, managers can more effectively oversee the species involved.

In Yellowstone, coyotes and wolves coexisted until the early 1920s, when gray wolves were extirpated as part of a predator management program. After wolves became locally extinct, coyote numbers increased, and coyotes partially filled the vacant ecological niche. After lengthy and heated debate,

## Prewolf Period, 1989–1995
*Pack extents in 1995*

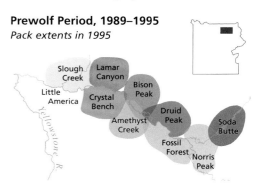

## Wolf Colonization Period, 1996–2002  *Coyote pack extents from 1998 to 2000 shown*

1998

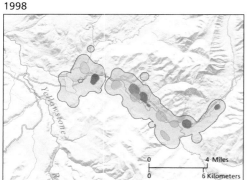

1999

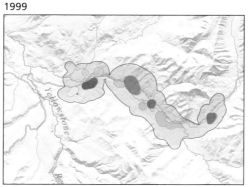

2000

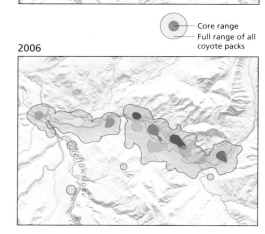

Core range
Full range of all
coyote packs

## Wolf Saturation Period, 2003–2007 *Coyote pack extents from 2004 to 2006 shown*

2004

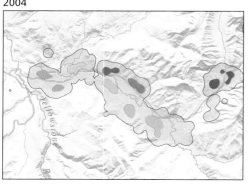

2005

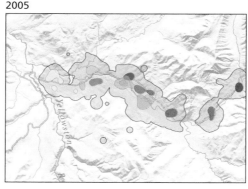

2006

## Coyote Research, 1937–2009

| Year | Study |
|---|---|
| 1937–1940 | A. Murie (*Ecology of the Coyote in the Yellowstone*) |
| 1945–1950 | Robinson and Cummings |
| 1951 | Craighead (Grand Teton) |
| 1970–1976 | Bekoff (Grand Teton) |
| 1970–1976 | Camenzind (Grand Teton) |
| 2009 | Twenty-year Northern Range study ends |

## Lamar Valley Coyote Population, 1990–2009

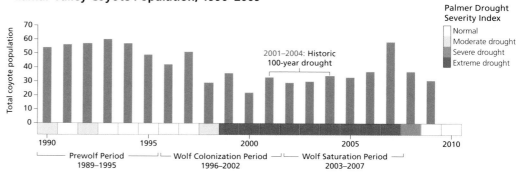

Palmer Drought
Severity Index

Normal
Moderate drought
Severe drought
Extreme drought

2001–2004: Historic
100-year drought

Prewolf Period
1989–1995
Wolf Colonization Period
1996–2002
Wolf Saturation Period
2003–2007

restoration of the gray wolf to Yellowstone occurred in 1995. In 1995–96, thirty-one wolves were transported from Canada and released in the park where they established a robust population. Wolf restoration provided an opportunity to evaluate the effects of this apex predator on its relative and competitor, the coyote. Prior to wolf restoration, a long-term ecological study of Yellowstone coyotes had established a clear understanding of coyote social structure, food habits, and general ecology. In the years from 1996 to 2002, as the colonizing wolf population rapidly grew, coyotes accommodated to the change. After 2003, the wolf population no longer increased, settling in as a reestablished component of the system. From a coyote perspective, wolf restoration can be characterized in three general phases: prerestoration (wolves absent); wolf colonization (rapid increase); and wolf saturation period (wolf population no longer increasing).

Wolves are dominant lethal competitors with coyotes, accounting for 40 percent of the adult coyote mortality in Yellowstone. Since wolf restoration, coyote numbers in Yellowstone have dropped; coyotes are largely absent from areas of highest wolf density, such as around wolf dens during summer. Yet the remaining coyotes also benefit from wolves, finding increased scavenging opportunities at wolf-killed ungulate carcasses. Coyotes also use areas of greater human concentration (trails, roads, structures) for refuges from wolves, which tend to avoid humans.

Coyotes are highly social. In Yellowstone, they occur in packs of as many as thirteen individuals, typically a mated pair and offspring from previous years. These large social groups sometimes hunt and travel together but often forage individually, regrouping daily for social interaction and howling sessions. This aspect of coyote ecology had not been well understood in studies from other regions where human exploitation disrupted coyote social structure. Interestingly, coyotes on the northern range of Yellowstone were the subject of one of the world's first truly ecological studies of predators, that of naturalist Adolf Murie. His study, published in 1940, presented a presciently modern overview of the role of predators serving an essential ecological function.

Across the continent, coyote range expansion in the past 120 years has been concurrent with broad wolf population reductions. It has also coincided with large-scale habitat alteration, loss of migratory ungulate populations, and expansion of human land-use practices such as agriculture that favor smaller generalist predators. In areas where wolf populations have been reduced or extirpated, coyote numbers typically increase.

Coyotes are avatars of adaptability, the ultimate generalist and survivor among the carnivores. From an ecological perspective, coyotes also serve as barometers of ecosystem disruption; as a rule, the more social, long-lived, and stable coyote populations are, the healthier their ecosystem. Take away their prey base, den sites, travel routes, and home ranges, and they will persist, though with adapted habits and social interactions. Coyotes forsake their highly social pack structure (typical of pristine settings where they are not killed by humans) for the transient, highly mobile, extremely opportunistic behaviors typical of the animal across most of North America.

## Lamar Valley Coyote Pack Extents, 2007

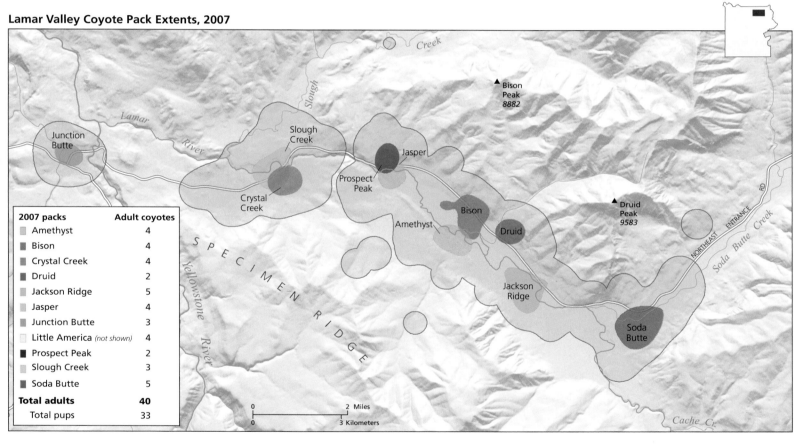

| 2007 packs | Adult coyotes |
|---|---|
| Amethyst | 4 |
| Bison | 4 |
| Crystal Creek | 4 |
| Druid | 2 |
| Jackson Ridge | 5 |
| Jasper | 4 |
| Junction Butte | 3 |
| Little America (not shown) | 4 |
| Prospect Peak | 2 |
| Slough Creek | 3 |
| Soda Butte | 5 |
| **Total adults** | **40** |
| Total pups | 33 |

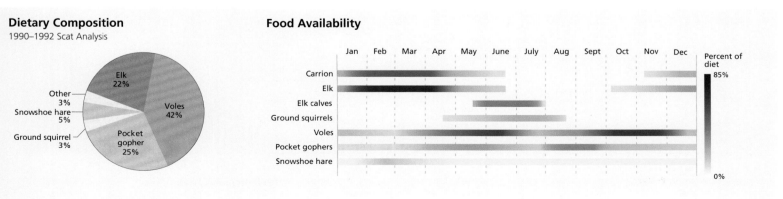

## Dietary Composition
1990–1992 Scat Analysis

Elk 22%
Voles 42%
Pocket gopher 25%
Ground squirrel 3%
Snowshoe hare 5%
Other 3%

## Food Availability

Percent of diet
85%
0%

# Bison

The ancestors of modern American bison migrated from Siberia to North America across the Bering Strait "land bridge" about 10,000 years ago. The continent was hospitable to the animals; their numbers grew to the tens of millions. During the westward settlement of North America in the mid-eighteenth century, these vast herds were decimated by harvest and slaughter. The bison population declined to mere thousands by the time Yellowstone National Park was founded in 1872. The devastation was so extensive that concerned individuals around the country began efforts to save the species.

Yellowstone was the first preserve created by the federal government where American bison were protected from extinction. However, early park managers had little legal authority with which to safeguard the animals. The legislation that established the park allowed hunting for subsistence and sport and bison numbers continued to decline for thirty years. When the Lacey Act of 1900 granted Yellowstone officials the authority to punish those who illegally killed wildlife in the park, the Yellowstone bison population had dwindled to fewer than two dozen, all concentrated in Pelican Valley. In 1902 Congress appropriated funds for Yellowstone to develop a bison restoration project. It began with the acquisition of twenty-one animals from the Goodnight (Texas) and Walking Coyote (Montana) captive populations, initially kept in an enclosure in Mammoth. By 1907 the Buffalo Ranch in Lamar Valley was constructed to manage these bison and increase their numbers, which by the 1930s had grown

## Bison Range 1500–1880

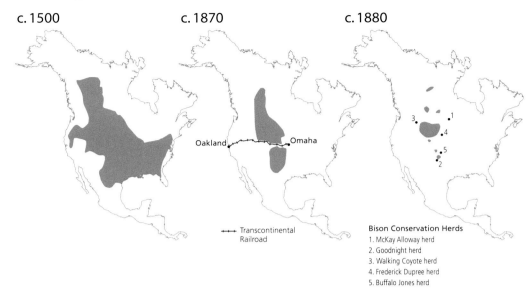

c. 1500    c. 1870    c. 1880

Oakland — Omaha

→← Transcontinental Railroad

**Bison Conservation Herds**
1. McKay Alloway herd
2. Goodnight herd
3. Walking Coyote herd
4. Frederick Dupree herd
5. Buffalo Jones herd

## Historic Events in Bison Population

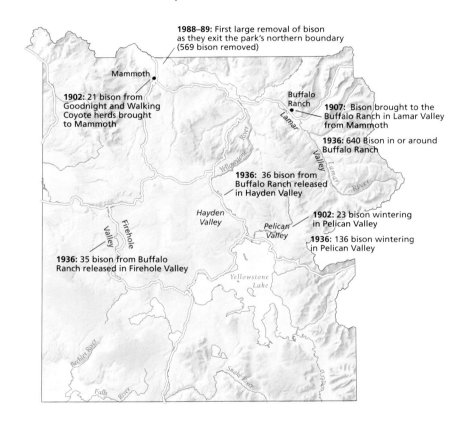

**1988–89:** First large removal of bison as they exit the park's northern boundary (569 bison removed)

Mammoth

**1902:** 21 bison from Goodnight and Walking Coyote herds brought to Mammoth

Buffalo Ranch

**1907:** Bison brought to the Buffalo Ranch in Lamar Valley from Mammoth

**1936:** 640 bison in or around Buffalo Ranch

**1936:** 36 bison from Buffalo Ranch released in Hayden Valley

Hayden Valley

Pelican Valley

**1902:** 23 bison wintering in Pelican Valley

**1936:** 136 bison wintering in Pelican Valley

Firehole Valley

**1936:** 35 bison from Buffalo Ranch released in Firehole Valley

Yellowstone Lake

## Bison Conservation Herds, 2003

Yellowstone Herd

● Stocked with Yellowstone bison
○ Not stocked with Yellowstone bison
*Not shown: 12 bison conservation herds in Canada and Alaska that are unrelated to Yellowstone*

## Bison Population in Yellowstone National Park, 1901–2010

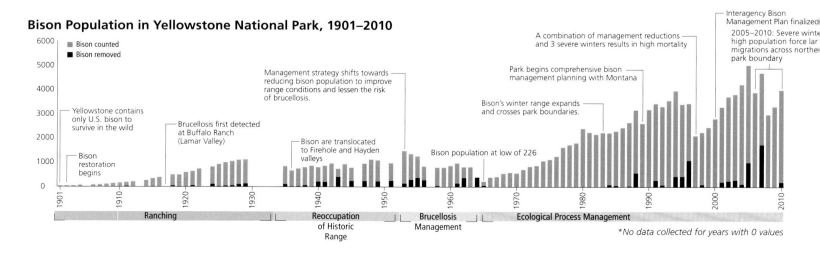

■ Bison counted
■ Bison removed

Yellowstone contains only U.S. bison to survive in the wild

Bison restoration begins

Brucellosis first detected at Buffalo Ranch (Lamar Valley)

Management strategy shifts towards reducing bison population to improve range conditions and lessen the risk of brucellosis.

Bison are translocated to Firehole and Hayden valleys

Bison population at low of 226

Bison's winter range expands and crosses park boundaries.

Park begins comprehensive bison management planning with Montana

A combination of management reductions and 3 severe winters results in high mortality

Interagency Bison Management Plan finalized

2005–2010: Severe winter, high population force large migrations across northern park boundary

Ranching | Reoccupation of Historic Range | Brucellosis Management | Ecological Process Management

*No data collected for years with 0 values*

sufficiently to allow shipment of bison both to unpopulated ranges within the park and across the country to establish conservation herds.

In 1970 the bison population in the park was about 600. During the following decade bison congregated for summer breeding in the Hayden Valley as well as atop the ridges of the Mirror Plateau and the west facing slopes of the Absaroka Range. As the accumulating snowpack made forage less available, they dispersed to three primary wintering areas: the Lamar Valley floor, the Pelican Valley, and the geothermal areas of western Hayden Valley and the Firehole Geyser Basin. The general pattern of winter movement on this latter area was a slow but steady westward migration over Mary Mountain and a reverse passage back to Hayden Valley for breeding season.

By the early 1980s the population had grown to more than 2,000 and the breeding areas on the Northern Range now included the Lamar Valley floor; a broadening of wintering ranges also began at this time. As the population has continued to grow—to a current level of about 4,000—winter ranges have expanded primarily downstream along the Yellowstone and Madison river corridors. A portion of the Hayden Valley breeding group now moves northward, along several migration routes, until autumn snows block passage. The park's present bison population comprises two sub-populations that congregate separately during breeding season. As winter progresses, these breeding groups spread out extensively in response to weather severity and population abundance. Some overlap of the groups has developed on the Northern Range since the mid-1990s.

Bison move over nearly a quarter of Yellowstone in the course of a year, although the core of their summer breeding range is less than 4 percent of the park. Winter ranges extend through numerous interconnected open valleys, parts of which are near large geothermal features that help moderate harsh winter conditions. The Northern Range also provides a relatively mild environment, with lower elevations, less snow accumulation, many south-facing slopes, and rain shadows to the east of tall peaks.

From the 1960s onward, park policy has evolved to support preservation of ecological processes, and thus to allow bison to resume their functional role in the ecosystem. Given the animals' massive size, growing numbers, and far-reaching range, this role—which includes serving as food for predators and scavengers and eating and otherwise affecting plants—is of increasing significance to the park's ecology. When sufficient numbers of bison migrate beyond the park boundary, a new set of challenges can arise as the animals become the subject of social and political conflict among humans. The current management strategy maintains the bison population between 2,500 to 4,500. This supports bison conservation needs while also minimizing conflicts on low-elevation winter ranges outside the park.

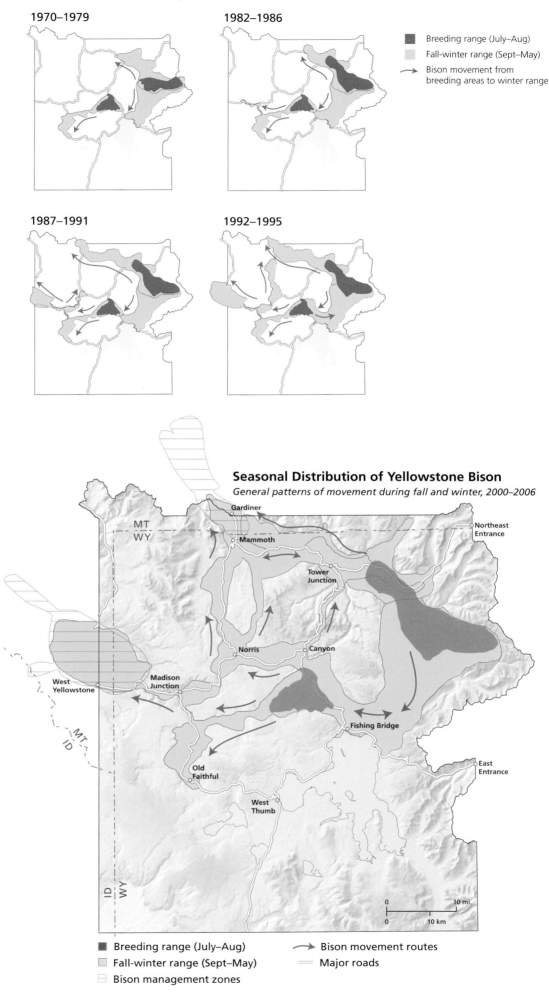

**Expansion of Bison Range, 1970–1995**

1970–1979

1982–1986

1987–1991

1992–1995

- Breeding range (July–Aug)
- Fall-winter range (Sept–May)
- Bison movement from breeding areas to winter range

**Seasonal Distribution of Yellowstone Bison**
*General patterns of movement during fall and winter, 2000–2006*

- Breeding range (July–Aug)
- Fall-winter range (Sept–May)
- Bison management zones
- Bison movement routes
- Major roads

# Bison Movement

Winter weather is an important regulating mechanism that affects the survival of all species associated with Yellowstone National Park. As the snowpack accumulates, food becomes increasingly difficult to procure; eventually, bison and other grazing animals must rely more on their stored fat reserves than on daily consumption. The choice becomes to stay and pay the consequences of the harsh winter conditions or to migrate to areas with less snow, where foraging is easier.

Bison are uniquely adapted to digging in the snowpack, using their large heads to sweep snow aside and expose food. They are known to migrate long distances in Yellowstone to find food. Some move early in the autumn to low-elevation ranges, expending energy at a time when food is more easily attainable and using little if any of their stored fat. They can then spend the whole winter in locations where less energy is required to procure food. The trade-off with this strategy is that the low-elevation ranges typically produce less useful vegetation.

Others migrate later, choosing to rely on their knowledge of the landscape and finding small patches of habitat with high forage production. This strategy, too, has associated costs—the energy expended digging through the deep snow and the fat reserves burned to supplement energy intake before seeking alternate ranges. Working in their favor is the animals' remarkable ability to plow through deep snow and to conserve energy by moving in single file along snow trails. The risks of a late departure from the interior ranges of Yellowstone include miscalculation of the amount of energy stored in body fats needed to get to the new location. Another risk: as the snowpack begins to freeze and thaw in late winter, the advantage shifts toward predators, animals less likely to sink in the snow or have to push snow aside to make a path.

In low-elevation winter ranges, bison face human-related risks as well. Bison migrating beyond negotiated conservation areas outside the park results in hazing, the forced relocation of animals by humans using horses, all-terrain vehicles, and helicopters. Hazing requires animals to expend energy that would otherwise be conserved. When numbers of bison on these boundary ranges reach a level where distribution can no longer be managed effectively, some of the migrants are captured and removed from the ecosystem to prevent dispersal and potential brucellosis transmission to domestic livestock on private lands surrounding the national park.

Animals move about on the landscape to avoid predators and severe climatic conditions. Many grazing species have evolved migratory patterns to more efficiently utilize food reserves that fluctuate seasonally in abundance and nutritional quality. The low-elevation winter ranges are the first locations to become free of snow in the spring and the first to produce new growth of vegetation—and thus attract many bison. As the seasons progress, bison abandon the lower elevation ranges as these areas move past peak productivity, traveling to higher, more productive elevations. This pattern of movement allows the bison to occupy highly productive ranges for a longer period and collect body fat as energy reserves to assist in survival during the following winter. The breeding season coincides with the peak of plant productivity in the high country areas of Lamar and Hayden valleys. This is the time of year that the bison are most concentrated.

The geothermal features of the park provide bison with opportunities to forage during winter in areas that would otherwise present an unacceptable survival risk (requiring too much effort to obtain the available food). Bison are able to maintain a network of trails between the geothermal basins on interior ranges. The large body mass and musculature of adult males is ideal for breaking trail following snowstorms. The smaller animals in the cow-and-calf groups work together to keep the trails packed. By passing over these trails in single file the bison conserve energy. Travel corridors are often closely associated with stream courses. Riparian vegetation provides bison easily accessible food as they migrate between ranges.

"Exploratory migration" may be a precursor to bison engaging in a dispersal movement or a change in migratory pattern. Exploratory movements allow animals to compare resource availability in current habitats to those in adjacent areas. Climatic changes that varied on a local scale throughout the Great Plains likely changed the quality of available food for bison centuries ago. Animals that adapted to changing conditions by exploring other options were likely better suited for survival when patterns of resource abundance changed.

## Individual Bison Movement Patterns, 2003–04

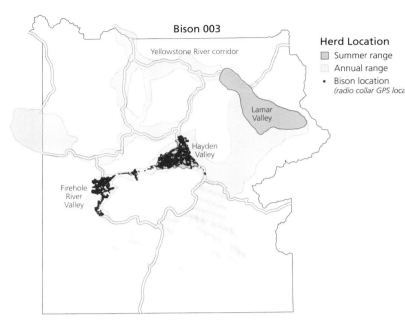

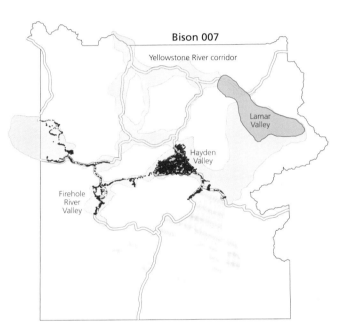

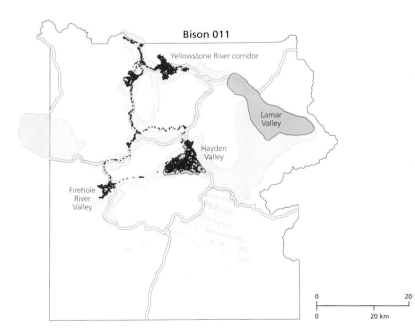

# Seasonal Bison Movement, 2003–04

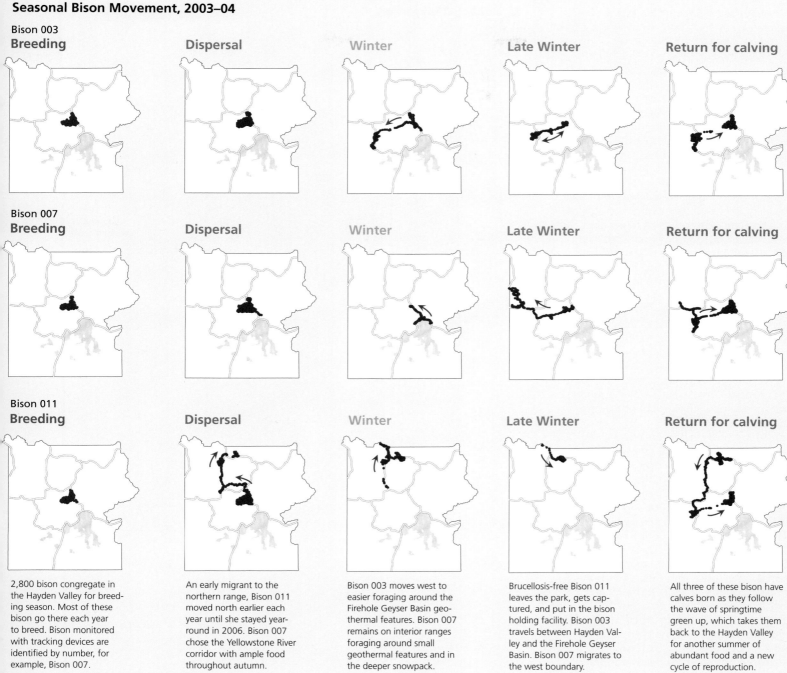

### Bison 003
**Breeding** · **Dispersal** · Winter · **Late Winter** · **Return for calving**

### Bison 007
**Breeding** · **Dispersal** · Winter · **Late Winter** · **Return for calving**

### Bison 011
**Breeding** · **Dispersal** · Winter · **Late Winter** · **Return for calving**

2,800 bison congregate in the Hayden Valley for breeding season. Most of these bison go there each year to breed. Bison monitored with tracking devices are identified by number, for example, Bison 007.

An early migrant to the northern range, Bison 011 moved north earlier each year until she stayed year-round in 2006. Bison 007 chose the Yellowstone River corridor with ample food throughout autumn.

Bison 003 moves west to easier foraging around the Firehole Geyser Basin geothermal features. Bison 007 remains on interior ranges foraging around small geothermal features and in the deeper snowpack.

Brucellosis-free Bison 011 leaves the park, gets captured, and put in the bison holding facility. Bison 003 travels between Hayden Valley and the Firehole Geyser Basin. Bison 007 migrates to the west boundary.

All three of these bison have calves born as they follow the wave of springtime green up, which takes them back to the Hayden Valley for another summer of abundant food and a new cycle of reproduction.

| **Breeding** | **Dispersal** | Winter | **Late Winter** | **Return for calving** |

| July | Aug | Sept | Oct | Nov | Dec | Jan | Feb | March | April | May | June |

---

## Snow Pack and Bison Movement

In Yellowstone there is no sanctuary from the effects of winter weather. Yellowstone's restored bison herds have reestablished seasonal migratory patterns that lead them to low-elevation areas outside the park. Habitat available to Yellowstone bison on these ranges provide food and space to roam during a critical time period in the annual cycle when most of the animals are thin from consuming body fat during winter. It is here also where they come into conflict with human society. Bison that migrate outside the park are subject to recurrent hazing and culling operations to meet brucellosis risk management obligations established in a court-supervised negotiated settlement with the State of Montana (tolerance for bison outside the park is the prerogative of the state). Numbers of bison allowed on habitats outside the park are limited due to disease as well as social and political concerns.

While the migratory process begins in late autumn, most of this migration occurs between mid-winter and mid-spring, with peak numbers moving to the north boundary in February and to the west boundary in April or early May. Natural migration back into the park follows the seasonally advancing abundance of vegetation at higher elevations. This typically begins in April on the northern range and June across the central range.

The migrations are predictable, but their magnitude changes in response to environmental conditions. The probability of migratory movements to low elevations increases rapidly when population numbers exceed 3,000, the previous summer's forage conditions fell well below average, or the winter snowpack is well above average. Snowpack conditions on the bison ranges within the park fluctuate greatly from year to year. Since 2000, when the Interagency Bison Management Plan was implemented, winter snowpacks have generally been below the twenty-year average.

Managers must understand the factors influencing movement patterns and distribution of Yellowstone bison to effectively oversee these animals—this is especially true on the boundary winter ranges.

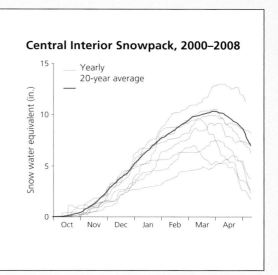

### Central Interior Snowpack, 2000–2008

# Elk

Yellowstone National Park provides summer range for 15,000 to 25,000 elk from eight herds, most of which winter at lower elevations outside the park. These world-renowned herds attract many visitors for wildlife viewing, photography, sport hunting, and other activities that add revenue to the local economy. Yellowstone was the major source of elk for restoration of the species throughout the nation during the first half of the twentieth century.

Elk play an important role in the Yellowstone ecosystem. For example, elk account for approximately 85 percent of kills made by wolves during winter and serve as an important source of protein for black and grizzly bears during spring and early summer. Mountain lions and at least twelve species of scavengers, including bald eagles and coyotes, also rely on elk for food. Competition with elk for resources can influence the diet, habitat selection, and population characteristics of other hoofed mammals (ungulates) such as bighorn sheep, bison, moose, mule deer, and pronghorn. In addition, elk browsing and excretions can have significant effects on vegetative production, soil fertility, and plant diversity. Changes in elk numbers and distribution affect surrounding plants and animals.

The largest elk herd in Yellowstone winters on the grasslands and shrub steppes of the northern range, which extends along the northern boundary of the park and into Montana. Northern Yellowstone elk inspired one of this country's most productive, if sometimes bitter, debates on wildland ecosystem management. For more than fifty years, this dialogue focused on whether there were too many elk—and too much damage associated with their grazing and browsing. Between 1930 and 1968, park staff members removed some 26,000 Yellowstone elk due to concerns about overgrazing, while hunters outside the park removed another 45,000 animals. As a result, elk counts decreased from approximately 12,000 to fewer than 4,000. After the culling program ceased in 1969, and with a much-reduced harvest outside the park (fewer than 210 elk per year), the population grew rapidly and rose to about 12,000 by the mid-1970s and 19,000 in 1994.

The recovery of grizzly bears and wolves shifted the debate from concerns about too many elk to speculation about a future with too few elk due to predation. The winter count of northern Yellowstone elk was around 17,000 when wolves were first reintroduced in 1995 and 1996. The count decreased significantly by 1998 following a substantial winter kill and harvest of more than 3,300 elk outside the park during the severe winter of 1997. Counts varied from 11,500 to 14,500 elk between 1999 and 2001, but decreased to fewer than 7,000 by 2007. The primary factors contributing to this trend were wolf and bear predation and, outside the park, hunting by humans.

## Seasonal Elk Migration

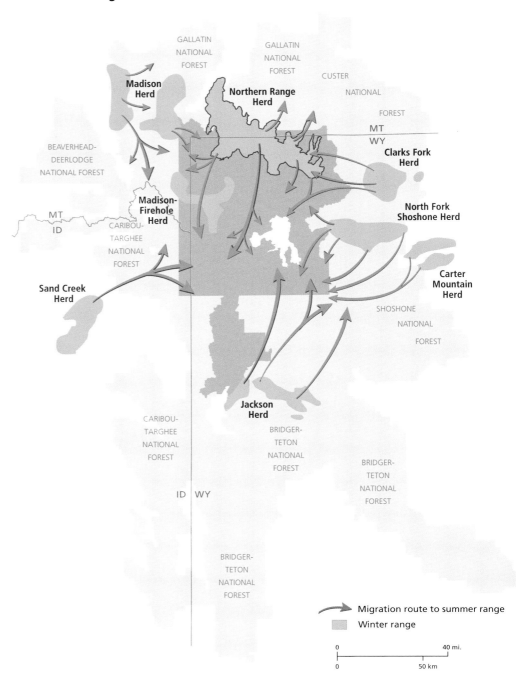

Migration route to summer range

Winter range

## Elk Location and Winter Severity

### Mild Winter, 1994

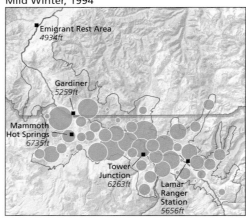

### Severe Winter, 1992

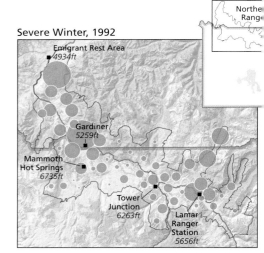

Elk per counting unit

1,000  500  100  •10

## Northern Range: Elk Habitat and Wolf Kills

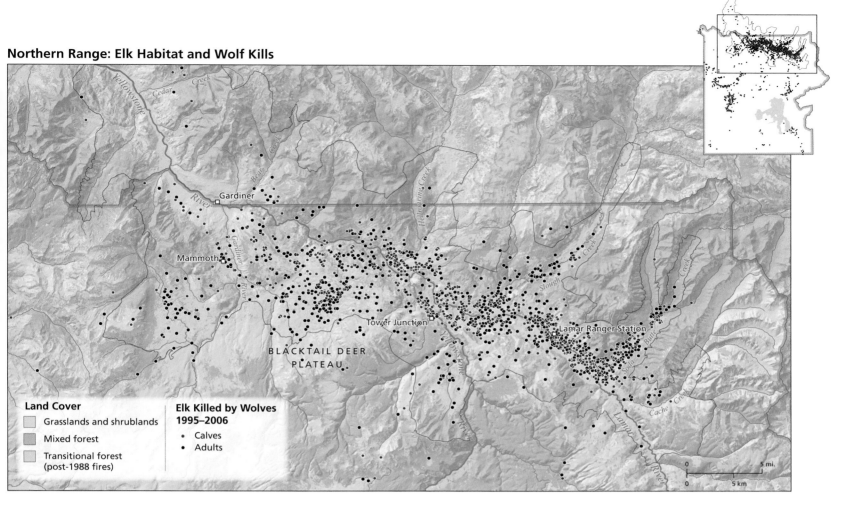

**Land Cover**
- Grasslands and shrublands
- Mixed forest
- Transitional forest (post-1988 fires)

**Elk Killed by Wolves 1995–2006**
- Calves
- Adults

The population dynamics of northern Yellowstone elk appear linked to snowpack. Snow levels influence human harvests of elk by affecting both the number of animals migrating outside the park and hunter success. Elk migration outside the park was proportional to snowpack between 1989 and 2006, and harvests reflected these fluctuations. Snowpack also influences elk vulnerability to wolves, with kill rates rising during severe snow conditions. Most of this predation occurs on the portion of the northern range inside the park, where wolf densities are higher than outside the park. Of the elk that migrated outside the park between 1995 and 2002, the number killed by hunters exceeded conservative estimates of the number killed by wolves. But this ratio reversed in the following six years—a trend likely to continue in the near future because

Montana Fish, Wildlife, and Parks reduced the number of hunting permits for antlerless elk—cows and calves (animals less than one year old)—by more than 99 percent during that period. This change essentially eliminated harvest of antlerless animals as a factor decreasing elk numbers.

Predation on newborns, however, is not so easily controlled. Intensive field studies conducted in recent years found that 69 percent of the calves under study died, most within thirty days after birth. Grizzly and black bears accounted for about 60 percent of these deaths of newborn elk, with wolves responsible for roughly 15 percent. Over the course of their first year, 40 percent of all elk calves are killed by bears, 10 percent are killed by wolves, with only 24 percent surviving.

### Elk Calf Mortality, 2003–2006
*Elk less than one year old*

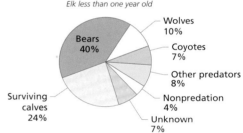

- Bears 40%
- Surviving calves 24%
- Wolves 10%
- Coyotes 7%
- Other predators 8%
- Nonpredation 4%
- Unknown 7%

### Age and Cause of Death, 1995–2002

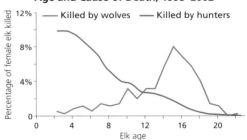

— Killed by wolves    — Killed by hunters

(y-axis: Percentage of female elk killed; x-axis: Elk age, 0 to 20)

## Northern Range Elk Counts and Removals, 1932–2009

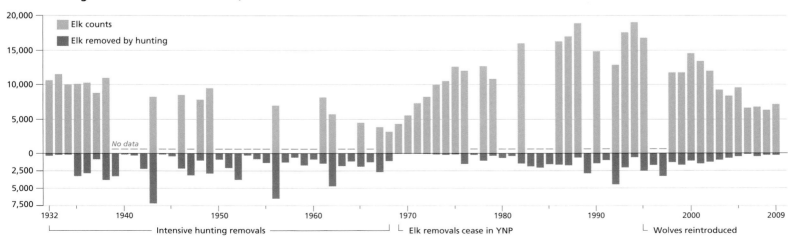

- Elk counts
- Elk removed by hunting

No data

Intensive hunting removals — Elk removals cease in YNP — Wolves reintroduced

159

# Fish

## Yellowstone National Park Fishery

When Yellowstone National Park was established in 1872, thirteen fish species populated 52 percent of the park's surface waters. Waterfalls or other barriers kept fish from the remaining waters. Early park managers placed a high value on visitor access to fishing. Noting the vast fishless waters of Yellowstone, managers asked the U.S. Fish Commission to "see that all waters are stocked so that the pleasure seeker can enjoy fine fishing within a few rods of any hotel or camp." Fish from outside the park were first planted in 1889 and 1890, including nonnative species such as brook trout in the upper Firehole River, rainbow trout in the upper Gibbon River, and brown and lake trout in Lewis and Shoshone lakes. Between 1881 and 1955 the stocking program planted more than 310 million native and nonnative fish, greatly reducing the park's fishless waters and establishing five nonnative fish species.

Nonnative fish have profoundly affected Yellowstone's ecology. The more serious consequences include displacement of natives such as westslope cutthroat trout and Arctic grayling, hybridization of Yellowstone and westslope cutthroat trout with nonnative rainbow trout, and predation of Yellowstone cutthroat trout by nonnative lake trout.

Preservation of native species is a top priority of current Yellowstone fisheries management. Restoring native fish and monitoring lake and stream ecosystem health are also key pursuits, as are tracking invasive species and providing early warning of habitat degradation. Angling is allowed in the park but closely controlled through catch-and-release regulations for native species and a variety of harvest regulations for nonnatives. Nonnative regulations range from two fish per angler, providing the opportunity for a fresh fish dinner, to "must kill" on lake trout in Yellowstone Lake, assisting with lake trout suppression.

## Selected Native Fish

### Yellowstone Cutthroat Trout

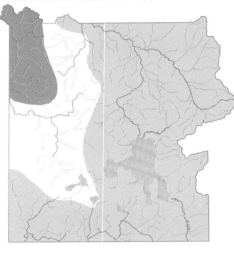

Originally Yellowstone cutthroat trout inhabited the park's Snake, Shoshone, Falls, and Yellowstone river drainages. Historic hatchery operations expanded this range by planting fish beyond their native waters. The western cutthroat trout, a close relative to the Yellowstone cutthroat, populates waters in the park's northwestern corner.

 Westslope Cutthroat Trout

Yellowstone Cutthroat Trout

### Mountain Whitefish

Mountain whitefish are native to the headwaters of the Missouri River, including the Yellowstone River up to the falls, the Gallatin and Madison rivers, and the Snake and Lewis rivers.

### Arctic Grayling

A native species that has been largely lost from the park, Arctic grayling are found in the Gibbon River and less commonly in the Madison and Firehole rivers.

**Yellowstone native fish not shown**

| | |
|---|---|
| Longnose dace | Snake River cutthroat trout |
| Longnose sucker | Speckled dace |
| Mottled sculpin | Utah chub |
| Mountain sucker | Utah sucker |
| Redside shiner | |

## Nonnative Fish

### Rainbow Trout

Due to historic stocking practices rainbow trout are widely distributed, although not found in Yellowstone Lake, the Yellowstone River above the falls, or the Snake River.

### Lake Trout

Lake trout are an extremely harmful species. They are currently in Heart, Lewis, Shoshone, and Yellowstone lakes as a result of both historic and illegal stocking.

### Brook Trout

Historic stocking widely distributed this species, though no brook trout inhabit Yellowstone Lake, the Yellowstone River above the falls, or the Gallatin River.

### Brown Trout

Brown trout were introduced to the Gallatin, Gibbon, Firehole, Madison, Lewis, and Snake rivers, and the Yellowstone River below Knowles Falls.

## Yellowstone Cutthroat Trout

Yellowstone National Park lies at the heart of the native Yellowstone cutthroat range in Idaho, Montana, and Wyoming. The fish is a keystone species in the park, providing an important source of food for an estimated forty-two species of birds and mammals and supporting a $36 million annual sport fishery. Genetically pure Yellowstone cutthroat trout populations have declined throughout their natural range in the Intermountain West, succumbing to habitat degradation and destruction by people, excessive fishing, competition with and predation by nonnative fish, and loss of genetic integrity through interbreeding with introduced fish species. Many of the remaining genetically pure Yellowstone cutthroat trout are found within Yellowstone National Park, particularly in Yellowstone Lake and its drainages.

Yellowstone cutthroat trout have disappeared altogether from some streams and suffered substantial population declines elsewhere in Yellowstone National Park. The greatest threat to Yellowstone cutthroats may be illegally introduced nonnative lake trout: one adult lake trout can consume forty or more cutthroat trout per year. Park biologists have removed more than 560,000 lake trout from Yellowstone Lake following its discovery there in 1994, with 115,000 of the total removal in 2010. Another 48,600

were removed by contract netting in 2009 and 2010. Hybridization is another major factor in the loss of Yellowstone cutthroat trout, leaving genetically pure fish in only a fraction of their historical range. Pathogens also pose a significant threat; in particular, *Myxobolus cerebralis*, an introduced parasite that causes whirling disease, has reduced fish populations in several areas of the Intermountain West. In Yellowstone Lake, up to

20 percent of juvenile and adult Yellowstone cutthroat trout are infected with the parasite. Since the late 1980s, periods of drought have decreased the volume of Yellowstone Lake. Water levels as low as those of the Dust Bowl years have exposed bars that block fish passage in the late summer and early autumn when fry migrate to the lake.

### Yellowstone Cutthroat Trout Distribution

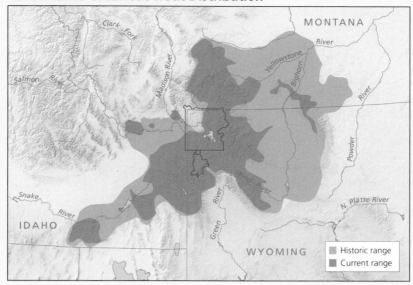

### Hybridization

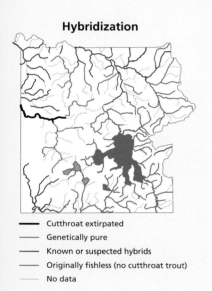

- Cutthroat extirpated
- Genetically pure
- Known or suspected hybrids
- Originally fishless (no cutthroat trout)
- No data

### Whirling Disease 1995–2004

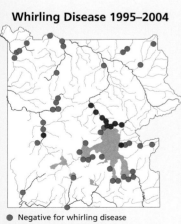

- Negative for whirling disease
- Positive for whirling disease

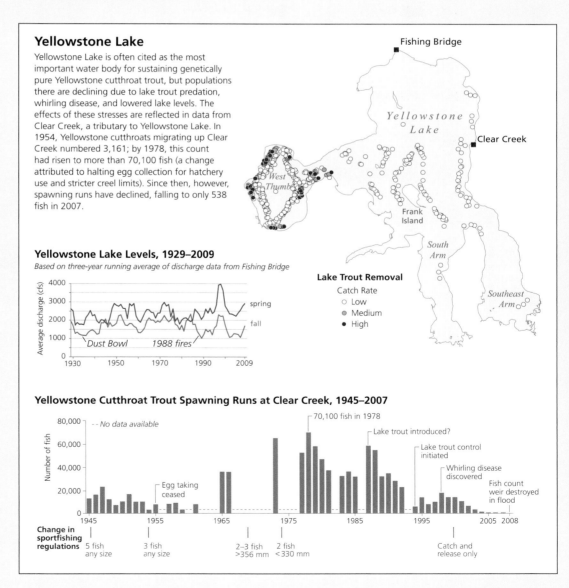

### Yellowstone Lake

Yellowstone Lake is often cited as the most important water body for sustaining genetically pure Yellowstone cutthroat trout, but populations there are declining due to lake trout predation, whirling disease, and lowered lake levels. The effects of these stresses are reflected in data from Clear Creek, a tributary to Yellowstone Lake. In 1954, Yellowstone cutthroats migrating up Clear Creek numbered 3,161; by 1978, this count had risen to more than 70,100 fish (a change attributed to halting egg collection for hatchery use and stricter creel limits). Since then, however, spawning runs have declined, falling to only 538 fish in 2007.

**Lake Trout Removal**
Catch Rate
- ○ Low
- ◔ Medium
- ● High

**Yellowstone Lake Levels, 1929–2009**
*Based on three-year running average of discharge data from Fishing Bridge*

**Yellowstone Cutthroat Trout Spawning Runs at Clear Creek, 1945–2007**

161

# Potential Wildlife Habitats

Strictly speaking, "habitat" refers to the set of environmental conditions that allow an organism to survive and reproduce. But the term often is used to describe the places where organisms live. People may speak broadly about Yellowstone National Park being good wildlife habitat or far more specifically of a single wetland as habitat for one population of boreal toads. Wildlife species are often described in terms of their patterns of habitat use. The coyote is a "habitat generalist" because it uses many different environments with no strong preference for any one type. In contrast, the water vole is a "habitat specialist," found only along the edges of subalpine and alpine streams.

Using computers, researchers can map multiple environmental characteristics across large landscapes. Estimated habitat distributions can be derived by combining those mapped environmental characteristics used by particular species. Landscape-scale habitat maps help answer questions about a species' current status; for example, do American martens occur in several small and disconnected populations in Yellowstone or is their habitat contiguous enough that they likely form one large population? The maps are also useful in forecasting the larger effects of local management actions.

A habitat map is only as good as the data and knowledge that underlie it. Maps for well-known species are more accurate than those for poorly understood ones. The maps shown here are based on vegetation types thought suitable for each species. The map for black rosy-finch (long known to be strongly specialized to alpine tundra) is an example of a reasonably accurate and reliable habitat map. The map for rubber boa, a rarely observed and little-studied species whose vegetation associations are more speculative, is less so.

## Major Habitat Types

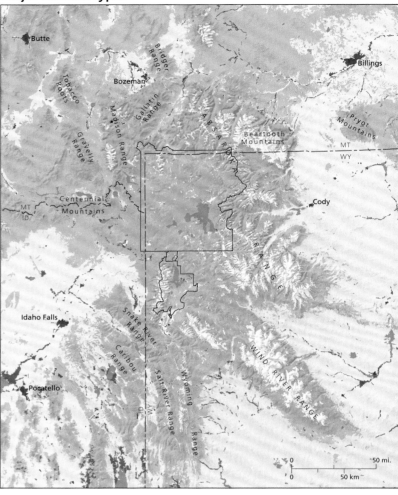

| | | |
|---|---|---|
| ☐ Alpine and subalpine | ☐ Grassland and shrubland | ■ Wetland and lakes |
| ▨ Forest | ■ Riparian | ■ Human development |
| ▨ Foothills | | |

## Alpine and Subalpine

Many wildlife species avoid cold, barren mountaintops, but some are adapted to these environments. Most of the world's black rosy-finches nest among Yellowstone's glaciers and tundra. American pika are at home in alpine rocks and cliffs. Wolverine—one of the rarest species in the area—range across several habitats but appear to require high, cold retreats for successful reproduction. Changes in mountaintop environments associated with global climate change will affect these animals.

**Wolverine**

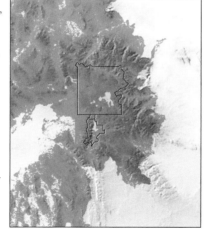

**Black rosy-finch** (breeding range)

**American pika**

## Forest

Yellowstone is home to many types of forest. Large stands of conifers (lodgepole pine, Engelmann spruce, subalpine fir) support well-known wildlife such as chipmunks and squirrels. Some animals common in eastern North America, such as the blue jay, reach their westernmost limits in low-elevation deciduous thickets surrounding the Yellowstone highlands. Wildfires, bark beetles, avalanches, and blowdowns constantly restructure the region's forests—and wildlife habitat.

**Uinta chipmunk**

**Fisher**

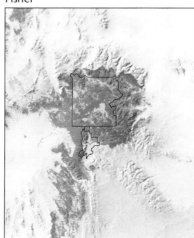

**Blue jay**

162

## Foothills

Foothills are well suited for many species: not as cold and snowy as the mountains above but not as hot and dry as the plains below. These conditions support varied and productive vegetation exploited by many wildlife species. Migratory birds are especially drawn to the seeds, fruits, and insects produced in foothills environments during the summer. The rubber boa—the northernmost true boa constrictor—is thought to occupy foothills throughout Yellowstone but is seldom seen.

Rubber boa
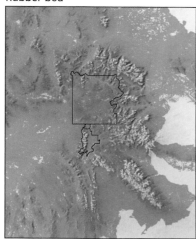

Pinyon jay

Least flycatcher *(breeding range)*

## Grassland and Shrubland

Yellowstone's abundant wildlife results partly from the intermingling of grasslands and shrublands (offering forage) with forests, canyons, and other environments (providing shelter). Some species are specialized to meadows and shrublands. The distribution of sage sparrows closely tracks with sagebrush, whereas Brewer's sparrows use all shrub types. Prairie rattlesnakes inhabit open plains in the highlands, extending upward only along major river valleys.

Prairie rattlesnake
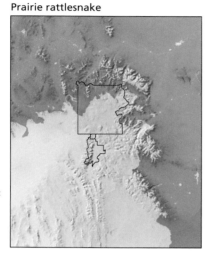

Sage sparrow *(breeding range)*

Brewer's sparrow

## Riparian

"Riparian" refers to the zone separating dry uplands from waterlogged lowlands—the productive environment that surrounds wetlands, streams, and lakes. Riparian soils are fertile and reliably moist, supporting vegetation that is often first to green up in spring and last to fade out in fall. Many insects and insect-eating vertebrates such as flycatchers thrive in these habitats. Most wildlife species use riparian zones during part of their life cycles. Yellowstone's amphibians are common riparian residents.

Willow flycatcher
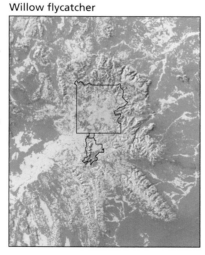

Western jumping mouse

Columbia spotted frog

## Wetland and Lakes

Bodies of permanent water in Yellowstone support a great variety of fish, mollusks, crustaceans, and other aquatic animals. Many of the area's most charismatic wildlife species, such as river otters and trumpeter swans, rely on aquatic habitats. White-faced ibis breed in heavily vegetated wetlands at lower elevations. Large mountain lakes, including Yellowstone Lake, support the continent's southernmost breeding pairs of the iconic common loon.

White-faced ibis *(breeding range)*

Double-crested cormorant *(breeding range)*
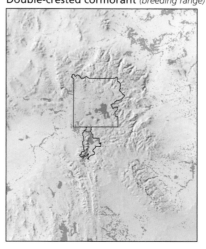

Common loon *(breeding range)*

163

# Sagebrush-Steppe Habitat

The sagebrush-steppe ecosystem once covered nearly half the American West. The term steppe refers to dry, treeless, generally level grassland; sagebrush identifies the dominant shrub species. The sagebrush-steppe ecosystem is among the largest native environments in the United States, but as it is being rapidly reduced, fragmented, and altered, it is also one of the most imperiled.

Sagebrush-steppe habitats support a wide variety of native wildlife, but they are not well suited for cattle. Land managers on private and public lands have used fire, chemicals, and mechanical treatments to remove sagebrush in favor of seeded grasses for livestock forage. Agriculture, oil and gas development, sprawl, and off-road recreation also have eliminated or altered vast areas of sagebrush-steppe landscapes.

Nonnative plants have taken advantage of human-caused disturbance to invade sagebrush habitat. Chief among these invaders is cheatgrass, introduced from Eurasia in the late nineteenth century. Cheatgrass out-competes native grasses in spring and becomes wildfire fuel in summer. Unlike native plants, cheatgrass is well adapted to surviving fire, thus perpetuating itself while eliminating competing species. Cheatgrass has invaded roughly half the existing sagebrush-steppe in western North America, and approximately 10 percent of these environments are now cheatgrass monocultures.

Well-developed sagebrush-steppe habitats occupy portions of the Lamar, Hayden, and Pelican valleys and Gardner's Hole in Yellowstone National Park; the glacial outwash plains and other lowlands of the Upper Snake River Basin, including Grand Teton National Park in Jackson Hole; and the Wyoming Basin, including the Bighorn, Wind River, and Upper Green River basins. These areas are dominated by sagebrush along with shrubs such as bitterbrush and rabbitbrush; various grasses such as bluebunch and western wheatgrasses, Idaho fescue, and needlegrass and thread grass; and other plants like larkspur, scarlet globemallow, and prickly pear cactus. Sagebrush-steppe habitats within the national parks remain largely unchanged except for effects of wildfire in some areas. While cheatgrass occurs throughout the region, it has not attained the dominant foothold in Yellowstone that it has in other places. Nonetheless, this species' range is expanding, especially in the southern part of the greater Yellowstone region.

Sagebrush is valuable to wildlife as forage and cover. In Wyoming, for example, sagebrush-steppe habitats support current counts of eighty-seven species of mammals, 297 species of birds, and sixty-three species of reptiles and amphibians. Sagebrush is a critical food source for several species, including pronghorn, elk, mule deer, and sage grouse. It also helps protect native grasses by creating a partial barrier to severe trampling and grazing by livestock. Several species of vertebrates are classified as sagebrush obligates (species that cannot survive without an adequate, intact expanse of sagebrush-steppe).

The greater sage-grouse is an icon of

**Western U.S. Sagebrush Distribution**

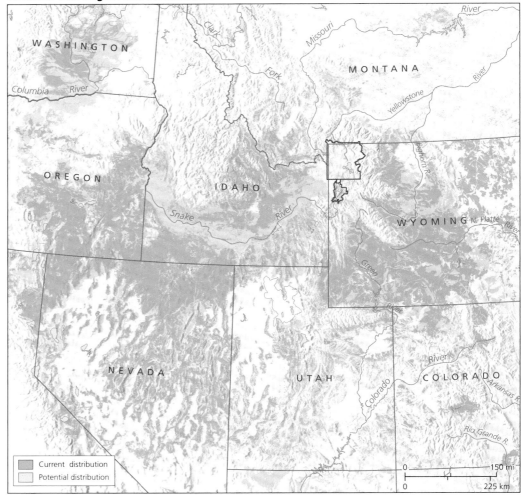

Current distribution
Potential distribution

**Greater Yellowstone Area Sagebrush Distribution**

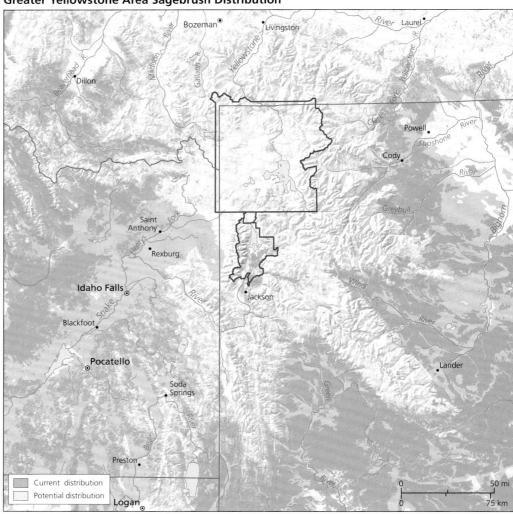

Current distribution
Potential distribution

the sagebrush-steppe. The largest grouse in North America (weighing up to seven pounds), its distribution is limited to sagebrush habitats. The species was once widespread and abundant, ranging across portions of thirteen western states and three western Canadian provinces. It has declined significantly during the past fifty to seventy years, currently inhabiting only eleven states and two provinces—60 percent of its original range. The species is a candidate for listing under the Endangered Species Act. In 2010, the U.S. Fish and Wildlife Service declared the greater sage-grouse "warranted but precluded" from federal protection, meaning it is imperiled but a lower priority than more critically threatened species. Habitat degradation is considered the primary cause of greater sage-grouse decline.

Sage grouse depend on sagebrush environments for different things in different seasons. In spring, males perform elaborate displays to attract females in shared open spaces called leks. The birds are vulnerable to predation while displaying on leks, and sagebrush provides ready cover for escape. Following mating, hens seek out protected nesting sites under sagebrush and other dense shrubs. Chicks require sagebrush cover with undergrowth to screen them from predators and to provide plants and insects for forage. During summer, hens lead their young to areas with more moisture but return to the sagebrush in autumn and begin to feed on sagebrush vegetation. During winter the birds depend almost exclusively on sagebrush for food and protection from predators and weather. Sagebrush is the most important component in this species' yearly diet, comprising 60 to 80 percent of all food consumed. Sage grouse generally maintain or gain weight during cold, snowy winters—testament to their ability to digest and assimilate nutrients contained in sagebrush.

The greater sage-grouse inhabits the Wyoming and Upper Snake River basins, including the southern portion of Grand Teton National Park and adjacent areas. The species is largely absent from higher elevations in the greater Yellowstone region. In

## Historic and Current Sage-Grouse Distribution

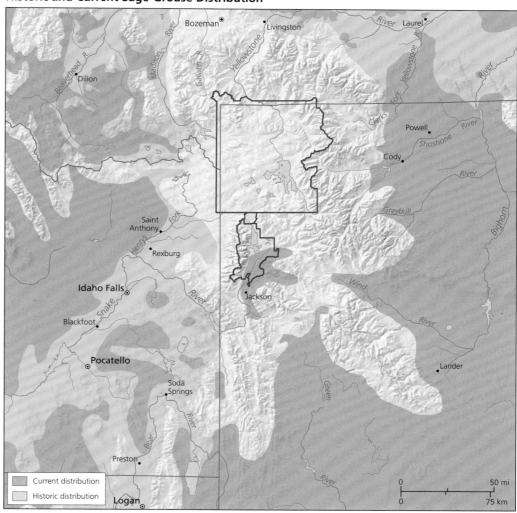

both the Upper Snake River and Wyoming basins, greater sage-grouse numbers declined for several decades, with population lows generally occurring in the mid-1990s. More recent lek counts indicate modest population increases. Several working groups around the greater Yellowstone region are focused on maintaining and improving sagebrush-steppe habitats, fostering research and monitoring, and informing the public, policymakers, and other stakeholders about sage-grouse and sagebrush-steppe conservation.

## Other Sagebrush-Dependent Species In the Greater Yellowstone Area

| Common Name | Scientific Name |
| --- | --- |
| Brewer's sparrow | *Spizella breweri* |
| Greater sage-grouse | *Centrocercus urophasianus* |
| Pronghorn | *Antilocapra americana* |
| Sage sparrow | *Amphispiza belli* |
| Sage thrasher | *Oreoscoptes montanus* |
| Sagebrush lizard | *Sceloporus graciosus* |
| Sagebrush vole | *Lemmiscus curtatus* |

## Greater Sage-Grouse in Grand Teton

Jackson Hole is largely protected from development, but local sage grouse are vulnerable because of their small numbers and relative isolation from surrounding populations. Wildlife biologists therefore have monitored the birds at leks since the 1940s. A 2008 survey included twelve active leks, two inactive leks that had been used in the previous ten years, four leks that were considered abandoned, and one that could not be classified. In total, 103 males were counted at these leks, more than in preceding years but still below historical averages. Identifying clear population trends is difficult, however, because of variations in surveys between years and discoveries of leks that may have been previously overlooked.

Population declines have been associated with predation as well as with habitat fragmentation and loss due to fire, land development, and livestock grazing. Sage grouse hunting in national forests around Jackson Hole ended in 2000. Grand Teton National Park now sets up seasonal protective zones around active leks to minimize human disturbance. Current studies are focusing on how predators, such as ravens, affect chick survival and sage grouse productivity.

## Grand Teton Sage-Grouse Survey, 2008

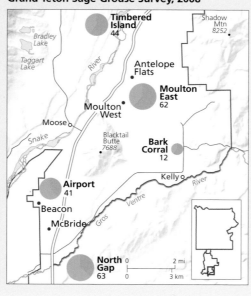

## Grand Teton Greater Sage-Grouse Population

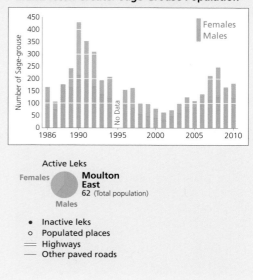

# Thermophiles

## Tree of Life

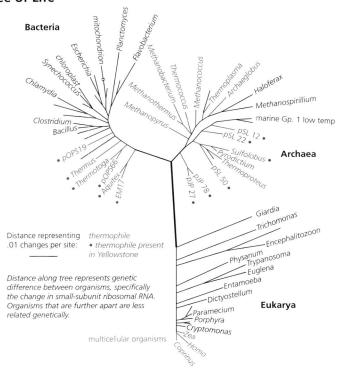

**Bacteria**

Archaea

Eukarya

Distance representing
.01 changes per site:    *thermophile*
   • *thermophile present*
   *in Yellowstone*

*Distance along tree represents genetic
difference between organisms, specifically
the change in small-subunit ribosomal RNA.
Organisms that are further apart are less
related genetically.*

multicellular organisms

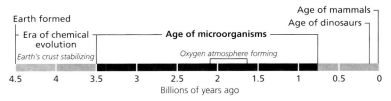

Age of mammals
Age of dinosaurs
Earth formed
Era of chemical
evolution
**Age of microorganisms**
*Earth's crust stabilizing*
*Oxygen atmosphere forming*

4.5    4    3.5    3    2.5    2    1.5    1    0.5    0
Billions of years ago

DNA sequence comparisons reveal that life on Earth evolved from a common ancestor into three major domains on the Tree of Life. Two (Bacteria and Archaea) and most of the third (Eukarya) are composed of microorganisms that evolved toward the present (ends of lines). Eukaryal microorganisms gave rise to multicellular life forms (blue lines) relatively recently. This phylogenetic record parallels the fossil record, which also suggests an "age of microorganisms" before the evolution of metazoan life. Most of life's diversity is among the microorganisms, which drive the cycling of elements upon which all life depends, but most of us know little about them because they are too small to see.

Microorganisms adapted to extreme environments are easily recognized in Yellowstone's geothermal features, however, because of the colorful communities they form. Individual microbial cells may be up to a thousand times smaller than the head of a pin, but billions of them make up communities large enough to see with the human eye. Many of these thermophiles (red lines) are the closest living relatives of life's common ancestor, possibly indicating that life arose in extreme environments.

The colors of microbial communities differ according to how geology variously influences the chemistry of the thermal fluids in which they live. Each of the three common types of Yellowstone hot springs exhibit a different progression of microorganisms as temperature and chemistry change along the effluent flow.

*Alkaline siliceous springs,* which result from hot water interactions with volcanic rocks, typically found within the Yellowstone caldera, range in pH from 7 to 11 and are rich in silica. Below 72°C (162°F), conspicuous green, orange, and brown mat communities are formed by the photosynthetic metabolism of cyanobacteria. Some springs contain dissolved inorganic chemicals that support the chemosynthetic metabolism of filamentous bacteria adapted to temperatures between 75 and 90°C (167 and 194°F).

Around the Yellowstone caldera, hydrogen sulfide-laden steam rises to the surface through fracture zones, where archaea and bacteria convert it to sulfuric acid. The resulting *acid sulfate chloride springs,* exemplified by Norris Geyser Basin and Mud Volcano, are extremely acidic, ranging in pH from 0 to 5. The hot source waters are rich in hydrogen gas, hydrogen sulfide, and iron, which are metabolized by chemosynthetic archaea and bacteria, forming yellow mats laden with elemental sulfur and rust-colored mats laden with iron. Photosynthetic microorganisms are only found below approximately 56°C (133°F), where algae form brilliant cyan-colored mats and brown filaments at lower temperatures.

## Chemistry

### Alkaline Siliceous

In alkaline siliceous hot springs, green cyanobacterial mats form a V-shape in the effluent channel below ~72°C as the water cools fastest at the edges. As the water continues to cool, colors change to orange then brown as cyanobacteria adapted to lower temperatures become predominant. Filamentous chemosynthetic bacteria are sometimes found closer to the source pool.

### Acid Sulfate Chloride

Waters from two acid sulfate chloride springs merge. The hotter flow contains a rust-colored mat constructed by iron-oxidizing archaea and bacteria. The cooler flow contains a cyan-colored mat constructed by algae. Near source pools, yellow mats may be found. Downstream from cyan mats, brown filamentous algae may be found.

### Sulfide Carbonate

Sulfide carbonate spring effluents deposit carbonate minerals forming a large mound. Orange and green colors staining the mound are from cyanobacteria that live only where sulfide has been removed by filamentous chemosynthetic bacteria that form white streamers and/or by green or purple photosynthetic bacteria that form mats in the source pools atop the mound.

### Thermal Area Chemistry

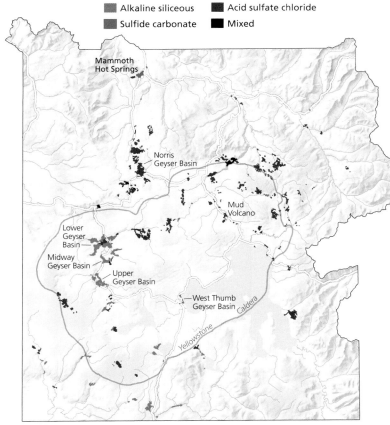

Alkaline siliceous    Acid sulfate chloride
Sulfide carbonate    Mixed

Mammoth Hot Springs

Norris Geyser Basin

Mud Volcano

Lower Geyser Basin

Midway Geyser Basin

Upper Geyser Basin

West Thumb Geyser Basin

Yellowstone Caldera

*Sulfide-carbonate springs* are restricted to locations, such as Mammoth Terraces, where geothermal fluids interacted with ancient carbonate-rich sedimentary rock formations. The water takes on the chemistry of the sediments, and as it rises to the surface, microorganisms convert some of the chemicals to hydrogen sulfide. Carbonate precipitation forms distinctive terraces, but it is the hydrogen sulfide that causes the difference in microbial communities. Above around 66°C (151°F), filamentous chemosynthetic bacteria grow by converting the hydrogen sulfide to sulfuric acid, but the carbonates buffer the acid, keeping the pH near neutral. Below approximately 66°C, photosynthetic green and purple bacteria use hydrogen sulfide, which is poisonous to cyanobacteria. Because the sulfide is rapidly consumed, these mats are usually restricted to small source pools. In downstream sulfide-free waters, cyanobacteria form green and orange mats.

## Thermophile Locations

### Upper, Midway, Lower, and West Thumb Geyser Basins

| Characteristics | Temperature | Community Color and Type | Organisms |
|---|---|---|---|
| Underlain by rhyolitic rock | 92°C (198°F) | Colorless | Bacteria and archaea |
| Water rich in silica, which forms sinter and geyserite deposits | 75–90°C (167–194°F) | Pink, yellow, orange, gray chemosynthetic streamers | Aquificales (*Thermocrinis*) |
| | 50–72°C (122–162°F) | Green cyanobacterial mats | Unicellular cyanobacteria (*Synechococcus*) and filamentous green photosynthetic bacteria (*Roseiflexus*) |
| pH 7–11, (alkaline) | 40–50°C (104–122°F) | Orange cyanobacterial mats | Filamentous (*Leptolyngbya*) and unicellular cyanobacteria (*Synechococcus*) |
| 0 ← 7 → 14  acid  neutral  alkaline | 30–40°C (86–104°F) | Brown cyanobacterial mats | Filamentous cyanobacteria (*Calothrix*) |

### Norris Geyser Basin and Mud Volcano Area

| Characteristics | Temperature | Community Color and Type | Organisms |
|---|---|---|---|
| Sulfidic steam rises to the surface | 70–90°C (158–194°F) | Gray muddy pools | *Sulfolobus* and other archaea |
| Archaea and bacteria convert sulfide to sulfuric acid | 60–85°C (140–185°F) Sulfide present | Yellow mats and streamers | *Hydrogenobaculum* and archaea |
| | 60–85°C (140–185°F) Sulfide absent | Rust-colored mats | *Metallosphaera* and other archaea |
| pH 0–5, (acidic) | 40–56°C (104–133°F) | Cyan-colored mats | Unicellular algae (*Galdierea*, *Cyanidioschyzon*) and fungi |
| 0 ← 7 → 14  acid  neutral  alkaline | 25–40°C (77–104°F) | Brown streamers | Filamentous algae (*Zygogonium*) |

### Mammoth Hot Springs

| Characteristics | Temperature | Community Color and Type | Organisms |
|---|---|---|---|
| Underlain by ancient limestone deposits | 55–75°C (131–167°F) Sulfide present | Cream-colored chemosynthetic streamers | Filamentous chemosynthetic bacteria (*Sulfurihydrogenibium*) |
| Water rich in calcium carbonate and sulfide | 55–66°C (131–151°F) Sulfide present | Green and purple photosynthetic mats | Filamentous green bacteria (*Chloroflexus*) and unicellular purple sulfur bacteria (*Chromatium*) |
| pH 6–8, (neutral to slightly acidic)  0 ← 7 → 14  acid  neutral  alkaline | 25–55°C (77–131°F) Sulfide absent | Green and orange cyanobacterial mats | Unicellular (*Synechococcus*) and filamentous cyanobacteria (*Spirulina*) |

## Selected Thermophiles

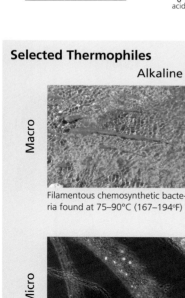

**Macro**

### Alkaline Siliceous Springs

Filamentous chemosynthetic bacteria found at 75–90°C (167–194°F)

~50°C cyanobacterial mat with entrained bubbles of oxygen produced during photosynthesis

### Acid Sulfate Chloride Springs

Cyan-colored algal mat

### Sulfide Carbonate Springs

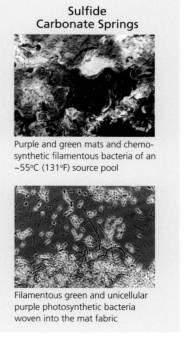

Purple and green mats and chemosynthetic filamentous bacteria of an ~55°C (131°F) source pool

**Micro**

Filamentous chemosynthetic bacteria entwined into streamers

Filamentous and unicellular cyanobacteria woven into mat fabric

Unicellular algae densely packed in mat

Filamentous green and unicellular purple photosynthetic bacteria woven into the mat fabric

# Dinosaurs

In the Greater Yellowstone Area and the surrounding area, fossils of dinosaurs come mainly from rocks deposited during the middle and late Jurassic period (160 to 145.5 million years ago) and the Cretaceous period (145.5 to 65.5 million years ago). These fossils include body fossils and trace fossils such as dinosaur tracks.

During the middle of the Jurassic period, much of the region was underwater, but some areas were periodically lifted out of the seas. Groups of dinosaurs wandered these Jurassic beaches, and today we see their footprints preserved in the sandstone that was once a sandy shore.

Later, during the late Jurassic and early Cretaceous periods, this part of the continent was mainly a seasonally dry plain with midsized rivers, small lakes, occasional dune fields, and swampy areas. This was the environment in which the sediments of the Morrison and Cloverly formations were deposited, and in these rocks today we find many different dinosaur fossils. Late Jurassic dinosaurs from the Morrison Formation include the big sauropods such as *Apatosaurus*, *Camarasaurus*, and *Brachiosaurus*; smaller plant eaters such as *Stegosaurus* and *Camptosaurus*; and the predator *Allosaurus*. Dinosaurs are not as common in the rocks of

the early Cretaceous Cloverly Formation, but they include the plant eaters *Sauropelta* and *Tenontosuarus* and the predator *Deinonychus*.

In the middle of the Cretaceous period, a broad interior seaway spread across the region, splitting North America in half and lasting in many areas through the end of the Cretaceous period. Large numbers of dinosaur fossils have been found in the sediments deposited along the shores of this seaway and in the rivers and deltas that fed into it. These dinosaurs include the plant-eating *Triceratops*, *Ankylosaurus*, and *Hadrosaurus* as well as the king predator, *Tyrannosaurus rex*.

## Outcrops of Mesozoic Rocks

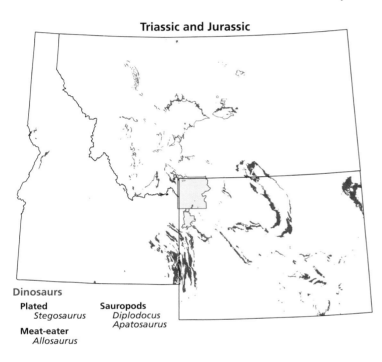

**Triassic and Jurassic**

**Dinosaurs**

**Plated**
*Stegosaurus*

**Meat-eater**
*Allosaurus*

**Sauropods**
*Diplodocus*
*Apatosaurus*

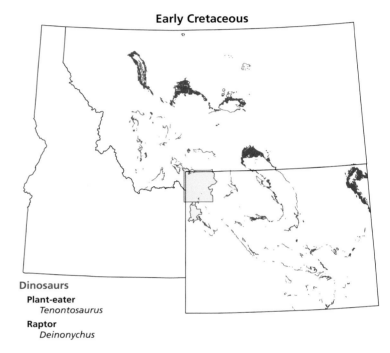

**Early Cretaceous**

**Dinosaurs**

**Plant-eater**
*Tenontosaurus*

**Raptor**
*Deinonychus*

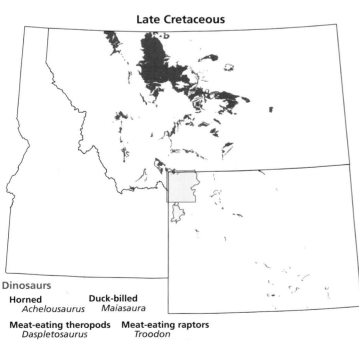

**Late Cretaceous**

**Dinosaurs**

**Horned**
*Achelousaurus*

**Duck-billed**
*Maiasaura*

**Meat-eating theropods**
*Daspletosaurus*
*Albertosaurus*

**Meat-eating raptors**
*Troodon*
*Saurornitholestes*

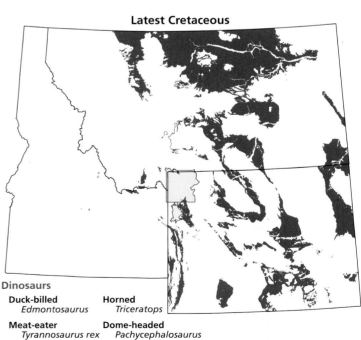

**Latest Cretaceous**

**Dinosaurs**

**Duck-billed**
*Edmontosaurus*

**Meat-eater**
*Tyrannosaurus rex*

**Horned**
*Triceratops*

**Dome-headed**
*Pachycephalosaurus*

## Mesozoic Seas and Landmasses

### Middle Triassic, 230 MYA

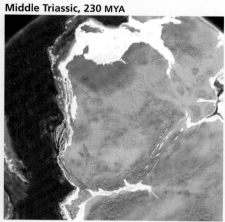

### Late Jurassic, 150 MYA

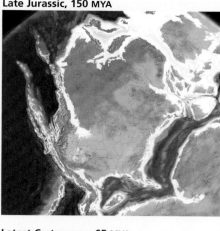

### Early Cretaceous, 115 MYA

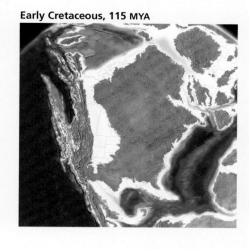

### Late Cretaceous, 75 MYA

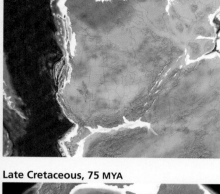

### Latest Cretaceous, 65 MYA

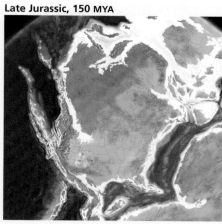

Over millions of years, climates changed and sea levels rose and fell around the globe. Throughout the Paleozoic and Mesozoic eras, North America was periodically exposed above sea level, flooded by seaways, then exposed again. Seawater covered parts of what is now the arid landscape of much of Wyoming and Montana, including the Greater Yellowstone Area. This map series shows the changing inland seas of North America during the Mesozoic era.

MYA = millions of years ago

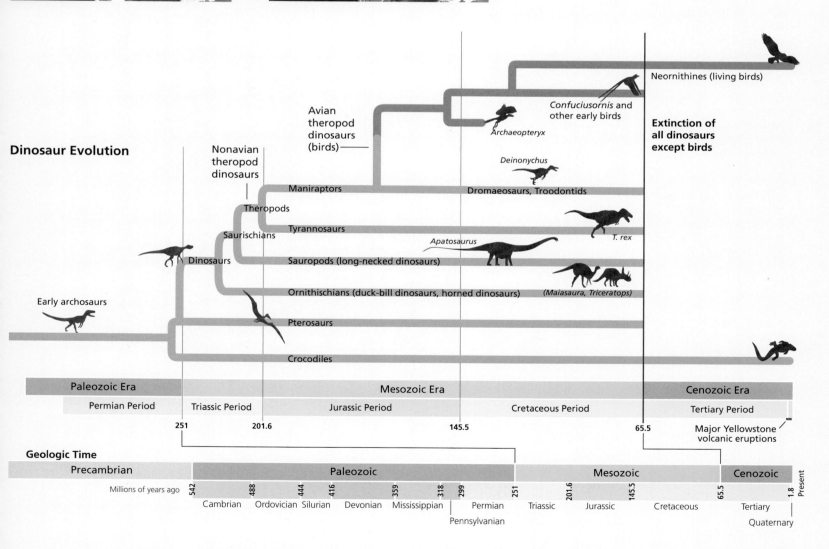

**Dinosaur Evolution**

- Avian theropod dinosaurs (birds)
- Nonavian theropod dinosaurs
- Maniraptors
- Theropods
- Saurischians
- Tyrannosaurs
- Dinosaurs
- Sauropods (long-necked dinosaurs)
- Early archosaurs
- Ornithischians (duck-bill dinosaurs, horned dinosaurs)
- Pterosaurs
- Crocodiles

Neornithines (living birds)

*Confuciusornis* and other early birds

*Archaeopteryx*

**Extinction of all dinosaurs except birds**

*Deinonychus*

Dromaeosaurs, Troodontids

*T. rex*

*Apatosaurus*

(*Maiasaura, Triceratops*)

| Paleozoic Era | Mesozoic Era | | Cenozoic Era |
|---|---|---|---|
| Permian Period | Triassic Period | Jurassic Period | Cretaceous Period | Tertiary Period |

251    201.6    145.5    65.5

Major Yellowstone volcanic eruptions

### Geologic Time

| Precambrian | Paleozoic | Mesozoic | Cenozoic |
|---|---|---|---|

Millions of years ago  542  488  444  416  359  318  299  251  201.6  145.5  65.5  1.8  Present

Cambrian  Ordovician  Silurian  Devonian  Mississippian  Permian  Triassic  Jurassic  Cretaceous  Tertiary

Pennsylvanian    Quaternary

169

# Vertebrate Species

An important function of the National Park System is to protect and maintain the level of biological diversity found within parks. Park managers, planners, and scientists require reliable information on the status of species as a basis for making decisions and working with other agencies, the scientific community, and the public for the long-term protection of park ecosystems. The following list of vertebrate species that occur in Yellowstone and Grand Teton national parks was compiled between 2000 and 2005 from existing species lists, evidence records (vouchers, scientific documents, and observation records that support the species occurrences), and targeted studies. The information was reviewed and certified by subject-matter experts in 2008. Species federally designated as threatened or endangered as of June 2011 are shown in red.

*Yellowstone*
*Grand Teton*

| Latin name | Common name |
|---|---|

## MAMMALS

### Rodents
| | |
|---|---|
| *Tamiasciurus hudsonicus* | Red squirrel |
| *Glaucomys sabrinus* | Northern pocket gopher |
| *Spermophilus armatus* | Uinta ground squirrel |
| *Spermophilus lateralis* | Golden-mantled ground squirrel |
| *Marmota flaviventris* | Yellow-bellied marmot |
| *Tamias umbrinus* | Uinta chipmunk |
| *Tamias amoenus* | Yellow-pine chipmunk |
| *Tamias minimus* | Least chipmunk |
| *Thomomys talpoides* | Northern pocket gopher |
| *Castor canadensis* | North American beaver |
| *Zapus princeps* | Western jumping mouse |
| *Microtus longicaudus* | Long-tailed vole |
| *Microtus richardsoni* | Richardson water vole |
| *Microtus montanus* | Montane vole |
| *Microtus pennsylvanicus* | Meadow vole |
| *Lemmiscus curtatus* | Sagebrush vole |
| *Myodes gapperi* | Southern red-backed vole |
| *Ondatra zibethicus* | Muskrat |
| *Phenacomys intermedius* | Western heather vole |
| *Neotoma cinerea* | Bushy-tailed woodrat |
| *Peromyscus maniculatus* | Deer mouse |
| *Peromyscus leucopus* | White-footed mouse |
| *Mus musculus* | House mouse |
| *Erethizon dorsatum* | North American porcupine |

### Hares, Rabbits, and Pikas
| | |
|---|---|
| *Ochotona princeps* | American pika |
| *Lepus americanus* | Snowshoe hare |
| *Lepus townsendii* | White-tailed jack rabbit |
| *Sylvilagus nuttallii* | Mountain cottontail |
| *Sylvilagus audubonii* | Desert cottontail |

### Shrews
| | |
|---|---|
| *Sorex hoyi* | Pygmy shrew |
| *Sorex cinereus* | Cinereus shrew |
| *Sorex vagrans* | Vagrant shrew |
| *Sorex palustris* | Water shrew |
| *Sorex preblei* | Preble's shrew |
| *Sorex nanus* | Dwarf shrew |
| *Sorex monticolus* | Montane shrew |

### Bats
| | |
|---|---|
| *Eptesicus fuscus* | Big brown bat |
| *Lasiurus cinereus* | Hoary bat |
| *Corynorhinus townsendii* | Townsend's big-eared bat |
| *Euderma maculatum* | Spotted bat |
| *Antrozous pallidus* | Pallid bat |
| *Lasionycteris noctivagans* | Silver-haired bat |
| *Myotis leibii* | Eastern small-footed myotis |
| *Myotis lucifugus* | Little brown bat |
| *Myotis thysanodes* | Fringed myotis |
| *Myotis yumanensis* | Yuma myotis |

*Yellowstone*
*Grand Teton*

| | |
|---|---|
| *Myotis volans* | Long-legged myotis |
| *Myotis evotis* | Long-eared myotis |
| *Myotis californicus* | California myotis |
| *Myotis ciliolabrum* | Western small-footed myotis |

### Felines
| | |
|---|---|
| *Lynx rufus* | Bobcat |
| *Lynx canadensis* | Canada lynx |
| *Puma concolor* | Cougar |

### Canines
| | |
|---|---|
| *Vulpes vulpes* | Red fox |
| *Canis latrans* | Coyote |
| *Canis lupus* | Gray wolf |

### Weasels, Badgers, and Otters
| | |
|---|---|
| *Mustela erminea* | Ermine |
| *Mustela nivalis* | Least weasel |
| *Mustela frenata* | Long-tailed weasel |
| *Gulo gulo* | Wolverine |
| *Martes pennanti* | Fisher |
| *Martes americana* | American marten |
| *Neovison vison* | American mink |
| *Taxidea taxus* | American badger |
| *Lontra canadensis* | Northern river otter |

### Bears
| | |
|---|---|
| *Ursus americanus* | American black bear |
| *Ursus arctos horribilis* | Grizzly bear |

### Skunks and Raccoons
| | |
|---|---|
| *Mephitis mephitis* | Striped skunk |
| *Spilogale gracilis* | Western spotted skunk |
| *Procyon lotor* | Northern raccoon |

### Even-toed Ungulates
| | |
|---|---|
| *Bison bison* | American bison |
| *Oreamnos americanus* | Mountain goat |
| *Ovis canadensis* | Bighorn sheep |
| *Alces alces* | Moose |
| *Odocoileus hemionus* | Mule deer |
| *Odocoileus virginianus* | White-tailed deer |
| *Cervus elaphus* | Elk |
| *Antilocapra americana* | Pronghorn |

## AMPHIBIANS
| | |
|---|---|
| *Bufo boreas boreas* | Boreal toad |
| *Bufo woodhousii* | Woodhouse's toad |
| *Pseudacris triseriata maculata* | Boreal chorus frog |
| *Rana catesbeiana* | American bullfrog |
| *Rana luteiventris* | Columbia spotted frog |
| *Rana pipiens* | Northern leopard frog |
| *Spea bombifrons* | Plains spadefoot |
| *Spea intermontana* | Great Basin spadefoot |
| *Ambystoma tigrinum melanostictum* | Blotched tiger salamander |

## REPTILES
| | |
|---|---|
| *Chrysemys picta bellii* | Western painted turtle |
| *Phrynosoma douglasii* | Short-horned lizard |
| *Phrynosoma hernandesi* | Greater short-horned lizard |
| *Sceloporus graciosus graciosus* | Northern sagebrush lizard |
| *Charina bottae* | Rubber boa |
| *Coluber constrictor* | Eastern racer |
| *Pituophis catenifer* | Gopher snake |
| *Pituophis catenifer sayi* | Bullsnake |
| *Thamnophis elegans vagrans* | Wandering garter snake |
| *Thamnophis sirtalis fitchi* | Valley garter snake |
| *Crotalus viridis viridis* | Prairie rattlesnake |

## FISH

| | |
|---|---|
| *Catostomus ardens* | Utah sucker |
| *Catostomus catostomus* | Longnose sucker |
| *Catostomus discobolus* | Bluehead sucker |
| *Catostomus platyrhynchus* | Mountain sucker |
| *Chasmistes liorus* | June sucker |
| *Couesius plumbeus* | Lake chub |
| *Gila atraria* | Utah chub |
| *Gila copei* | Leatherside chub |
| *Rhinichthys cataractae* | Longnose dace |
| *Rhinichthys osculus* | Speckled dace |
| *Richardsonius balteatus* | Redside shiner |
| *Snyderichthys copei* | Leatherside chub |
| *Poecilia reticulata* | Guppy |
| *Xiphophorus helleri* | Green swordtail |
| *Cichlasoma nigrofasciatum* | Convict cichlid |
| *Oncorhynchus clarki behnkei* | Snake River fine-spotted cutthroat trout |
| *Oncorhynchus clarkii* | Cutthroat trout |
| *Oncorhynchus clarkii bouvieri* | Yellowstone cutthroat trout |
| *Oncorhynchus clarkii lewisi* | Westslope cutthroat trout |
| *Oncorhynchus mykiss* | Rainbow trout |
| *Prosopium williamsoni* | Mountain whitefish |
| *Salmo trutta* | Brown trout |
| *Salvelinus fontinalis* | Brook trout |
| *Salvelinus namaycush* | Lake trout |
| *Thymallus arcticus* | Arctic grayling |
| *Cottus bairdii* | Mottled sculpin |
| *Cottus beldingii* | Paiute sculpin |

## BIRDS

### Ducks, Geese, and Swans

| | |
|---|---|
| *Anser albifrons* | Greater white-fronted goose |
| *Chen caerulescens* | Snow goose |
| *Chen rossii* | Ross's goose |
| *Branta bernicla* | Brant |
| *Branta hutchinsii* | Cackling goose |
| *Branta canadensis* | Canada goose |
| *Cygnus buccinator* | Trumpeter swan |
| *Cygnus columbianus* | Tundra swan |
| *Cygnus cygnus* | Whooper swan |
| *Aix sponsa* | Wood duck |
| *Anas strepera* | Gadwall |
| *Anas penelope* | Eurasian wigeon |
| *Anas americana* | American wigeon |
| *Anas rubripes* | American black duck |
| *Anas platyrhynchos* | Mallard |
| *Anas discors* | Blue-winged teal |
| *Anas cyanoptera* | Cinnamon teal |
| *Anas clypeata* | Northern shoveler |
| *Anas acuta* | Northern pintail |
| *Anas crecca* | Green-winged teal |
| *Aythya valisineria* | Canvasback |
| *Aythya americana* | Redhead |
| *Aythya collaris* | Ring-necked duck |
| *Aythya marila* | Greater scaup |
| *Aythya affinis* | Lesser scaup |
| *Histrionicus histrionicus* | Harlequin duck |
| *Melanitta perspicillata* | Surf scoter |
| *Melanitta fusca* | White-winged scoter |
| *Melanitta americana* | Black scoter |
| *Clangula hyemalis* | Long-tailed duck |
| *Bucephala albeola* | Bufflehead |
| *Bucephala clangula* | Common goldeneye |
| *Bucephala islandica* | Barrow's goldeneye |
| *Lophodytes cucullatus* | Hooded merganser |
| *Mergus merganser* | Common merganser |
| *Mergus serrator* | Red-breasted merganser |
| *Oxyura jamaicensis* | Ruddy duck |

### Partridges, Pheasants, and Grouse

| | |
|---|---|
| *Alectoris chukar* | Chukar |
| *Perdix perdix* | Gray partridge |
| *Centrocercus urophasianus* | Greater sage-grouse |
| *Phasianus colchicus* | Ring-necked pheasant |
| *Bonasa umbellus* | Ruffed grouse |
| *Dendragapus obscurus* | Dusky grouse |
| *Meleagris gallopavo* | Wild turkey |

### Loons

| | |
|---|---|
| *Gavia stellata* | Red-throated loon |
| *Gavia pacifica* | Pacific loon |
| *Gavia immer* | Common loon |

### Grebes

| | |
|---|---|
| *Podilymbus podiceps* | Pied-billed grebe |
| *Podiceps auritus* | Horned grebe |
| *Podiceps grisegena* | Red-necked grebe |
| *Podiceps nigricollis* | Eared grebe |
| *Aechmophorus occidentalis* | Western grebe |
| *Aechmophorus clarkii* | Clark's grebe |

### Storks

| | |
|---|---|
| *Mycteria americana* | Wood stork |

### Cormorants

| | |
|---|---|
| *Phalacrocorax auritus* | Double-crested cormorant |

### Pelicans

| | |
|---|---|
| *Pelecanus erythrorhynchos* | American white pelican |
| *Pelecanus occidentalis* | Brown pelican |

### Herons

| | |
|---|---|
| *Botaurus lentiginosus* | American bittern |
| *Ardea herodias* | Great blue heron |
| *Ardea alba* | Great egret |
| *Egretta thula* | Snowy egret |
| *Egretta caerulea* | Little blue heron |
| *Egretta tricolor* | Tricolored heron |
| *Bubulcus ibis* | Cattle egret |
| *Butorides virescens* | Green heron |
| *Nycticorax nycticorax* | Black-crowned night-heron |

### Ibises

| | |
|---|---|
| *Plegadis falcinellus* | Glossy ibis |
| *Plegadis chihi* | White-faced ibis |

### Vultures

| | |
|---|---|
| *Cathartes aura* | Turkey vulture |

### Osprey

| | |
|---|---|
| *Pandion haliaetus* | Osprey |

### Eagles, Harriers, and Hawks

| | |
|---|---|
| *Haliaeetus leucocephalus* | Bald eagle |
| *Circus cyaneus* | Northern harrier |
| *Accipiter striatus* | Sharp-shinned hawk |
| *Accipiter cooperii* | Cooper's hawk |
| *Accipiter gentilis* | Northern goshawk |
| *Buteo lineatus* | Red-shouldered hawk |
| *Buteo platypterus* | Broad-winged hawk |
| *Buteo swainsoni* | Swainson's hawk |
| *Buteo jamaicensis* | Red-tailed hawk |
| *Buteo regalis* | Ferruginous hawk |
| *Buteo lagopus* | Rough-legged hawk |
| *Aquila chrysaetos* | Golden eagle |

# Vertebrate Species

*Yellowstone*
*Grand Teton*

| • • *Latin name* | Common name |
|---|---|

### Falcons
| | |
|---|---|
| • *Caracara cheriway* | Crested caracara |
| • • *Falco sparverius* | American kestrel |
| • • *Falco columbarius* | Merlin |
| • • *Falco rusticolus* | Gyrfalcon |
| • • *Falco peregrinus* | Peregrine falcon |
| • • *Falco mexicanus* | Prairie falcon |

### Rails, Crakes, and Coots
| | |
|---|---|
| • • *Coturnicops noveboracensis* | Yellow rail |
| • • *Rallus limicola* | Virginia rail |
| • • *Porzana carolina* | Sora |
| • • *Fulica americana* | American coot |

### Cranes
| | |
|---|---|
| • • *Grus canadensis* | Sandhill crane |

### Plovers
| | |
|---|---|
| • • *Pluvialis squatarola* | Black-bellied plover |
| • *Charadrius alexandrinus* | Snowy plover |
| • • *Charadrius semipalmatus* | Semipalmated plover |
| • • *Charadrius vociferus* | Killdeer |

### Avocets and Stilts
| | |
|---|---|
| • • *Himantopus mexicanus* | Black-necked stilt |
| • • *Recurvirostra americana* | American avocet |

### Sandpipers
| | |
|---|---|
| • • *Actitis macularius* | Spotted sandpiper |
| • • *Tringa solitaria* | Solitary sandpiper |
| • • *Tringa incana* | Wandering tattler |
| • • *Tringa melanoleuca* | Greater yellowlegs |
| • • *Tringa semipalmata* | Willet |
| • *Tringa flavipes* | Lesser yellowlegs |
| • *Bartramia longicauda* | Upland sandpiper |
| • • *Numenius americanus* | Long-billed curlew |
| • *Limosa haemastica* | Hudsonian godwit |
| • *Limosa fedoa* | Marbled godwit |
| • *Arenaria interpres* | Ruddy turnstone |
| • *Calidris canutus* | Red knot |
| • *Calidris alba* | Sanderling |
| • *Calidris pusilla* | Semipalmated sandpiper |
| • *Calidris mauri* | Western sandpiper |
| • *Calidris minutilla* | Least sandpiper |
| • *Calidris fuscicollis* | White-rumped sandpiper |
| • *Calidris bairdii* | Baird's sandpiper |
| • *Calidris melanotos* | Pectoral sandpiper |
| • *Calidris alpina* | Dunlin |
| • *Calidris himantopus* | Stilt sandpiper |
| • *Limnodromus griseus* | Short-billed dowitcher |
| • • *Limnodromus scolopaceus* | Long-billed dowitcher |
| • *Gallinago delicata* | Wilson's snipe |
| • • *Gallinago gallinago* | Common snipe |
| • • *Phalaropus tricolor* | Wilson's phalarope |
| • • *Phalaropus lobatus* | Red-necked phalarope |

### Gulls
| | |
|---|---|
| • *Xema sabini* | Sabine's gull |
| • *Chroicocephalus philadelphia* | Bonaparte's gull |
| • *Leucophaeus atricilla* | Laughing gull |
| • *Leucophaeus pipixcan* | Franklin's gull |
| • *Larus canus* | Mew gull |
| • *Larus delawarensis* | Ring-billed gull |
| • *Larus californicus* | California gull |
| • *Larus argentatus* | Herring gull |
| • *Sternula antillarum* | Least tern |
| • *Hydroprogne caspia* | Caspian tern |

| | |
|---|---|
| • • *Chlidonias niger* | Black tern |
| • • *Sterna hirundo* | Common tern |
| • *Sterna paradisaea* | Arctic tern |
| • *Sterna forsteri* | Forster's tern |

### Jaegers
| | |
|---|---|
| • • *Stercorarius parasiticus* | Parasitic jaeger |

### Murrelets
| | |
|---|---|
| • *Brachyramphus perdix* | Long-billed murrelet |

### Pigeons and Doves
| | |
|---|---|
| • • *Columba livia* | Rock pigeon |
| • • *Patagioenas fasciata* | Band-tailed pigeon |
| • *Streptopelia decaocto* | Eurasian collared-dove |
| • • *Zenaida asiatica* | White-winged dove |
| • • *Zenaida macroura* | Mourning dove |

### Cuckoos
| | |
|---|---|
| • *Coccyzus americanus* | Yellow-billed cuckoo |
| • • *Coccyzus erythropthalmus* | Black-billed cuckoo |

### Owls
| | |
|---|---|
| • • *Tyto alba* | Barn owl |
| • *Otus flammeolus* | Flammulated owl |
| • *Megascops kennicottii* | Western screech-owl |
| • *Megascops asio* | Eastern screech-owl |
| • • *Bubo virginianus* | Great horned owl |
| • • *Bubo scandiacus* | Snowy owl |
| • • *Glaucidium gnoma* | Northern pygmy-owl |
| • • *Athene cunicularia* | Burrowing owl |
| • *Strix varia* | Barred owl |
| • • *Strix nebulosa* | Great gray owl |
| • • *Asio otus* | Long-eared owl |
| • • *Asio flammeus* | Short-eared owl |
| • • *Aegolius funereus* | Boreal owl |
| • • *Aegolius acadicus* | Northern saw-whet owl |

### Nighthawks
| | |
|---|---|
| • • *Chordeiles minor* | Common nighthawk |
| • *Phalaenoptilus nuttallii* | Common poorwill |

### Swifts
| | |
|---|---|
| • *Chaetura vauxi* | Vaux's swift |
| • • *Aeronautes saxatalis* | White-throated swift |

### Hummingbirds
| | |
|---|---|
| • *Archilocus alexandri* | Black-chinned hummingbird |
| • • *Stellula calliope* | Calliope hummingbird |
| • • *Selasphorus platycercus* | Broad-tailed hummingbird |
| • • *Selasphorus rufus* | Rufous hummingbird |

### Kingfishers
| | |
|---|---|
| • • *Megaceryle alcyon* | Belted kingfisher |

### Woodpeckers
| | |
|---|---|
| • • *Melanerpes lewis* | Lewis's woodpecker |
| • • *Melanerpes erythrocephalus* | Red-headed woodpecker |
| • *Melanerpes carolinus* | Red-bellied woodpecker |
| • • *Sphyrapicus thyroideus* | Williamson's sapsucker |
| • • *Sphyrapicus varius* | Yellow-bellied sapsucker |
| • • *Sphyrapicus nuchalis* | Red-naped sapsucker |
| • • *Picoides pubescens* | Downy woodpecker |
| • • *Picoides villosus* | Hairy woodpecker |
| • • *Picoides albolarvatus* | White-headed woodpecker |
| • • *Picoides dorsalis* | American three-toed woodpecker |
| • • *Picoides arcticus* | Black-backed woodpecker |
| • • *Colaptes auratus* | Northern flicker |
| • • *Dryocopus pileatus* | Pileated woodpecker |

## Flycatchers

- • *Contopus cooperi* — Olive-sided flycatcher
- • *Contopus sordidulus* — Western wood-pewee
- • *Empidonax traillii* — Willow flycatcher
- • *Empidonax minimus* — Least flycatcher
- • *Empidonax hammondii* — Hammond's flycatcher
- • *Empidonax wrightii* — Gray flycatcher
- • *Empidonax oberholseri* — Dusky flycatcher
- • *Empidonax difficilis* — Pacific-slope flycatcher
- • *Empidonax occidentalis* — Cordilleran flycatcher
- • *Myiarchus tuberculifer* — Dusky-capped flycatcher
- • *Sayornis saya* — Say's phoebe
- • *Myiarchus cinerascens* — Ash-throated flycatcher
- • *Tyrannus verticalis* — Western kingbird
- • *Tyrannus tyrannus* — Eastern kingbird
- • *Tyrannus forficatus* — Scissor-tailed flycatcher

## Shrikes

- • *Lanius ludovicianus* — Loggerhead shrike
- • *Lanius excubitor* — Northern shrike

## Vireos

- *Vireo flavifrons* — Yellow-throated vireo
- *Vireo plumbeus* — Plumbeous vireo
- *Vireo solitarius* — Blue-headed vireo
- *Vireo gilvus* — Warbling vireo
- *Vireo philadelphicus* — Philadelphia vireo
- *Vireo olivaceus* — Red-eyed vireo

## Corvids

- • *Perisoreus canadensis* — Gray jay
- *Gymnorhinus cyanocephalus* — Pinyon jay
- • *Cyanocitta stelleri* — Steller's jay
- • *Cyanocitta cristata* — Blue jay
- • *Nucifraga columbiana* — Clark's nutcracker
- • *Pica hudsonia* — Black-billed magpie
- • *Corvus brachyrhynchos* — American crow
- • *Corvus corax* — Common raven

## Larks

- • *Eremophila alpestris* — Horned lark

## Swallows

- • *Tachycineta bicolor* — Tree swallow
- • *Tachycineta thalassina* — Violet-green swallow
- • *Stelgidopteryx serripennis* — Northern rough-winged swallow
- • *Riparia riparia* — Bank swallow
- • *Petrochelidon pyrrhonota* — Cliff swallow
- • *Hirundo rustica* — Barn swallow

## Chickadees

- • *Poecile atricapillus* — Black-capped chickadee
- • *Poecile gambeli* — Mountain chickadee

## Nuthatches

- • *Sitta canadensis* — Red-breasted nuthatch
- • *Sitta carolinensis* — White-breasted nuthatch
- • *Sitta pygmaea* — Pygmy nuthatch

## Creepers

- • *Certhia americana* — Brown creeper

## Wrens

- • *Salpinctes obsoletus* — Rock wren
- • *Catherpes mexicanus* — Canyon wren
- • *Troglodytes aedon* — House wren
- • *Troglodytes hiemalis* — Winter wren
- • *Cistothorus platensis* — Sedge wren
- • *Cistothorus palustris* — Marsh wren

## Gnatcatchers

- • • *Polioptila caerulea* — Blue-gray gnatcatcher

## Dippers

- • • *Cinclus mexicanus* — American dipper

## Kinglets

- • • *Regulus satrapa* — Golden-crowned kinglet
- • • *Regulus calendula* — Ruby-crowned kinglet

## Thrushes

- • • *Sialia mexicana* — Western bluebird
- • • *Sialia currucoides* — Mountain bluebird
- • • *Myadestes townsendi* — Townsend's solitaire
- • • *Catharus fuscescens* — Veery
- • • *Catharus ustulatus* — Swainson's thrush
- • • *Catharus guttatus* — Hermit thrush
- • • *Turdus migratorius* — American robin
- • • *Ixoreus naevius* — Varied thrush

## Mimids

- • • *Dumetella carolinensis* — Gray catbird
- • • *Mimus polyglottos* — Northern mockingbird
- • • *Oreoscoptes montanus* — Sage thrasher
- • • *Toxostoma rufum* — Brown thrasher

## Starlings

- • • *Sturnus vulgaris* — European starling

## Pipits

- • • *Anthus rubescens* — American pipit
- • • *Anthus spragueii* — Sprague's pipit

## Waxwings

- • • *Bombycilla garrulus* — Bohemian waxwing
- • • *Bombycilla cedrorum* — Cedar waxwing

## Silky-flycatchers

- • *Phainopepla nitens* — Phainopepla

## Longspurs and Snow Buntings

- • • *Calcarius lapponicus* — Lapland longspur
- • *Rhynchophanes mccownii* — McCown's longspur
- • • *Plectrophenax nivalis* — Snow bunting

## Wood-Warblers

- • • *Oreothlypis peregrina* — Tennessee warbler
- • • *Oreothlypis celata* — Orange-crowned warbler
- • • *Oreothlypis ruficapilla* — Nashville warbler
- • *Oreothlypis virginiae* — Virginia's warbler
- • • *Dendroica petechia* — Yellow warbler
- • • *Dendroica pensylvanica* — Chestnut-sided warbler
- • • *Dendroica tigrina* — Cape May warbler
- • *Dendroica caerulescens* — Black-throated blue warbler
- • • *Dendroica coronata* — Yellow-rumped warbler
- • *Dendroica nigrescens* — Black-throated gray warbler
- • • *Dendroica townsendi* — Townsend's warbler
- • • *Dendroica fusca* — Blackburnian warbler
- • • *Dendroica dominica* — Yellow-throated warbler
- • • *Dendroica discolor* — Prairie warbler
- • • *Dendroica palmarum* — Palm warbler
- • *Dendroica castanea* — Bay-breasted warbler
- • *Dendroica striata* — Blackpoll warbler
- • • *Mniotilta varia* — Black-and-white warbler
- • • *Setophaga ruticilla* — American redstart
- • *Protonotaria citrea* — Prothonotary warbler
- • • *Seiurus aurocapilla* — Ovenbird
- • • *Parkesia noveboracensis* — Northern waterthrush
- • • *Oporornis tolmiei* — MacGillivray's warbler
- • • *Geothlypis trichas* — Common yellowthroat

# Vertebrate Species

**BIRDS** *(continued from previous page)*

| Latin name | Common name |
|---|---|
| • • *Wilosnia citrina* | Hooded warbler |
| • • *Wilsonia pusilla* | Wilson's warbler |
| • • *Icteria virens* | Yellow-breasted chat |

### Sparrows
| | |
|---|---|
| • • *Pipilo chlorurus* | Green-tailed towhee |
| • *Pipilo maculatus* | Spotted towhee |
| • *Pipilo erythrophthalmus* | Eastern sowhee |
| • • *Spizella arborea* | American tree sparrow |
| • • *Spizella passerina* | Chipping sparrow |
| • *Spizella pallida* | Clay-colored sparrow |
| • *Spizella breweri* | Brewer's sparrow |
| • *Spizella pusilla* | Field sparrow |
| • *Spizella atrogularis* | Black-chinned sparrow |
| • • *Pooecetes gramineus* | Vesper sparrow |
| • • *Chondestes grammacus* | Lark sparrow |
| • • *Amphispiza bilineata* | Black-throated sparrow |
| • • *Amphispiza belli* | Sage sparrow |
| • • *Calamospiza melanocorys* | Lark bunting |
| • • *Passerculus sandwichensis* | Savannah sparrow |
| • *Ammodramus savannarum* | Grasshopper sparrow |
| • *Ammodramus leconteii* | Le Conte's sparrow |
| • *Passerella iliaca* | Fox sparrow |
| • *Melospiza melodia* | Song sparrow |
| • *Melospiza lincolnii* | Lincoln's sparrow |
| • *Melospiza georgiana* | Swamp sparrow |
| • *Zonotrichia albicollis* | White-throated sparrow |
| • *Zonotrichia querula* | Harris's sparrow |
| • *Zonotrichia leucophrys* | White-crowned sparrow |
| • *Zonotrichia atricapilla* | Golden-crowned sparrow |
| • *Junco hyemalis* | Dark-eyed junco |

### Cardinals
| | |
|---|---|
| • *Piranga olivacea* | Scarlet tanager |
| • *Piranga ludoviciana* | Western tanager |
| • *Pheucticus ludovicianus* | Rose-breasted grosbeak |
| • *Pheucticus melanocephalus* | Black-headed grosbeak |
| • *Passerina amoena* | Lazuli bunting |
| • *Passerina cyanea* | Indigo bunting |

### Blackbirds and Orioles
| | |
|---|---|
| • • *Dolichonyx oryzivorus* | Bobolink |
| • • *Agelaius phoeniceus* | Red-winged blackbird |
| • • *Sturnella neglecta* | Western meadowlark |
| • • *Xanthocephalus xanthocephalus* | Yellow-headed blackbird |
| • • *Euphagus cyanocephalus* | Brewer's blackbird |
| • • *Quiscalus quiscula* | Common grackle |
| • • *Molothrus ater* | Brown-headed cowbird |
| • • *Icterus bullockii* | Bullock's oriole |
| • • *Icterus galbula* | Baltimore oriole |

### Cardueline Finches
| | |
|---|---|
| • • *Leucosticte tephrocotis* | Gray-crowned rosy-finch |
| • • *Leucosticte atrata* | Black rosy-finch |
| • • *Pinicola enucleator* | Pine grosbeak |
| • *Carpodacus purpureus* | Purple finch |
| • • *Carpodacus cassinii* | Cassin's finch |
| • • *Carpodacus mexicanus* | House finch |
| • • *Loxia curvirostra* | Red crossbill |
| • • *Loxia leucoptera* | White-winged crossbill |
| • • *Acanthis flammea* | Common redpoll |
| • • *Acanthis hornemanni* | Hoary redpoll |
| • • *Spinus pinus* | Pine siskin |
| • *Spinus psaltria* | Lesser goldfinch |
| • • *Spinus tristis* | American goldfinch |
| • • *Coccothraustes vespertinus* | Evening grosbeak |

### Old World Sparrows
| | |
|---|---|
| • *Passer domesticus* | House sparrow |

The NPSpecies list (the National Park Service Biodiversity Database) provides information on the presence of species in America's national parks. Although the data are reviewed by experts in taxonomy using the best information available at the time of disclosure, lists of species found to be occurring in parks are not exhaustive; the absence of a species from the database does not mean the species is absent from the park. Varying degrees of effort spent surveying species or mining historical reference information may have resulted in data gaps. Accepted taxonomy changes over time; species names shift accordingly.

# Reference Maps

# Greater Yellowstone Reference Maps

## Legend

Maps in this section are at a scale of 1:500,000.
One inch equals approximately eight miles on the ground.

### Settlements

◉ County seat
○ Incorporated city or town
○ Unincorporated city or town

### Boundaries

━━━━━━━ National Park Service unit boundary
━━━ ━ ━ State boundary
━━━ ━ ━ County boundary

### Roads and Transportation

🛡15 Interstate route
⬭191 U.S. Highway route
◯390 State route

━━━━━ Interstate freeway
━━━━━ Highway
━━━━━ Paved road
────── Unpaved road
═════⬎ Runway
┣━━━┫ Railroad track
─ ─ ─ ─ Trail
─ ─ ─ ─ National trail

### Water Features

⌒ Perennial stream
⌒ Intermittent stream
⌒ Canal
⌢ Dam

⬭ Perennial water body
⬭ Intermittent pond
⬭ Glacier
❀ Wetland

### Other Symbols

· · · · · · · Continental Divide
*el 6290* Elevation in feet
• Peak, summit, or point (elevation in feet)
⌣ Pass, gap, or saddle
✈ Airport with commercial service
✛ Airport

## Hypsometric Tints (Elevation Colors)

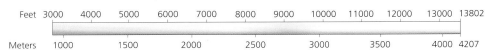

| Feet | 3000 | 4000 | 5000 | 6000 | 7000 | 8000 | 9000 | 10000 | 11000 | 12000 | 13000 | 13802 |
|---|---|---|---|---|---|---|---|---|---|---|---|---|
| Meters | 1000 | | 1500 | | 2000 | | 2500 | | 3000 | | 3500 | 4000 | 4207 |

## Reference Map Index

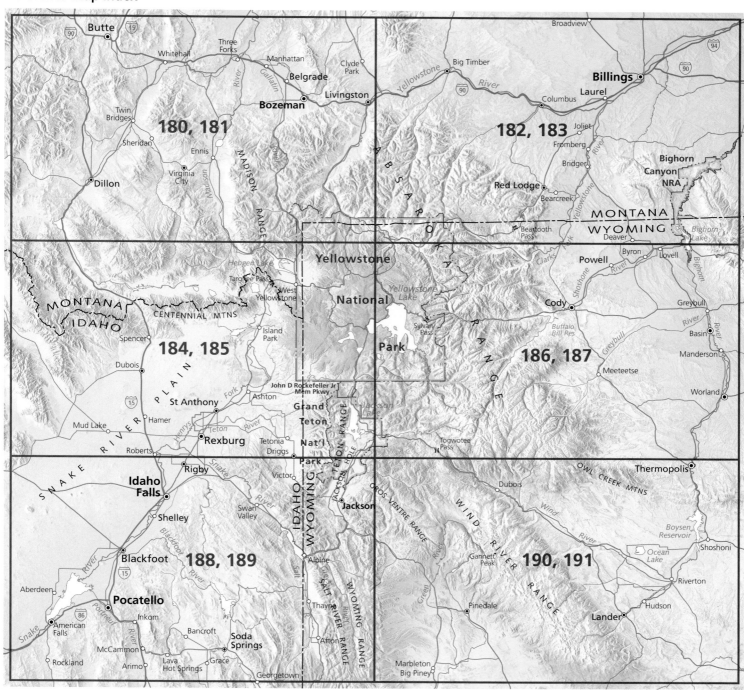

Reference maps provide a comprehensive overview of areas by displaying place names, elevations, roads, rivers, and other key information. The maps immediately following this page have a scale of 1:500,000, meaning that a measurement of one unit on the map is equal to 500,000 of those same units in the real world. A distance of one inch on these maps equals 500,000 inches or about eight miles.

The 1:500,000 scale maps are small-scale maps, because one divided by 500,000 is a small number. The later maps at a scale of 1:100,000 are larger scale, because one divided by 100,000 is larger than one divided by 500,000. Counter-intuitively, the smaller scale maps show larger areas. The 1:500,000 maps provide broad overviews of the Greater Yellowstone Area but at a reduced level of detail. Small streams are not shown at all and larger streams are exaggerated in width so they remain legible. Stream representations are simplified; compare the Snake River through Jackson Hole as it is shown on the 1:500,000 and 1:100,000 maps. Roads, too, are shown selectively: important routes are included with their widths exaggerated, but local roads and most backcountry routes are dropped altogether.

The Yellowstone regional maps highlight variations in landforms using a combination of relief shading and elevation tinting, both derived from extremely detailed digital elevation databases. The colors do not have meaning in and of themselves; for example, these maps do *not* use green to indicate vegetation. Rather, the colors are ones that have proven to be readily distinguished from one another. The maps here use a double shift, meaning that the shade varies within colors, progressing from dark to light greens at the lowest elevations to light grays to white at the highest elevations. This approach sacrifices forceful depiction of high elevations in favor of overall brightening and a clearer representation of relief shading, emphasizing the detailed forms of high ridges.

Slope is indicated by shading. These maps are illuminated from the left, or due west, which more clearly shows the important northwest-trending landforms like the Lamar Valley—a contrast to the northwest illumination traditionally used for most topographic shading. The elevation tints and shading are built on top of digital elevation data with ten-meter resolution. This resolution yields 10,000 data points in every square kilometer, far more than can be shown in the two-millimeter square that represents a square kilometer at 1:500,000. The digital elevation data are therefore thinned on the regional maps to a twenty-five-meter resolution, or 1,600 data points per square kilometer. The resulting shading is more detailed than is commonly shown. Elevation tints are lightened to maintain shading visibility.

# Bozeman

See Page 184

180

SCALE 1:500,000

5 Miles 0 5 10 15 20 25 30 Miles

**MILES**

Elevations are in feet

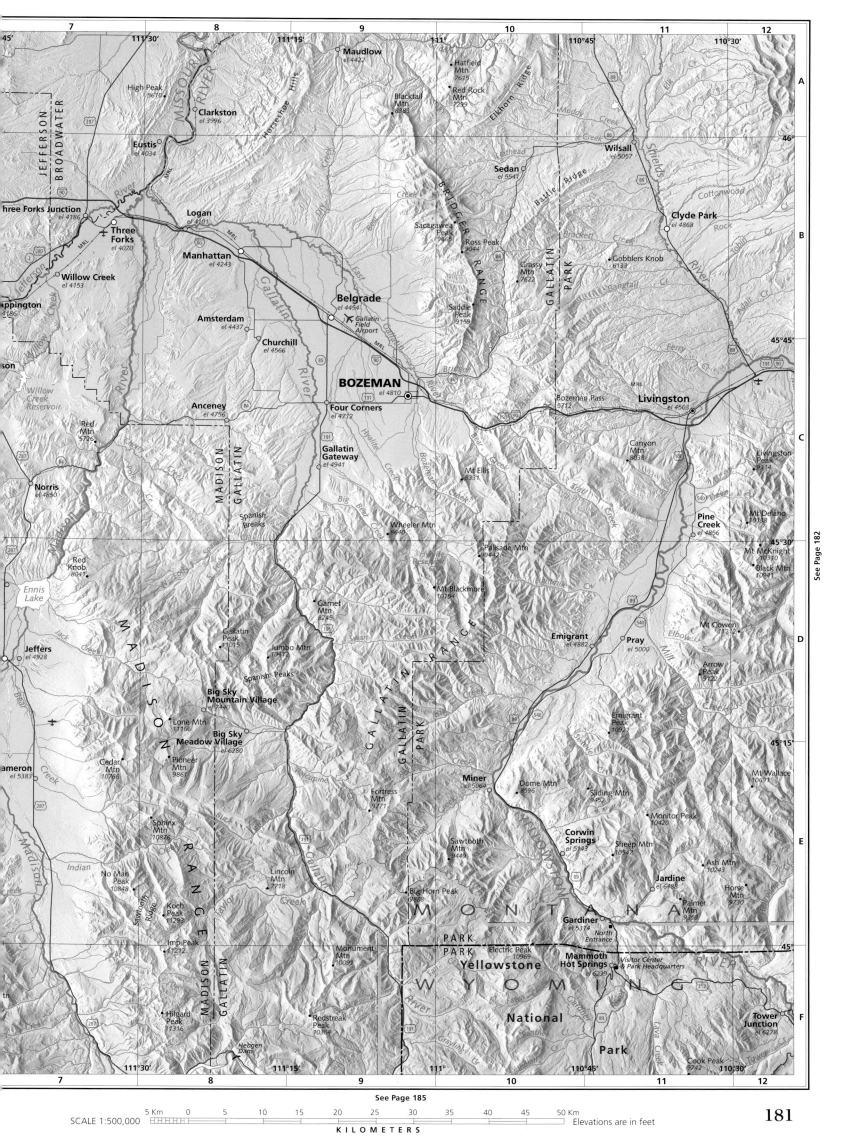

# Billings

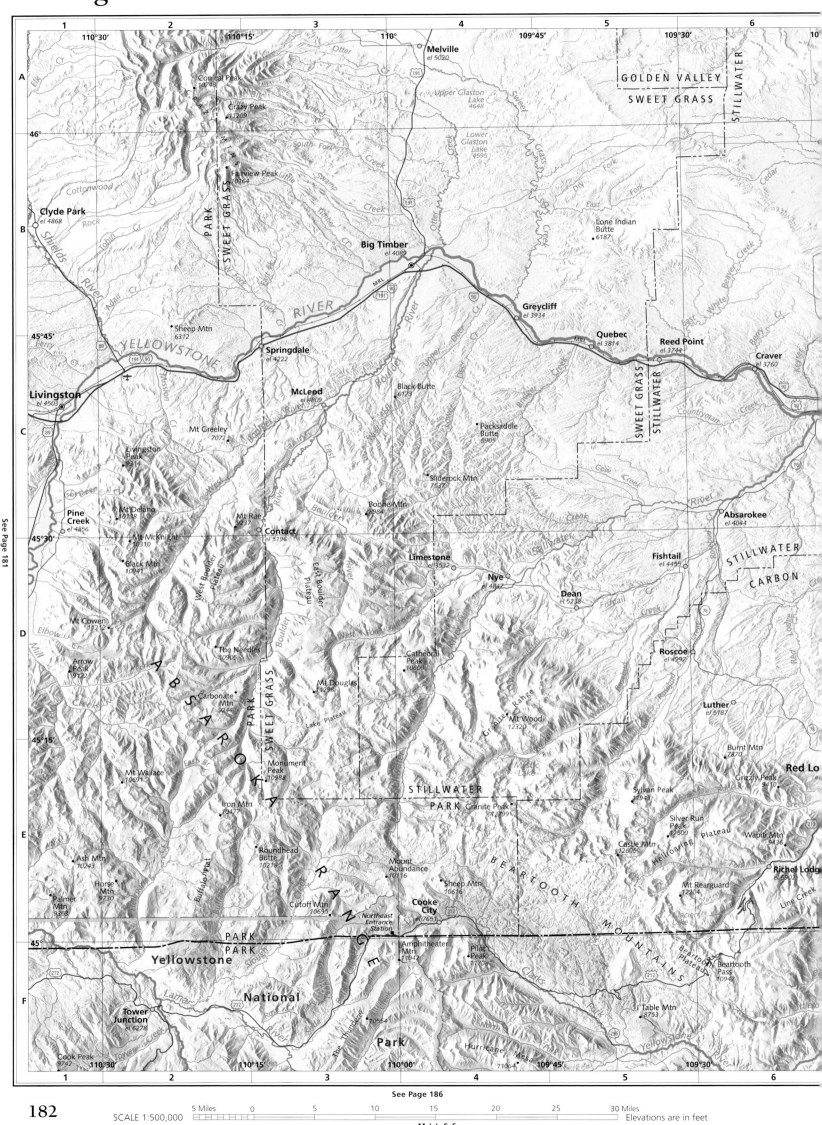

110°30'     110°15'     110°     109°45'     109°30'     10

A

46°

B

45°45'

C

45°30'

45°15'

D

E

45°

F

Melville
el 5020

GOLDEN VALLEY
SWEET GRASS
STILLWATER

Conical Peak
10748

Crazy Peak
11209

Fairview Peak
10764

PARK

SWEET GRASS

Upper Glaston
Lake
4648

Lower
Glaston
Lake
4595

Lone Indian
Butte
6187

Clyde Park
el 4868

Big Timber
el 4089

RIVER

Greycliff
el 3934

Quebec
el 3814

Reed Point
el 3744

Craver
el 3760

YELLOWSTONE

Sheep Mtn
6312

Springdale
el 4222

McLeod
el 4809

Black Butte
6173

SWEET GRASS

STILLWATER

Livingston
el 4503

Mt Greeley
7072

Packsaddle
Butte
5905

Absarokee
el 4044

Livingston
Peak
9314

Sliderock Mtn
7537

STILLWATER

CARBON

Pine
Creek
el 4856

Mt Delano
10138

Mt Rae
9237

Contact
el 5196

Boone Mtn
8984

Limestone
el13532

Nye
el 4847

Dean
el 5238

Fishtail
el 4455

Mt McKnight
10310

West Boulder Plateau

East Boulder Plateau

Roscoe
el 4997

Black Mtn
10941

Luther
el 5187

Mt Cowen
11212

Boulder

The Needles
10905

Mt Douglas
11298

Cathedral
Peak
10600

Granite Range

Mt Wood
12320

Arrow
Peak
9722

ABSAROKA

SWEET GRASS

Lake Plateau

Burnt Mtn
7870

Carbonate
Mtn
9244

Red Lo

Mt Wallace
10691

Monument
Peak
10988

PARK

STILLWATER

Sylvan Peak
11943

Grizzly Peak
9410

Iron Mtn
10477

PARK

Granite Peak
12799

Silver Run
Peak
12500

Heliroaring Plateau

Wapiti Mtn
9436

Castle Mtn
12605

Ash Mtn
10243

Roundhead
Butte
10218

Mount
Abundance
10116

Sheep Mtn
10616

BEARTOOTH

Richel Lodg
el 5503

Horse
Mtn
9730

Cutoff Mtn
10695

Cooke
City
el 7651

Mt Rearguard
12204

Palmer
Mtn
9363

Northeast
Entrance
Station

212

RANGE

PARK

PARK

Amphitheater
Mtn
11042

Pilot
Peak
11708

MOUNTAINS

Beartooth
Plateau

Beartooth
Pass
10947

Yellowstone

Tower
Junction
el 6278

National

Park

10554

Table Mtn
8753

Cook Peak
9742

Hurricane
11064

110°30'     110°15'     110°00'     109°45'     109°30'

See Page 181

See Page 186

SCALE 1:500,000     5 Miles     0     5     10     15     20     25     30 Miles

MILES     Elevations are in feet

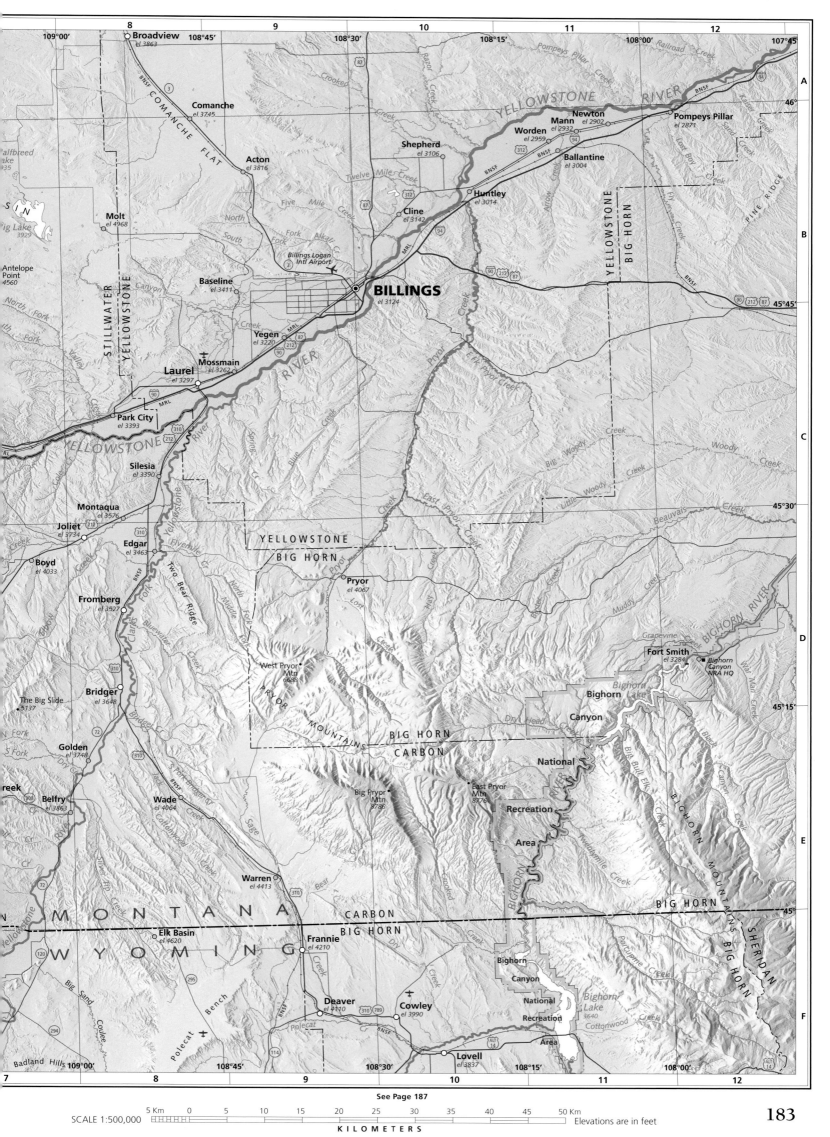

Broadview
el 3863

Comanche
el 3745

Acton
el 3816

Molt
el 4968

Antelope
Point
4560

Baseline
el 3411

Billings Logan
Intl Airport

Shepherd
el 3106

Worden
el 2959

Mann
el 2932

Newton
el 2902

Pompeys Pillar
el 2871

Ballantine
el 3004

Huntley
el 3014

Cline
el 3142

BILLINGS
el 3124

Yegen
el 3220

Mossmain
el 3262

Laurel
el 3297

Park City
el 3393

Silesia
el 3390

Montaqua
el 3576

Joliet
el 3734

Edgar
el 3463

Boyd
el 4033

Pryor
el 4067

Fromberg
el 3527

West Pryor
Mtn
6685

The Big Slide
5137

Bridger
el 3648

Fort Smith
el 3284

Bighorn
Canyon
NRA HQ

Bighorn
Lake

Bighorn

Dry Head

Canyon

Golden
el 3748

National

Belfry
el 3863

Wade
el 4064

Big Pryor
Mtn
8786

East Pryor
Mtn
8776

Recreation

Area

Warren
el 4413

CARBON
BIG HORN

BIG HORN
CARBON

YELLOWSTONE
BIG HORN

YELLOWSTONE
BIG HORN

PRYOR
MOUNTAINS

MONTANA
WYOMING

Elk Basin
el 4620

Frannie
el 4210

Bighorn

Canyon

National

Recreation

Area

Bighorn
Lake
3640

Deaver
el 4410

Cowley
el 3990

Lovell
el 3837

BIG HORN
SHERIDAN

BIG HORN MOUNTAINS

See Page 187

SCALE 1:500,000

5 Km    0    5    10    15    20    25    30    35    40    45    50 Km

KILOMETERS

Elevations are in feet

183

# Rexburg

See Page 180

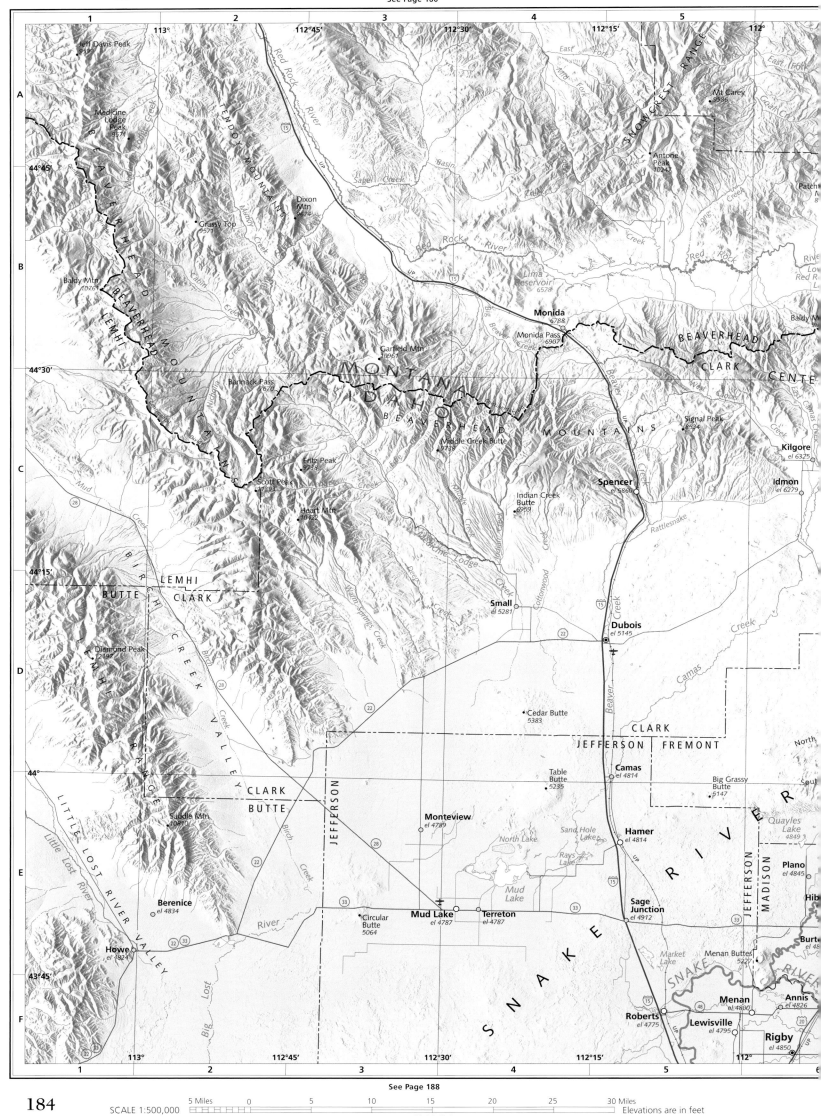

See Page 188

184

SCALE 1:500,000

5 Miles 0 5 10 15 20 25 30 Miles

MILES

Elevations are in feet

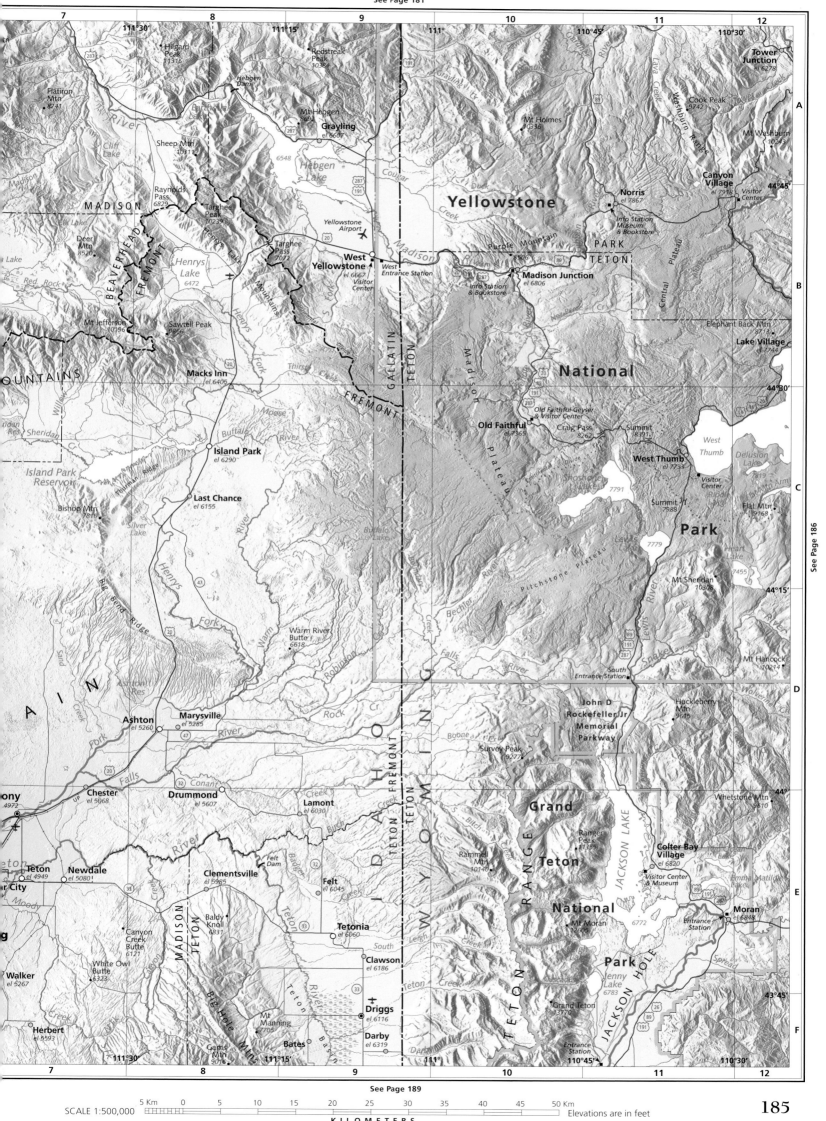

SCALE 1:500,000

KILOMETERS

Elevations are in feet

# Cody

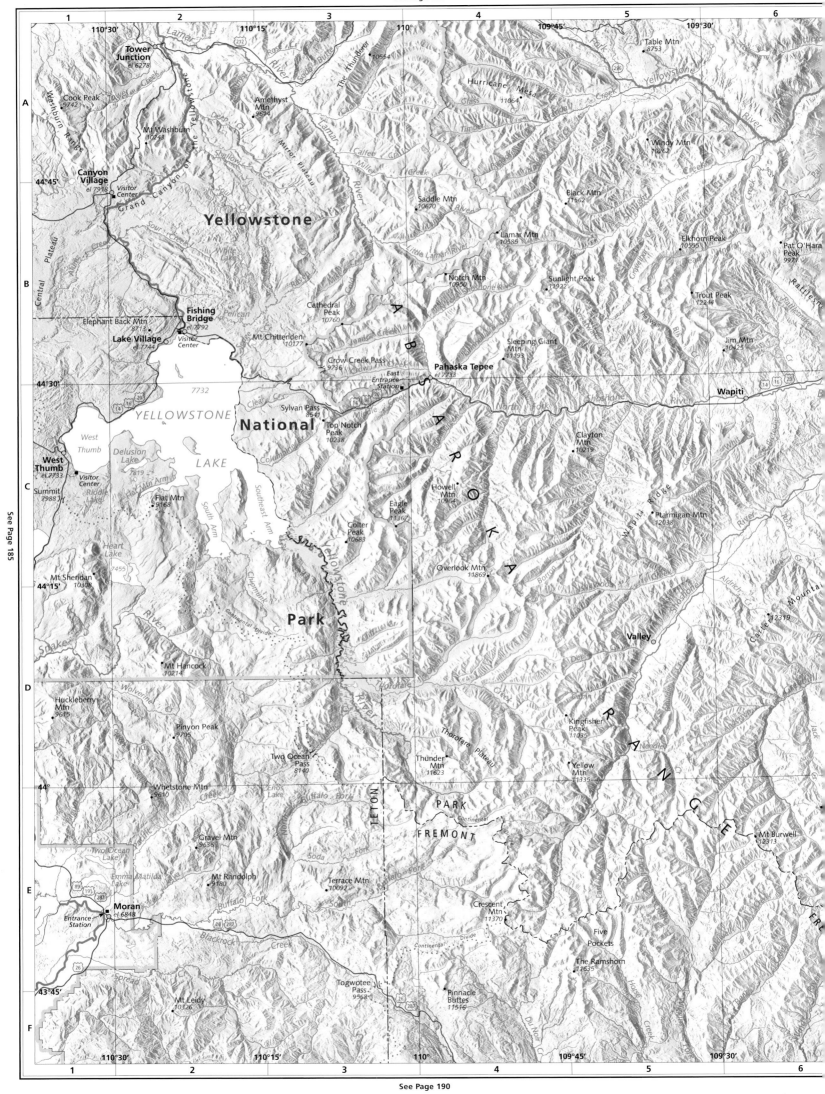

See Page 185

SCALE 1:500,000

5 Miles   0   5   10   15   20   25   30 Miles

MILES

Elevations are in feet

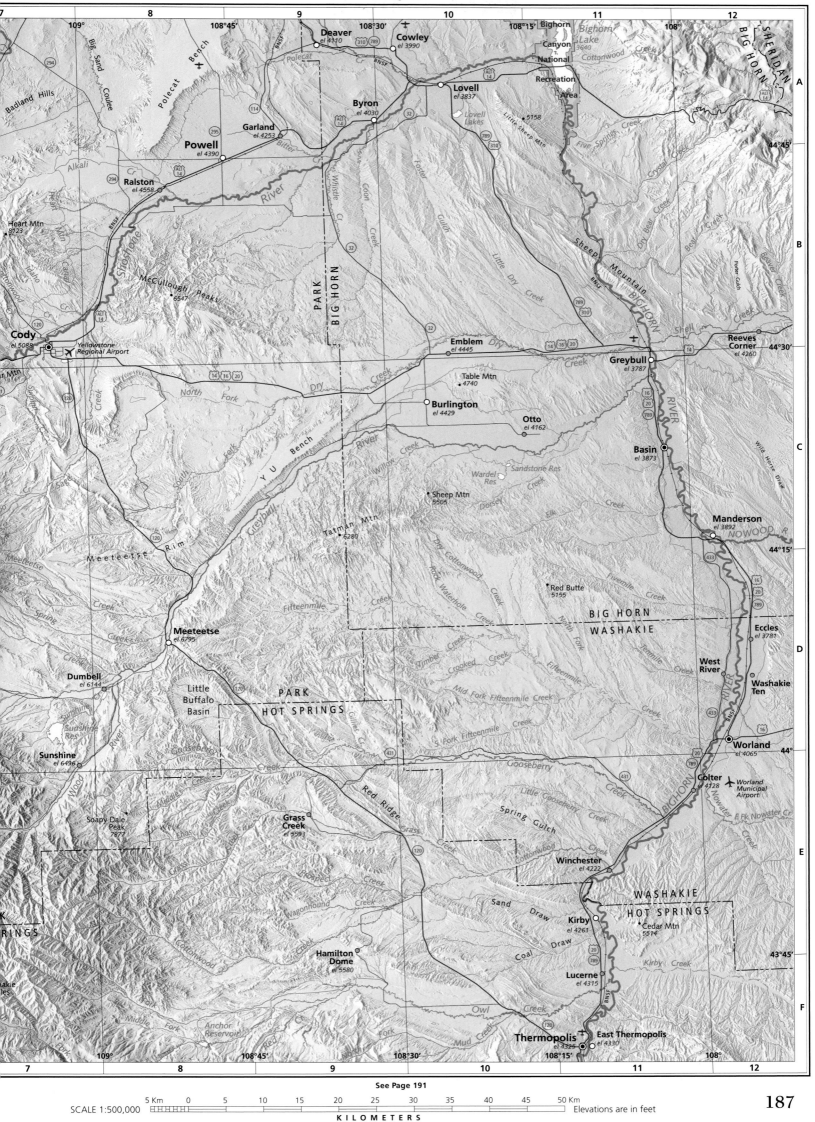

SCALE 1:500,000

5 Km 0 5 10 15 20 25 30 35 40 45 50 Km

Elevations are in feet

K I L O M E T E R S

**187**

# Pocatello

See Page 184

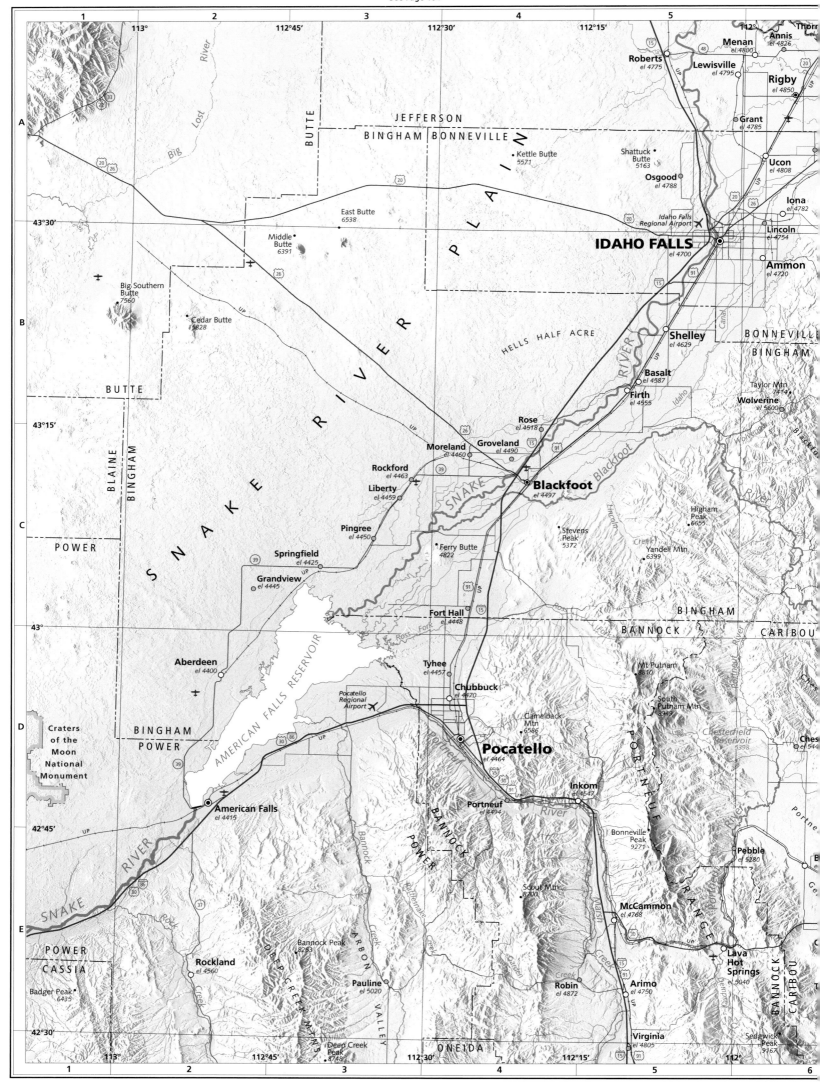

JEFFERSON

BINGHAM  BONNEVILLE

BUTTE

*Kettle Butte*
5571

*Shattuck*
*Butte*
5163

**Roberts**
el 4775

**Menan**
el 4800

**Annis**
el 4826

**Thorn**

**Lewisville**
el 4795

**Rigby**
el 4850

**Grant**
el 4785

**Ucon**
el 4808

*Osgood*
el 4788

**Iona**
el 4782

**IDAHO FALLS**
el 4700

*Idaho Falls*
*Regional Airport*

**Lincoln**
el 4754

43°30'

*East Butte*
6538

*Middle*
*Butte*
6391

**Ammon**
el 4720

*Big Southern*
*Butte*
7560

BONNEVILLE

HELLS HALF ACRE

BINGHAM

BUTTE

*Cedar Butte*
5828

**Shelley**
el 4629

43°15'

BLAINE

BINGHAM

**Basalt**
el 4587

*Taylor Mtn*
7414

**Firth**
el 4555

**Wolverine**
el 5600

S N A K E   R I V E R   P L A I N

**Rose**
el 4518

**Moreland**
el 4460

**Groveland**
el 4490

**Rockford**
el 4463

**Blackfoot**
el 4497

*Higham*
*Peak*
6655

**Liberty**
el 4459

POWER

**Pingree**
el 4450

*Ferry Butte*
4822

*Stevens*
*Peak*
5372

*Yandell Mtn*
6399

43°

**Springfield**
el 4425

**Grandview**
el 4445

S N A K E

**Fort Hall**
el 4448

BINGHAM

BANNOCK

CARIBOU

**Aberdeen**
el 4400

*Mt Putnam*
8810

*South*
*Putnam Mtn*
9440

**Tyhee**
el 4457

AMERICAN FALLS RESERVOIR

*Pocatello*
*Regional*
*Airport*

**Chubbuck**
el 4470

*Camelback*
*Mtn*
6586

*Chesterfield*
*Reservoir*
5398

**Ches**

BINGHAM

POWER

**Pocatello**
el 4464

Craters
of the
Moon
National
Monument

**Inkom**
el 4547

**Portneuf**
el 4494

*Portneuf River*

**Pebble**
el 5280

*Bonneville*
*Peak*
9271

**American Falls**
el 4415

BANNOCK

POWER

*Scout Mtn*
8700

**McCammon**
el 4768

SNAKE RIVER

42°45'

*Bannock Peak*
8263

CARBON

CREEK

MTNS

**Lava**
**Hot**
**Springs**
el 5040

POWER

CASSIA

**Rockland**
el 4560

**Pauline**
el 5020

VALLEY

**Robin**
el 4872

**Arimo**
el 4750

BANNOCK

CARIBOU

42°30'

*Badger Peak*
6435

DEEP CREEK

*Deep Creek*
*Peak*
8748

ONEIDA

**Virginia**
el 4805

*Sedgwick*
*Peak*
9167

SCALE 1:500,000

5 Miles        0        5        10        15        20        25        30 Miles

M I L E S

Elevations are in feet

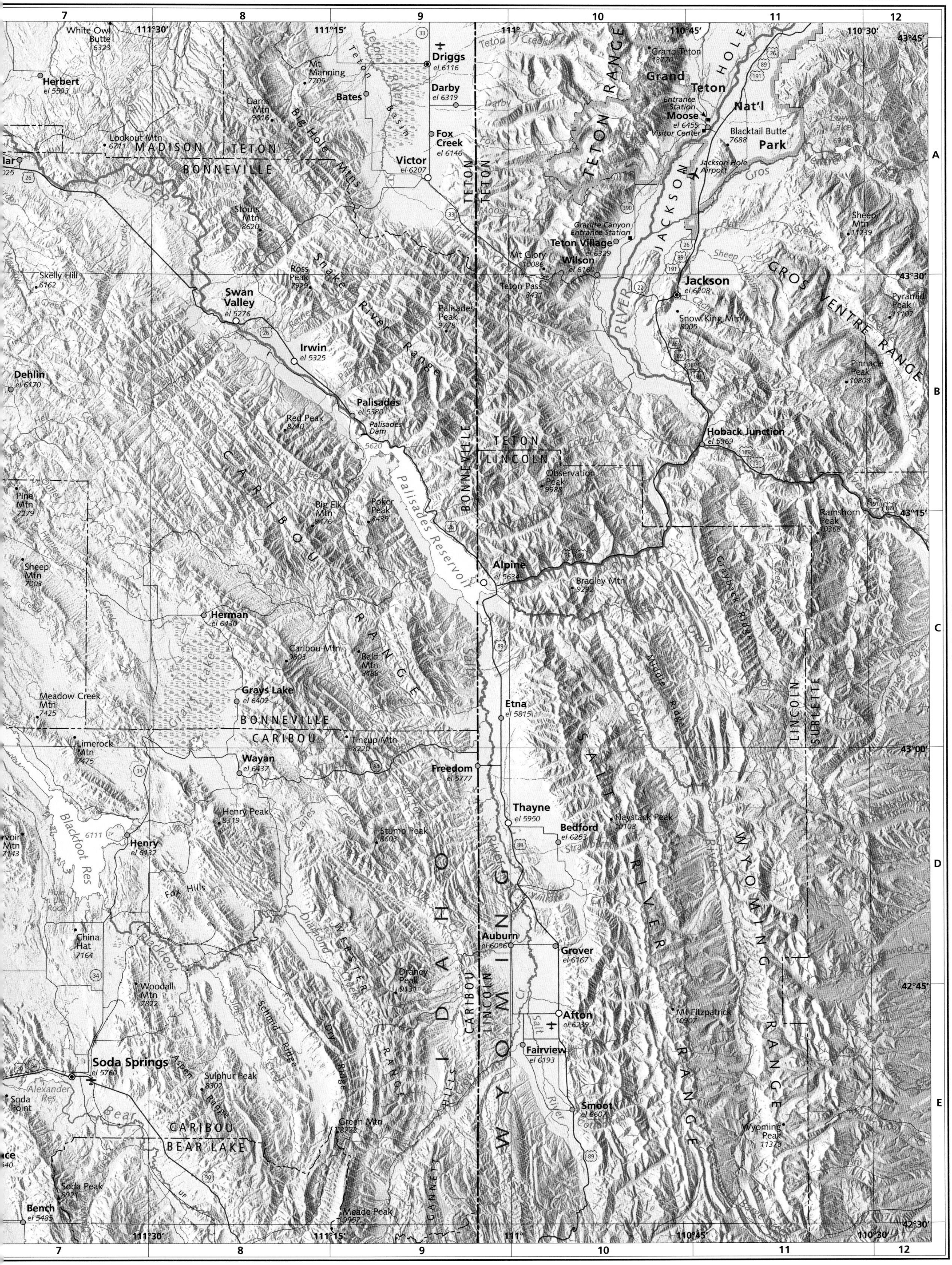

7   8   9   10   11   12

111°30'   111°15'   111°   110°45'   110°30'   43°45'

White Owl
Butte
6323
Herbert
el 5593
Lookout Mtn
6211
Skelly Hill
6162

MADISON   TETON
BONNEVILLE

Garns
Mtn
9016
Mt
Manning
7705
Bates

Driggs
el 6116
Darby
el 6319

Fox
Creek
el 6146

Victor
el 6207

Stouts
Mtn
8620

Ross
Peak
7929

Swan
Valley
el 5276

Irwin
el 5325

Dehlin
el 6170

Red Peak
8240

Palisades
el 5380

Palisades
Dam

Big Elk
Mtn
9476

Poker
Peak
8439

Pine
Mtn
7279

Sheep
Mtn
7003

Herman
el 6440

Caribou Mtn
9803

Bald
Mtn
8884

Grays Lake
el 6402

Meadow Creek
Mtn
7425

Limerock
Mtn
7475

BONNEVILLE
CARIBOU

Tincup Mtn
8220

Wayan
el 6437

Freedom
el 5777

Henry Peak
8319

Henry
el 6132

Fox Hills

Stump Peak
8601

China
Hat
7164

Woodall
Mtn
7822

Drancy
Peak
9137

Soda Springs
el 5760

Sulphur Peak
8901

Green Mtn
6772

CARIBOU
BEAR LAKE

Soda Peak
8321

Bench
el 5485

Meade Peak
9957

TETON RANGE
Grand Teton
13770

Grand Teton Nat'l Park

Entrance
Station
Moose
el 6459
Visitor Center

Blacktail Butte
7688

Granite Canyon
Entrance Station
Teton Village
el 6329

Wilson
el 6160

Mt Glory
10086

Teton Pass
8431

Jackson
el 6208

Snow King Mtn
8005

JACKSON HOLE

Sheep
Mtn
11239

GROS VENTRE RANGE

Pyramid
Peak
11107

Pinnacle
Peak
10808

Hoback Junction
el 5969

Observation
Peak
9588

Ramshorn
Peak
10368

Alpine
el 5634

Bradley Mtn
9292

Etna
el 5815

Thayne
el 5950

Bedford
el 6253

Haystack Peak
10108

Auburn
el 6056

Grover
el 6167

Afton
el 6239

Mt Fitzpatrick
10907

Fairview
el 6193

Smoot
el 6607

Wyoming
Peak
11378

SALT RIVER RANGE

WYOMING RANGE

CARIBOU RANGE

WEBSTER RANGE

SNAKE RANGE

GANNETT HILLS

IDAHO   WYOMING

LINCOLN   SUBLETTE

TETON   LINCOLN

BONNEVILLE   LINCOLN

Palisades Reservoir

Blackfoot Res

Alexander Res

Snake River

43°30'   43°15'   43°00'   42°45'   42°30'

111°30'   111°15'   111°   110°45'   110°30'

A   B   C   D   E

See Page 190

# Lander

See Page 189

**1**    110°30'    **2**    110°15'    **3**    110°    **4**    109°45'    **5**    109°30'    **6**

GTNP

Mt Leidy
10326

Togwotee
Pass
9658

Pinnacle
Buttes
11516

Sheep
Mtn
11239

Dubois
el 6940

Table Mtn
8120

Upper
8135

43°30'

Pyramid
Peak
11107

The
Six
Lakes

Sportsman
Ridge

FREMONT
SUBLETTE

Union
Pass
9210

Union Peak
11491

Whiskey
Mtn
11132

Lower
Table
7878

Wi

TETON

Pinnacle
Peak
10808

GROS VENTRE RANGE

Doubletop
Peak
11682

Elk Ridge

Little Sheep
Mtn
10223

Seven
Lakes

Simpson
Lake

Green River
Lakes

Continental
Glacier

WIND

Ross
Lake

Dinwoody
Lakes

43°15'

Ramshorn
Peak
10368

Bondurant
el 6588

Big Sheep
Mtn
11618

Saltlick Mtn
11045

Gannett Peak
13804
(Highest Point
in Wyoming)

Mammoth
Glacier

RIVER

Bold
10679

LINCOLN
SUBLETTE

Fremont
Peak
13745

Big
Lake

43°

New Fork
Lakes

7698

Willow
Lake

Fremont
Lake

FREMONT
SUBLETTE

Alpine
Lakes

Merna
el 7678

Cora
el 7341

Half
Lake
7595

Daniel
el 7193

Pinedale
el 7176

Boulder
Lake

Ra
Pea
125

42°45'

Boulder
el 7016

Silver

Big Sandy
el 7240

Wyoming
Peak
11378

Sixty-Seven
Res

The Mesa

East Fork River

Muddy Cr

Big

Marbleton
el 6869

351

Big Piney
el 6820

42°30'    110°30'    110°15'    110°    109°45'    109°30'

**1**    **2**    **3**    **4**    **5**    **6**

SCALE 1:500,000    5 Miles   0   5   10   15   20   25   30 Miles

**MILES**    Elevations are in feet

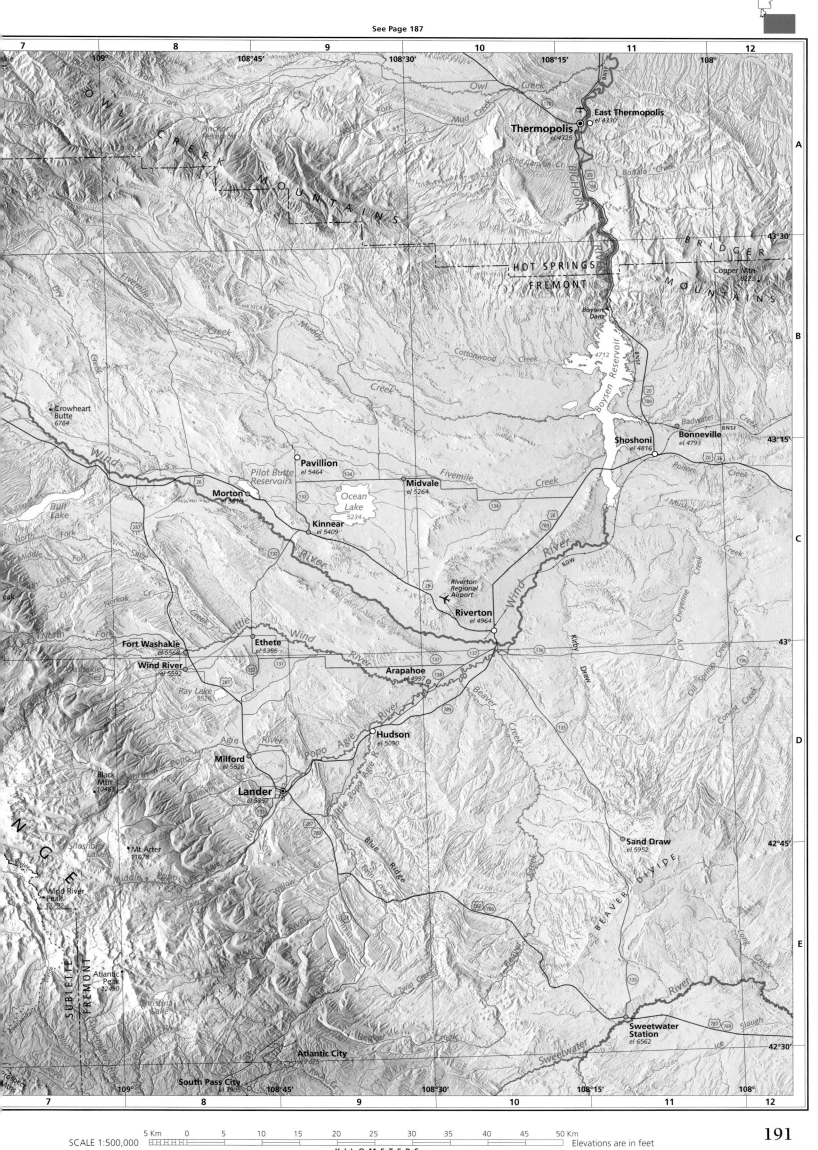

7     109°     8     108°45'     9     108°30'     10     108°15'     11     108°     12

OWL CREEK MOUNTAINS

East Thermopolis
el 4330

Thermopolis
el 4325

BIGHORN RIVER

BRIDGER

A

43°30'

HOT SPRINGS
FREMONT

Copper Mtn
8223

MOUNTAINS

Boysen
Dam

4712

B

Crowheart
Butte
6764

Cottonwood   Creek

Badwater

Shoshoni
el 4816

Bonneville
el 4793

43°15'

Wind

Pavillion
el 5464

Pilot Butte
Reservoir

Fivemile   Creek

Polson

Muskrat   Creek

134

Bull
Lake

Morton
el 5610

133

Ocean
Lake
5234

Midvale
el 5264

134

River

C

North Fork

Kinnear
el 5409

132

River

28

Riverton
Regional
Airport

Wind

BDW

Riverton
el 4964

Fort Washakie
el 5558

Ethete
el 5355

Little   Wind

137

137

136

Kirby

43°

Wind River
el 5592

132

137

Arapahoe
el 4997

138

River

789

135

Oil Spring Creek

Dry Cheyenne Creek

Conant Creek

Ray Lake
5526

287

Beaver

D

Black Mtn
10463

Popo Agie Rivey

Agie

Hudson
el 5090

789

Creek

Milford
el 5526

Agie R.

Beaver Creek

RANGE

Little Popo

Lander
el 5347

287

Sand Draw
el 5952

42°45'

Mt Arter
11078

Shoshone
Lake

789

Blue Ridge

BEAVER DIVIDE

135

E

SUBLETTE
FREMONT

Wind River
Peak
13192

Atlantic
Peak
12490

Christina Lake

Willow Creek

Beaver Creek

Twin Creek

135

Long Creek

Sweetwater
Station
el 6562

287 789

Atlantic City
el 7675

Sweetwater

South Pass City
el 7905     108°45'

108°30'

108°15'

108°

7     109°     8     9     10     11     12

42°30'

SCALE 1:500,000     5 Km   0   5   10   15   20   25   30   35   40   45   50 Km

KILOMETERS     Elevations are in feet

191

# National Park Reference Maps

## Legend

Maps in this section are at a scale of 1:100,000.
One inch equals approximately 1.58 miles on the ground.

### Settlements

◉ County seat
○ Incorporated city or town
◉ Unincorporated city or town

### Boundaries

▬▬▬ National park boundary
— - — State boundary
— - - — County boundary

### Roads and Transportation

(15) Interstate route
(191) U.S. highway route
(390) State route

—— Interstate freeway
—— Highway
— Paved road
— Unpaved road
▭▭▭ Runway
|—•—| Railroad track
– – – – Trail
— — — National trail

### Water Features

⌒ Perennial stream
⌒ Intermittent stream
⌒ Canal
⌒ Dam
◯ Perennial water body
◯ Intermittent pond
◯ Glacier
▨ Wetland
◦ Springs
○ Geysers or fumaroles
/ Falls

### Land Cover

▨ Forest
▭ Shrub, bare land, or regenerating forest
▭ Wetland
▨ Urban area

### Other Symbols

········ Continental Divide
*el 6290* Elevation in feet
• Peak, summit, or point (elevation in feet)
⌣ Pass, gap, or saddle
✈ Airport with commercial service
✦ Airport

The 1:100,000 scale national park maps on the following pages differ markedly in appearance from the preceding 1:500,000 series of regional maps. The larger scale of the 1:100,000 maps allows for abundant detail relative to the regional maps. Even the smallest roads are shown. All mapped streams and rivers are shown, and many more of them are named. Canals and ditches are differentiated (by their green color) from streams. Even small glaciers are identified. The Snake River in Jackson Hole, shown on the greater Yellowstone region maps as a single dark line, appears here in all its braided complexity.

The national park map series emphasizes land cover over elevation (the feature that is prioritized on the regional maps). The colors on the 1:100,000 maps parallel real world colors, as when green is used for forests. The land-cover classes for these maps are simplified from the original 2001 National Land Cover Data, a data set maintained by the U.S. Geological Survey that portrays twenty-nine classes

of land cover. For this application, the classes have been generalized into just four categories. Multiple forest types and wetland communities, for example, have been grouped into the single categories of "Forest" and "Wetland." The category "Shrub, bare land, and regenerating forest" includes areas that burned in recent decades that are still markedly evident as such on the landscape. These categories accurately portray overall patterns of land cover but may be incorrect at individual sites.

Elevations on the national park land cover maps are based on the same ten-meter elevation data set used for the 1:500,000 regional maps. However, the elevation data were not thinned out as they were for the regional maps. The resulting landforms depiction thus is more detailed than on the regional maps. The shading is lightened to accommodate the land cover coloring, which is given priority on these maps.

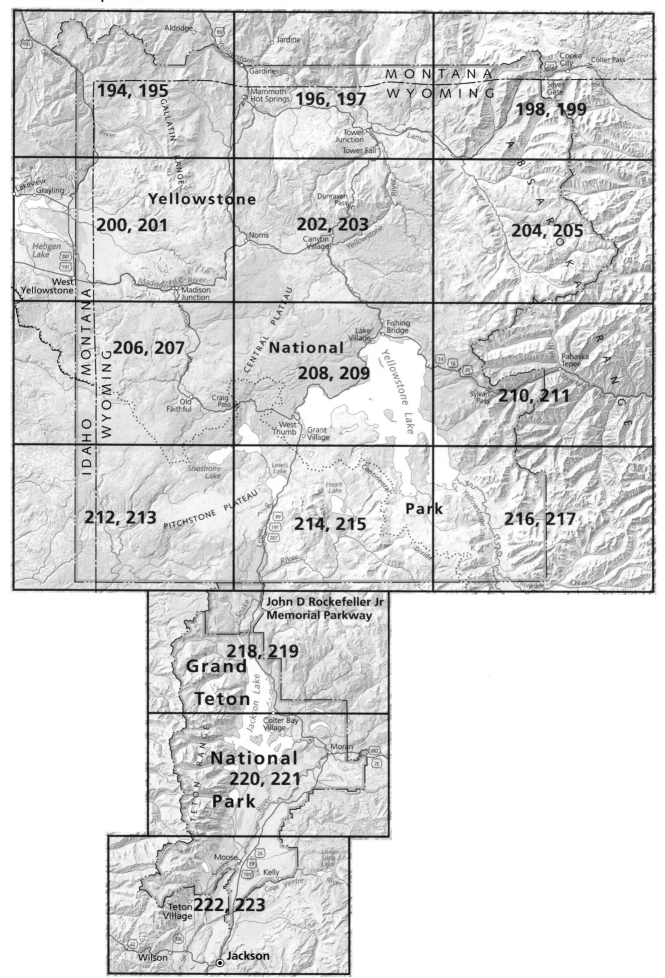

# Electric Peak

See Page 200

SCALE 1:100,000

0    1    2    3    4    5 Miles

M I L E S

Elevations are in feet

Burnt Top 8721
Grouse Mtn 8428
Tepee Pass 7644
Daly Pass 8350
Canary Bird Peak 7941
Sawtooth Mtn 9449
Sheep Mtn 9855
Marble Point 8044
Pulpit Rock
Sunshine Point 8235
Cameron Point 7606
Crown Butte 8059
King Butte 9315
Big Horn Peak 9888
Lava Butte 7904
Meldrum Mtn 9553
Crescent Lake 9577
Sedge Lake
Leech Lake 8764
Snowflake Springs
Black Butte 8410
Sportsman Lake
Monument Meadow
Monument Mtn 10091
Snowslide Mtn 10037
Red Mtn 9994
Fawn Pass
Cone Peak 9678
Redstreak Peak 10384
White Peak 10353
Upper Tepee Basin
Tepee Point 9423
Divide Lake
Tepee Basin
Bighorn Pass
Cabin Creek Cabin
Grayling

Gallatin
Taylor
Snowflake Ridge
Big Spring Creek
Sage Creek
Little Sage Creek
Skyline Ridge
Gallatin Petrified Forest
Gallatin River
Bacon Rind Creek
Snowslide Creek
Monument Creek
Specimen Creek
North Fork
East Fork
YELLOWSTONE NATIONAL PARK
MONTANA
WYOMING
Fan Creek
Gallatin River
Tepee East Fork Creek

Yellowstone National Park

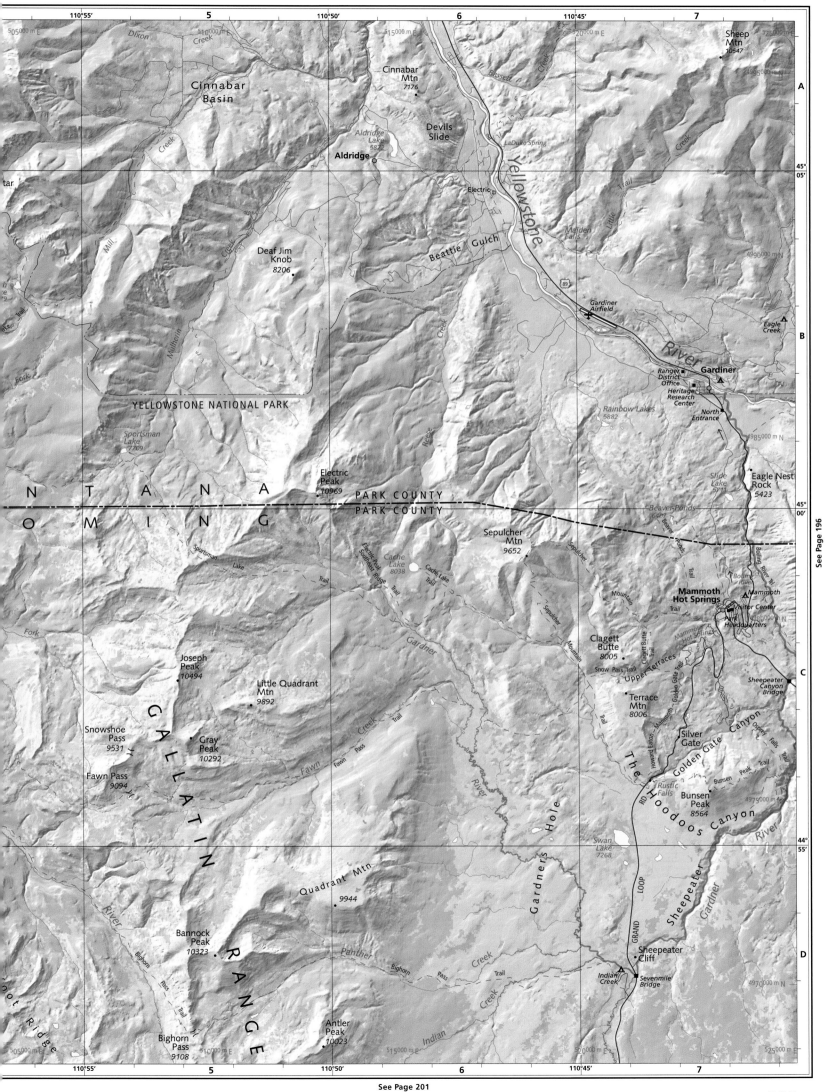

Sheep
Mtn
*10547*

505000 m E
510000 m E
*515000 m E*
*520000 m E*
*525000 m E*

Dixon Creek

Cinnabar
Basin

Cinnabar
Mtn
*7176*

4965000 m N

Aldridge
Lake
*5872*

Devils
Slide

Bassett Creek

**A**

LaDuke Spring

**Aldridge**

Electric

Yellowstone

4990000 m N

Little Creek

Maiden
Falls

**45°
05'**

Mill Creek

Deaf Jim
Knob
*8206*

Mulherin Creek

Beattie Gulch

Reese Creek

89

Gardiner
Airfield

**B**

River

Ranger
District
Office

Eagle
Creek

**Gardiner**

YELLOWSTONE NATIONAL PARK

Heritage
Research
Center

4985000 m N

Fork

Sportsman
Lake
*7709*

Rainbow Lakes
*5882*

North
Entrance

Slide
Lake
*5711*

Eagle Nest
Rock
*5423*

**45°
00'**

**M O N T A N A**
**W Y O M I N G**

Electric
Peak
*10969*

**PARK COUNTY**
**PARK COUNTY**

Sportsman Lake
Trail

Electric Peak Southeast Ridge Trail

Cache
Lake
*8038*

Cache Lake Trail

Sepulcher
Mtn
*9652*

Sepulcher

Sepulcher

Beaver Ponds
Trail

Beaver Ponds

Boiling River Trail

See Page 196

Mountain

**Mammoth
Hot Springs**

Bunsen
Mountain

Mammoth

Visitor Center

Park
Headquarters

4980000 m N

Fork

Gardner

Joseph
Peak
*10494*

**G A L L A T I N**

Little Quadrant
Mtn
*9892*

Clagett
Butte
*8005*

Clagett Butte

Mammoth

Upper Terraces

Snow Pass Trail

Terrace
Mtn
*8006*

Mammoth Golden Gate Trail

Silver
Gate

Golden Gate

Sheepeater
Canyon
Bridge

**C**

Oppet Falls

Canyon

Trail

**R A N G E**

Snowshoe
Pass
*9531*

Gray
Peak
*10292*

Creek

Howard Eaton Trail

Trail

Fawn Pass Trail

Fawn Pass
*9094*

Fawn

Fawn

River

Gardner's Hole

Rustic
Falls

RD

The Hoodoos Canyon

Bunsen
Peak
*8564*

Bunsen Peak Trail

4975000 m N

GRAND

Swan
Lake
*7268*

**44°
55'**

Quadrant Mtn
*9944*

LOOP

Sheepeater

Gardner

Bannock
Peak
*10323*

Bighorn

Panther

Bighorn Pass

Creek

Trail

Indian
Creek

Sheepeater
Cliff

**D**

Antler
Peak
*10023*

Bighorn
Pass
*9108*

Indian

Sevenmile
Bridge

4970000 m N

nut Ridge

River

505000 m E
510000 m E
*515000 m E*
*520000 m E*
525000 m E

See Page 201

SCALE 1:100,000

0　1　2　3　4　5　　KILOMETERS　　10 Km

Elevations are in feet

**195**

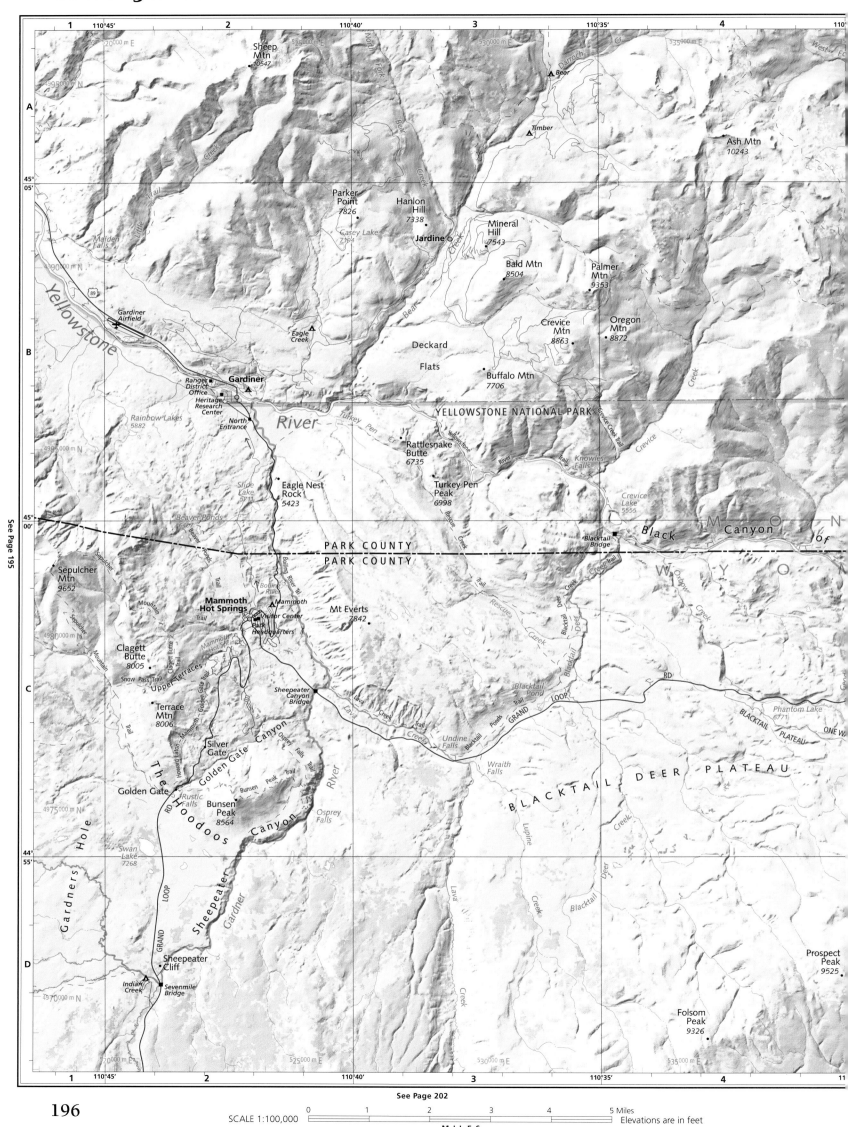

See Page 202

SCALE 1:100,000

MILES

5 Miles

Elevations are in feet

See Page 195

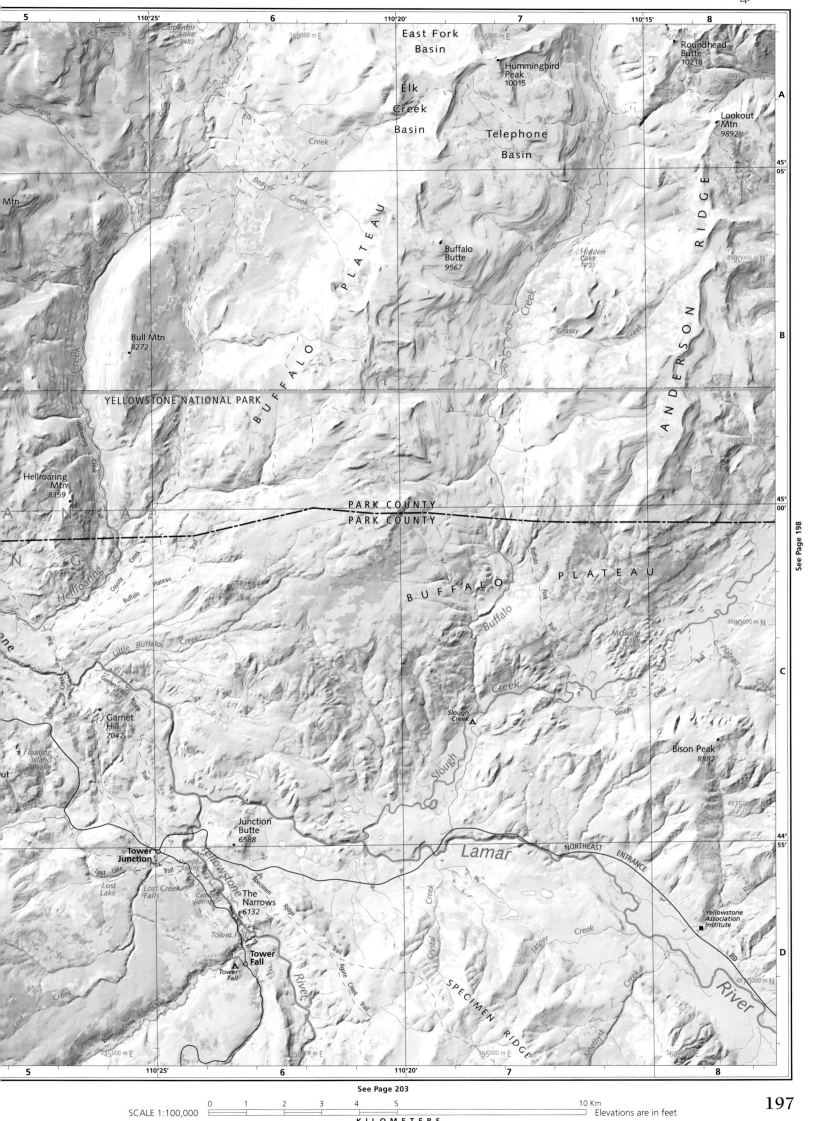

5    110°25'    6    110°20'    7    110°15'    8

A

45°
05'

B

YELLOWSTONE NATIONAL PARK

PARK COUNTY
PARK COUNTY

45°
00'

C

D

44°
55'

Carpenter
Lake
7482

East Fork
Basin

Hummingbird
Peak
10015

Roundhead
Butte
10218

Lookout
Mtn
9892

Elk

Creek

Basin

Telephone

Basin

Buffalo
Butte
9567

Hidden
Lake
7723

ANDERSON RIDGE

Grassy

Bull Mtn
8272

BUFFALO

Hellroaring
Mtn
8359

Hellroaring

BUFFALO PLATEAU

McBride
Lake
6585

Garnet
Hill
7047

Slough
Creek

Bison Peak
8882

Floating
Island
Lake

Junction
Butte
6588

Lamar

NORTHEAST ENTRANCE

Tower
Junction

Lost
Lake

The
Narrows
6132

Yellowstone
Association
Institute

Tower
Fall

SPECIMEN RIDGE

River

SCALE 1:100,000    0   1   2   3   4   5     10 Km
KILOMETERS    Elevations are in feet

# Silver Gate

See Page 204

SCALE 1:100,000

0  1  2  3  4  5 Miles

MILES

Elevations are in feet

See Page 205

SCALE 1:100,000

KILOMETERS

Elevations are in feet

# West Yellowstone

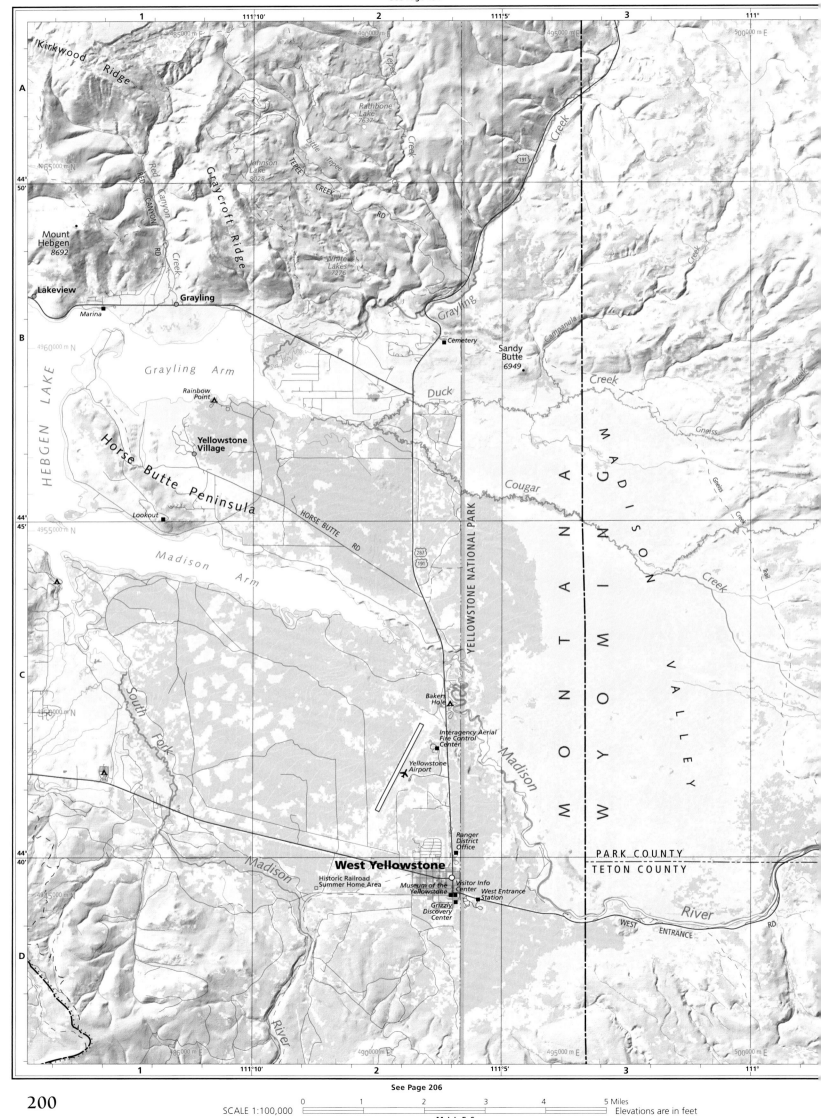

Kirkwood Ridge

Red Canyon RD

Graycroft Ridge

TEPEE CREEK

Tepee Creek

Little Tepee Creek

Rathbone Lake
7637

Johnson Lake
8028

Creek

191

Creek

Mount Hebgen
8692

White Lakes
7276

RD

Grayling

Lakeview

Marina

Grayling

Cemetery

Grayling

Campanula Creek

Sandy Butte
6949

Duck

Creek

Gneiss Creek

HEBGEN LAKE

Grayling Arm

Rainbow Point

Yellowstone Village

Horse Butte Peninsula

Lookout

Cougar

M A D I S O N

MADISON VALLEY

HORSE BUTTE RD

Madison Arm

287
191

South Fork

YELLOWSTONE NATIONAL PARK

Madison

Gneiss Creek

Trail

Bakers Hole

Interagency Aerial Fire Control Center

Yellowstone Airport

M O N T A N A
W Y O M I N G

Ranger District Office

PARK COUNTY
TETON COUNTY

West Yellowstone

Historic Railroad Summer Home Area

Museum of the Yellowstone

Visitor Info Center

West Entrance Station

Grizzly Discovery Center

Madison

River

WEST ENTRANCE RD

River

SCALE 1:100,000

0   1   2   3   4   5 Miles

MILES

Elevations are in feet

See Page 202

SCALE 1:100,000

0    1    2    3    4    5                              10 Km     Elevations are in feet

KILOMETERS

# Canyon Village

See Page 201

Willow
Park

Winter      Creek

Obsidian

Apollinaris
Spring

Horseshoe
Hill
*8314*

Cook Peak
*9742*

W
A
S
H
B
U
R
N

R
A
N
G
E

Tower

Obsidian
Lake
*7736*

Beaver
Lake

Crystal
Spring

Obsidian
Cliff
*7526*

The
Landmark
*8275*

Creek

Beaver
Lake
*7585*

Grizzly     Lake

Trail

Solfatara

Lake of the
Woods

Amphitheater
Springs

GRAND

Clearwater
Springs

Semi-Centennial
Geyser

Roaring
Mtn.
*8152*

Creek

Arrow

Canyon

Observation
Peak
*9397*

Observation

LOOP

Twin
Lakes
*9826*

Porcelain

Basin

Sulfatara

Trail

Peak     Trail

Cascade

Grebe
Lake

*8021*

Cascade
Lake
*7990*

Nymph
Lake
*7496*

Wolf Lake
*7958*

S
O
L
F
A
T
A
R
A

P
L
A
T
E
A
U

RD.

River

Gibbon

Trail

Grebe Lake Trail

Norris

Museum of the
National Park Ranger

Howard  Eaton

Cascade Lake – Norris

Norris
Geyser
Basin

Info Station
Museum &
Bookstore

**Norris**

Ice Lake

Wolf Lake Tr.

Virginia
Cascade

NORRIS     CANYON     RD.

River

Gibbon River
Rapids

Otter      Creek

Artists
Paintpots

Gibbon
Hill
*8601*

C
E
N
T
R
A
L

P
L
A
T
E
A
U

Monument
Geyser

Beryl
Spring

Paintpot
Hill
*8055*

Cygnet Lakes   Trail

Creek

PARK COUNTY
TETON COUNTY

Gibbon

Cygnet
Lakes
*8290*

Violet Springs

Gibbon
Falls

Canyon        Creek

Magpie        Creek

Mary
Mtn
*8611*

Alum        Creek

H
A
Y
D
E
N

SCALE 1:100,000

0        1        2        3        4        5 Miles

**M I L E S**

Elevations are in feet

See Page 204

**5**    110°25'    **6**    110°20'    **7**    110°15'    **8**

545 000m E    550 000m E    555 000m E    560 000m E

A

44°50'

4965 000m N

B

4960 000m N

44°45'

4955 000m N

C

4950 000m N

44°40'

4945 000m N

D

YELLOWSTONE

SPECIMEN RIDGE

Agate Creek
Agate Creek

GRAND LOOP RD
CHITTENDEN RD

Camellian Creek

Antelope Creek

Amethyst Creek

Chalcedony Creek

Amethyst Mtn
9614

Specimen Ridge Trail

Deep Creek

Burnt Creek

Mt Washburn
10243

Dunraven Pass
8859

Dunraven Peak
9904

Mt Washburn Trail
Mt Washburn Spur Trail

Broad Creek

Shallow Creek

Creek

Rainbow Springs

Wrong Creek

Washburn Hot Springs
Mt Washburn

Inkpot Spring

Seven Mile Hole Trail

Sulphur Creek

GRAND CANYON OF THE

Yellowstone River

Moss Creek

Coffee Pot Hot Springs

Josephs Coat Springs

Wapiti Lake

Wapiti Lake Trail

Broad Creek

Ribbon Lake

Artist Point

Ribbon Lake Loop

Forest Spgs

Wrangler Lake

Sour Creek

Wrangler Lake

Pelican Creek Trail

Fern Lake
8241

Fern Lake Trail

Pelican Creek

Dewdrop Lake

Sour Creek

Ponuntpa Springs

Tern Lake
8219

Astringent Creek

Cottongrass Creek

Yellowstone River

Hills
7985

Sour Mtn
37

Howard Eaton Trail

Bluff Creek

Creek

White Lake
8230

Astringent Creek

Trail

Elk Antler Creek

Sulphur Caldron

Dragons Mouth Spring

Mud Volcano

Stonetop Mtn
9042

The Mushpots

545 000m E    550 000m E    555 000m E    560 000m E

**5**    110°25'    **6**    110°20'    **7**    110°15'    **8**

SCALE 1:100,000    0   1   2   3   4   5    10 Km

KILOMETERS    Elevations are in feet

# Lamar Valley

See Page 203

SCALE 1:100,000

0  1  2  3  4  5 Miles

MILES

Elevations are in feet

The Needle
*9908*

S A R O K A

Creek

Canoe
Lake
*9160*

YELLOWSTONE NATIONAL PARK

Bootjack Gap

Trail

R A N G E

Saddle
Mtn
*10670*

Saddle

Hague Mtn
*10565*

Lamar

Little Lamar River

Castor
Peak
*10854*

Pollux
Peak
*11067*

Grant Peak
*10858*

Notch
Mtn
*10950*

Parker
Peak
*10203*

Hoodoo Basin

Trail

Lamar
Mtn
*10585*

Moose
Mtn
*10922*

Land
Mtn
*11268*

HURRICANE MESA

*11064*

Closed

Creek

Timber

Creek

Indian Peak
*10923*

Papoose

Creek

Hoodoo

Hoodoo
Peak
*10563*

Hoodoo Basin

One Hunt

Creek

Dike
Mtn
*10697*

Hoodoo Basin Trail

Painter
Cabin

Sunlight Creek

Brown Bear
Spring

N Fk Crandall Cr

Crandall Creek

Creek

Hoodoo

Creek

Black
Mtn
*11562*

Stinkingwater
Peak
*11597*

Copper
Lakes

110°
580000 m E
585000 m E
590000 m E
595000 m E
5
109°55′
6
109°50′
7
109°45′

44°50′
4965000 m N

44°45′
4955000 m N

44°40′
4945000 m N

A

B

C

D

SCALE 1:100,000

0  1  2  3  4  5        10 Km

K I L O M E T E R S        Elevations are in feet

205

# Old Faithful

See Page 200

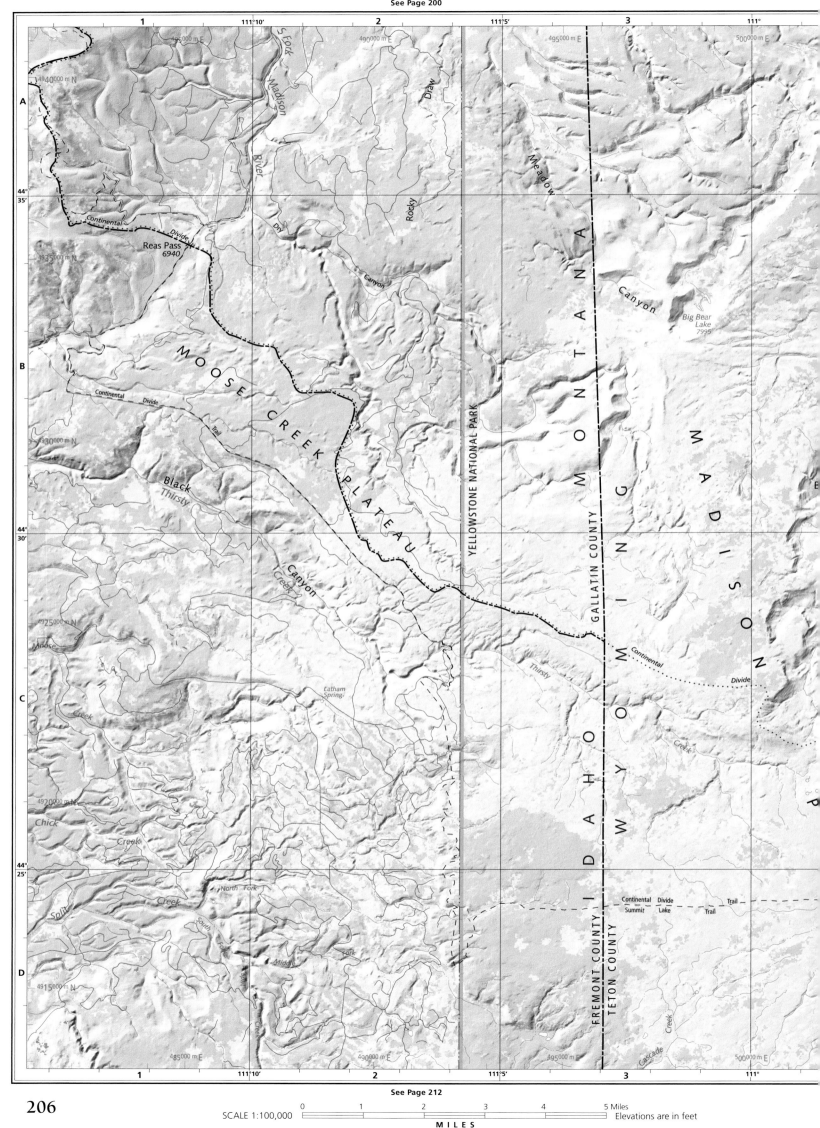

See Page 212

SCALE 1:100,000

MILES

0  1  2  3  4  5 Miles

Elevations are in feet

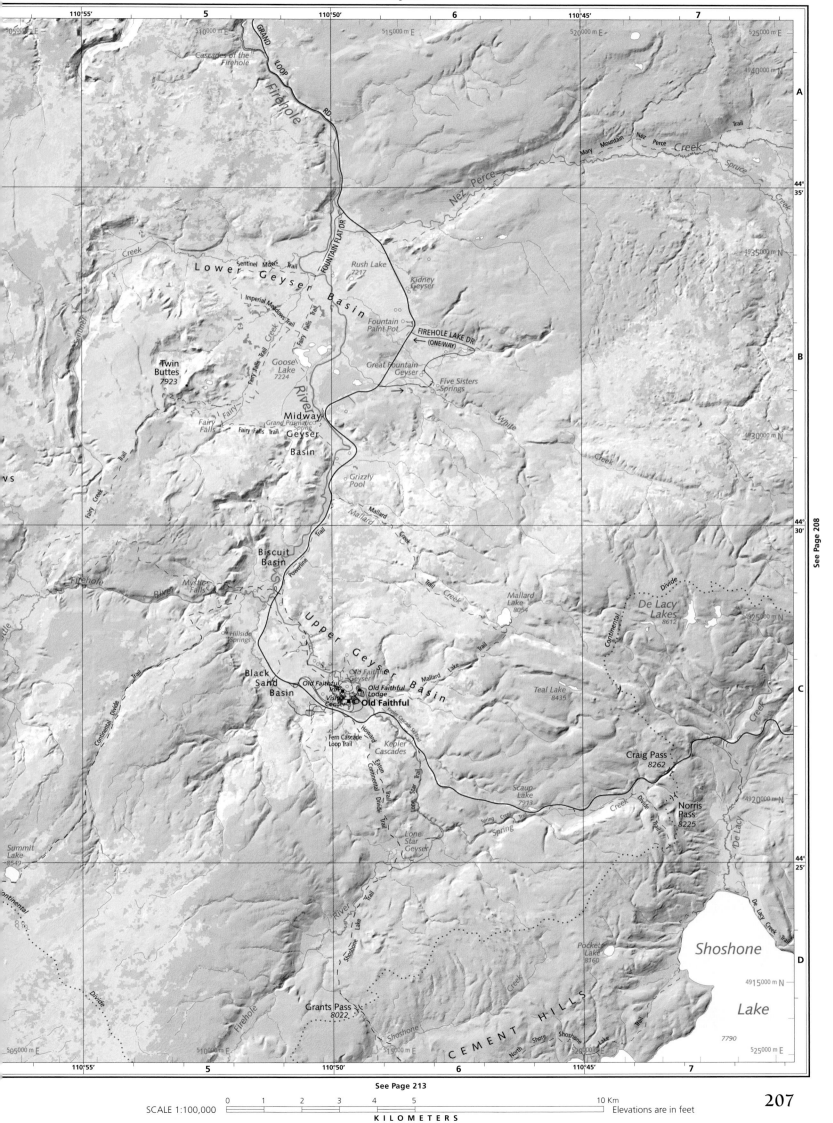

See Page 208

SCALE 1:100,000

0  1  2  3  4  5                          10 Km
KILOMETERS                    Elevations are in feet

# Lake Village

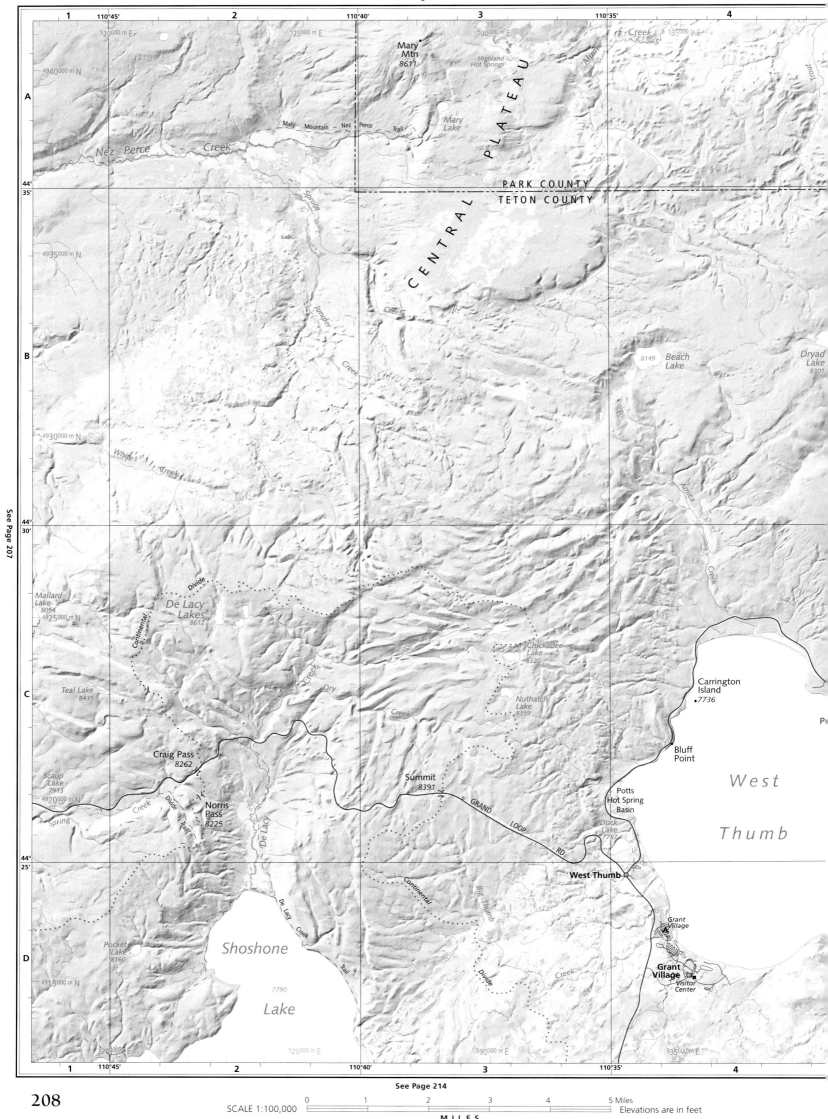

See Page 207

Mary
Mtn
*8611*

*Highland
Hot Springs*

Mary
*Mountain — Nez   Perce   Trail*

Mary
*Lake*

C E N T R A L     P L A T E A U

**PARK COUNTY**
**TETON COUNTY**

*Nez   Perce      Creek*

*Spruce*

*Juniper*

*Creek*

*Creek*

*White*     *Creek*

8149    *Beach
Lake*

*Dryad
Lake
8301*

*Arnica*

*Creek*

*Divide*

*De Lacy
Lakes
8612*

*Continental*

Mallard
Lake
8054

*Chickadee
Lake
8320*

Teal Lake
8435

*Creek*

*Dry*

*Creek*

Nuthatch
Lake
8359

Carrington
Island
•*7736*

Craig Pass
8262

Scaup
Lake
7913

*Spring*

*Creek*

*Divide     Trail*

Norris
Pass
8225

*De Lacy*

Summit
8391

Bluff
Point

Potts
Hot Spring
Basin

GRAND     LOOP

RD

*Duck
Lake
7787*

*West*

*Thumb*

**West Thumb**

*Continental*

*Big   Thumb*

Grant
Village

Pocket
Lake
8160

*De Lacy*

*Creek*

*Trail*

*Shoshone*

7790

*Lake*

*Divide*

*Creek*

**Grant
Village**

Visitor
Center

SCALE 1:100,000

0     1     2     3     4     5 Miles

M I L E S

Elevations are in feet

See Page 203

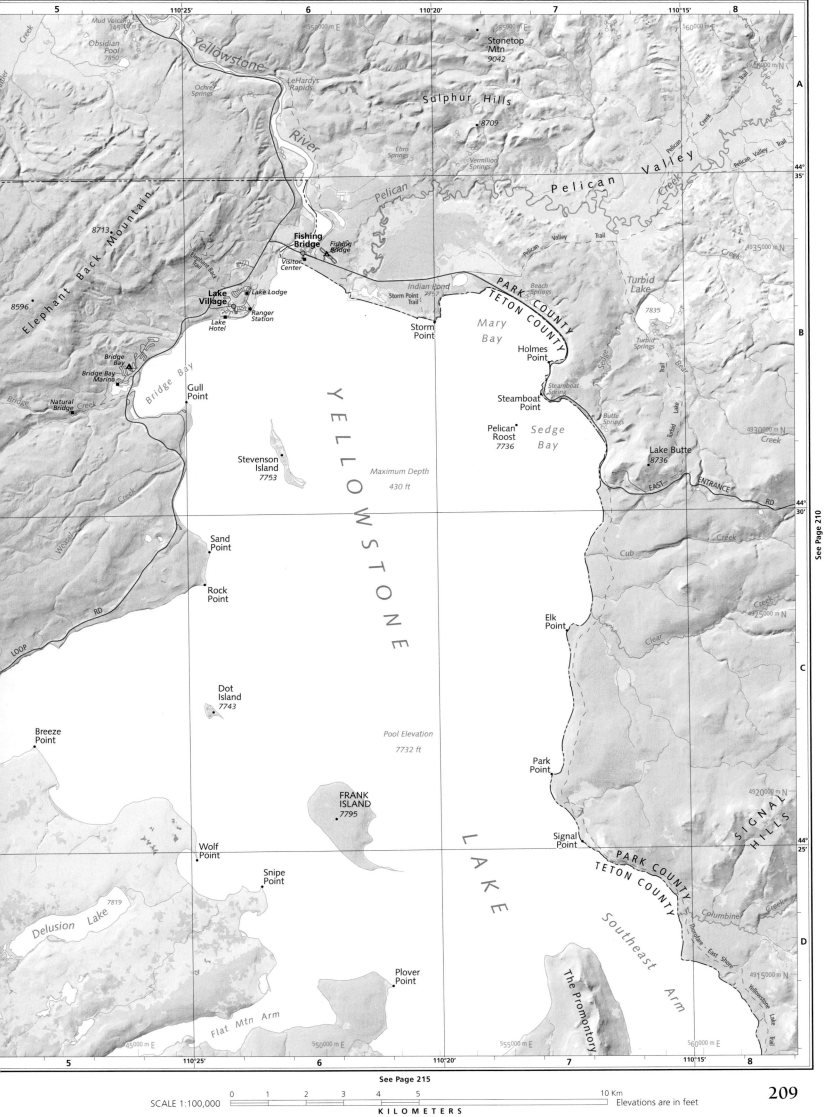

See Page 210

5     110°25'     6     110°20'     7     110°15'     8

44°35'

44°30'

44°25'

A

B

C

D

Mud Volcano
545000 m E

Obsidian
Pool
7850

Yellowstone

Ochre
Springs

LeHardys
Rapids

River

550000 m E

Stonetop
Mtn
9042

Sulphur Hills

8709

4940000 m N

Ebro
Springs

Vermilion
Springs

555000 m E

Pelican

Pelican    Valley

Pelican Valley Trail

Creek

560000 m E

8713

Elephant Back Mountain

Elephant Back Trail

8596

Fishing
Bridge

Fishing
Bridge

Visitor
Center

Lake Lodge

Lake
Village

Ranger
Station

Lake
Hotel

Indian Pond
7757

Storm Point Trail

Storm
Point

Mary
Bay

Holmes
Point

PARK COUNTY
TETON COUNTY

Beach
Springs

4935000 m N

Valley   Trail

Pelican

Turbid
Lake

7835

Turbid
Springs

Trail

Steamboat
Spring

Steamboat
Point

Butte
Springs

Sedge
Bay

Sedge   Lake

Turbid

Creek

4930000 m N

Bridge Bay

Bridge Bay Marina

Bridge

Natural
Bridge

Creek

Gull
Point

Bridge Bay

Weasel

Creek

Pelican
Roost
7736

Lake Butte
8736

EAST     ENTRANCE

RD

44°30'

Stevenson
Island
7753

Y E L L O W S T O N E

Maximum Depth

430 ft

Cub   Creek

4925000 m N

Sand
Point

Rock
Point

RD

LOOP

Dot
Island
7743

Pool Elevation

7732 ft

Elk
Point

Clear   Creek

C

Breeze
Point

Park
Point

4920000 m N

FRANK
ISLAND
7795

SIGNAL   HILLS

Wolf
Point

Snipe
Point

7819

L A K E

Signal
Point

PARK COUNTY
TETON COUNTY

44°25'

Delusion   Lake

Plover
Point

Southeast   Arm

The Promontory

Columbine   Creek

Thorofare   East Shore

4915000 m N

Yellowstone   Lake   Trail

D

Flat Mtn Arm

545000 m E

550000 m E

555000 m E

560000 m E

5     110°25'     6     110°20'     7     110°15'     8

See Page 215

SCALE 1:100,000

0   1   2   3   4   5       10 Km

K I L O M E T E R S

Elevations are in feet

**209**

# East Entrance

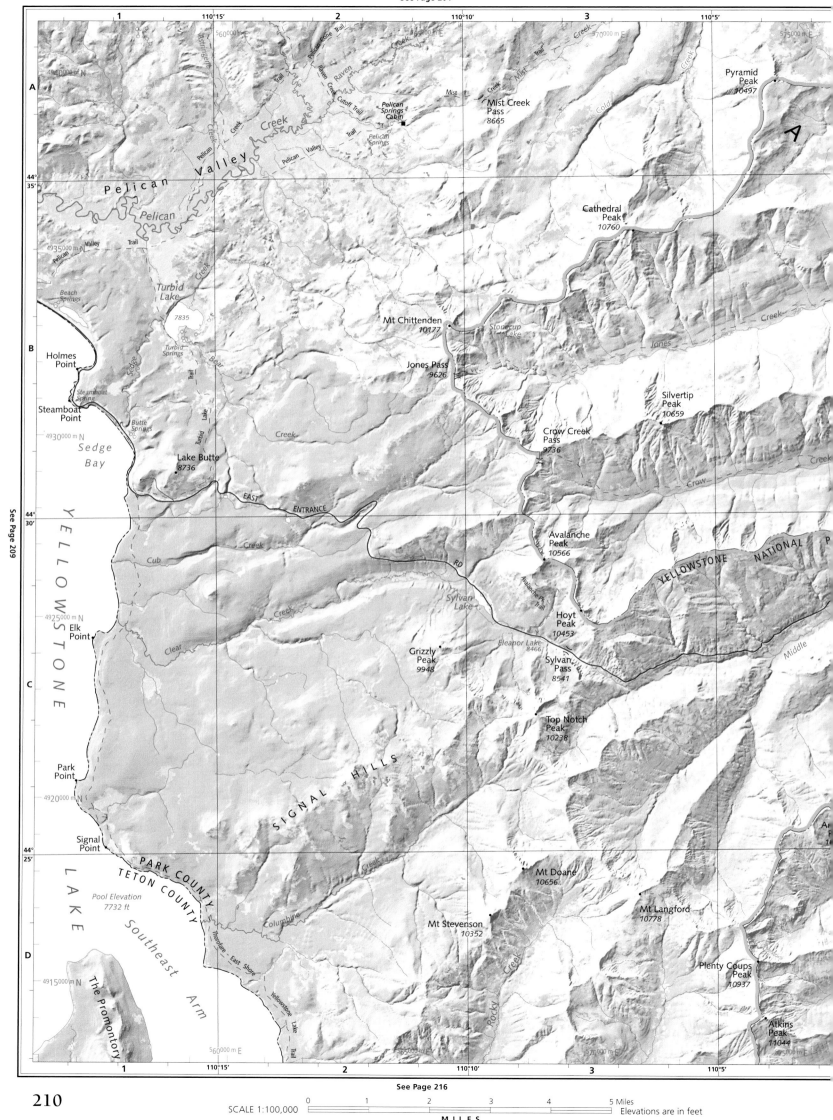

See Page 209

**A**

Pyramid
Peak
*10497*

Mist Creek
Pass
*8665*

Cathedral
Peak
*10760*

110°15′  110°10′  110°5′

560000 m E  565000 m E  570000 m E  575000 m E

4940000 m N

Pelican Cone Trail

Raven Creek

Raven Creek Cutoff Trail

Pelican
Springs
Cabin

Mist

Pelican Springs

Pelican Valley

Pelican Valley

Creek

Pelican

Valley

**44°35′**

Pelican

Pelican Valley Trail

4935000 m N

Beach
Springs

Turbid
Lake

*7835*

Turbid
Springs

Mt Chittenden
*10177*

Stonecup
Lake

Jones

Jones Pass
*9626*

Silvertip
Peak
*10659*

**B**

Holmes
Point

Sedge

Bear

Steamboat
Spring

Turbid Lake Trail

Creek

Crow Creek
Pass
*9736*

Steamboat
Point

Butte
Springs

4930000 m N

*Sedge
Bay*

Lake Butte
*8736*

EAST   ENTRANCE

Crow

Creek

**YELLOWSTONE**

Creek

Cub

RD

Avalanche
Peak
*10566*

**YELLOWSTONE NATIONAL P**

**44°30′**

Creek

*Sylvan
Lake*

Avalanche Pk Trail

4925000 m N

Elk
Point

Creek

Clear

Eleanor Lake
*8466*

Hoyt
Peak
*10453*

Middle

Grizzly
Peak
*9948*

Sylvan
Pass
*8541*

**C**

Top Notch
Peak
*10238*

Park
Point

4920000 m N

**S I G N A L   H I L L S**

Signal
Point

**44°25′**

Creek

Mt Doane
*10656*

**PARK COUNTY**

**TETON COUNTY**

*Pool Elevation
7732 ft*

Columbine

Mt Langford
*10778*

Mt Stevenson
*10352*

**L A K E**

*Southeast*

Thorofare East Shore

Plenty Coups
Peak
*10937*

**D**

4915000 m N

*The Promontory*

*Arm*

Yellowstone Lake

Rocky

Creek

Atkins
Peak
*11044*

560000 m E  570000 m E

110°15′  110°10′  110°5′

1  2  3

SCALE 1:100,000

0  1  2  3  4  5 Miles

**M I L E S**

Elevations are in feet

110°

5

109°55'

6

109°50'

7

109°45'

580000 m E

*Creek*

585000 m E

590000 m E

595000 m E

Sunlight
Peak
*11922*

4940000 m N

A

44°
35'

*Sweetwater
Creek
Falls*

Nipple
Mesa
*11492*

4935000 m N

Whirlwind
Peak
*10981*

Grinnell Meadows

*Clearwater*

B

R
O
K
A

Giant Castle
Mtn
*10161*

Sleeping
Giant
Mtn
*11193*

Monument
Mtn
*10914*

4930000 m N

Crow
Peak
*9348*

**Pahaska
Tepee**

44°
30'

NORTH  FORK  HWY

East
Entrance
Station

14
16
20

*North Fork*

Eagle
Creek

Wayfarer's
Chapel

Elephant
Head
Rock
*7018*

Chimney
Rock
*6490*

Lodge

Lodge

Newton
Springs

Newton
Creek

C

Shoshone

*River*

Canfield

R

A

N

Creek

Eagle

Kitty

Creek

G

44°
25'

Neva

Norris

Fishhawk

E

D

West

Sheep

Blackwater

4915000 m N

*Creek*

Coxcomb
Mtn
*11096*

580000 m E

Howell
Mtn
*10964*

585000 m E

Blackwater
Natural Bridge

110°

5

109°55'

6

109°50'

7

109°45'

SCALE 1:100,000

KILOMETERS

0  1  2  3  4  5

10 Km

Elevations are in feet

# Bechler Meadows

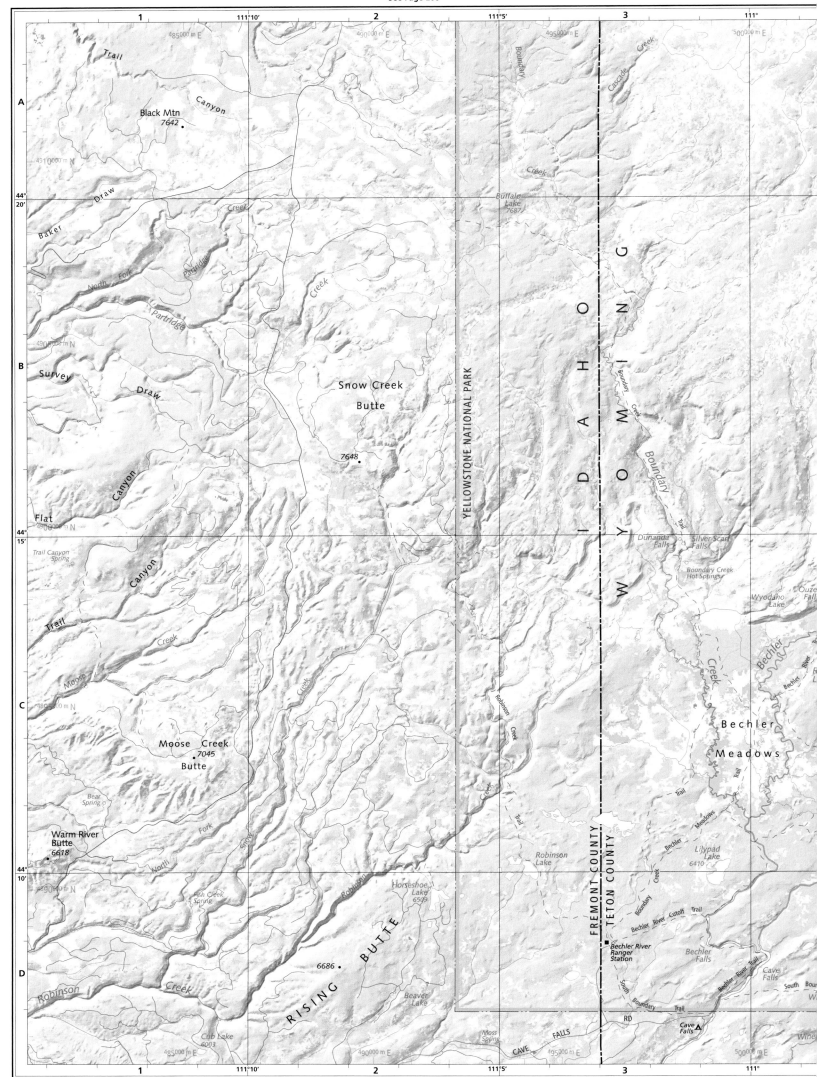

Trail

Canyon

Black Mtn
7642

Baker

Draw

North Fork

Partridge

Partridge

Creek

Creek

Survey

Draw

Snow Creek
Butte

7648

Canyon

Flat

Trail Canyon
Spring

Canyon

Trail

Creek

Moose

Moose   Creek
7045
Butte

Bear
Spring

Warm River
Butte
6618

North

Fork

Snow

Robinson

Creek

Fish Creek
Spring

Horseshoe
Lake
6509

RISING        BUTTE

6686

Beaver
Lake

Robinson   Creek

Cub Lake
6003

Moss
Spring

CAVE      FALLS

Robinson

Creek

Robinson
Lake

YELLOWSTONE NATIONAL PARK

Buffalo
Lake
7687

Boundary

Creek

Cascade        Creek

I  D  A  H  O

W  Y  O  M  I  N  G

Boundary

Creek

Boundary

Trail

Dunanda
Falls

Silver Scarf
Falls

Boundary Creek
Hot Springs

Wyodaho
Lake

Ouzel
Fall

Bechler

Creek

Bechler    River

Bechler

B  e  c  h  l  e  r

M  e  a  d  o  w  s

Trail

Bechler

Meadows

Lilypad
Lake
6410

Creek

Bechler River Cutoff   Trail

FREMONT COUNTY

TETON COUNTY

Bechler River
Ranger
Station

Bechler
Falls

Bechler

River

Trail

Cave
Falls

South     Boundary

South   Bou

RD

Cave
Falls

W

500000 m E

Wind

SCALE 1:100,000

0          1          2          3          4          5 Miles

M I L E S

Elevations are in feet

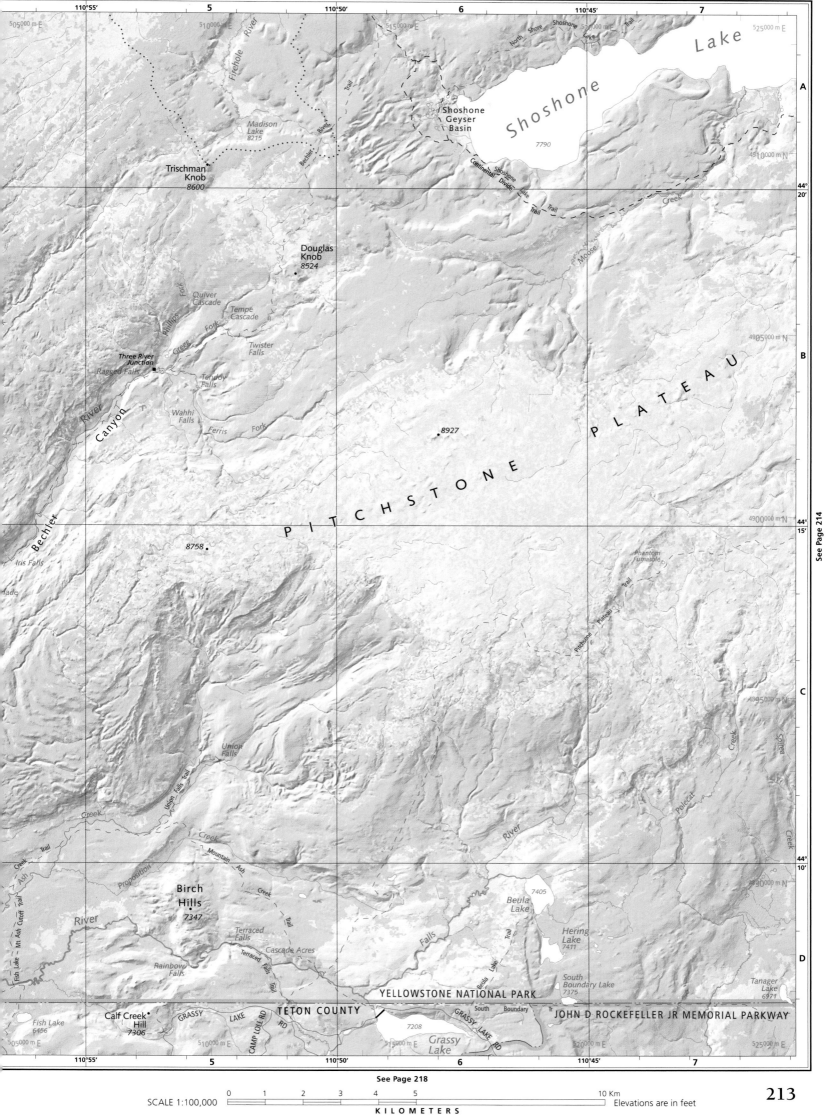

110°55'　　　　　5　　　　110°50'　　　　　6　　　　110°45'　　　　　7

505000 m E　　510000 m E　　515000 m E　　520000 m E　　525000 m E

Shoshone　Lake

North Shore Shoshone Lake Trail

4910000 m N

A

Firehole River

Madison
Lake
8215

Bechler　River

Shoshone
Geyser
Basin

7790

Continental Divide

Shoshone Lake Trail

Trail

44°
20'

Trischman
Knob
8600

Moose

Creek

Douglas
Knob
8524

Phillips Fork

Quiver
Cascade

Tempe
Cascade

Gregg Fork

4905000 m N

B

Three River
Junction

Ragged Falls

Twister
Falls

Tendoy
Falls

P　I　T　C　H　S　T　O　N　E

River

Canyon

Wahhi
Falls

Ferris　Fork

•8927

P　L　A　T　E　A　U

Bechler

4900000 m N

44°
15'

8758 •

Phantom
Fumarole

Iris Falls

ade

Pitchstone Plateau Trail

See Page 214

4895000 m N

C

Union
Falls

River

Union Falls Trail

Polecat

Spitzer Creek

Creek

Creek

Creek

Mountain Ash

Creek

44°
10'

Proposition

Birch
Hills
7347

Ash

Creek

Trail

Beula
Lake

7405

4890000 m N

Falls

Hering
Lake
7411

River

Terraced
Falls

Cascade Acres

Beula Lake Trail

South
Boundary Lake
7375

Tanager
Lake
6971

D

Fish Lake – Mt Ash Cutoff Trail

Rainbow
Falls

Terraced Falls Trail

YELLOWSTONE NATIONAL PARK

South Boundary

Creek

Fish Lake
6456

Calf Creek
Hill
7306

GRASSY
LAKE

CAMP LOLL RD

TETON COUNTY

7208

GRASSY LAKE RD

JOHN D ROCKEFELLER JR MEMORIAL PARKWAY

505000 m E

510000 m E

Grassy
Lake

515000 m E

520000 m E

525000 m E

110°55'　　　　　5　　　　110°50'　　　　　6　　　　110°45'　　　　　7

SCALE 1:100,000

0　1　2　3　4　5　　　　　　　10 Km

K I L O M E T E R S

Elevations are in feet

# Lewis Lake

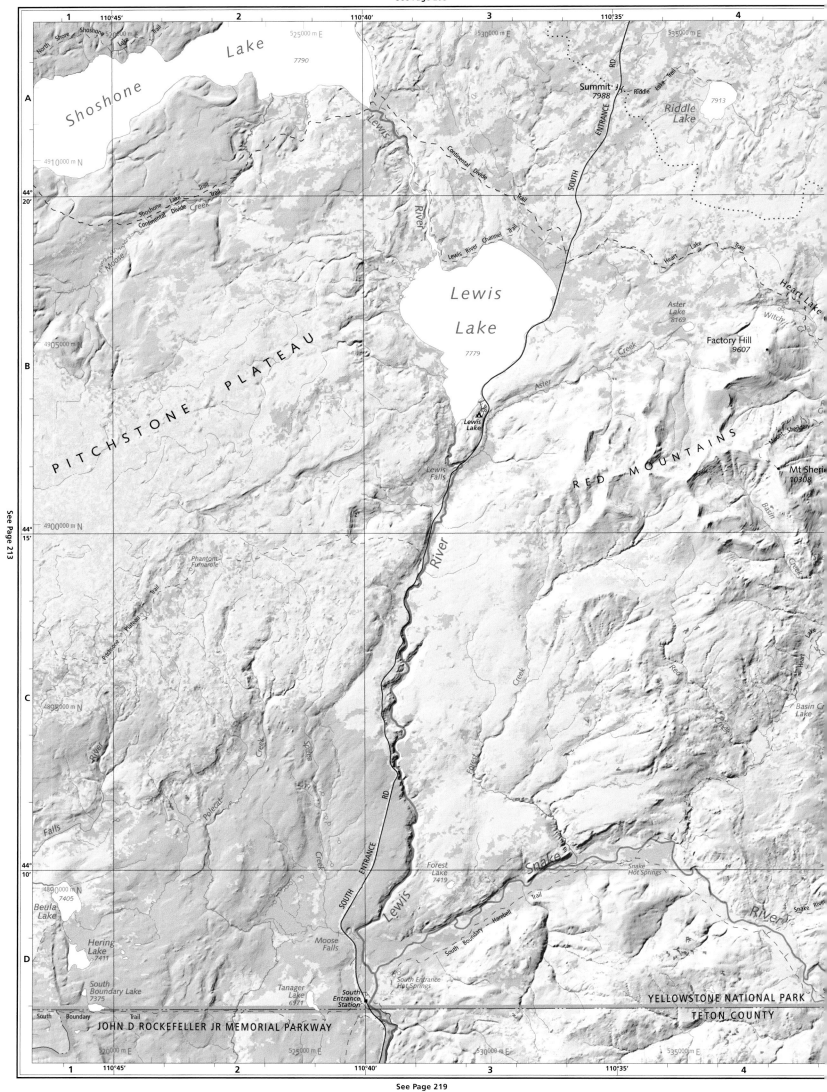

See Page 213

See Page 219

214

SCALE 1:100,000

0   1   2   3   4   5 Miles

M I L E S

Elevations are in feet

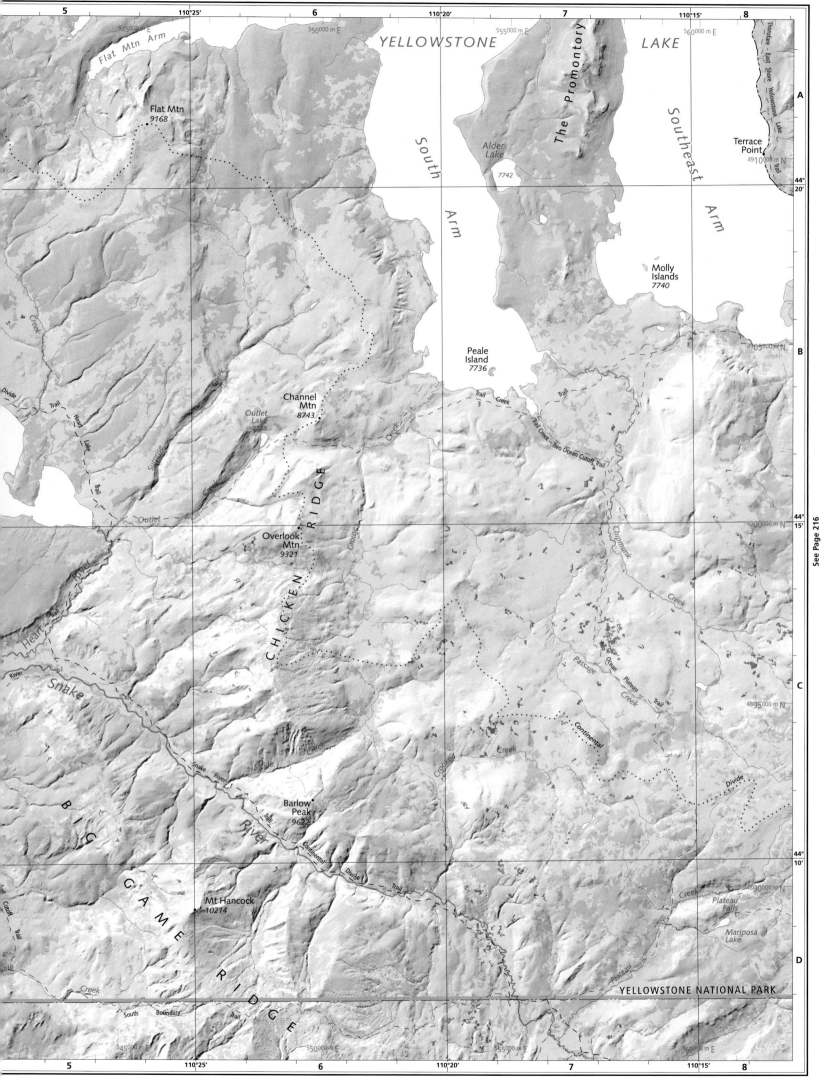

See Page 216

SCALE 1:100,000

KILOMETERS

Elevations are in feet

# Thorofare

See Page 210

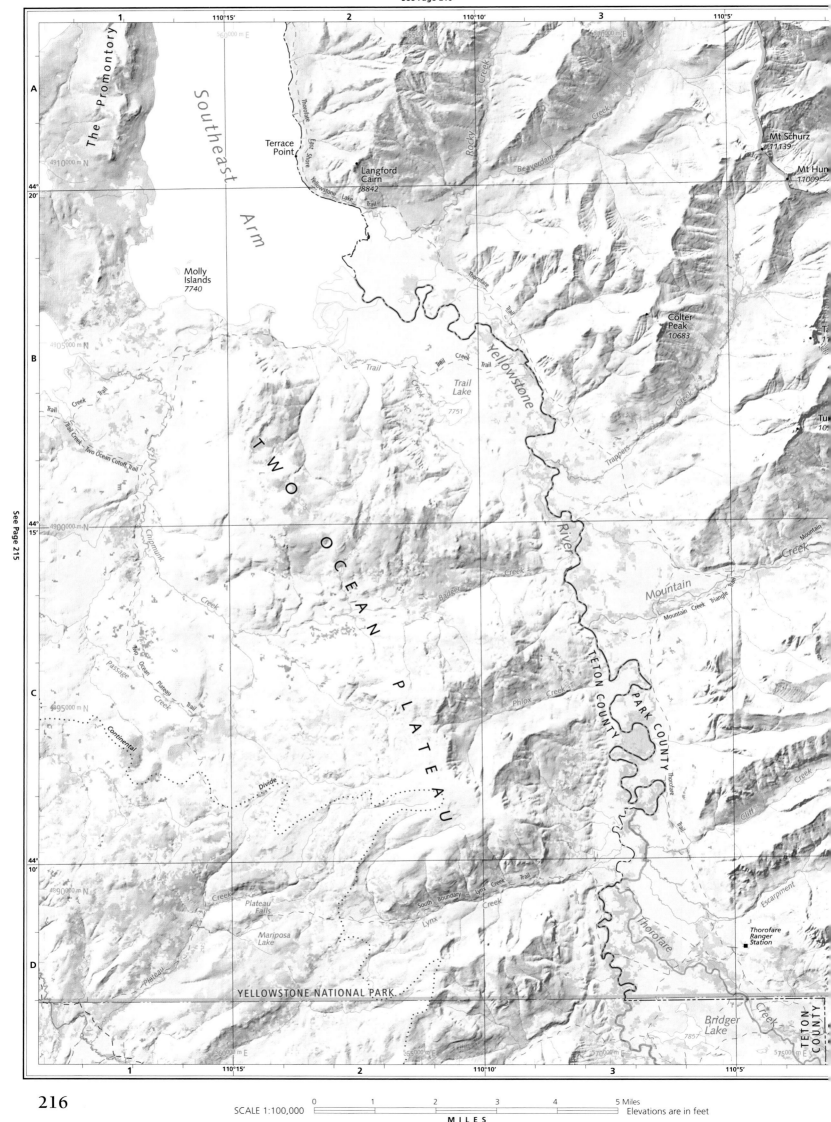

The Promontory

Southeast Arm

Thorofare Creek

Terrace Point

Langford Cairn
*8842*

Yellowstone Lake Trail

Rocky Creek

Beaverdam Creek

Mt Schurz
*11139*

Mt Hum
*11009*

Molly Islands
*7740*

Thorofare Trail

Yellowstone

Colter Peak
*10683*

Trail

Creek

Trail Lake
*7751*

Trail

TWO

Creek

Ta

Tu
*10*

Trail

Creek

Trappers

Creek

Trail Creek

Two Ocean Cutoff Trail

OCEAN

River

Mountain

Creek

Chipmunk

Creek

Badger Creek

Mountain Creek Triangle

Mountain

PLATEAU

Creek

Two Ocean

Passage

Plateau Creek Trail

Phlox Creek

TETON COUNTY

PARK COUNTY

Thorofare Trail

Cliff Creek

Continental

Divide

Lynx Creek Trail

South Boundary

Creek

Lynx

Creek

Escarpment

Creek

Plateau Falls

Thorofare Ranger Station

Mariposa Lake

Plateau

YELLOWSTONE NATIONAL PARK

Bridger Lake
*7857*

TETON COUNTY

See Page 215

## 216

SCALE 1:100,000

0  1  2  3  4  5 Miles

MILES

Elevations are in feet

ABSAROKA RANGE

Howell Mtn
10964

Blackwater
Natural Bridge

Sheep Mesa

11300

Fortress Mtn
12085

Eagle Nest
10798

Pinnacle Mtn
11466

Eagle Pass
9628

Seclusion Creek

Battlement Mtn
11813

Rampart

Chaos Mtn
11403

Fishhawk Glacier

Glacier Basin

Overlook Mtn
11869

Rampart Pass
10719

Ishawooa Cone
11853

Ishawooa Pass
9915

YELLOWSTONE NATIONAL PARK

Silver Tip

PETRIFIED RIDGE

Deer Creek Pass
10289

11444

Thorofare Buttes

Thorofare

SCALE 1:100,000

0  1  2  3  4  5          10 Km

KILOMETERS          Elevations are in feet

# Flagg Ranch

See Page 213

TETON COUNTY

510 000 m E

YELLOWSTONE NATIONAL PARK

515 000 m E

520 000 m E

525 000 m E

Tanager Lake
6971

South
Entrance
Station

JOHN D ROCKEFELLER JR MEMORIAL PARKWAY

Calf Creek
Hill
7306

GRASSY LAKE

CAMP LOLL RD

South Boundary

GRASSY

LAKE

Trail

7208

Grassy
Lake

Flagg
Ranch

4885000 m N

Tillery Lake
7312

Glade

RD

A

Camp Loll
(Boy Scouts of America)

7372

Lake of the
Woods

44°
05'

4880000 m N

Mt Berry
8951

Creek

Lewis

River

JOHN D ROCKEFELLER JR

South Boone Creek

Survey Peak

Trail

Hechtman
Lake
7864

Glade Creek

Steamboat
Mtn
7872

B

Survey
Peak
9277

Jackass
Pass
8640

Jackass Pass

Berry

JOHN D ROCKEFELLER JR MEMORIAL PARKWAY

GRAND TETON NATIONAL PARK

Creek

Creek

Berry

Creek

Trail

89
191
287

4875000 m N

Forellen Divide

Trail

Berry Owl Cutoff Trl

Harem
Hill
7330

Carrot
Knoll
8810

Conant
Pass
8831

Conant Pass Trail

Forellen
Peak
9776

Creek

Trail

Owl Creek

ELK Ridge

Lizard
Creek

44°
00'

Youngs
Point
9272

Owl

Creek

Fonda
Point

Lizard Creek

4870000 m N

North

Bitch

Creek

Red Mtn
10177

Elk Mtn
10760

Canyon

Creek

Wilcox
Point

JOHN D

C

South

Bitch

Moose Basin Divide

Webb

Canyon

Webb

Colter

Canyon

JACKSON LAKE

Bitch Creek
Narrows
7333

Mt Nord
9720

Moose
Mtn
10054

Moose
Basin

Moose

Canyon

4865000 m N

Nord
Pass
9419

TETON COUNTY

GRAND TETON NATIONAL PARK

Ranger
Peak
11355

Columbine Cascades

43°
55'

Camp
Lake

Creek

9046

Doane
Peak
11355

Wilderness
Falls

Waterfalls

Canyon

Pool Elevation
6769 ft

9732

Talus
Lake

D

Rammell
Mtn
10140

4860000 m N

Eagles Rest
Peak
1257

JACKSON LAKE

510 000 m E

515 000 m E

520 000 m E

525 000 m E

See Page 220

SCALE 1:100,000

0  1  2  3  4  5 Miles

MILES

Elevations are in feet

See Page 221

SCALE 1:100,000

0  1  2  3  4  5                    10 Km

KILOMETERS                          Elevations are in feet

**219**

# Grand Teton

See Page 218

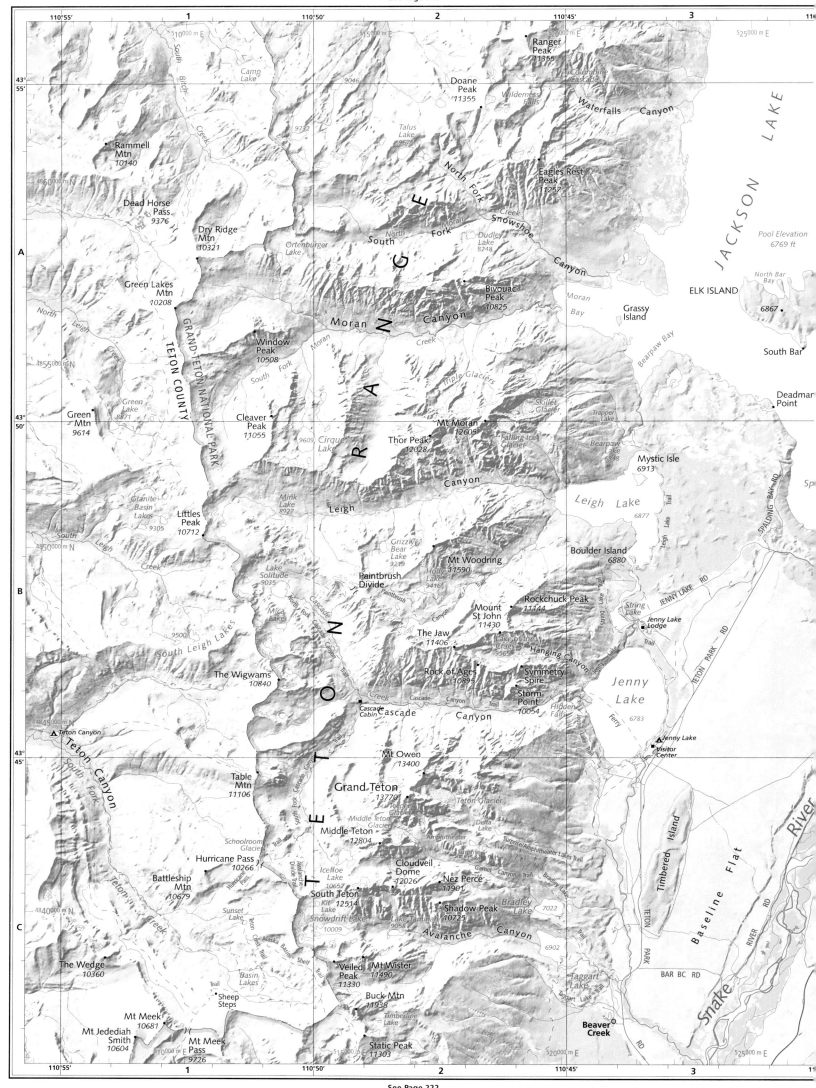

See Page 222

220

SCALE 1:100,000

MILES

Elevations are in feet

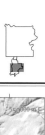

See Page 223

**221**

SCALE 1:100,000

0  1  2  3  4  5  10 Km

KILOMETERS

Elevations are in feet

# Moose

**1**    111°    110°55'    **2**    110°50'    **3**

500000 m E
4840000 m N

Sunset
Lake

Kit
Lake
Snowdrift Lake
10009

Lake Taminah
9058

Shadow Peak
10725

Bradley
Lake

Teton
Creek
Trail

Alaska
Basin
Trail

The Wedge
10360

Basin
Lakes

Veiled
Peak
11330

Mt Wister
11490

Buck Mtn
11938

Taggart
Lake

Avalanche
Canyon

**A**

Sheep
Steps

Mt Meek
10681

Mt Jedediah
Smith
10604

Mt Meek
Pass
9726

Static Peak
11303

Timberline
Lake

Stewarts
Draw

43°
40'

4835000 m N

Darby Camp
(Latter-Day Saints)

Mt Bannon
10966

Death

Alaska Basin Trail

Canyon

White Grass
Ranch

DARBY CANYON RD

Darby Creek

Fossil Mtn
10916

Prospectors
Mtn
11241

Rimrock
Lake
9516

Death Canyon
Ranger Station

Phelps
Lake
6633

RANGE

Fox Creek
Pass
9544

Forget-me-not
Lakes
9727

Coyote Lake

Open Canyon

Indian
Lake
9805

RD

Moose-Wilson RD

Sol

**B**

4830000 m N

Fox Creek

Housetop
Mtn
10537

Marion
Lake
9238

N Fk Granite Canyon Trail

Mt Hunt
10783

Granite Canyon
Creek

Granite Canyon Trail

Laurence Rockefeller
Preserve Center

River

Game Creek Pass

Mid Fk Granite Canyon Trail

GRAND TETON NATIONAL PARK

S Fk Granite Canyon Trail

Granite

Granite

Apres Vous
Peak
8426

Granite Canyon
Entrance Station

ZENITH

Game

Moose
Lake

TETON COUNTY

Rendezvous Mtn Trail

Mountain

Jackson Hole
Mountain Resort

43°
35'

4825000 m N

TETON

Rendezvous

10753

Teton
Village

SAGEBRUSH DR

Moose

Moose
Meadows

Rendezvous
Peak
10927

Creek

MOOSE-WILSON RD

PRINCE PL

**C**

4820000 m N

22

Phillips
Pass
8932

Canyon

Ski Lake
8654

Phillips

PHILLIPS RIDGE

Teton Pines
Golf Course

390

Teton Pines

West Gros Ventre Butte

SPRING CREEK RD

East Gros Ventre Butte

Taylor
Mtn
10352

RIDGE

FISH CREEK

6890

43°
30'

Coal
Creek

Mt Glory
10086

Creek

Fish Creek

22

Wilson

JACKSON

22

Journeys
School

Bridger-Teton
National Forest
Headquarters &
Visitor Center

**D**

4815000 m N

Teton
Pass
8431

Lake
7148

Trail

Canyon

FALL CREEK RD

ELY SPRS RD

Skyline
Ranch

Boyles Hill
6623

HOLE

Jackson Hole
Museum

Ja

Black

FALL CREEK RD

500000 m E

515000 m E

BOYLES HILL RD

Snow K

Snow

**1**    111°    110°55'    **2**    110°50'    **3**

SCALE 1:100,000    0   1   2   3   4   5 Miles

**MILES**

Elevations are in feet

4  110°40'  5  110°35'  6  110°30'  7

**Baseline Flat**

TETON
PARK

BAR BC RD

Blacktail
Ponds
6476

Thomas-
overy &
Center

**Moose**

Moose
Junction

Rock Face
6985

Blacktail
Butte
7688

26
89
191

W RD

s Ventre
tion

Gros
Ventre

Flat

son National
Hatchery

SHEEP

Sheep

rs-Butte  6775

Antelope Flats

ANTELOPE FLATS

FLATS RD

MORMON ROW

RD

Kelly Warm
Spring

**Kelly**

Gros
Ventre

GROS VENTRE

Long Hollow

Flat
RD

FLAT CREEK

CREEK

Creek

RD

Curtis
Canyon

North

Creek

SHADOW MTN

Antelope
Spring

Shadow
Mtn
8252

Hardeman
Reservoir
7401

South Fork

Creek

Ditch

Teton
Science
School

GROS VENTRE

River

GRAND TETON NATL PARK

TETON COUNTY

Timpa

Creek

Atherton
Creek

6908

Lower
Slide
Lake

Red Hills

Grizzly
Lake
7398

Bitter Creek

Crystal

West Miners

Creek

Blue Miner
Lake
9406

GROS VENTRE RANGE

Sheep Mountain

11239

Flat Creek
Ranch

Table Mtn
9987

Goodwin
Lake
9504

Creek

Jackson Peak
10741

4  110°40'  5  110°35'  6  110°30'  7

**223**

SCALE 1:100,000
0  1  2  3  4  5  10 Km
Elevations are in feet
**K I L O M E T E R S**

# Gazetteer

This gazetteer gives the page number and coordinates for the most important features named on the 1:100,000 and 1:500,000 scale reference maps. Features appearing on more than one map or at both scales are listed only once; the given listing portrays the location or feature in best context at the more appropriate scale.

## Cities, Towns, Locales and Sites

### A

| | | |
|---|---|---|
| Aberdeen | 188 | D2 |
| Absarokee | 182 | C6 |
| Acton | 183 | B9 |
| Afton | 189 | E10 |
| Alder | 180 | D5 |
| Aldridge | 195 | A6 |
| Alpine | 189 | C9 |
| American Falls | 188 | D2 |
| Ammon | 188 | B6 |
| Amsterdam | 181 | B8 |
| Anceney | 181 | C8 |
| Annis | 188 | A6 |
| Arapahoe | 191 | D10 |
| Arimo | 188 | E5 |
| Ashton | 185 | D8 |
| Atlantic City | 191 | E9 |
| Auburn | 189 | D9 |

### B

| | | |
|---|---|---|
| Ballantine | 183 | B11 |
| Bancroft | 188 | E6 |
| Basalt | 188 | B5 |
| Baseline | 183 | B9 |
| Basin | 187 | C11 |
| Bates | 189 | A9 |
| Bearcreek | 183 | E7 |
| Beaver Creek | 223 | A4 |
| Bedford | 189 | D10 |
| Belfry | 183 | E7 |
| Belgrade | 181 | B9 |
| Bench | 189 | E7 |
| Berenice | 184 | E2 |
| Big Piney | 190 | E3 |
| Big Sandy | 190 | E6 |
| Big Sky Meadow Village | 181 | D8 |
| Big Sky Mountain Village | 181 | D8 |
| Big Timber | 182 | B4 |
| Billings | 183 | B10 |
| Blackfoot | 188 | C4 |
| Bondurant | 190 | C2 |
| Bone | 188 | B6 |
| Bonneville | 191 | B11 |
| Boulder | 190 | D5 |
| Boyd | 183 | D7 |
| Bozeman | 181 | C9 |
| Bridger | 183 | D8 |
| Broadview | 183 | A8 |
| Burlington | 187 | C10 |
| Burris | 190 | B6 |
| Burton | 184 | E6 |
| Butte | 180 | A3 |
| Byron | 187 | A9 |

### C

| | | |
|---|---|---|
| Camas | 184 | D5 |
| Cameron | 181 | E7 |
| Canyon Village | 203 | C5 |
| Canyon Village | 185 | A12 |
| Cardwell | 180 | B6 |
| Central | 188 | E6 |
| Chester | 185 | E7 |
| Chesterfield | 188 | D6 |
| Chubbuck | 188 | D4 |
| Churchill | 181 | C8 |
| Clarkston | 181 | A8 |
| Clawson | 185 | E9 |
| Clementsville | 185 | E8 |
| Cline | 183 | B10 |
| Clyde Park | 181 | B11 |

| | | |
|---|---|---|
| Cody | 187 | B7 |
| Colter | 187 | E12 |
| Colter Bay Village | 220 | A4 |
| Colter Bay Village | 185 | E11 |
| Columbus | 182 | C6 |
| Comanche | 183 | A8 |
| Contact | 182 | C3 |
| Cooke City | 199 | B5 |
| Cooke City | 182 | E4 |
| Cora | 190 | D3 |
| Corwin Springs | 181 | E10 |
| Cowley | 183 | F10 |
| Craver | 182 | C6 |
| Crowheart | 191 | B7 |

### D

| | | |
|---|---|---|
| Daniel | 190 | D3 |
| Darby | 185 | F9 |
| Dean | 182 | D5 |
| Deaver | 183 | F9 |
| Dehlin | 189 | B7 |
| Dewey | 180 | B2 |
| Dillon | 180 | E3 |
| Divide | 180 | B3 |
| Driggs | 185 | F9 |
| Drummond | 185 | E8 |
| Dubois (Clark Co.) | 184 | D5 |
| Dubois (Fremont Co.) | 190 | A5 |
| Dumbell | 187 | D8 |

### E

| | | |
|---|---|---|
| East Thermopolis | 191 | A11 |
| Eccles | 187 | D12 |
| Edgar | 183 | D8 |
| Egin | 184 | E6 |
| Elk Basin | 183 | F8 |
| Emblem | 187 | B10 |
| Emigrant | 181 | D10 |
| Ennis | 181 | D7 |
| Ethete | 191 | C8 |
| Etna | 189 | C9 |
| Eustis | 181 | B7 |

### F

| | | |
|---|---|---|
| Fairview | 189 | E10 |
| Felt | 185 | E9 |
| Firth | 188 | B5 |
| Fishing Bridge | 209 | B6 |
| Fishing Bridge | 186 | B2 |
| Fishtail | 182 | D5 |
| Fishtrap | 180 | B1 |
| Fort Hall | 188 | C4 |
| Fort Smith | 183 | D12 |
| Fort Washakie | 191 | C8 |
| Four Corners | 181 | C9 |
| Fox Creek | 189 | A9 |
| Frannie | 183 | F9 |
| Freedom | 189 | D9 |
| Fromberg | 183 | D8 |

### G

| | | |
|---|---|---|
| Gallatin Gateway | 181 | C9 |
| Gardiner | 196 | B2 |
| Gardiner | 181 | E11 |
| Garland | 187 | A9 |
| Golden | 183 | E8 |
| Grace | 189 | E7 |
| Grandview | 188 | C2 |
| Grant | 188 | A5 |
| Grant Village | 208 | D4 |
| Grass Creek | 187 | E9 |
| Grayling | 200 | B1 |

| | | |
|---|---|---|
| Grayling | 185 | A9 |
| Grays Lake | 189 | C8 |
| Greybull | 187 | C11 |
| Greycliff | 182 | B4 |
| Groveland | 188 | C4 |
| Grover | 189 | D10 |

### H

| | | |
|---|---|---|
| Hamer | 184 | E5 |
| Hamilton Dome | 187 | E9 |
| Harrison | 180 | C6 |
| Hatch | 188 | D6 |
| Henry | 189 | D7 |
| Herbert | 189 | A7 |
| Herman | 189 | C8 |
| Hibbard | 184 | E6 |
| Hoback Junction | 189 | B11 |
| Howe | 184 | E1 |
| Hudson | 191 | D9 |
| Huntley | 183 | B10 |

### I

| | | |
|---|---|---|
| Idaho Falls | 188 | B5 |
| Idmon | 184 | C6 |
| Inkom | 188 | D4 |
| Iona | 188 | A6 |
| Irwin | 189 | B8 |
| Island Park | 185 | C8 |

### J

| | | |
|---|---|---|
| Jackson | 223 | D4 |
| Jackson | 189 | B10 |
| Janney | 180 | B4 |
| Jardine | 196 | B3 |
| Jardine | 181 | E11 |
| Jeffers | 181 | D7 |
| Jefferson Island | 180 | B6 |
| Joliet | 183 | D8 |

### K

| | | |
|---|---|---|
| Kelly | 223 | B5 |
| Kilgore | 184 | C6 |
| Kinnear | 191 | C9 |
| Kirby | 187 | E11 |

### L

| | | |
|---|---|---|
| Lake Village | 209 | B6 |
| Lake Village | 186 | B2 |
| Lakeview | 200 | B1 |
| Lakeview | 184 | B6 |
| Lamont | 185 | E9 |
| Lander | 191 | D9 |
| Last Chance | 185 | C8 |
| Laurel | 183 | C8 |
| Laurin | 180 | D5 |
| Lava Hot Springs | 188 | E5 |
| Lewisville | 188 | A5 |
| Liberty | 188 | C3 |
| Limestone | 182 | D4 |
| Lincoln | 188 | A6 |
| Livingston | 181 | C11 |
| Logan | 181 | B8 |
| Lorenzo | 184 | F6 |
| Lovell | 187 | A10 |
| Lucerne | 187 | F11 |
| Luther | 182 | D6 |

### M

| | | |
|---|---|---|
| Macks Inn | 185 | B8 |
| Madison Junction | 201 | D5 |
| Madison Junction | 185 | B10 |
| Mammoth Hot Springs | 196 | C2 |
| Mammoth Hot Springs | 181 | F11 |
| Manderson | 187 | C12 |
| Manhattan | 181 | B8 |
| Mann | 183 | B11 |
| Marbleton | 190 | E3 |
| Marysville | 185 | D8 |
| Maudlow | 181 | A9 |
| McAllister | 180 | D6 |
| McCammon | 188 | E5 |
| McLeod | 182 | C3 |
| Meeteetse | 187 | D8 |
| Melrose | 180 | C3 |
| Melville | 182 | A4 |
| Menan | 188 | A6 |
| Merna | 190 | D2 |
| Midvale | 191 | C9 |
| Milford | 191 | D8 |
| Milo | 188 | A6 |

| | | |
|---|---|---|
| Miner | 181 | E10 |
| Molt | 183 | B8 |
| Monida | 184 | B4 |
| Montaqua | 183 | C8 |
| Monteview | 184 | E3 |
| Moose | 223 | B4 |
| Moose | 189 | A11 |
| Moran | 221 | A5 |
| Moran | 185 | E11 |
| Moreland | 188 | C4 |
| Morton | 191 | C8 |
| Mossmain | 183 | C9 |
| Mud Lake | 184 | E4 |

### N

| | | |
|---|---|---|
| Newdale | 185 | E7 |
| Newton | 183 | B11 |
| Niter | 189 | E7 |
| Norris (Madison Co.) | 181 | C7 |
| Norris (Park Co.) | 202 | C2 |
| Norris (Park Co.) | 185 | B11 |
| Nye | 182 | D4 |

### O

| | | |
|---|---|---|
| Old Faithful | 207 | C6 |
| Old Faithful | 185 | C10 |
| Opportunity | 180 | A2 |
| Osgood | 188 | A5 |
| Otto | 187 | C10 |

### P

| | | |
|---|---|---|
| Pahaska Tepee | 211 | B5 |
| Pahaska Tepee | 186 | B4 |
| Palisades | 189 | B9 |
| Park City | 183 | C8 |
| Parker | 185 | E7 |
| Pauline | 188 | E3 |
| Pavillion | 191 | C9 |
| Pebble | 188 | E6 |
| Pine Creek | 181 | C11 |
| Pinedale | 190 | D4 |
| Pingree | 188 | C3 |
| Plano | 184 | E6 |
| Pocatello | 188 | D4 |
| Pompeys Pillar | 183 | B12 |
| Pony | 180 | C6 |
| Poplar | 189 | A7 |
| Portneuf | 188 | D4 |
| Powell | 187 | A8 |
| Pray | 181 | D11 |
| Pryor | 183 | D9 |

### Q

| | | |
|---|---|---|
| Quebec | 182 | C5 |

### R

| | | |
|---|---|---|
| Ralston | 187 | B8 |
| Rapelje | 182 | B6 |
| Red Lodge | 182 | E6 |
| Reed Point | 182 | C5 |
| Reeves Corner | 187 | B12 |
| Renova | 180 | B5 |
| Rexburg | 184 | E6 |
| Richel Lodge | 182 | E6 |
| Rigby | 188 | A6 |
| Ririe | 188 | A6 |
| Riverton | 191 | C10 |
| Roberts (Carbon Co.) | 183 | D7 |
| Roberts (Jefferson Co.) | 188 | A5 |
| Robin | 188 | E5 |
| Rockford | 188 | C3 |
| Rockland | 188 | E2 |
| Roscoe | 182 | D5 |
| Rose | 188 | B4 |

### S

| | | |
|---|---|---|
| Sage Junction | 184 | E5 |
| Salem | 184 | E6 |
| Sand Draw | 191 | D11 |
| Sappington | 180 | B6 |
| Sedan | 181 | B10 |
| Selmes | 183 | D7 |
| Shelley | 188 | B5 |
| Shepherd | 183 | B10 |
| Sheridan | 180 | D5 |
| Shoshoni | 191 | B11 |
| Silesia | 183 | C8 |
| Silver Gate | 199 | B5 |
| Silver Star | 180 | C4 |
| Skyline Ranch | 222 | D3 |

Small 184 D4
Smoot 189 E10
Soda Springs 189 E7
Solitude 222 B3
South Pass City 191 R8
Spencer 184 C5
Springdale 182 C3
Springfield 188 C3
St Anthony 185 E7
Sugar City 185 E7
Sunshine 187 D7
Swan Valley 189 B8
Sweetwater Station 191 E11

T

Terreton 184 E4
Teton 185 E7
Teton Pines 222 C2
Teton Village 222 B2
Teton Village 189 A10
Tetonia 185 E9
Thayne 189 D10
Thermopolis 191 A11
Thornton 184 E6
Three Forks 181 B7
Three Forks Junction 181 B7
Tower Fall 197 D6
Tower Junction 197 D5
Tower Junction 182 F2
Turner 188 E6
Twin Bridges 180 C4
Tyhee 188 D4

U

Ucon 188 A6

V

Valley 186 D5
Victor 189 A9
Virginia 188 E5
Virginia City 180 D6

W

Wade 183 E8
Walker 185 E7
Walkerville 180 A3
Wapiti 186 C6
Warren 183 E9
Washakie Ten 187 D12
Wayan 189 D8
West River 187 D12
West Thumb 208 D4
West Thumb 185 C11
West Yellowstone 200 D2
West Yellowstone 185 B9
Whitehall 180 B5
Wilderness 190 B6
Willow Creek 181 B7
Wilsall 181 B11
Wilson 222 D2
Wilson 189 A10
Winchester 187 E11
Wind River 191 D8
Wise River 180 B2
Wolverine 188 B6
Worden 183 B11
Worland 187 D12

Y

Yegen 183 C9
Yellowstone Village 200 B1

## Physical and Hydrologic Features

A

Abiathar Peak 198 C4
Absaroka Range 186 C4
Adair Creek 182 B2
Agate Creek 203 A6
Alder Lake 215 A7
Aldrich Creek 186 D6
Aldridge Lake 195 A6
Alexander Reservoir 189 E7
Alkali Creek (Park Co.) 187 B7
Alkali Creek (Sublette Co.) 190 E4
Alp Rock 199 A6
Alpine Lake 190 C6
Alum Creek 202 D4
American Falls Reservoir 188 D2
Amethyst Creek 198 D1
Amethyst Mountain 203 B7
Amphitheater Creek 198 C3
Amphitheater Lake 220 C2
Amphitheater Mountain 198 C4
Amphitheater Springs 202 B2
Anchor Reservoir 191 A8
Anderson Creek 186 D6
Anderson Ridge 198 B2
Antelope Creek (Bonneville Co.) 189 B7
Antelope Creek (Park Co.) 203 B6
Antelope Flats 220 C4
Antelope Point 183 B7
Antelope Spring 223 A5
Antler Peak 201 A5
Antone Peak 184 A5
Apollinaris Spring 202 A2
Apperson Creek 190 E1
Apres Vous Peak 222 B3
Aquarius Lake 199 B7
Arbon Valley 188 E3
Arizona Creek 219 C4
Arizona Island 219 C4
Arizona Lake 219 C4
Arnica Creek 208 B4
Arrow Canyon Creek 202 B3
Arrow Creek 183 B11
Arrow Peak 181 D11
Arthur Peak 210 C4
Artist Point 203 C5
Artists Paintpots 201 C7
Ash Mountain 196 A4
Ashton Reservoir 185 D8
Aspen Range 189 E8
Aster Creek 214 B3
Aster Lake 214 B4
Astringent Creek 210 A2
Atkins Peak 210 D4
Atlantic Creek 186 D3
Atlantic Peak 191 E7
Avalanche Canyon 220 C2
Avalanche Creek 217 B5
Avalanche Peak 210 C3

B

Bachelor Mountain 180 E1
Bacon Creek 190 A3
Bacon Rind Creek 194 C2
Badger Creek (Park Co.) 216 C2
Badger Creek (Teton Co.) 185 E9
Badger Island 220 A4
Badger Pass 180 E2
Badger Peak 188 E1
Badland Hills 187 A7
Badwater Creek 191 B11
Bailey Creek 219 C4
Baker Draw 212 B1
Bald Knob 199 B7
Bald Mountain (Bonneville Co.) 196 B3
Bald Mountain (Park Co.) 189 C9
Baldwin Creek 191 D8
Baldy Knoll 185 E8
Baldy Mountain (Beaverhead Co.) 180 D1
Baldy Mountain (Beaverhead/Clark Co.) 184 B6
Baldy Mountain (Beaverhead/Lemhi Co.) 184 B1
Baldy Mountain (Madison Co.) 180 E6
Bangtail Creek 181 B11
Bannack Pass 184 C2
Bannock Creek 188 D3
Bannock Peak (Park Co.) 188 E3
Bannock Peak (Power Co.) 195 D5
Barlow Peak 215 C6
Barronette Peak 198 C3
Baseline Flat 220 C3
Basin Creek (Beaverhead Co.) 184 A3
Basin Creek (Silver Bow Co.) 180 B3
Basin Creek (Teton Co.) 214 B4
Basin Creek Lake 214 C4
Basin Lakes 220 C1
Bassett Creek 195 A6
Battle Ridge 181 B10
Battlement Mountain 217 B7
Battleship Mountain 220 C1
Beach Lake 208 B4
Beach Springs 209 B7
Bear Creek (Big Horn Co.) 187 B12
Bear Creek (Bonneville Co.) 189 B8
Bear Creek (Carbon Co. east) 183 E9
Bear Creek (Carbon Co. west) 183 E7
Bear Creek (Fremont Co.) 190 A5
Bear Creek (Gallatin Co. north) 181 B9
Bear Creek (Gallatin Co. south) 181 C10
Bear Creek (Park Co. Montana) 196 B3
Bear Creek (Park Co. Wyoming east) 210 A4
Bear Creek (Park Co. Wyoming west) 210 B1
Bear Peak 191 C7
Bear Spring 212 C1
Bearpaw Bay 220 A3
Bearpaw Lake 220 B3
Beartooth Mountains 182 E4
Beartooth Pass 182 F6
Beartooth Plateau 182 E5
Beattie Gulch 195 B6
Beauvais Creek 183 D11
Beaver Creek (Big Horn Co.) 187 B12
Beaver Creek (Clark Co.) 184 D5
Beaver Creek (Fremont Co.) 191 D10
Beaver Creek (Park Co. Montana) 197 B6
Beaver Creek (Teton Co. Wyoming) 215 B5
Beaver Divide 191 E11
Beaver Lake (Fremont Co.) 212 D2
Beaver Lake (Park Co. east) 201 B7
Beaver Lake (Park Co. west) 201 B6
Beaver Ponds 196 B2
Beaverdam Creek 216 A3
Beaverhead Mountains 184 C3
Beaverhead River 180 E2
Beaverhead Rock 180 D4
Bechler Canyon 212 C4
Bechler Falls 212 D3
Bechler Meadows 212 C3
Bechler River 212 C4
Bennett Creek 182 F6
Berry Creek 218 B1
Beryl Spring 201 C6
Beula Lake 213 D6
Bierer Creek 223 B6
Big Bear Creek 181 C9
Big Bear Lake 206 B3
Big Beaver Creek 184 B4
Big Bend Ridge 185 C7
Big Bull Elk Creek 183 E11
Big Creek (Park Co. Montana) 181 D9
Big Creek (Park Co. Wyoming) 186 B5
Big Elk Creek 189 B9
Big Elk Mountain 189 B8
Big Game Ridge 215 C5
Big Grassy Butte 184 D5
Big Hole Mountains 189 A8
Big Hole River 180 B1
Big Horn Mountain 180 F6
Big Horn Peak 194 B3
Big Lake 183 B7
Big Lost River 188 A2
Big Milky Lake 190 C5
Big Moose Lake 199 C7
Big Pipestone Creek 180 B4
Big Pryor Mountain 183 E10
Big Sand Coulee 187 A8
Big Sandy River 190 E6
Big Sheep Creek 184 B2
Big Sheep Mountain 190 B4
Big Southern Butte 188 B1
Big Spring Creek 194 B1
Big Thumb Creek 208 D3
Big Timber Creek 182 A3
Big Twin Creek 190 C3
Big Woody Creek 183 C11
Bighorn Lake 187 A11
Bighorn Lake (Montana) 183 D11
Bighorn Lake (Wyoming) 183 F11
Bighorn Mountains 183 E11
Bighorn Pass 195 D5
Bighorn River 187 B11
Birch Creek (Beaverhead Co.) 180 D2
Birch Creek (Bonneville Co.) 189 A7
Birch Creek (Clark Co.) 184 D2
Birch Creek (Teton Co.) 185 E9
Birch Creek Valley 184 C1
Birch Hills 213 D5
Biscuit Basin 207 C5
Bishop Mountain 185 C7
Bison Peak 198 C2
Bitch Creek Narrows 218 C1
Bitter Creek 187 A9
Bivouac Peak 220 A2
Black Butte (Gallatin Co.) 194 B2
Black Butte (Madison Co.) 180 F6
Black Butte (Sweet Grass Co.) 182 C4
Black Butte Creek 194 B2
Black Canyon (Fremont Co.) 206 B1
Black Canyon (Teton Co.) 222 D1
Black Canyon Creek 183 E12
Black Canyon of the Yellowstone 196 C4
Black Mountain (Fremont Co. Idaho) 212 A1
Black Mountain (Fremont Co. Wyoming north) 191 D7
Black Mountain (Fremont Co. Wyoming south) 191 A7
Black Mountain (Park Co. Montana) 181 D12
Black Mountain (Park Co. Wyoming) 205 C7
Black Sand Basin 207 C5
Blackfoot Mountains 188 B6
Blackfoot Reservoir 189 D7
Blackfoot River 188 C5
Blackrock Creek 186 E2
Blacktail Butte 223 B4
Blacktail Deer Creek (Beaverhead Co.) 180 E3
Blacktail Deer Creek (Park Co.) 196 D3
Blacktail Deer Plateau 196 C3
Blacktail Mountain 181 A9
Blacktail Mountains 180 E3
Blacktail Pond 196 C3
Blacktail Ponds 223 A4
Blackwater Creek 211 D7
Blackwater Natural Bridge 217 A7
Bliss Pass 198 C3
Blucher Creek 191 E7
Blue Creek 183 C9
Blue Miner Lake 223 C6
Blue Ridge 191 D9
Bluewater Creek 183 D8
Bluff Creek 203 D6
Bluff Point 208 C4
Bob Creek 190 C6
Bobcat Mountain 180 C1
Bobcat Ridge 219 B6
Boiling River 196 C2
Bold Mountain 190 C6
Bonneville Peak 188 D5
Boone Creek 185 D10
Boone Mountain 182 C3
Bootjack Gap 205 B6
Borron Creek 186 D4
Boulder Creek (Park Co.) 186 D5
Boulder Creek (Sublette Co. north) 190 C4
Boulder Creek (Sublette Co. south) 190 D5
Boulder Island 220 B3
Boulder Lake 190 D5
Boulder River (Jefferson Co.) 180 B6
Boulder River (Sweet Grass Co.) 182 D3
Boulder Valley 180 A6
Boundary Creek 212 B3
Boundary Creek Hot Springs 212 C3
Boyles Hill 222 D3
Boysen Reservoir 191 B11
Bozeman Creek 181 C9
Bozeman Pass 181 C10
Brackett Creek 181 B10
Bradley Lake 220 C2
Bradley Mountain 189 C10
Breeze Point 209 C5
Breteche Creek 186 C6
Bridge Bay 209 B5
Bridge Creek 209 B5
Bridger Creek (Carbon Co.) 183 D8
Bridger Creek (Gallatin Co.) 181 C10
Bridger Creek (Sweet Grass Co.) 182 C4
Bridger Lake 216 D3
Bridger Mountains 191 A11
Bridger Range 181 B9

Broad Creek 203 B6
Broadwater Lake 199 B6
Broadwater River 199 B6
Brown Bear Spring 205 B7
Brundage Creek 197 A5
Brush Creek 188 C6
Buck Creek 181 E8
Buck Lake 198 D3
Buck Mountain 222 A3
Buffalo Bill Reservoir 187 C7
Buffalo Butte 197 B7
Buffalo Creek (Beaverhead Co.) 180 E1
Buffalo Creek (Hot Springs Co.) 191 A11
Buffalo Creek (Park Co.) 182 E2
Buffalo Fork Blackrock Creek 186 E2
Buffalo Fork Snake River 221 A5
Buffalo Fork Timothy Creek 204 C3
Buffalo Horn Creek 194 A1
Buffalo Lake 212 B3
Buffalo Meadows 206 B4
Buffalo Mountain 196 B3
Buffalo Plateau 197 B6
Buffalo River 185 C8
Bugle Lake 199 C7
Bull Creek (Park Co. Montana) 198 A2
Bull Creek (Park Co. Wyoming) 186 C6
Bull Lake 191 C7
Bull Lake Creek 190 C6
Bull Mountain (Jefferson Co.) 180 A5
Bull Mountain (Park Co.) 197 B5
Bunsen Peak 196 C2
Burned Ridge 220 B4
Burns Mountain 180 E2
Burnt Creek 203 B6
Burnt Lake 190 D5
Burnt Mountain (Carbon Co.) 182 E6
Burnt Mountain (Silver Bow Co.) 180 B2
Burnt Top 194 A1
Buster Creek 183 D11
Butte Creek 217 D6
Butte Springs 209 B7

## C

Cabin Creek (Beaverhead Co.) 184 B2
Cabin Creek (Park Co. central) 217 A7
Cabin Creek (Park Co. east) 186 D5
Cabin Creek (Park Co. west) 210 D4
Cache Creek (Park Co.) 204 B3
Cache Creek (Teton Co.) 189 B11
Cache Lake 195 C6
Cache Mountain 199 D5
Calcite Springs 197 D6
Caldwell Creek 186 E5
Calf Creek Hill 218 A1
Calfee Creek (Park Co. north) 204 B3
Calfee Creek (Park Co. south) 204 B4
Calvert Hill 180 B1
Camas Creek 184 D5
Camelback Mountain 188 D4
Cameron Point 194 B1
Camp Creek 180 C3
Camp Lake 218 C1
Campanula Creek 200 B3
Canary Bird Peak 194 A3
Canfield Creek 211 C5
Canoe Lake 205 B5
Canyon Creek (Beaverhead Co.) 180 C2
Canyon Creek (Gallatin Co.) 194 B1
Canyon Creek (Madison Co.) 185 E8
Canyon Creek (Park Co.) 186 C5
Canyon Creek (Teton Co.) 201 D6
Canyon Creek (Yellowstone Co.) 183 B8
Canyon Creek Butte 185 E7
Canyon Lake 199 B7
Canyon Mountain 181 C11
Carbonate Mountain 182 D2
Caribou Mountain 189 C8
Caribou Range 189 B8
Carmichael Fork Slate Creek 221 C6
Carnelian Creek 203 B5
Carpenter Lake 197 A6
Carrington Island 208 C4
Carrot Knoll 218 B1
Carter Creek (Beaverhead Co.) 180 D3
Carter Creek (Park Co.) 187 C7
Carter Mountain 186 D6
Cascade Acres 213 D5
Cascade Canyon 220 B2
Cascade Creek (Teton Co. north) 212 A3
Cascade Creek (Teton Co. south) 220 B1
Cascade Lake 202 B4
Cascade Mountain 184 A6
Cascades of the Firehole 207 A5
Casey Lake 196 B2
Castle Mountain 182 E5
Castor Peak 205 D5
Cathedral Peak (Park Co.) 210 B3
Cathedral Peak (Stillwater Co.) 182 D4
Cave Falls 212 D4
Cedar Butte (Bingham Co.) 188 B2
Cedar Butte (Clark Co.) 184 D4
Cedar Creek 182 B6
Cedar Mountain (Hot Springs Co.) 187 E11
Cedar Mountain (Madison Co.) 181 E7
Cedar Mountain (Park Co.) 187 C7
Cement Hills 207 D6
Centennial Mountains 184 B5
Central Plateau 208 B3
Chalcedony Creek 204 A2
Channel Mountain 215 B6
Chaos Mountain 217 B5
Cherry Creek (Beaverhead Co.) 180 C2
Cherry Creek (Madison Co. east) 181 C7
Cherry Creek (Madison Co. south) 181 E6
Cherry Creek (Madison Co. west) 180 C4
Chesterfield Range 188 D6
Chesterfield Reservoir 188 D6
Chick Creek 206 C1
Chickadee Lake 208 C3
Chicken Ridge 215 C6
Chief Mountain 186 E6
Chimney Rock 211 C7
China Hat 189 D7
Chipmunk Creek 215 C7
Christian Pond 221 A5
Christina Lake 191 E8
Cinnabar Basin 195 A5
Cinnabar Mountain 195 A6
Circular Butte 184 E3
Cirque Lake 220 B2
Clagett Butte 195 C7
Clark Canyon Reservoir 180 F2
Clarks Fork Yellowstone River 183 D8
Clayton Mountain 186 C5
Clear Creek (Park Co.) 210 C1
Clear Creek (Teton Co.) 190 B3
Clear Lake 203 C5
Clearwater Creek 211 B7
Clearwater Springs 201 B7
Cleaver Peak 220 A1
Cliff Creek (Park Co.) 216 C4
Cliff Creek (Sublette Co.) 189 C11
Cliff Lake 185 A7
Closed Creek 205 A6
Cloudburst Creek 211 D5
Cloudveil Dome 220 C2
Clover Creek 184 B4
Coal Draw 187 E10
Coburn Creek 189 B10
Cody Peak 210 B4
Coffee Pot Hot Springs 203 B7
Cold Creek 210 A3
Cole Creek 183 C7
Colonnade Falls 212 C4
Colter Bay 220 A4
Colter Canyon 218 C2
Colter Pass 199 B6
Colter Peak 216 B3
Columbine Cascade 218 C3
Columbine Creek 210 D2
Comanche Flat 183 A8
Conant Creek (Fremont Co. Idaho) 185 D8
Conant Creek (Fremont Co. Wyoming) 191 D11
Conant Pass 218 B1
Cone Peak 194 C1
Conical Peak 182 A2
Continental Glacier 190 B5
Cook Peak 202 A4
Coon Creek 187 B9
Cooney Reservoir 183 D7
Copper Lakes 205 D7
Copper Mountain (Fremont Co.) 191 B11
Copper Mountain (Madison Co.) 180 D5
Corner Lake 199 B6
Cottongrass Creek 203 C5
Cottonwood Creek (Big Horn Co.) 183 F11
Cottonwood Creek (Carbon Co.) 183 E8
Cottonwood Creek (Clark Co.) 184 D4
Cottonwood Creek (Fremont Co.) 191 B10
Cottonwood Creek (Lincoln Co.) 189 E10
Cottonwood Creek (Park Co. Montana) 181 B11
Cottonwood Creek (Park Co. Wyoming) 187 B7
Cottonwood Creek (Sublette Co.) 190 D3
Cottonwood Creek (Teton Co.) 190 A3
Cottonwood Creek (Washakie Co.) 187 E10
Cougar Creek 200 B3
Coulter Creek 219 A5
Countryman Creek 182 C5
County Peak 186 E6
Cow Creek 182 C5
Cow Island 219 C4
Cow Lake 221 B5
Coxcomb Mountain 211 D7
Coyote Lake 222 B2
Cradle Lake 199 B7
Crag Lake 194 B4
Craig Pass 207 C7
Crandall Creek 205 B7
Cranes Creek 189 C7
Crater Hills 202 D4
Crater Lake 222 D1
Crazy Creek 199 C7
Crazy Mountains 182 A2
Crazy Peak 182 A2
Crescent Hill 197 C5
Crescent Lake 194 B4
Crescent Mountain 186 E4
Crevice Creek 196 B4
Crevice Lake 196 B4
Crevice Mountain 196 B3
Crooked Creek (Carbon Co.) 183 E10
Crooked Creek (Teton Co.) 215 C6
Crooked Creek (Washakie Co.) 187 D10
Crooked Creek (Yellowstone Co.) 183 A9
Crouch Creek 210 D4
Crow Creek (Caribou Co.) 189 E9
Crow Creek (Fremont Co.) 191 B7
Crow Creek (Park Co.) 210 B3
Crow Creek Pass 210 B3
Crow Peak 211 B5
Crowfoot Ridge 194 D4
Crowheart Butte 191 B7
Crown Butte (Gallatin Co.) 194 B2
Crown Butte (Park Co.) 199 B5
Crystal Creek (Big Horn Co.) 187 B11
Crystal Creek (Park Co.) 197 D7
Crystal Creek (Teton Co.) 190 A2
Crystal Falls 202 C4
Crystal Spring 201 B7
Cub Creek 210 C1
Cub Lake 212 D1
Curl Lake 199 B6
Cutoff Creek 198 B2
Cutoff Mountain 198 B3
Cygnet Lakes 202 D3
Cygnet Pond 221 A4

## D

Dailey Creek 194 B2
Daisy Pass 199 B5
Daly Pass 194 A2
Darby Creek 189 A9
Darroch Creek 196 A3
Davis Hill 221 A6
De Lacy Creek 208 C2
De Lacy Lakes 208 C2
Dead Horse Pass 220 A1
Dead Indian Creek 186 B5
Deadman Point 220 A3
Deadmans Bar 220 B4
Deaf Jim Knob 195 B5
Death Canyon 222 A2
Deckard Flats 196 B3
Deep Creek (Clark Co.) 184 C3
Deep Creek (Deer Lodge Co.) 180 B1
Deep Creek (Fremont Co.) 191 E9
Deep Creek (Park Co. Montana) 181 C11
Deep Creek (Park Co. Wyoming) 203 B6
Deep Creek Mountains 188 E2
Deep Creek Peak 188 F3
Deer Creek (Park Co. north) 187 B8
Deer Creek (Park Co. south) 186 D5
Deer Creek Pass 217 D7
Deer Mountain 185 B7
Delmoe Lake 180 A4
Delta Lake 220 C2
Delusion Lake 209 D5
Dempsey Creek 188 E5
Devils Slide 195 A6
Dewdrop Lake 203 C6
Dewey Lake 199 A7
Diamond Creek 189 D8
Diamond Peak 184 D1
Dickie Peak 180 B2
Dike Creek 217 B5
Dike Mountain 205 C7
Dinwoody Creek 190 C5
Dinwoody Lakes 190 B6
Ditch Creek 223 A5
Divide Creek 180 B3
Divide Lake 194 D3
Dixon Creek 195 A5
Dixon Mountain 184 B2
Doane Peak 218 D2
Doherty Mountain 180 B6
Dollar Island 220 A4
Dome Mountain (Park Co. Montana) 181 E10
Dome Mountain (Park Co. Wyoming) 201 A5
Donoho Point 220 A4
Dorsey Creek 187 C10
Dot Island 209 C6
Doubletop Peak 190 B2
Douglas Knob 213 B5
Dragons Mouth Spring 203 D5
Draney Peak 189 D9
Druid Peak 198 D2
Dry Bear Creek 187 B11
Dry Canyon 206 B2
Dry Cheyenne Creek 191 C11
Dry Cottonwood Creek 187 C10
Dry Creek (Big Horn Co. north) 183 F10
Dry Creek (Big Horn Co. south) 187 B10
Dry Creek (Carbon Co.) 183 E7
Dry Creek (Fremont Co. north) 191 B7
Dry Creek (Fremont Co. south) 190 B6
Dry Creek (Gallatin Co.) 181 B9
Dry Creek (Lincoln Co.) 189 E10
Dry Creek (Madison Co.) 180 B4
Dry Creek (Park Co.) 187 B7
Dry Creek (Teton Co.) 208 C2
Dry Fork Sweet Grass Creek 182 B5
Dry Head Creek 183 E10
Dry Lake 221 B5
Dry Mountain 180 B5
Dry Ridge 189 E8
Dry Ridge Mountain 220 A1
Dryad Lake 208 B4
Du Noir Creek 190 A4
Duck Creek (Gallatin Co.) 200 B2
Duck Creek (Park Co.) 200 C4
Duck Creek (Sublette Co.) 190 D4
Duck Lake 208 C4
Dudley Lake 220 A2
Dunanda Falls 212 C3
Dunlap Creek 180 E1
Dunraven Pass 203 B5
Dunraven Peak 203 B5

## E

Eagle Creek 211 C5
Eagle Nest 217 A5
Eagle Nest Rock 196 B2
Eagle Pass 216 B4
Eagle Peak 216 B4
Eagles Rest Peak 220 A2
Earthquake Lake 185 A8
East Boulder Plateau 182 D3
East Boulder River 182 C3
East Branch Hellroaring Creek 182 E2
East Butte 188 A3
East Camas Creek 184 C6
East Fork Basin 197 A7
East Fork Blacktail Deer Creek 180 F4
East Fork Creek 186 E5
East Fork Duck Creek 182 B3
East Fork Fan Creek 194 C4
East Fork Mill Creek 181 D11
East Fork Nowater Creek 187 E12
East Fork Pilgrim Creek 219 C5

East Fork Pryor Creek 183 C10
East Fork River 190 E5
East Fork Ruby River 180 F6
East Fork Specimen Creek 194 B3
East Fork Sweet Grass Creek 182 B5
East Fork Tepee Creek 194 D2
East Fork Wind River 190 B6
East Gallatin River 181 B9
East Gros Ventre Butte 222 D3
East Pryor Creek 183 C10
East Pryor Mountain 183 E10
East Rosebud Creek 182 D5
East Sweetwater River 191 E8
East White Beaver Creek 182 B6
Ebro Springs 209 A6
Echo Peak 201 A5
Edie Creek 184 C3
Eightmile Creek (Bear Lake Co.) 189 F7
Eightmile Creek (Park Co.) 181 D10
Elbow Creek (Carbon Co.) 183 D7
Elbow Creek (Park Co.) 181 D11
Eleanor Lake 210 C3
Electric Peak 195 B5
Elephant Back Mountain 209 B5
Elephant Head Rock 211 C7
Elk Antler Creek 203 D5
Elk Creek (Big Horn Co.) 187 C11
Elk Creek (Park Co. Montana north) 181 A11
Elk Creek (Park Co. Montana south) 197 A6
Elk Creek (Park Co. Wyoming) 197 C5
Elk Creek Basin 197 A7
Elk Fork 186 C5
Elk Island 220 A3
Elk Lake 185 B7
Elk Mountain 218 C2
Elk Park Pass 180 A4
Elk Point 209 C7
Elk Ranch Reservoir 221 B6
Elk Ridge (Sublette Co.) 190 B3
Elk Ridge (Teton Co.) 218 B2
Elk River 184 A6
Elk Tongue Creek 198 C2
Elkhorn Peak 186 B5
Elkhorn Ridge 181 A10
Emigrant Peak 181 D11
Emma Matilda Lake 221 A5
Ennis Lake 181 D7
Enos Creek 187 E8
Enos Lake 186 E3
Escarpment Creek 216 D4
Eye of the Needle 204 A4
Eynon Draw 221 B5
Eyrie Creek 217 B5

# F

Factory Hill 214 B4
Fairview Peak 182 B2
Fairy Creek 207 B5
Fairy Falls 207 B5
Fall Creek (Bonneville Co.) 189 B8
Fall Creek (Sublette Co.) 190 D5
Fall Creek (Teton Co.) 189 B10
Falling Ice Glacier 220 B2
Falls Creek 217 C5
Falls River 185 D7
Fan Creek 194 C3
Fawn Creek 195 C5
Fawn Pass 195 C5
Fern Lake 203 C7
Ferris Fork Bechler River 213 B5
Ferry Butte 188 C4
Ferry Creek 181 C11
Fifteenmile Creek 187 D11
Firehole Falls 201 D5
Firehole River 207 A5
Fish Creek (Silver Bow Co.) 180 B4
Fish Creek (Sublette Co.) 190 E1
Fish Creek (Teton Co. east) 222 C2
Fish Creek (Teton Co. west) 190 A3
Fish Creek Spring 212 D1
Fish Lake 212 D4
Fisherman Creek 190 C2
Fishhawk Creek 211 D6
Fishhawk Glacier 217 B6
Fishtail Creek 182 D5
Five Mile Creek 183 B9
Five Pockets 186 E5
Five Sisters Springs 207 B6
Five Springs Creek 187 A11
Fivemile Creek (Big Horn Co.) 187 D11
Fivemile Creek (Carbon Co.) 183 D8
Fivemile Creek (Fremont Co.) 191 C10
Fizzle Lake 199 A7
Flat Canyon 212 B1
Flat Creek 223 C4
Flat Mountain 215 A5
Flat Mtn Arm, Yellowstone Lake 209 D6
Flathead Creek 181 B10
Flatiron Mountain 185 A7
Flint Creek 204 B3
Floating Island Lake 197 C5
Fly Creek 183 B12
Folsom Peak 196 D4
Fonda Point 218 C3
Forellen Peak 218 B2
Forest Creek 214 C3
Forest Lake 214 C3
Forest Springs 203 C5
Forget-me-not Lakes 222 B2
Fork Creek 181 D11
Fortress Mountain (Park Co. Montana) 181 E9
Fortress Mountain (Park Co. Wyoming) 217 A7
Fossil Lake 199 A7
Fossil Mountain 222 B1
Foster Gulch 187 B10
Fountain Paint Pot 207 B6
Fox Creek (Park Co.) 199 C6
Fox Creek (Teton Co.) 189 A9
Fox Creek Pass 222 B2
Fox Hills 189 D8
Fox Lake 199 B7
Francs Peak 186 E6
Frank Island 209 C6
Frederick Peak 198 C2
Fremont Creek 190 C5
Fremont Lake 190 D4
Fremont Peak 190 C5
Frenchy Meadow 198 A3
Fritz Peak 184 C3
Frontier Creek 186 E5
Frost Lake 204 D4

# G

Gallagher Mountain 180 E3
Gallatin Lake 201 A5
Gallatin Peak 181 D8
Gallatin Petrified Forest 194 A2
Gallatin Range (Gallatin Co.) 181 E9
Gallatin Range (Park Co.) 201 A5
Gallatin River 181 B8
Game Creek 222 B1
Gannett Hills 189 E9
Gannett Peak 190 C5
Garden Creek 188 E4
Gardner River 181 F10
Gardners Hole 195 D6
Garfield Mountain 184 B3
Garnet Hill 197 C5
Garnet Mountain 181 D9
Garns Mountain 189 A8
Gem Valley 188 E6
Geode Creek 196 C4
Giant Castle Mountain 210 B4
Gibbon Falls 201 D6
Gibbon Geyser Basin 201 C6
Gibbon Hill 202 C2
Gibbon River 185 B10
Gibbon River Rapids 201 C7
Gilbert Creek 199 C7
Gillies Creek 187 D9
Glacier Basin 217 B6
Glacier Creek 217 B6
Glacier Lake 182 E5
Glade Creek 218 A2
Gneiss Creek 200 B3
Goal Creek 184 A6
Gobblers Knob 181 B11
Goff Creek 211 C6
Golden Gate 196 C2
Golden Gate Canyon 196 C2
Goodwin Lake 223 D5
Goose Lake (Park Co.) 199 A6
Goose Lake (Teton Co.) 207 B5
Gooseberry Creek 187 E10
Grand Canyon of the Yellowstone 203 C5
Grand Prismatic Spring 207 B5
Grand Teton 220 C2
Grand View Point 221 A5
Granite Basin Lakes 220 B1
Granite Canyon 222 B2
Granite Creek (Madison Co.) 180 D6
Granite Creek (Teton Co. east) 222 B2
Granite Creek (Teton Co. west) 190 B2
Granite Peak (Madison Co.) 180 C5
Granite Peak (Park Co.) 182 E4
Granite Range 182 E4
Grant Peak 205 D5
Grants Pass 207 D6
Grapevine Creek 183 D11
Grass Creek 187 E10
Grasshopper Creek 180 E1
Grasshopper Valley 180 D1
Grassy Creek 197 B7
Grassy Island 220 A3
Grassy Lake 213 D6
Grassy Mountain 181 B10
Grassy Top 184 B2
Gravel Creek 219 B7
Gravel Mountain 186 E2
Gravelbar Creek 186 B5
Gravelly Range 180 E6
Gray Peak 195 C5
Grayback Ridge 189 C11
Graycroft Ridge 200 A1
Grayling Arm, Hegben Lake 200 B1
Grayling Creek 200 B2
Grays Lake 189 C8
Great Fountain Geyser 207 B6
Grebe Lake 202 B4
Green Lake 220 A1
Green Lakes Mountain 220 A1
Green Mountain (Caribou Co.) 189 E9
Green Mountain (Teton Co.) 220 A1
Green River 190 D3
Green River Lakes 190 B4
Greenhorn Range 180 E5
Gregg Fork Bechler River 213 B5
Greybull River 187 C9
Greys River 189 C10
Grinnell Creek 211 B6
Grinnell Meadows 211 B6
Grizzly Bear Lake 220 B2
Grizzly Lake (Park Co.) 201 B6
Grizzly Lake (Teton Co.) 223 B7
Grizzly Peak (Caribou Co.) 182 E6
Grizzly Peak (Park Co.) 210 C2
Grizzly Pool 207 B6
Gros Ventre Range 223 C6
Gros Ventre River 223 C4
Grouse Creek 215 C6
Grouse Mountain 194 A1
Grove Creek 183 E7
Guitar Lake 199 C5
Gull Point 209 B6
Gunbarrel Creek 211 C7
Gypsum Creek 190 C4

# H

Hague Mountain 205 C5
Hailstone Lake 183 A7
Half Moon Bay 220 A4
Half Moon Lake 190 D4
Halfbreed Lake 183 B7
Hanging Canyon 220 B2
Hanlon Hill 196 B3
Hardeman Reservoir 223 A6
Hardpan Creek 186 C5
Harebell Creek 215 D5
Harem Hill 218 B3
Harlequin Lake 201 D5
Hatfield Mountain 181 A10
Hay Creek 183 D10
Hayden Valley 202 D4
Haystack Peak 189 D10
Heart Lake 215 B5
Heart Lake Geyser Basin 214 B4
Heart Mountain (Clark Co.) 184 C3
Heart Mountain (Park Co.) 187 B7
Heart Mountain Canal 187 B7
Heart River 215 C5
Hebgen Lake 200 B1
Hechtman Lake 218 B2
Hedges Peak 203 B5
Hedrick Pond 220 B4
Hell Creek 189 B7
Hellroaring Creek 197 C5
Hellroaring Mountain 197 B5
Hellroaring Plateau 182 E5
Hells Canyon Creek 180 C4
Hells Half Acre 188 B4
Henderson Mountain 199 B5
Henneberry Ridge 180 E1
Henry Peak 189 D8
Henrys Fork 185 E7
Henrys Lake 185 B8
Henrys Lake Mountains 185 B8
Hering Lake 213 D6
Hermitage Point 220 A4
Heron Pond 221 A4
Hidden Falls 220 B2
Hidden Lake 197 B7
High Lake 194 B4
High Peak 181 A8
Higham Peak 188 C5
Highland Hot Springs 208 A3
Highland Mountains 180 B3
Hilgard Peak 185 A8
Hillside Springs 207 C5
Hoback River 189 B11
Hole in the Rock 189 D7
Holly Lake 220 B2
Holmes Point 209 B7
Homer Creek 189 C7
Homestake Creek 180 B4
Honeymoon Bay 219 C4
Hoodoo Basin 205 C6
Hoodoo Creek (Park Co. east) 205 C7
Hoodoo Creek (Park Co. west) 187 C7
Hoodoo Peak 205 C6
Horse Butte Peninsula 200 B1
Horse Creek (Fremont Co.) 190 A5
Horse Creek (Park Co. Montana east) 197 A5
Horse Creek (Park Co. Montana west) 194 A4
Horse Creek (Sublette Co.) 190 D2
Horse Creek (Teton Co.) 189 B11
Horse Mountain 197 B5
Horseshoe Hill 202 A2
Horseshoe Hills 181 B8
Horseshoe Lake 212 D2
Housetop Mountain 222 B1
Howell Creek 216 B4
Howell Fork Mountain Creek 217 C5
Howell Mountain 211 D5
Hoyt Peak 210 C3
Huckleberry Lake 199 A5
Huckleberry Mountain 219 A4
Huckleberry Ridge 219 B4
Hughes Cr. 205 D7
Humbolt Mountain 180 D2
Hummingbird Peak 197 A7
Hurricane Mesa 205 A6
Hurricane Pass 220 C1
Hyalite Creek 181 C9
Hyalite Reservoir 181 D10

# I

Ice Lake 202 C3
Ice Slough 191 E11
Icefloe Lake 220 C2
Idaho Canal 188 B5
Idaho Creek 187 B7
Imp Peak 181 E8
Index Peak 199 C6
Indian Creek (Madison Co.) 181 E7
Indian Creek (Park Co.) 195 D6
Indian Creek (Teton Co.) 184 C4
Indian Creek Butte 184 C4
Indian Lake 222 B2
Indian Peak 205 B6
Indian Pond 209 B6
Inkpot Spring 203 B6
Inspiration Point 203 C5
Iris Falls 212 C4
Iron Mountain 182 E2
Ishawooa Cone 217 C7
Ishawooa Creek 186 D5
Ishawooa Pass 217 C7
Island Park Reservoir 185 C7
Ivy Lake 199 C7

# J

Jack Creek (Caribou Co.) 183 E8
Jack Creek (Madison Co.) 181 D7
Jack Creek (Park Co.) 186 D6

Jackass Pass ........................... 218 B1
Jackknife Creek ...................... 189 C9
Jackson Creek ........................ 181 C10
Jackson Hole........................... 189 A10
Jackson Lake .......................... 185 E11
Jackson Peak .......................... 223 D5
Jakeys Fork Wind River............ 190 B5
Jasper Creek .......................... 197 D7
Jeff Davis Peak ....................... 180 F1
Jefferson River....................... 180 C4
Jenny Lake ............................ 220 B3
Jerry Creek............................. 180 B2
Jim Creek............................... 190 C4
Jim Mountain.......................... 186 B6
Jim Smith Peak ....................... 199 C7
Johnson Lake .......................... 200 A1
Jones Creek ........................... 210 B3
Jones Pass .............................. 210 B2
Joseph Peak ........................... 195 C5
Josephs Coat Springs ............... 203 C7
Jumbo Mountain ...................... 181 D8
Junco Lake ............................. 212 D4
Junction Butte ........................ 197 C6
Juniper Creek ......................... 208 B2

**K**

Kaiser Creek .......................... 183 A12
Kelly Warm Spring.................... 223 B5
Kepler Cascades ...................... 207 C6
Kersey Lake ............................ 199 B6
Kettle Butte ........................... 188 A4
Keyser Creek .......................... 182 C6
Kidney Geyser......................... 207 B6
Kilgore Creek.......................... 190 C1
King Butte .............................. 194 B3
Kingfisher Peak ....................... 186 D5
Kirby Creek............................ 187 F11
Kirby Draw ............................. 191 C10
Kirkwood Ridge ....................... 200 A1
Kit Lake.................................. 220 C2
Kitty Creek............................. 211 D6
Knowles Falls.......................... 196 B3
Koch Peak............................... 181 E8

**L**

La Barge Creek ....................... 189 E11
Laduke Spring.......................... 195 A6
Lady of the Lake ...................... 199 B6
Lake Abundance ...................... 198 A4
Lake Abundance Creek .............. 198 A3
Lake Basin.............................. 182 B6
Lake Butte.............................. 209 B7
Lake Fork Rock Creek ............... 182 E5
Lake of the Clouds ................... 199 B7
Lake of the Crags .................... 220 B2
Lake of the Winds.................... 199 B7
Lake of the Woods (Park Co.) ..... 202 B2
Lake of the Woods (Teton Co.).... 218 A1
Lake Plateau .......................... 182 D3
Lake Reno .............................. 199 C7
Lake Solitude.......................... 220 B1
Lake Taminah .......................... 220 C2
Lamar Mountain....................... 205 C6
Lamar River ........................... 182 F2
Land Mountain......................... 205 D6
Lander Creek .......................... 191 E7
Lanes Creek ........................... 189 D8
Langford Cairn........................ 216 A2
Latham Spring ........................ 206 C2
Lava Butte .............................. 194 B2
Lava Creek (Park Co.)................ 196 C2
Lava Creek (Teton Co.) ............. 186 E2
Ledford Creek ......................... 180 F5
Leech Lake ............................. 194 B4
Lehardys Rapids ...................... 209 A6
Leidy Creek ............................ 221 C7
Leidy Lake .............................. 221 C7
Leigh Canyon........................... 220 B1
Leigh Lake .............................. 220 B3
Lemhi Range ........................... 184 D1
Lewis Falls.............................. 214 B3
Lewis Lake ............................. 214 B3
Lewis River............................. 185 D11
Lilypad Lake ........................... 212 C3
Lima Reservoir........................ 184 B4
Limerock Mountain................... 189 C7
Lincoln Creek ......................... 188 C5
Lincoln Mountain ..................... 181 E8
Line Creek ............................. 183 F7
Line Creek Plateau .................. 182 E6

Little Lost River Valley.............. 184 D1
Little Buffalo Basin .................. 187 D8
Little Buffalo Creek.................. 197 C5
Little Dry Creek (Big Horn Co.)... 187 B10
Little Dry Creek (Fremont Co.) ... 191 A7
Little Firehole River ................. 206 C4
Little Gooseberry Creek............. 187 E10
Little Greys River .................... 189 C10
Little Lamar River ................... 205 D5
Little Lost River....................... 184 E1
Little Mackinaw Bay ................. 220 A4
Little Pipestone Creek .............. 180 B4
Little Popo Agie River .............. 191 D9
Little Quadrant Mountain .......... 195 C5
Little Saddle Mountain .............. 204 C4
Little Sage Creek...................... 194 C1
Little Sheep Creek ................... 184 B3
Little Sheep Mountain
   (Big Horn Co.)...................... 187 A10
Little Sheep Mountain
   (Sublette Co.)....................... 190 B4
Little Tepee Creek .................. 200 A2
Little Timber Creek .................. 182 B3
Little Trail Creek ..................... 195 B7
Little Twin Creek ..................... 190 C3
Little Wind River ..................... 191 C8
Little Woody Creek ................... 183 C11
Littlerock Creek....................... 182 F6
Littles Peak............................ 220 B1
Livingston Peak ....................... 181 C12
Lizard Creek ........................... 218 B3
London Hills............................ 180 B6
Lone Indian Butte .................... 182 B5
Lone Mountain......................... 181 D8
Lone Star Geyser...................... 207 C6
Long Creek (Beaverhead Co.) ..... 184 B5
Long Creek (Fremont Co.) ......... 191 E11
Long Hollow............................ 223 C5
Looking Glass Lake ................... 199 A7
Lookout Mountain (Madison Co.) 198 A2
Lookout Mountain (Park Co.)...... 189 A7
Lost Boy Creek ........................ 183 B12
Lost Creek (Big Horn Co.) ......... 183 D9
Lost Creek (Park Co. Montana) ... 198 B3
Lost Creek (Park Co. Wyoming).. 197 D5
Lost Creek (Teton Co.) .............. 220 C4
Lost Creek Falls....................... 197 D5
Lost Lake ............................... 197 D5
Lovell Lakes............................ 187 A10
Lovely Pass ............................ 204 D3
Lower Aero Lake ...................... 199 A6
Lower Deer Creek .................... 182 C4
Lower Falls............................. 203 C5
Lower Geyser Basin................... 207 B5
Lower Glaston Lake................... 182 B4
Lower Red Rock Lake ................ 184 B6
Lower Slide Lake ..................... 223 B6
Lower Table ........................... 190 A6
Lozier Hill............................... 221 A5
Lupine Creek ........................... 196 C3
Lynx Creek ............................. 216 D2

**M**

Madison Arm, Hegben Lake ....... 200 C1
Madison Canyon....................... 200 D4
Madison Lake .......................... 213 A5
Madison Plateau ...................... 206 B3
Madison Range ........................ 181 D7
Madison River ......................... 181 C7
Madison Valley ........................ 200 B3
Magpie Creek.......................... 202 D2
Maiden Falls ........................... 195 B6
Mallard Creek ......................... 207 B6
Mallard Lake ........................... 207 C6
Mammoth Glacier ..................... 190 C5
Mammoth Hot Springs............... 196 C2
Maple Creek ........................... 200 B4
Marble Point ........................... 194 B1
Mariane Lake .......................... 199 B7
Marie Island ........................... 220 A4
Marion Lake ........................... 222 B1
Mariposa Lake......................... 216 D2
Market Lake ........................... 184 E5
Marmot Mountain.................... 180 F6
Marquette Creek...................... 186 C6
Marsh Creek ........................... 188 E5
Mary Bay ............................... 209 B7
Mary Lake .............................. 208 A3
Mary Mountain ....................... 208 A3
Maurice Mountain.................... 180 C1

Maverick Mountain .................. 180 D1
McBride Lake .......................... 197 C7
McCartney Mountain ................ 180 C3
McCoy Creek .......................... 189 C8
McCullough Peaks .................... 187 B8
McHessor Creek ...................... 180 D4
Meade Peak ............................ 189 E9
Meadow Canyon....................... 206 A3
Meadow Canyon Creek .............. 190 E2
Meadow Creek ........................ 189 A7
Meadow Creek Mountain........... 189 C7
Medicine Lodge Creek
   (Beaverhead Co.)................... 180 F1
Medicine Lodge Creek
   (Clark Co.) ........................... 184 C3
Medicine Lodge Peak ................ 184 A1
Meeteetse Creek ...................... 187 C7
Meeteetse Rim ........................ 187 C7
Meldrum Mountain ................... 194 B3
Menan Buttes.......................... 184 E6
Meridan Peak .......................... 198 B4
Mesa Creek ............................ 211 D7
Mica Lake .............................. 220 B1
Middle Beaver Creek................. 190 C2
Middle Butte .......................... 188 B3
Middle Creek (Clark Co.) .......... 184 C4
Middle Creek (Gallatin Co.) ....... 181 A9
Middle Creek (Hot Springs Co.).. 187 E8
Middle Creek (Park Co.)............. 210 C4
Middle Creek Butte .................. 184 C3
Middle Fork Blacktail Deer Creek 180 F4
Middle Fork Ditch Creek ........... 221 C5
Middle Fork Fifteenmile Creek ... 187 D10
Middle Fork Fivemile Creek ....... 183 D9
Middle Fork Hellroaring Creek ... 182 E2
Middle Fork Owl Creek.............. 191 A8
Middle Fork Sage Creek ............ 191 C7
Middle Fork South Fork
   Split Creek........................... 206 D2
Middle Fork Wood River ........... 187 E7
Middle Mountain...................... 199 B6
Middle Piney Creek .................. 190 E1
Middle Popo Agie River............. 191 E7
Middle Ridge .......................... 189 C10
Middle Teton .......................... 220 C2
Middle Teton Glacier ................ 220 C2
Midway Geyser Basin................. 207 B5
Mill Creek (Deer Lodge Co.)....... 180 A1
Mill Creek (Park Co. north)........ 181 D11
Mill Creek (Park Co. south)........ 195 B5
Miller Creek ........................... 204 B4
Miller Mountain ...................... 199 B5
Millers Butte .......................... 223 D4
Mineral Hill ........................... 196 B3
Mineral Mountain .................... 199 B5
Mink Lake .............................. 220 B1
Mirror Fork Timothy Creek ....... 204 C3
Mirror Lake ............................ 204 C3
Mirror Plateau ........................ 204 B2
Mission Creek ......................... 182 C2
Missouri River......................... 181 A8
Mist Creek ............................. 204 D3
Mist Creek Pass ...................... 210 A3
Molly Islands .......................... 216 B1
Monida Pass ........................... 184 B4
Monitor Peak .......................... 181 E11
Monument Creek...................... 194 C2
Monument Geyser..................... 201 C6
Monument Meadow .................. 194 B2
Monument Mountain
   (Gallatin Co.) ....................... 194 C1
Monument Mountain
   (Park Co.) ............................ 211 B7
Monument Peak ...................... 182 E3
Moody Creek .......................... 185 E7
Moore Creek .......................... 180 D6
Moose Basin ........................... 218 C1
Moose Basin Divide .................. 218 C1
Moose Creek (Fremont Co.
   north)................................. 185 C8
Moose Creek (Fremont Co.
   south)................................. 212 C1
Moose Creek (Silver Bow Co.)..... 180 C3
Moose Creek (Teton Co. central) 218 C2
Moose Creek (Teton Co. north)... 213 B6
Moose Creek (Teton Co. south).... 189 A9
Moose Creek Butte ................... 212 C1
Moose Creek Plateau ................ 206 B1
Moose Falls ............................ 214 D2
Moose Island .......................... 219 C4

Moose Lake ............................ 222 B1
Moose Meadows ....................... 222 C1
Moose Mountain (Park Co.) ....... 205 D6
Moose Mountain (Teton Co.) ...... 218 C1
Moran Bay .............................. 220 A2
Moran Canyon ......................... 220 A2
Mormon Creek ........................ 211 C6
Mosquito Creek ....................... 189 B10
Moss Creek ............................ 203 B6
Moss Spring ........................... 212 D2
Mount Abundance ..................... 198 B4
Mount Arter ........................... 191 D8
Mount Baldy ........................... 191 D7
Mount Bannon ......................... 222 A2
Mount Berry ........................... 218 B2
Mount Blackmore ..................... 181 D10
Mount Burwell ........................ 186 E6
Mount Carey........................... 184 A5
Mount Chittenden .................... 210 B2
Mount Cowen .......................... 181 D12
Mount Delano ......................... 181 C12
Mount Doane .......................... 210 D3
Mount Douglas........................ 182 D3
Mount Ellis............................. 181 C10
Mount Evans .......................... 180 A1
Mount Everts .......................... 196 C3
Mount Fitzpatrick ..................... 189 E10
Mount Fleecer ........................ 180 B2
Mount Glory ........................... 222 C1
Mount Greeley ........................ 182 C2
Mount Haggin ......................... 180 A1
Mount Hancock........................ 215 D6
Mount Haynes ........................ 200 D4
Mount Hebgen ........................ 200 B1
Mount Holmes ........................ 201 B5
Mount Hornaday ...................... 198 C3
Mount Humbug ........................ 180 B3
Mount Humphreys.................... 216 A4
Mount Hunt ............................ 222 B2
Mount Jedediah Smith .............. 222 A2
Mount Jefferson ...................... 185 B7
Mount Langford ...................... 210 D3
Mount Leidy ........................... 221 C6
Mount Manning ....................... 185 F8
Mount Maurice........................ 182 E6
Mount McKnight ...................... 181 D12
Mount Meek ........................... 222 A2
Mount Meek Pass ..................... 222 A2
Mount Moran.......................... 220 B2
Mount Nord ............................ 218 C1
Mount Norris .......................... 198 D3
Mount Owen ........................... 220 B2
Mount Putnam ........................ 188 D5
Mount Rae ............................. 182 C2
Mount Randolph ...................... 186 E2
Mount Rearguard ..................... 182 E5
Mount Reid ............................ 219 C4
Mount Rosebud ........................ 199 A7
Mount Saint John ..................... 220 B2
Mount Schurz ......................... 216 A4
Mount Sheridan....................... 214 B4
Mount Stevenson ..................... 210 D2
Mount Tahepia ........................ 180 C2
Mount Wallace ........................ 182 E2
Mount Washakie....................... 191 D7
Mount Washburn...................... 203 B5
Mount Wister .......................... 222 A3
Mount Wood ........................... 182 D4
Mount Woodring ...................... 220 B2
Mountain Ash Creek.................. 212 D4
Mountain Creek ....................... 216 C3
Mud Creek (Hot Springs Co.) ..... 191 A10
Mud Creek (Lemhi Co.) ............. 184 C1
Mud Lake (Jefferson Co.)........... 184 E4
Mud Lake (Park Co.) ................. 199 B6
Mud Volcano .......................... 203 D5
Muddy Creek (Beaverhead Co.)... 184 B2
Muddy Creek (Big Horn Co.)...... 183 D11
Muddy Creek (Fremont Co.) ...... 191 B9
Muddy Creek (Park Co. Montana) 181 A10
Muddy Creek (Sublette Co. east) . 190 E5
Muddy Creek (Sublette Co.
   north).................................. 190 C2
Muddy Creek (Sublette Co. west) 190 E3
Mulherin Creek ....................... 195 B5
Muskrat Creek ........................ 191 C11
Mystic Falls ............................ 207 C5
Mystic Isle ............................. 220 B3
Mystic Lake ........................... 182 E4

## N

National Park Mountain ........... 201 D5
Natler Peak ........... 195 D5
Needle Creek ........... 186 D5
Neva Creek ........... 211 D5
New Fork Lakes ........... 190 C4
New Fork River ........... 190 E4
Newton Spring ........... 211 C7
Nez Perce ........... 220 C2
Nez Perce Creek ........... 207 B6
Nicholia Creek ........... 184 C2
Nipple Mesa ........... 211 B7
No Man Peak ........... 181 E7
Nord Pass ........... 218 C1
Norkok Creek ........... 191 C7
Norris Creek ........... 211 D6
Norris Geyser Basin ........... 201 C7
Norris Pass ........... 207 C7
North Bar Bay ........... 220 A3
North Beaver Creek ........... 190 C3
North Bitch Creek ........... 218 C1
North Buffalo Fork Buffalo Fork Blackrock Creek ........... 186 E3
North Cottonwood Creek ........... 190 D2
North Fork Alkali Creek ........... 183 B9
North Fork Bear Creek ........... 196 A3
North Fork Butte Creek ........... 217 D7
North Fork Crandall Creek ........... 199 D6
North Fork Ditch Creek ........... 221 C5
North Fork Dry Creek (Carbon Co.) ........... 183 E7
North Fork Dry Creek (Park Co.) ........... 187 C8
North Fork Fan Creek ........... 194 B4
North Fork Fifteenmile Creek ........... 187 D11
North Fork Fish Creek (Fremont Co.) ........... 212 C1
North Fork Fish Creek (Teton Co.) ........... 190 A3
North Fork Fivemile Creek ........... 183 D9
North Fork Little Wind River ........... 191 C7
North Fork Moody Creek ........... 185 F7
North Fork Mud Creek ........... 191 A10
North Fork Owl Creek ........... 191 A7
North Fork Partridge Creek ........... 212 B1
North Fork Piney Creek ........... 189 A8
North Fork Rodent Creek ........... 219 A5
North Fork Sage Creek ........... 191 C7
North Fork Shoshone River ........... 211 C5
North Fork Snowshoe Canyon ........... 220 A2
North Fork Spanish Creek ........... 181 D8
North Fork Specimen Creek ........... 194 B3
North Fork Split Creek ........... 206 D1
North Fork Valley Creek ........... 183 B7
North Fork Willow Creek ........... 190 C6
North Fork Wood River ........... 186 E6
North Horse Creek ........... 190 D2
North Junipers ........... 184 D6
North Lake ........... 184 E4
North Leigh Creek ........... 220 A1
North Moran Creek ........... 220 A2
North Piney Creek ........... 190 E1
North Popo Agie River ........... 191 D8
North Willow Creek ........... 180 C6
Notch Mountain ........... 205 D5
Nowater Creek ........... 187 E12
Nowlin Creek ........... 223 D4
Nowood River ........... 187 C12
Nuthatch Lake ........... 208 C3
Nymph Lake (Park Co. Montana) ........... 199 B7
Nymph Lake (Park Co. Wyoming) ........... 201 B7
Nymph Spring ........... 197 D6

## O

Observation Peak (Lincoln Co.) ........... 202 B4
Observation Peak (Park Co.) ........... 189 B10
Obsidian Cliff ........... 201 B7
Obsidian Creek ........... 201 A7
Obsidian Lake ........... 202 A2
Obsidian Pool ........... 209 A5
Ocean Lake ........... 191 C9
Ochre Springs ........... 209 A6
Oil Springs Creek ........... 191 D11
Old Faithful Geyser ........... 207 C6
One Hunt Creek ........... 205 C7
Onemile Creek ........... 199 D7
Opal Creek ........... 204 B2
Open Canyon ........... 222 B2
Open Creek ........... 217 D5
Oregon Mountain ........... 196 B4
Ortenburger Lake ........... 220 A1
Osprey Falls ........... 196 C2
Otter Creek (Park Co.) ........... 202 C4
Otter Creek (Sweet Grass Co.) ........... 182 B4
Otter Lake ........... 199 B7
Outlet Creek ........... 215 B5
Outlet Lake ........... 215 B6
Ouzel Creek ........... 212 B4
Ouzel Falls ........... 212 C4
Overlook Mountain (Park Co.) ........... 217 B6
Overlook Mountain (Teton Co.) ........... 215 C6
Ovis Lake ........... 199 B5
Owl Creek (Hot Springs Co.) ........... 187 F10
Owl Creek (Teton Co.) ........... 218 C2
Owl Creek Mountains ........... 191 A7
Oxbow Bend ........... 221 A5
Oxbow Creek ........... 196 C4
Oxide Mountain ........... 199 B6

## P

Pacific Creek ........... 221 A5
Packsaddle Butte ........... 182 C4
Paint Creek ........... 186 B6
Paintbrush Divide ........... 220 B2
Paintpot Hill ........... 201 C7
Palisade Mountain ........... 181 D10
Palisades Creek ........... 189 B9
Palisades Peak ........... 189 B9
Palisades Reservoir ........... 189 B9
Palmer Mountain ........... 196 B3
Panther Creek ........... 195 D6
Papoose Creek ........... 205 B7
Park Point ........... 209 C7
Parker Peak ........... 205 C6
Parker Point ........... 196 B3
Partridge Creek ........... 212 B1
Pass Creek ........... 217 D6
Passage Creek ........... 215 C7
Pat O'Hara Creek ........... 187 B7
Pat O'Hara Peak ........... 186 B6
Patchtop Mountain ........... 184 B6
Peale Island ........... 215 B7
Pebble Creek ........... 198 C3
Pelican Cone ........... 204 D2
Pelican Creek ........... 209 B6
Pelican Roost ........... 209 B7
Pelican Springs ........... 210 A2
Pelican Valley ........... 209 B7
Petrified Ridge ........... 217 D6
Pettengill Creek ........... 180 C1
Pevah Creek ........... 191 C7
Phantom Fumarole ........... 214 C2
Phantom Lake ........... 196 C4
Phelps Lake ........... 222 B3
Phillips Canyon ........... 222 C2
Phillips Fork Bechler River ........... 213 B5
Phillips Pass ........... 222 C1
Phillips Ridge ........... 222 C1
Phlox Creek ........... 216 C3
Pickett Creek ........... 186 D6
Pilgrim Creek ........... 221 A4
Pilgrim Mountain ........... 219 C5
Pilot Butte Reservoir ........... 191 C8
Pilot Creek ........... 199 C6
Pilot Peak ........... 199 C6
Pine Creek ........... 189 B8
Pine Mountain ........... 189 B7
Pine Ridge ........... 183 B12
Pinnacle Buttes ........... 186 F4
Pinnacle Mountain ........... 217 B5
Pinnacle Peak ........... 189 B11
Pinyon Peak ........... 219 B7
Pioneer Mountain ........... 181 E8
Pioneer Mountains ........... 180 B1
Pipestone Pass ........... 180 B4
Pitchstone Plateau ........... 213 C5
Plateau Creek (Park Co.) ........... 198 C2
Plateau Creek (Teton Co.) ........... 215 D7
Plateau Falls ........... 215 D8
Plenty Coups Peak ........... 210 D4
Plover Point ........... 209 D6
Pocket Lake ........... 207 D7
Poison Creek ........... 191 C11
Poker Peak ........... 189 B9
Pole Creek (Madison Co.) ........... 181 C7
Pole Creek (Sublette Co.) ........... 190 D4
Polecat Bench ........... 187 A8
Polecat Creek (Big Horn Co.) ........... 187 A9
Polecat Creek (Park Co.) ........... 214 C2
Pollux Peak ........... 205 D5
Pompeys Pillar Creek ........... 183 A11
Ponuntpa Springs ........... 203 C7
Popo Agie River ........... 191 D9
Porcelain Basin ........... 202 B2
Porcupine Creek (Big Horn Co.) ........... 183 F11
Porcupine Creek (Gallatin Co.) ........... 181 E9
Porter Gulch ........... 187 B12
Portneuf Range ........... 188 D5
Portneuf River ........... 188 E3
Portneuf Valley ........... 188 D6
Potts Hot Spring Basin ........... 208 C4
Proposition Creek ........... 213 D5
Prospect Creek ........... 187 E9
Prospect Mountains ........... 191 E7
Prospect Peak ........... 196 D4
Prospectors Mountain ........... 222 B2
Pryor Creek ........... 183 C10
Pryor Mountains ........... 183 D9
Ptarmigan Mountain ........... 186 C5
Pulpit Rock ........... 194 A1
Pumice Point ........... 208 C4
Purple Mountain ........... 201 D5
Pyramid Peak (Park Co.) ........... 210 A4
Pyramid Peak (Teton Co.) ........... 190 B2

## Q

Quadrant Mountain ........... 195 D5
Quayles Lake ........... 184 E6
Quiver Cascade ........... 213 B5

## R

Raft Lake ........... 190 D6
Ragged Falls ........... 213 B5
Raid Peak ........... 190 D6
Railroad Creek ........... 183 A12
Rainbow Falls ........... 213 D5
Rainbow Lakes ........... 195 B7
Rainbow Springs ........... 203 B7
Rainey Creek ........... 189 B8
Rammell Mountain ........... 220 A1
Rampart Creek ........... 217 B7
Rampart Mountain ........... 180 A4
Rampart Pass ........... 217 C6
Ramshorn Peak ........... 189 C11
Ranger Lake ........... 212 C4
Ranger Peak ........... 218 C2
Rathbone Lake ........... 200 A2
Ratio Mountain ........... 180 A5
Rattlesnake Butte ........... 196 B3
Rattlesnake Creek (Beaverhead Co.) ........... 180 D2
Rattlesnake Creek (Clark Co.) ........... 184 C5
Rattlesnake Creek (Park Co.) ........... 186 B6
Rattlesnake Creek (Power Co.) ........... 188 E3
Rattlesnake Mountain ........... 186 B6
Raven Creek ........... 210 A2
Rawhide Creek ........... 187 D7
Ray Lake ........... 191 D8
Raynolds Pass ........... 185 B8
Rays Lake ........... 184 E4
Razor Creek ........... 183 A10
Reas Pass ........... 206 B1
Red Butte ........... 187 D11
Red Canyon Creek (Gallatin Co.) ........... 200 A1
Red Canyon Creek (Hot Springs Co.) ........... 191 A10
Red Creek (Hot Springs Co.) ........... 191 A9
Red Creek (Park Co.) ........... 211 A5
Red Creek (Teton Co.) ........... 214 C4
Red Hills ........... 223 B7
Red Knob ........... 181 D7
Red Lodge Creek ........... 183 D7
Red Mountain (Gallatin Co.) ........... 194 C2
Red Mountain (Madison Co.) ........... 181 C7
Red Mountain (Teton Co.) ........... 218 C1
Red Mountains ........... 214 B3
Red Peak ........... 189 B8
Red Ridge ........... 187 E9
Red Rock Creek ........... 185 B7
Red Rock Lakes ........... 199 A7
Red Rock Mountain ........... 181 A10
Red Rock River ........... 180 F2
Redmond Creek ........... 223 B6
Redstreak Peak ........... 194 C1
Reese Creek ........... 195 B6
Rendezvous Mountain ........... 222 C2
Rendezvous Peak ........... 222 C2
Republic Creek ........... 199 C5
Republic Mountain ........... 199 C5
Republic Pass ........... 199 C5
Republic Peak ........... 199 C5
Rescue Creek ........... 196 C3
Reservation Peak ........... 210 C4
Reservoir Mountain ........... 189 D7
Ribbon Lake ........... 203 C5
Riddle Lake ........... 214 A4
Rimrock Lake ........... 222 B2
Ririe Reservoir ........... 189 A7
Rising Butte ........... 212 D2
Roaring Mountain ........... 201 B7
Robb Creek ........... 180 F5
Robinson Creek ........... 185 D9
Robinson Lake ........... 212 C3
Rochester Creek ........... 180 C4
Rock Creek (Carbon Co.) ........... 183 E7
Rock Creek (Fremont Co. north) ........... 185 D9
Rock Creek (Fremont Co. south) ........... 191 E9
Rock Creek (Park Co. Montana north) ........... 181 B11
Rock Creek (Park Co. Montana south ) ........... 181 E9
Rock Creek (Park Co. Wyoming) ........... 186 C6
Rock Creek (Power Co.) ........... 188 E2
Rock Creek (Sublette Co.) ........... 190 B3
Rock Face ........... 223 B4
Rock Island Butte ........... 199 B7
Rock Island Lake ........... 199 B7
Rock Lake ........... 212 D4
Rock of Ages ........... 220 B2
Rock Point ........... 209 C6
Rock Tree Lake ........... 199 B7
Rock Waterhole Creek ........... 187 D10
Rockchuck Peak ........... 220 B2
Rocky Creek ........... 216 A2
Rocky Draw ........... 206 B2
Rodent Creek ........... 219 B5
Rose Creek (Park Co. Montana) ........... 187 D7
Rose Creek (Park Co. Wyoming) ........... 198 D2
Rosebud Creek ........... 182 D6
Ross Fork, American Falls Reservoir ........... 188 D3
Ross Lake ........... 190 B5
Ross Peak (Bonneville Co.) ........... 189 A8
Ross Peak (Gallatin Co.) ........... 181 B10
Round Lake ........... 199 B6
Round Top Mountain ........... 180 B1
Roundhead Butte ........... 198 A2
Ruby Creek ........... 180 E6
Ruby Range ........... 180 E4
Ruby River ........... 180 D5
Ruby River Reservoir ........... 180 E5
Rush Lake ........... 207 B6
Russell Lake ........... 199 B7
Rustic Falls ........... 196 C2
Rustic Geyser ........... 214 B4
Ruth Creek ........... 217 A6

## S

Sacagawea Peak ........... 181 B10
Saddle Mountain (Butte Co.) ........... 184 E2
Saddle Mountain (Park Co.) ........... 205 C5
Saddle Peak ........... 181 B10
Sage Creek (Beaverhead Co.) ........... 184 B3
Sage Creek (Big Horn Co.) ........... 183 E9
Sage Creek (Fremont Co.) ........... 191 C8
Sage Creek (Gallatin Co.) ........... 181 E9
Sage Creek (Madison Co.) ........... 180 E4
Sage Creek (Park Co) ........... 187 C7
Salt River ........... 189 C9
Salt River Range ........... 189 C10
Saltlick Mountain ........... 190 C4
Sand Creek (Fremont Co.) ........... 185 D7
Sand Creek (Yellowstone Co.) ........... 183 B12
Sand Draw ........... 187 E10
Sand Hole Lake ........... 184 E5
Sand Point ........... 209 C6
Sandstone Reservoir ........... 187 C10
Sandy Butte ........... 200 B3
Sargents Bay ........... 219 C4
Sawmill Creek ........... 191 E8
Sawtell Peak ........... 185 B8
Sawtooth Mountain ........... 194 A4
Sawtooth Ridge ........... 181 E8
Scatter Creek ........... 217 D5
Scaup Lake ........... 207 C6
Schmid Ridge ........... 189 E8
Schoolroom Glacier ........... 220 C1

Scotch Bonnet Mountain ........... 199 B5
Scott Peak ........... 184 C2
Scout Mountain ........... 188 E4
Seclusion Creek ........... 217 B7
Secret Valley ........... 201 C6
Sedge Bay ........... 209 B7
Sedge Creek ........... 210 B1
Sedge Lake ........... 194 B4
Sedgwick Peak ........... 188 E6
Semicentennial Geyser ........... 201 B7
Sentinel Creek ........... 207 B5
Sepulcher Mountain ........... 195 C6
Seven Lakes ........... 190 B4
Seymore Mountain ........... 180 C1
Seymour Creek ........... 180 B1
Shadow Mountain ........... 221 C5
Shadow Peak ........... 220 C2
Shallow Creek ........... 203 B7
Shane Creek ........... 182 C6
Shattuck Butte ........... 188 A5
Shaw Mountain ........... 180 C1
Sheep Creek (Park Co.) ........... 211 D7
Sheep Creek (Teton Co.) ........... 223 C4
Sheep Mesa ........... 217 A7
Sheep Mountain (Beaverhead Co.) ........... 180 C2
Sheep Mountain (Big Horn Co. north) ........... 187 B11
Sheep Mountain (Big Horn Co. south) ........... 187 C10
Sheep Mountain (Bingham Co.) ........... 189 C7
Sheep Mountain (Madison Co.) ........... 185 A8
Sheep Mountain (Park Co. Montana central) ........... 195 A7
Sheep Mountain (Park Co. Montana east) ........... 199 B5
Sheep Mountain (Park Co. Montana north) ........... 182 B2
Sheep Mountain (Park Co. Montana west) ........... 194 A4
Sheep Mountain (Teton Co.) ........... 223 C6
Sheep Steps ........... 222 A2
Sheepeater Canyon ........... 196 D2
Sheepeater Cliff ........... 196 D2
Sheffield Creek ........... 219 A4
Sheffield Island ........... 220 A4
Shell Creek ........... 187 B11
Sheridan Creek ........... 185 C7
Sheridan Lake ........... 214 C4
Sheridan Res. ........... 185 C7
Shields River ........... 182 B1
Shooting Star Mountain ........... 194 B4
Shoshone Creek ........... 207 D6
Shoshone Geyser Basin ........... 213 A6
Shoshone Lake (Fremont Co.) ........... 191 D7
Shoshone Lake (Teton Co.) ........... 185 C10
Shoshone River ........... 187 B8
Sickle Creek ........... 215 C6
Siggins Fork Open Creek ........... 217 C5
Signal Hills ........... 210 D2
Signal Mountain ........... 221 A5
Signal Peak ........... 184 C5
Signal Point ........... 209 C7
Silver Bow Creek ........... 180 B3
Silver Creek ........... 190 E5
Silver Gate ........... 196 C2
Silver Lake ........... 185 C8
Silver Run Peak ........... 182 E5
Silver Scarf Falls ........... 212 C3
Silver Tip Creek (Carbon Co.) ........... 183 E8
Silver Tip Creek (Park Co.) ........... 217 C6
Silvertip Peak ........... 210 B3
Simpson Lake ........... 190 B5
Sixmile Creek ........... 181 D10
Sixty-Seven Reservoir ........... 190 E3
Skelly Hill ........... 189 B7
Ski Lake ........... 222 C1
Skillet Glacier ........... 220 A2
Skull Creek ........... 221 B7
Sky Top Creek ........... 199 A6
Skyline Ridge ........... 194 C1
Sleeping Giant Mountain ........... 211 B6
Slide Lake ........... 196 B2
Sliderock Mountain ........... 182 C4
Sliding Mountain ........... 181 E11
Slough Creek ........... 197 C7
Slug Creek ........... 189 E8
Smoke Jumper Hot Springs ........... 206 C4
Snake Hot Springs ........... 214 D4
Snake River Plain ........... 188 C2
Snake River Range ........... 189 A8
Snake River ........... 188 C4
Snipe Point ........... 209 D6
Snow Creek ........... 212 C1
Snow Creek Butte ........... 212 B2
Snow King Mountain ........... 222 D3
Snowcrest Mountain ........... 180 E5
Snowcrest Range ........... 180 F5
Snowdrift Lake ........... 220 C1
Snowflake Ridge ........... 194 B1
Snowflake Springs ........... 194 B1
Snowshoe Canyon ........... 220 A2
Snowshoe Pass ........... 195 C5
Snowslide Creek ........... 194 C2
Snowslide Mountain ........... 194 C1
Soapy Dale Peak ........... 187 E8
Soda Butte Creek ........... 198 D3
Soda Fork Buffalo Fork Blackrock Creek ........... 186 E3
Soda Lake ........... 190 D5
Soda Peak ........... 189 E7
Soda Point ........... 189 E7
Sodalite Lake ........... 199 A7
Solfatara Plateau ........... 202 C3
Solution Creek ........... 208 D4
Sour Creek ........... 203 C5
South Arm, Yellowstone Lake ........... 215 A6
South Baldy Mountain ........... 180 D6
South Bar ........... 220 A3
South Beaver Creek ........... 190 D2
South Bitch Creek ........... 185 E10
South Boone Creek ........... 218 B1
South Boulder River ........... 180 C5
South Boundary Lake ........... 213 D6
South Cache Creek ........... 204 A4
South Cottonwood Creek ........... 190 D1
South Entrance Hot Springs ........... 214 D3
South Fork Alkali Creek ........... 183 B9
South Fork Big Timber Creek ........... 182 B3
South Fork Bridger Creek ........... 183 E8
South Fork Buffalo Fork Blackrock Creek ........... 186 E3
South Fork Butte Creek ........... 198 A2
South Fork Ditch Creek ........... 223 A6
South Fork Dry Creek (Caribou Co.) ........... 183 E7
South Fork Dry Creek (Park Co.) ........... 187 C8
South Fork Fifteenmile Creek ........... 187 D10
South Fork Fish Creek ........... 190 A3
South Fork Hoback River ........... 190 D1
South Fork Indian Creek ........... 189 C9
South Fork Little Wind River ........... 191 D7
South Fork Madison River ........... 200 C1
South Fork Moody Creek ........... 185 F7
South Fork Moran Creek ........... 220 A1
South Fork Mud Creek ........... 191 A9
South Fork Owl Creek ........... 191 A7
South Fork Sage Creek ........... 191 C7
South Fork Shoshone River ........... 186 D5
South Fork Snowshoe Canyon ........... 220 A2
South Fork Spanish Creek ........... 181 D8
South Fork Split Creek ........... 206 D1
South Fork Teton Creek ........... 220 C1
South Fork Tincup Creek ........... 189 D9
South Fork Valley Creek ........... 183 B7
South Fork Warm Spring Creek ........... 190 A4
South Fork Willow Creek ........... 190 C6
South Fork Wood River ........... 187 E7
South Horse Creek ........... 190 D2
South Junipers ........... 184 E6
South Leigh Creek ........... 185 E9
South Leigh Lakes ........... 220 B1
South Piney Creek ........... 190 E2
South Putnam Mountain ........... 188 D5
South Teton ........... 220 C2
South Teton River ........... 184 E6
South Willow Creek ........... 180 C6
Southeast Arm, Yellowstone Lake ........... 209 D7
Spalding Bay ........... 220 B4
Spanish Breaks ........... 181 C8
Spanish Peaks ........... 181 D8
Specimen Creek ........... 194 B3
Specimen Falls ........... 197 B5
Specimen Ridge ........... 203 A7
Sphinx Mountain ........... 181 E8
Spirea Creek ........... 214 C2
Split Creek ........... 206 D1
Sportman Ridge ........... 190 B2
Sportsman Lake ........... 195 B5
Spread Creek ........... 221 B5
Spring Creek (Clark Co.) ........... 184 C6
Spring Creek (Madison Co.) ........... 180 D4
Spring Creek (Park Co.) ........... 187 D7
Spring Creek (Teton Co.) ........... 207 C6
Spring Creek (Yellowstone Co.) ........... 183 C9
Spring Gulch ........... 187 E10
Spruce Creek ........... 208 B2
Standard Creek ........... 180 F6
Static Peak ........... 222 A3
Steamboat Mountain ........... 218 B3
Steamboat Point ........... 209 B7
Steamboat Spring ........... 209 B7
Stevens Peak ........... 188 C4
Stevenson Island ........... 209 B6
Stewarts Draw ........... 222 A3
Stillwater River ........... 182 E4
Stine Mountain ........... 180 C1
Stinkingwater Peak ........... 205 D7
Stone Creek ........... 180 D3
Stonecup Lake ........... 210 B3
Stonetop Mountain ........... 209 A7
Storm Peak ........... 180 C2
Storm Point (Teton Co. north) ........... 209 B6
Storm Point (Teton Co. south) ........... 220 B2
Stouts Mountain ........... 189 A8
Straight Creek ........... 201 B6
Strawberry Creek ........... 189 D10
String Lake ........... 220 B3
Stump Creek ........... 189 D9
Stump Peak ........... 189 D9
Sugarloaf Mountain (Beaverhead Co.) ........... 180 C2
Sugarloaf Mountain (Park Co.) ........... 198 B3
Sulfatra Creek ........... 202 B2
Sulpher Creek ........... 203 B5
Sulphur Caldron ........... 203 D5
Sulphur Creek ........... 187 C7
Sulphur Hills ........... 209 A6
Sulphur Mountain ........... 203 D5
Sulphur Peak ........... 189 E8
Sulphur Spring ........... 203 D5
Summit Lake ........... 206 C4
Summit Lake Hot Springs ........... 206 D4
Sunlight Creek ........... 186 B5
Sunlight Peak ........... 211 A7
Sunset Lake ........... 220 C1
Sunset Peak ........... 199 B5
Sunshine Creek ........... 187 D7
Sunshine Point ........... 194 B1
Sunshine Reservoir ........... 187 D7
Surprise Creek ........... 215 B5
Surprise Lake ........... 220 C2
Survey Draw ........... 212 B1
Survey Peak ........... 218 B1
Swamp Creek ........... 182 B3
Swamp Lake ........... 199 B6
Swan Creek ........... 181 D9
Swan Lake (Beaverhead Co.) ........... 185 B7
Swan Lake (Park Co.) ........... 195 C7
Swan Lake (Teton Co.) ........... 220 A4
Sweet Grass Creek ........... 182 A4
Sweetwater Creek (Madison Co.) ........... 180 E4
Sweetwater Creek (Park Co.) ........... 186 B5
Sweetwater Creek Falls ........... 211 A7
Sweetwater River ........... 191 E7
Sylvan Lake ........... 210 C3
Sylvan Pass ........... 210 C3
Sylvan Peak ........... 182 E5
Symmetry Spire ........... 220 B2


## T


Table Butte ........... 184 D4
Table Mountain (Big Horn Co.) ........... 187 C10
Table Mountain (Fremont Co.) ........... 190 A5
Table Mountain (Park Co. north) ........... 182 F5
Table Mountain (Park Co. south) ........... 216 B4
Table Mountain (Silver Bow/ Madison Co.) ........... 180 C4
Table Mountain (Teton Co. north) ........... 220 C1
Table Mountain (Teton Co. south) ........... 223 C6
Taggart Lake ........... 222 A3
Talus Lake ........... 220 A2
Tanager Lake ........... 214 D2
Tappan Creek ........... 190 A5
Targhee Pass ........... 185 B8
Targhee Peak ........... 185 B8
Tash Peak ........... 180 D1
Tatman Mountain ........... 187 C9
Taylor Creek ........... 181 E8
Taylor Mountain (Bingham Co.) ........... 188 B6
Taylor Mountain (Teton Co.) ........... 222 C1
Teal Lake ........... 207 C6
Teepee Glacier ........... 220 C2
Telephone Basin ........... 197 A7
Tempe Cascade ........... 213 B5
Tendoy Falls ........... 213 B5
Tendoy Mountains ........... 184 A2
Tenmile Creek ........... 187 D11
Tenmile Pass ........... 188 D6
Tepee Basin ........... 194 D2
Tepee Creek (Gallatin Co. north) ........... 194 B1
Tepee Creek (Gallatin Co. south) ........... 200 A2
Tepee Creek (Park Co.) ........... 199 D7
Tepee Creek (Sublette Co.) ........... 190 B3
Tepee Pass ........... 194 A2
Tepee Point ........... 194 D2
Tern Lake ........... 203 D7
Terrace Mountain (Park Co.) ........... 186 E3
Terrace Mountain (Teton Co.) ........... 195 C7
Terrace Point ........... 216 A2
Terraced Falls ........... 213 D5
Teton Basin ........... 185 E9
Teton Canyon ........... 220 C1
Teton Creek ........... 185 E9
Teton Glacier ........... 220 C2
Teton Pass ........... 222 D1
Teton Range ........... 220 C2
Teton River ........... 185 E7
The Big Slide ........... 183 D7
The Crags ........... 200 A4
The Cut ........... 197 C5
The Hoodoos ........... 195 C7
The Jaw ........... 220 B2
The Landmark ........... 202 B2
The Mesa ........... 190 E4
The Mushpots ........... 204 D2
The Narrows ........... 197 D6
The Needle ........... 204 A4
The Needles ........... 182 D2
The Potholes ........... 220 B4
The Promontory ........... 209 D7
The Ramshorn ........... 186 E5
The Six Lakes ........... 190 B2
The Thunderer ........... 198 D4
The Trident ........... 216 C4
The Wall ........... 186 B5
The Wedge ........... 222 A1
The Wigwams ........... 220 B1
Thirsty Creek ........... 206 B1
Thor Peak ........... 220 B2
Thorofare Buttes ........... 217 D7
Thorofare Creek ........... 186 D3
Thorofare Plateau ........... 186 D4
Three Rivers Peak ........... 201 A5
Thunder Mountain ........... 186 D4
Thurman Ridge ........... 185 C7
Tillery Lake ........... 218 A1
Timber Creek (Park Co north) ........... 205 B6
Timber Creek (Park Co. south) ........... 187 D7
Timber Creek (Washakie Co.) ........... 187 D10
Timbered Island ........... 220 C3
Timberline Lake ........... 222 A3
Timothy Creek ........... 204 C3
Tincup Creek ........... 189 D8
Tincup Mountain ........... 189 C9
Tobacco Root Mountains ........... 180 C5
Tobin Creek ........... 182 B2
Togwotee Pass ........... 186 E3
Tom Creek ........... 190 B5
Tom Miner Creek ........... 181 E9
Top Notch Peak ........... 210 C3
Toppings Lakes ........... 221 C6
Torrent Creek ........... 205 D6
Torrey Mountain ........... 180 D2
Tosi Creek ........... 190 B3
Tough Creek ........... 199 D6
Tower Creek ........... 202 A4
Tower Fall ........... 197 D6
Tower Mountain ........... 180 D2
Trail Canyon (Fremont Co. north) ........... 212 A1
Trail Canyon (Fremont Co. south) ........... 212 C1
Trail Canyon Spring ........... 212 C1
Trail Creek (Park Co. Montana) ........... 181 C10
Trail Creek (Park Co. Wyoming) ........... 186 B6
Trail Creek (Teton Co. north) ........... 216 B2

Trail Creek (Teton Co. south)...... 189 A9
Trail Lake................................ 216 B2
Trapper Creek........................ 180 C2
Trapper Lake........................... 220 A3
Trappers Creek....................... 216 B3
Trilobite Lakes........................ 201 B5
Trilobite Point ....................... 201 B5
Triple Glaciers....................... 220 A2
Trischman Knob ..................... 213 A5
Trout Creek (Fremont Co.)........ 191 D7
Trout Creek (Park Co. east) ....... 186 B6
Trout Creek (Park Co. west) ...... 202 D4
Trout Creek (Stillwater Co.)....... 182 C4
Trout Lake............................ 198 D3
Trout Peak............................. 186 B5
Tucker Creek ......................... 198 B2
Turbid Lake........................... 209 B7
Turbid Springs ...................... 209 B7
Turkey Pen Creek.................... 196 B2
Turkey Pen Peak ..................... 196 B3
Turpin Creek .......................... 223 B6
Turret Mountain ..................... 216 B4
Tweedy Mountain ................... 180 D2
Twelve Mile Creek.................. 183 B9
Twentymile Creek ................... 183 E11
Twin Buttes........................... 207 B5
Twin Creek............................ 191 E9
Twin Island Lake..................... 199 B7
Twin Lakes............................ 201 B7
Twister Falls.......................... 213 B5
Two Bear Ridge....................... 183 D8
Two Ocean Lake ..................... 221 A5
Two Ocean Pass...................... 186 D3
Two Ocean Plateau ................. 216 B2

U

Uhl Draw.............................. 221 B6
Uhl Hill ................................ 221 B6
Undine Falls .......................... 196 C3
Union Falls............................ 213 C5
Union Pass ............................ 190 A4
Union Peak ........................... 190 B4
Upper Deer Creek................... 182 C4
Upper Falls............................ 203 C5
Upper Geyser Basin ................ 207 C5
Upper Glaston Lake................. 182 A4
Upper Red Rock Lake .............. 184 B6
Upper Table ........................... 190 A6
Upper Tepee Basin .................. 194 D1
Upper Terraces....................... 196 C2

V

Valley Creek .......................... 183 C8
Veiled Peak............................ 220 C2
Venus Creek .......................... 186 E5
Vermilion Springs ................... 209 A7
Violet Springs ........................ 202 D4
Virginia Cascade ..................... 202 C3

W

Wagon Creek.......................... 190 B3
Wagonhound Creek ................. 187 E9
Wahb Springs (Sulphur)........... 204 B3
Wahhi Falls ........................... 213 B5
Wapiti Lake............................ 203 C7
Wapiti Mountain .................... 182 E6
Wapiti Ridge.......................... 186 C5
War Man Creek ...................... 183 D12
Wardel Reservoir .................... 187 C10
Warm River ........................... 185 D8
Warm River Butte ................... 212 C1
Warm Spring Creek................. 190 A4
Warm Springs Creek................ 184 D3
Washakie Needles.................... 187 F7
Washakie Pass........................ 190 D6
Washakie Reservoir ................. 191 D7
Washburn Hot Springs............. 203 B5
Washburn Range ..................... 202 A3
Waterfalls Canyon................... 218 D3
Weasel Creek ......................... 209 C5
Webb Canyon ........................ 218 C2
Webber Creek......................... 184 C3
Webster Range ....................... 189 D8
West Boulder Plateau .............. 182 D2
West Camas Creek .................. 184 C5
West Fishtail Creek ................. 182 D4
West Fork Blacktail Deer Creek... 184 A4
West Fork Blackwater Creek...... 211 D7
West Fork Duck Creek.............. 182 B2
West Fork Hoodoo Creek .......... 205 C7

West Fork Horse Creek ............. 196 A4
West Fork Madison River.......... 184 B6
West Fork Mill Creek............... 181 E11
West Fork Rock Creek .............. 182 E5
West Fork Ruby River .............. 180 F5
West Fork Stillwater River......... 182 D3
West Fork Upper Deer Creek..... 182 C3
West Grinnell Creek ................ 211 B5
West Gros Ventre Butte............. 222 D3
West Miners Creek .................. 223 C6
West Pryor Mountain................ 183 D9
West Rosebud Creek................ 182 D5
West Thumb, Yellowstone Lake .. 208 C4
Wheeler Mountain .................. 181 C9
Whetstone Creek ..................... 219 B7
Whetstone Mountain ............... 219 C6
Whirlwind Peak ..................... 211 B6
Whiskey Mountain................... 190 B5
Whistle Creek ........................ 187 B9
Whit Creek ............................ 186 C6
White Creek ........................... 207 B6
White Lake............................ 203 D7
White Lakes ........................... 200 B2
White Owl Butte...................... 185 E7
White Peak ............................ 194 D1
White Peaks ........................... 201 B5
Whitetail Creek ...................... 180 A5
Whitetail Peak........................ 180 A4
Whitetail Reservoir.................. 180 A4
Widewater Lake...................... 199 B7
Wiggins Fork East Fork
    Wind River ...................... 190 A5
Wigwam Creek....................... 180 E6
Wilcox Point.......................... 218 C3
Wild Horse Draw.................... 187 C12
Wildcat Peak ......................... 219 B5
Wildcat Ridge ........................ 219 B4
Wilderness Falls ..................... 218 D2
Willow Creek (Beaverhead Co.)... 180 D2
Willow Creek (Big Horn Co.)...... 187 C9
Willow Creek (Carbon Co.)........ 182 D6
Willow Creek (Clark Co.) .......... 185 C7
Willow Creek (Fremont Co.
    north)............................. 190 C6
Willow Creek (Fremont Co.
    south)............................. 191 E9
Willow Creek (Gallatin Co.) ....... 181 C7
Willow Creek (Lincoln Co.) ....... 189 D10
Willow Creek (Park Co.) .......... 204 D3
Willow Creek (Teton Co. Idaho).. 189 B7
Willow Creek (Teton Co.
    Wyoming) ........................ 189 B11
Willow Creek Reservoir............ 181 C7
Willow Flats .......................... 220 A4
Willow Lake .......................... 190 D4
Willow Park .......................... 201 A6
Wind River............................ 191 C7
Wind River Peak .................... 191 E7
Wind River Range ................... 190 A4
Window Peak ......................... 220 A1
Windy Mountain .................... 186 A5
Winegar Creek ....................... 212 D4
Winegar Lake......................... 212 D4
Winter Creek ......................... 201 B6
Wise River............................. 180 C1
Witch Creek .......................... 214 B4
Wolf Creek ............................ 183 E7
Wolf Lake ............................. 202 C3
Wolf Point............................. 209 C6
Wolverine Creek (Bingham Co.).. 188 C5
Wolverine Creek (Park Co.)........ 198 B3
Wolverine Creek (Teton Co.) ...... 219 A6
Wolverine Pass ...................... 198 B4
Wolverine Peak (Fremont Co.).... 198 B4
Wolverine Peak (Park Co.) ......... 190 D6
Woodall Mountain ................... 189 E7
Woody Creek.......................... 183 C12
Woody Ridge ......................... 199 C5
Wraith Falls .......................... 196 C3
Wrangler Lake........................ 203 C5
Wrong Creek ......................... 203 B7
Wyodaho Lake ....................... 212 C4
Wyoming Peak ....................... 189 E11
Wyoming Range ..................... 189 C11

Y

Y U Bench............................. 187 C9
Yandell Mountain.................... 188 C5
Yellow Mountain..................... 186 D5
Yellowstone Lake ................... 186 C2

Yellowstone River .................... 183 C7
Youngs Point.......................... 218 C1

NORRIS  1:24,000-scale series

HEBGEN LAKE  1:100,000-scale series

National Park Service unit

Urban area

State boundary

County boundary

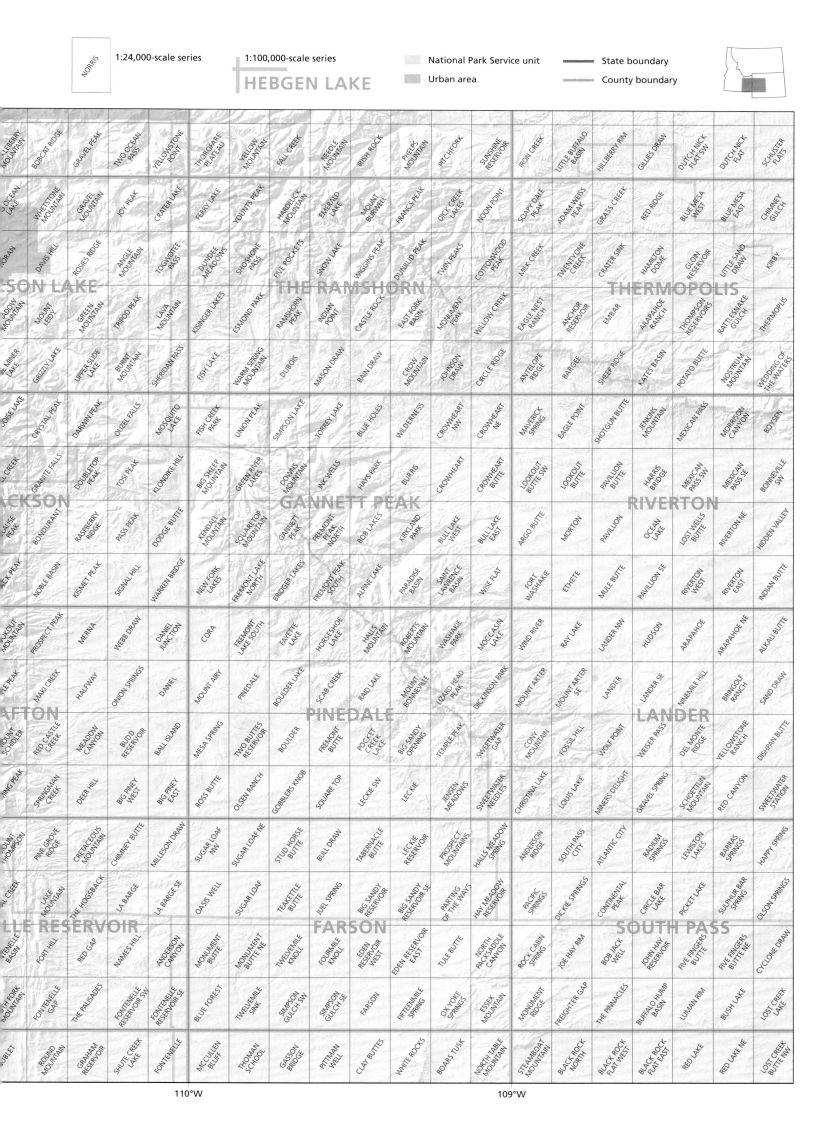

110°W          109°W

235

# Counties

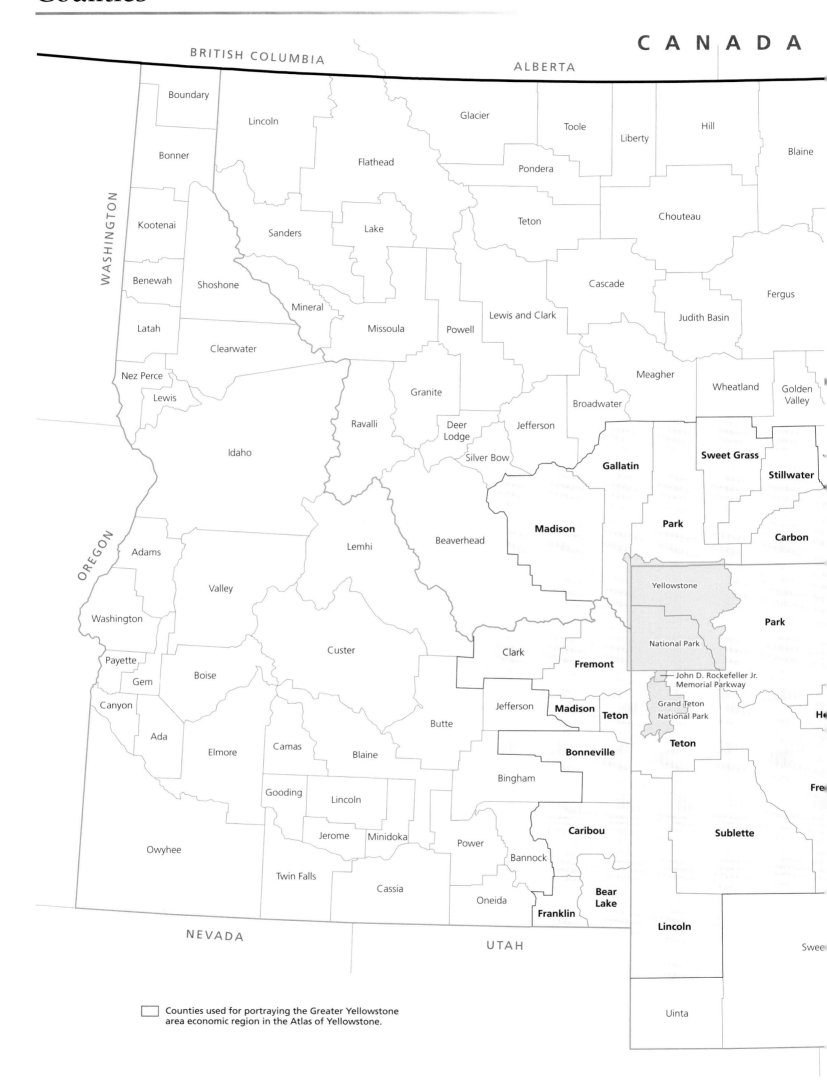

CANADA

BRITISH COLUMBIA

ALBERTA

WASHINGTON

Boundary

Lincoln

Glacier

Toole

Liberty

Hill

Blaine

Bonner

Flathead

Pondera

Kootenai

Sanders

Lake

Teton

Chouteau

Benewah

Shoshone

Cascade

Fergus

Latah

Mineral

Missoula

Lewis and Clark

Judith Basin

Clearwater

Powell

Nez Perce

Meagher

Wheatland

Golden Valley

Lewis

Granite

Broadwater

Ravalli

Deer Lodge

Jefferson

**Gallatin**

**Sweet Grass**

**Stillwater**

Idaho

Silver Bow

**Madison**

**Park**

**Carbon**

Adams

Lemhi

Beaverhead

Yellowstone

**Park**

OREGON

Valley

National Park

Washington

Custer

Clark

**Fremont**

John D. Rockefeller Jr. Memorial Parkway

Payette

**Madison**

**Teton**

Gem

Boise

Jefferson

Grand Teton National Park

**He**

Canyon

Butte

**Bonneville**

**Teton**

Ada

Camas

Blaine

Elmore

Bingham

**Caribou**

**Sublette**

**Fre**

Gooding

Lincoln

Owyhee

Jerome

Minidoka

Power

Bannock

**Bear Lake**

Twin Falls

Cassia

Oneida

**Franklin**

**Lincoln**

NEVADA

UTAH

Swee

Uinta

Counties used for portraying the Greater Yellowstone area economic region in the Atlas of Yellowstone.

KATCHEWAN

Daniels    Sheridan

Valley    Roosevelt

NORTH DAKOTA

Richland

McCone

Dawson

Garfield

Prairie    Wibaux

Rosebud    Fallon

Treasure    Custer

Carter

Big Horn    Powder River

Sheridan    Crook

SOUTH DAKOTA

Campbell

Johnson    Weston

Vashakie

Natrona    Converse    Niobrara

NEBRASKA

Platte    Goshen

Carbon    Albany

Laramie

COLORADO

# Greater Yellowstone Area Cultural Place Names

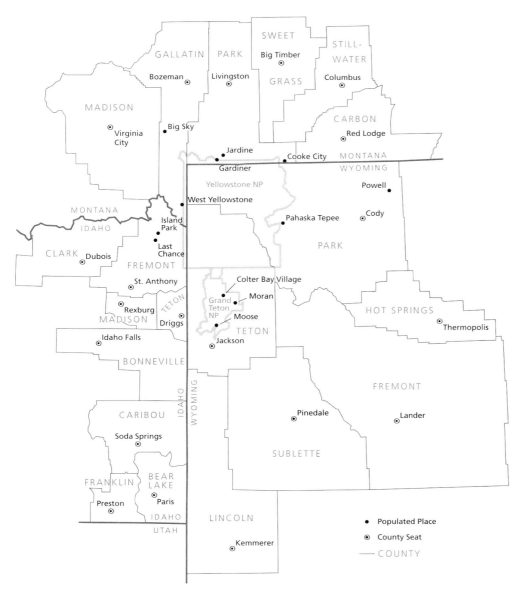

**Bear Lake County**, IDAHO Named for nearby Bear Lake, the county was established in 1875 after Mormons began colonizing the region in 1863. The county is in the state's southeast corner. Paris is the county seat.

**Big Sky**, MONTANA A resort complex, the creation of which was spearheaded by native Montanan and famed television newsman Chet Huntley (1911–74).

**Big Timber**, MONTANA The county seat of Sweet Grass County, Big Timber was named for the creek-side cottonwood trees in the area, which were, at the time, among the largest in the state.

**Bonneville County**, IDAHO Bordering Wyoming in the southeast part of the state, the county, created in 1911, is named after trapper and trailblazer B.L.E. Bonneville, who led an expedition that explored many areas in the West, including the Snake River region.

**Bozeman**, MONTANA The county seat of Gallatin County, Bozeman was named for John M. Bozeman, a pioneer who, in 1864, guided the first group of pioneers into the Gallatin Valley.

**Carbon County**, MONTANA Created in 1895 from sections of Yellowstone and Park counties, Carbon County took its name from the area's rich coal deposits.

**Caribou County**, IDAHO A participant in the1860 gold rush in the Cariboo region of British Columbia, Cariboo Fairchild is the namesake for Idaho's Caribou Mountains. The mountain range, in turn, inspired the naming of the county in the southeast part of the state.

**Clark County**, IDAHO Named for Samuel K. Clark, an early settler on Medicine Lodge Creek, who found great success as a rancher and businessman and was elected the first state senator from Clark County.

**Cody**, WYOMING Platted in 1895 and incorporated in 1902, Cody, the county seat of Park County, was named after William "Buffalo Bill" Cody.

**Colter Bay**, WYOMING Named for mountain man and Lewis and Clark expedition veteran John Colter.

**Columbus**, MONTANA The town of Stillwater was named for the Northern Pacific railroad station, built in 1882. Another town, also named Stillwater, this one on the railroad's main line in Minnesota, led to confusion and, eventually, the renaming of the Montana town, which incorporated in 1907.

**Cooke City**, MONTANA Named after mining investor Jay Cooke Jr., who was interested in the New World Mining District.

**Driggs**, IDAHO Mormon settlers established a colony in northeastern Idaho around 1888. Originally called Alpine, the area was renamed Driggs by the Post Office Department due to the large number of people surnamed Driggs who signed the petition requesting a local post office. The county seat of Teton County.

**Dubois**, IDAHO The county seat of Clark County was named to honor U.S. Senator Fred T. Dubois, an early settler (1880), U.S. marshal, and influential figure in the state's early political history.

**Franklin County**, IDAHO Mormon settlers came to the area in 1860, establishing the first permanent settlement in Idaho and naming it Franklin, for Franklin Richards, a church apostle.

**Fremont County**, IDAHO Named for John C. Fremont who led an expedition through the region in 1843.

**Fremont County**, WYOMING Established in 1884, Fremont County is named for John C. Fremont, who led an expedition passing through the area in 1842 (with Kit Carson serving as one of its guides).

**Gallatin County**, MONTANA In the state's southwest, this county roughly follows the contours of the Gallatin River, named by Lewis and Clark in 1805 to honor U.S. Secretary of the Treasury Albert Gallatin. Created by the territorial legislature in 1865, Gallatin was one of the territory's nine original counties.

**Gardiner**, MONTANA This northern entrance to Yellowstone park is named for mountain man Johnson Gardiner who trapped in the area of the upper Yellowstone River.

**Grand Teton National Park** Named for the Teton Range.

**Hot Springs County**, WYOMING The county, organized in 1913, takes its name from the hot springs at Thermopolis.

**Idaho** No certain origin is known for the word Idaho. It is popularly said to be taken from an Indian word Ee-da-how ("light on the mountain" or "gem of the mountains") though this origin is suspect and the name may have been coined. Early appearances of the word include the name of Idaho Springs, Colorado (1859), and the *Idaho*, a Columbia River steamer (1860).

**Idaho Falls**, IDAHO The county seat of Bonneville County had a number of names before 1891, when it became Idaho Falls, descriptive of nearby rapids in the Snake River.

**Island Park**, IDAHO The town, according to early resort proprietor Charlie Pond, took its name from the way a stand of timber rose from the surrounding plain of sagebrush.

**Jackson**, WYOMING The county seat of Teton County was named for Jackson Hole. In fur trappers' parlance, a hole was a mountain-protected valley, often named after the trapper most notably associated with the area. Trapper William Sublette named this place after David Jackson, his partner, in 1829 as Jackson's Hole, subsequently simplified to Jackson Hole.

**Jardine,** MONTANA  Named for A.C. Jardine, an executive with the Bear Gulch Mining Company.

**John D. Rockefeller Jr. Memorial Parkway**  Set aside by Congress in 1972 and managed by the National Park Service, this 24,000-acre scenic corridor between Yellowstone and Grand Teton national parks honors the conservationist and philanthropist whose efforts helped preserve the region's natural areas.

**Kemmerer,** WYOMING  The county seat of Lincoln County is named for M.S. Kemmerer, president of the Kemmerer Coal Company. It is home of the first J.C. Penney store, opened in 1902.

**Lander,** WYOMING  The county seat of Fremont County, Lander was named to honor Gen. F.W. Lander, who conducted surveys and improved passage along the Oregon Trail—work that led to the Lander Cut-Off. Local landowner B.F. Lowe renamed the town (previously known as Push Root and Camp Brown) after his friend in 1869.

**Last Chance,** IDAHO  The final resort (until Ashton, about forty miles) for those headed south on Highway 191. Previously known as Dewiner Inn and Ripleys Ford.

**Lincoln County,** WYOMING  Named after Abraham Lincoln.

**Livingston,** MONTANA  In 1822 railroad surveyors named the site Clark City to honor William Clark, who had explored the Yellowstone River and passed near the town's present location in mid-July of 1806. Later, the town, with its significant railroad-related assets, was renamed after Crawford Livingston, an executive of the Northern Pacific railroad.

**Madison County,** IDAHO  This county in the eastern part of the state is named for James Madison, fourth president of the United States.

**Madison County,** MONTANA  One of the original nine territorial counties established in 1865, Madison County is named for the Madison River, which, in turn, was named by Lewis and Clark to honor U.S. Secretary of State James Madison.

**Montana**  The state took its name from the Spanish word *montaña,* meaning mountain or mountainous region.

**Moose,** WYOMING  Named for the animals who inhabit the local area, Moose is home to the headquarters of Grand Teton National Park.

**Moran,** WYOMING  Named for Mount Moran, which, in turn, was named for Hayden expedition artist Thomas Moran, whose large-scale painting *Grand Canyon of*

*the Yellowstone* and other works contributed to the establishment of Yellowstone as the first national park.

**Pahaska Teepee,** WYOMING  A structure that served as a hunting lodge, built in 1901 by William "Buffalo Bill" Cody. Sioux Indians called Cody *Pahaska* (long hair).

**Park County,** MONTANA  Proximity to Yellowstone National Park accounts for this name, which was established in May 1887 to create a county seat nearer than Bozeman for the inhabitants of rapidly growing Clark City (Livingston).

**Park County,** WYOMING  Proximity to Yellowstone National Park accounts for this county's name, which was established in 1909.

**Paris,** IDAHO  The county seat of Bear Lake County in the state's southeast corner, Paris was settled by Mormon pioneers and named for Frederick Perris, who platted the site.

**Pinedale,** WYOMING  The county seat of Sublette County, Pinedale is located on and named for Pine Creek.

**Powell,** WYOMING  The federal government established Camp Colter (after John Colter) as headquarters for the Shoshone Project and related dam and canal work. It was later renamed Powell after explorer and geologist John Wesley Powell.

**Preston,** IDAHO  The seat of Franklin County, Preston was first known as Worm Creek and later renamed to honor William B. Preston, an early resident and prominent member of the Mormon Church.

**Red Lodge,** MONTANA  The county seat of Carbon County is believed to be named for the way Crow Indian lodges were hued by the area's red clay soils.

**Rexburg,** IDAHO  Directed by John Taylor, third president of the Church of Jesus Christ of Latter-Day Saints, to establish a settlement, Thomas Ricks traveled to this site in eastern Idaho. Rex and Ricks are etymologically related.

**Soda Springs,** IDAHO  The county seat of Caribou County is named after the area's soda deposits, mineral springs, and unusual rock formations.

**St. Anthony,** IDAHO  Noticing a resemblance between cascades on Henrys Fork of the Snake River and St. Anthony Falls on the Mississippi, local homesteader and postmaster Charles H. Moon suggested the name for what would become the county seat of Fremont County.

**Stillwater County,** MONTANA  Named for the Stillwater River in 1913, the county was established on lands previously part of Carbon, Sweet Grass, and Yellowstone counties.

**Sublette County,** WYOMING  Named for trapper, pioneer, and mountain man William Sublette, the state's first dude wrangler.

**Sweet Grass County,** MONTANA  Mrs. Paul Van Cleve Sr. suggested the name of this county after the local prairies near her home in Melville.

**Teton County,** IDAHO  Named for the Teton Range.

**Teton County,** WYOMING  Named for the Teton Range.

**Thermopolis,** WYOMING  The county seat of Hot Springs County, Thermopolis is named for the area's hot springs complex. Its name is a combination of the Latin word for hot baths (thermae) and the Greek word for city (polis).

**Virginia City,** MONTANA  After prospectors struck gold here in 1863, a gold rush ensued, bringing rapid growth (the population of the city shot to 18,000 by the fall of 1864) and leading to the incorporation the following year of Montana's first town. Originally named Varina, after the wife of Confederacy president Jefferson Davis, the town changed names when Judge G.G. Bissell, with no fond feelings for the Confederacy, refused to use the name on a legal document, writing instead "Virginia," later expanded to Virginia City.

**Wapiti Ranger Station,** WYOMING  The name of this 1903 structure, the oldest ranger station in the United States, is taken from the Indian name for elk.

**West Yellowstone,** MONTANA  The western entrance to Yellowstone National Park.

**Wyoming**  Congress created the Wyoming Territory in 1868, named for the Wyoming Valley in Pennsylvania. Other names had been considered, including Cheyenne, Shoshone, and Lincoln. The territory became the nation's forty-fourth state (with the same boundaries) in June 1890.

**Yellowstone National Park**  Named for the Yellowstone River. The river's name is derived from an Indian name, *Mi-tsi-a-da-zi,* Yellow Rock River. French trappers and explorers called it *Roche Jaune* (yellow rock) for the color of some geological features in the canyon. Lewis and Clark crossed the river in 1805 and used the name Yellowstone in their accounts.

# Greater Yellowstone Area Physical Place Names

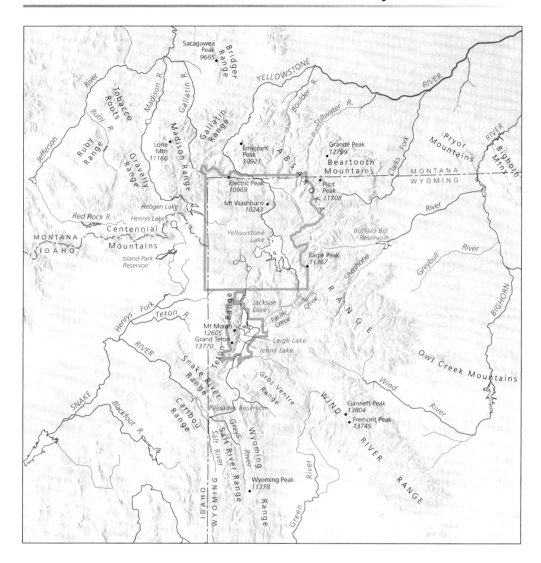

rise to the name for this peak (10,921 feet) as well as the creek and gulch.

**Fremont Peak,** WYOMING This is the peak (13,745 feet) John C. Fremont famously climbed in 1842 and unfurled an American flag, the first to fly over present-day Wyoming—or perhaps not. Analysis of Fremont's account of the climb has led investigators to believe he actually climbed the peak today known as Mount Woodrow Wilson. The Hayden party likely climbed this peak in 1878.

**Gallatin Range,** MONTANA A mountain range named for the Gallatin River, which, in turn, Lewis and Clark named for U.S. Secretary of the Treasury Albert Gallatin.

**Gallatin River,** MONTANA An honor bestowed on U.S. Secretary of the Treasury Albert Gallatin by Lewis and Clark in 1805.

**Gannett Peak,** WYOMING The highest point in Wyoming, Gannett Peak (13,804 feet) was named in 1906 after American geographer Henry Gannett (1846–1914), a participant in the Hayden explorations.

**Grand Teton,** WYOMING The Doane expedition of 1870 knew this feature (13,770 feet) as Mount Hayden. It is the high point of the Teton Range and second highest mountain in Wyoming. It received its name early (probably before 1800) from French-Canadian fur trappers of the Hudson's Bay Company, who named the main three peaks "*Les Trois Tetons*," meaning "the three nipples."

**Granite Peak,** MONTANA The highest point in Montana, this peak (12,799 feet) is named for its prominent granite formations.

**Gravelly Range,** MONTANA Course gravel deposits in the area give name to this range.

**Green River,** WYOMING In 1823 William Ashley named the river for its color, which resulted from erosion of its green soapstone banks.

**Greybull River,** WYOMING Indians named this river for an albino buffalo seen along its banks.

**Greys River,** WYOMING Named for trapper John Gray (or Grey)—previously known as John Day's River, after the hunter, trapper, and Astorian party member.

**Gros Ventre Mountains,** WYOMING These mountains got their name from the Gros Ventres Indian tribe. The tribe's name is French for "big bellies." Some historians believe that sign language used by local Indians was misinterpreted by the French to mean big belly. Others think that the name referred to the swollen stomachs of local Indians who had too little to eat.

**Hebgen Lake,** MONTANA The waters behind Hebgen Dam, named to honor hydroelectric engineer Max Hebgen, form Hebgen Lake.

**Henrys Fork River,** IDAHO Major Andrew Henry, a fur trader, built Fort Henry on Henrys Fork in 1810.

**Absaroka Range,** WYOMING Named for Absaroka Indians by the U.S. Geological Survey in 1885.

**Atlantic Creek,** WYOMING Flowing from a single water source that bifurcates at Two Ocean Pass on the continental divide, water from this creek reaches the Atlantic Ocean by way of the Yellowstone, Missouri, and Mississippi rivers. See Pacific Creek.

**Beartooth Mountains,** WYOMING Named after a peak that has the appearance of a bear's tooth.

**Bighorn River,** WYOMING Lewis and Clark named this river in 1803 from a translation of the Indian name, *Ah-sah-ta*, indicating the large numbers of bighorn sheep that could be found near the river.

**Big Horn Mountains,** MONTANA Named after the Bighorn River.

**Blackfoot River,** IDAHO Named in 1819 by Donald McKenzie after Indians in the area who called themselves *Siksika* (black) and *Kah* (foot).

**Boulder River,** MONTANA Named for the area's large rocks.

**Bridger Range,** MONTANA A mountain range named after mountain man, trapper, scout, and explorer Jim Bridger, the first white man to explore Yellowstone Park.

**Buffalo Bill Reservoir,** WYOMING The body of water behind Buffalo Bill Dam, completed in 1910, is named for William "Buffalo Bill" Cody.

**Caribou Range,** IDAHO A mountain range named after Cariboo Fairchild, a participant in the 1860 gold rush in British Columbia's Cariboo region.

**Centennial Mountains,** MONTANA A mountain range named for the Centennial Valley in southwestern Montana, itself named in 1876 to honor the nation's centennial.

**Clarks Fork,** WYOMING Named for William Clark.

**Eagle Peak,** WYOMING This peak (11,367 feet) of the Absaroka Range on the park's east boundary is the tallest mountain in Yellowstone National Park. It was named in 1885 because it was "shaped like a spread eagle."

**Electric Peak,** WYOMING Surveyor Henry Gannett was leading a team from the Hayden survey up the peak (10,969 feet) in 1872 when an intense electrical storm arose, resulting in shocks, hair standing on end, roaring in the climbers' ears, and pain in their hands.

**Emigrant Peak,** MONTANA Gold miners such as Thomas Curry, David Shorthill, and others came to the area in 1863–64, giving

**Henrys Lake,** IDAHO  Major Andrew Henry, a fur trader, built Fort Henry on Henrys Fork in 1810.

**Island Park Reservoir,** IDAHO  Island Park Dam is named for the town in northeast Idaho. The water behind the dam forms the reservoir. The town, according to early resort owner Charlie Pond, took its name from the way a stand of timber rose from the surrounding plain of sagebrush.

**Jackson Lake,** WYOMING  Named for Jackson Hole (which in turn got its name from David Jackson). Clark's early maps call this body of water Lake Biddle, after Nichols Biddle, publisher of the Lewis and Clark papers in 1809.

**Jefferson River,** MONTANA  Lewis and Clark named this river to honor Thomas Jefferson.

**Jenny Lake,** WYOMING  Richard "Beaver Dick" Leigh, a trapper and Hayden expedition guide, was married to an Indian woman whose name now graces this lake.

**Leigh Lake,** WYOMING  Named for Richard "Beaver Dick" Leigh, a trapper and Hayden expedition guide.

**Lone Mountain,** MONTANA  Surrounded by far lower peaks, Lone Mountain (11,166 feet) stands out prominently.

**Madison Range,** MONTANA  A mountain range named for the Madison River, which Lewis and Clark named for U.S. Secretary of State James Madison, later the fourth president of the United States.

**Madison River,** MONTANA  Named by Lewis and Clark for U.S. Secretary of State James Madison, later the fourth president of the United States.

**Mount Moran,** WYOMING  This feature (12,605 feet) was named for Hayden expedition artist Thomas Moran, whose large-scale painting *Grand Canyon of the Yellowstone* and other works contributed to the establishment of Yellowstone as the first national park.

**Mount Washburn,** WYOMING  Named by Hayden (1871) after Washburn party leader Henry Dana Washburn, who had climbed the mountain (10,243 feet) alone to find Yellowstone Lake and a suitable passage to it.

**Owl Creek Mountains,** WYOMING  Named for Owl Creek, itself named by Indians for the local abundance of owls.

**Pacific Creek,** WYOMING  Flowing from a single water source that bifurcates at Two Ocean Pass on the continental divide, water from this creek reaches the Pacific Ocean by way of the Snake and Columbia rivers. See Atlantic Creek.

**Palisades Reservoir,** IDAHO  Cliffs along the Snake River and tributaries gave Palisades Creek, and, in turn, this reservoir, their names.

**Pilot Peak,** WYOMING  A landmark (11,708 feet) useful for orienting travelers.

**Pryor Mountains,** MONTANA  Lewis and Clark honored expedition member Sergeant Nathaniel Hale Pryor with the naming of this range.

**Red Rock River,** MONTANA  Nearby clays and sandstones are red in color.

**Ruby Range,** MONTANA  A mountain range named after the garnets found in the region, mistakenly believed to be rubies.

**Ruby River,** MONTANA  The name is taken from the Ruby Mountains, where garnets, thought to be rubies, were found.

**Sacagawea Peak,** MONTANA  This peak (9,665 feet) is named for the Shoshone woman—whose name is spelled variously—who traveled with Lewis and Clark.

**Salt River,** WYOMING  Distinctive salt ledges and saline springs inspired the name of this river.

**Salt River Range,** WYOMING  Named for the Salt River.

**Shoshone River,** WYOMING  In 1807 John Colter named this "Stinking Water" after the sulfur smell associated with the geologically active springs near Cody—and it appears as such on William Clark's map of 1814. Preferring a less descriptive name, citizens persuaded the Wyoming legislature in 1902 to rename it Shoshone, after the Indian tribe.

**Snake River,** WYOMING  Upon discovering the river in 1805, William Clark named it the Lewis River, and that name appears on early maps. However, trappers were more descriptive, calling it *"la maudite riviere enrage"* (the accursed mad river). The present name was adopted later and is for the Snake Indians who lived there. The sign language gesture for the Snake Indians was moving the hand in a serpentine fashion, like the weaving of grass—woven grass being a component of their lodges.

**Snake River Range,** IDAHO  Named for the river, which takes its name from the Snake Indians.

**Stillwater River,** MONTANA  A curious name for a generally energetic body of water, it is attributed to an Indian story in which a pair of ill-fated lovers perish, their bodies washed away in a flood. As the floodwaters receded, a calm section of stream resulted.

**Teton Range,** WYOMING  French explorers and trappers used *"Les Trois Tetons"* (the three teats) to describe some of the mountains in the range.

**Teton River,** IDAHO  Named for the Teton Range.

**Tobacco Root Mountains,** MONTANA  This name is believed to refer to a root, native to these mountains, that could be mixed with or, in combination with other substances, used instead of tobacco.

**Wind River,** WYOMING  Named for the strong winds that blow through its canyon.

**Wind River Range,** WYOMING  A mountain range named for the river, itself named for the strong winds that blow through its canyon.

**Wyoming Peak,** WYOMING  The highest peak (11,378 feet) in the Wyoming Range.

**Wyoming Range,** WYOMING  Originally these mountains were known as the Bear River Range, but were renamed after the creation of the Wyoming Territory in 1869.

**Yellowstone Lake,** WYOMING  Named for the Yellowstone River. William Clark's early maps call this body of water Lake Eustis, after William Eustis, the U.S. Secretary of War (1809–13).

**Yellowstone River,** WYOMING  The river's name is derived from an Indian name, *Mitsi-a-da-zi,* Yellow Rock River, which French trappers and explorers called *"Roche Jaune"* (yellow rock) for the color of geological features that were probably not those of the Grand Canyon in the park but rather yellowish rocks far downstream (possibly the rim rocks at Billings, Montana, which appear bright yellow in the sunshine after a rain). In 1805, Lewis and Clark crossed the river and used the name Yellowstone in their accounts.

# Yellowstone National Park Place Names

All place names listed are located in Wyoming unless otherwise noted.

**Absaroka Range** This range, forming the eastern boundary of Yellowstone, was called the "Yellowstone Range" in fur trade days, but geologist Arnold Hague changed the name to Absaroka in 1885 for the Indian name of the Crow Indians, who lived to the northeast of the park. Absaroka means "children of the large beaked bird" or, according to Crow chief Jim Beckwourth, "the Sparrow-Hawk People."

**Alum Creek** Flowing to the Yellowstone River from Hayden Valley and the west, Alum Creek was named in 1870 because it tasted as if it were "strongly impregnated with alum."

**Amethyst Mountain** Located in the northeast park, this mountain (9,614 feet) received its name from a purple or violet-colored variety of quartz.

**Antler Peak** This peak (10,023 feet) of the Gallatin Range was named for elk and deer antlers that are commonly found shed in the park.

**Apollinaris Spring** This cold-water spring of calcium bicarbonate served as a drinking spot for many years for thirsty park visitors aboard stagecoaches. It was named by these travelers before 1890 because it tasted like "Apollinaris water," the commercially bottled German product (still available today).

**Arthur Peak** This peak (10,428 feet) of the Absaroka Range was named for President Chester A. Arthur, who, in 1883, became the first U.S. president to visit Yellowstone National Park.

**Artist Point** This point on the south rim of the Grand Canyon of the Yellowstone was named because it was thought—mistakenly—to be the spot where artist Thomas Moran drew sketches for his famous oil painting of that canyon.

**Artists' Paintpots** This plural-possessive name was inspired by this geothermal area's brightly colored muds, which reminded geologists of an artist's palette.

**Atkins Peak** This peak (11,044 feet) of the Absaroka Range was named for John D.C. Atkins, Commissioner of Indian Affairs (1885–87), who was so honored for his role in establishing the U.S. Geological Survey (1879).

**Bannock Ford** A ford of the Yellowstone River near Tower Fall, so named for the Bannock Indian tribe who used it (at least from 1840–78) on a trail that crossed the center of present Yellowstone National Park known as the Bannock Trail.

**Bannock Peak** This peak (10,323 feet) of the Gallatin Range was named in 1885 for the Bannock Indian tribe, who lived west of the present park, and for the Bannock Trail that passed a few miles to its south.

**Barlow Peak** Located southeast of the South Arm of Yellowstone Lake, Barlow Peak (9,622 feet) was named for U.S. Army Capt. John W. Barlow (1838–1914), an engineer who led a party on the first official exploration to the headwaters of the Snake River in 1871 and helped F.V. Hayden explore Yellowstone that same year.

**Barronette Peak** This peak (10,404 feet) of the Absaroka Range was named for Collins John H. ("Yellowstone Jack") Baronett (1829–1901), an important early prospector, guide, and assistant superintendent in Yellowstone who built the first bridge across the Yellowstone River in 1871. The name of the peak is officially misspelled, in connection with an early park map.

**Bechler River** This river in the park's southwest corner was so named in 1872 by a fellow topographer for Gustavus R. Bechler, the chief topographer of the Hayden expedition that year. Bechler drew several of the maps of the park from that survey party.

**Beehive Geyser** This geyser of Upper Geyser Basin was named in 1870 from its resemblance to "an old-fashioned straw beehive with the top cut off." It erupted that year for Washburn party members to heights of more than 200 feet.

**Beryl Spring** Beryl (pronounced "burl") is a blue-green mineral, and this spring was named by scientists who thought that its color resembled that of the gem. It is one of the park's hottest springs.

**Biscuit Basin** So named from the biscuit-shaped rock formations that formerly surrounded Sapphire Pool, as shown in early photographs. The 1959 Yellowstone earthquake broke off many of these "biscuits" and washed them into Firehole River.

**Boiling River** This feature—the largest discharging hot spring in the park—has probably never been hot enough to deserve its official name, which was given in the early 1880s for its turbulent action rather than its temperature.

**Bridge Bay** Park superintendent P.W. Norris named this bay on Yellowstone Lake in 1880 after a natural bridge of stone near the bay and a stream that had earlier been named Bridge Creek.

**Bumpus Butte** A small, square-shaped butte located on the west bank of Yellowstone River near Calcite Springs. It was named for Hermon Carey Bumpus (1862–1943), who made large contributions to the park's museum and interpretive programs during the 1920s.

**Bunsen Peak** F.V. Hayden named this peak (8,564 feet) in 1872. He had studied and been inspired by the theories about geysers promoted by German physicist Robert von Bunsen (1811–99).

**Cache Creek** This stream, which flows southwest to Lamar River from the Absaroka Range, was named in 1864 by a party of prospectors that included George Huston, Adam Miller, and Bill Hamilton. Indians stole their horses, leaving the men with only two donkeys, so they cached some of their belongings here and split up.

**Calfee Creek** This stream, flowing west from the Absaroka Range to the Lamar River, was named in 1880 for Henry Bird Calfee, a stereopticon photographer who was then earning some regional fame by producing and selling photographs of the new Yellowstone National Park.

**Carrington Island** Located two miles northeast of Bluff Point on Yellowstone Lake, this island was named in 1871 for E. Campbell Carrington, a zoologist with the Hayden survey who traveled in a boat around the lake and mapped its shores.

**Cascade Corner** This name for the southwest (Bechler) corner of Yellowstone National Park was given in 1921 because the region was noted for its many beautiful waterfalls and cascades.

**Castle Geyser** Located in the Castle Group of Upper Geyser Basin, this geyser was named in 1870 from its resemblance "to an old feudal tower partially in ruins." At times it erupts 100 feet in the air.

**Chalcedony Creek** Flowing northeast from Amethyst Mountain to Lamar River, this stream was named for chalcedony, a waxy quartz found in the park.

**Clagett Butte** This small butte (8,005 feet), located west of Mammoth Hot Springs, was named for William Horace Clagett, a Montana territorial delegate who introduced the bill into the U.S. House of Representatives in 1871 that designated Yellowstone as the world's first national park.

**Cleopatra Spring** This hot spring of the Mammoth Hot Springs, also known as Diana Spring, was named in about 1882 by photographers who sold pictures of it. Like many of the Mammoth Hot Springs, its name was confused among several springs. The name was probably intended to be a suggestion for classic beauty, the ancient queen of Egypt being the namesake.

**Clepsydra Geyser** This geyser of the Fountain Group of Lower Geyser Basin was named in 1873 because it seemed regular, marking the passage of time "by the discharge of water." The name comes from a metal vessel that was used as a water clock in ancient Greece.

**Cody Peak** This peak (10,267 feet) of the Absaroka Range was named for William F. "Buffalo Bill" Cody (1846–1917), who helped found the town of Cody, Wyoming, in about 1896 and who ran his famous wild west shows beginning in 1883.

**Colter Peak** This peak (10,683 feet) of the Absaroka Range was named for John Colter, "the first white man of whom we have any record who penetrated this rough and rugged country." Colter has been credited with being the white discoverer of the Yellowstone area in the winter of 1807–08 and its earliest explorer of record.

**Cook Peak**  This peak (9,742 feet) of the Washburn Range was named for Charles W. Cook (1839–1927), a leader of the Cook-Folsom-Peterson party that explored Yellowstone in 1869.

**Coulter Creek**  Flowing northwest to Snake River from Teton National Forest, Coulter Creek was named by members of the 1872 Hayden survey for their botanist, John Merle Coulter (1851–1928).

**Cowan Creek**  This stream, which flows southwest to Nez Perce Creek from Mary Mountain, was named in honor of George F. Cowan, a park tourist who was captured by Nez Perce Indians in 1877, shot, and left for dead. He survived, crawling nine miles in about sixty hours to reach the Lower Geyser Basin, where U.S. soldiers rescued him.

**Craig Pass**  This pass (8,262 feet) of the Continental Divide west of Old Faithful was named, likely by the army engineer who surveyed the road through the pass in 1891, for Ida M. Craig Wilcox, "the first tourist to cross the pass."

**Cummings Creek**  Flowing west from Jones Pass to Bear Creek, this stream was named for Lt. Joseph F. Cummings (1851–1912) who traveled through the area in 1881 on a military expedition with Wyoming governor John Hoyt. The trip was made to find a feasible wagon route into Yellowstone Park from the east, a route the party did not find.

**Dailey Creek**  This stream flows southwest to the Gallatin River from the park's northwest boundary. It was named for Andrew J. Dailey, one of the earliest settlers of Paradise Valley, Montana, just over the divide from the creek's headwaters. The name for Dailey Lake comes from this same family.

**DeLacy Creek**  Flowing south from DeLacy Lakes to Shoshone Lake, DeLacy Creek was named for Walter W. DeLacy (1819–92), the leader of a prospecting expedition that passed through Yellowstone in 1863. Although DeLacy made a map of the region and saw geysers erupting, he did not publish his findings until 1876 and so failed to receive credit as the first white to document his discoveries in Yellowstone National Park.

**Delusion Lake**  Named in 1878 after surveyors' discovered the "delusion" of earlier explorers, who mistakenly mapped it as an arm of Yellowstone Lake.

**Douglas Knob**  This low hill (8,524 feet) near the head of Little's Fork of Bechler River was named for Joseph O. Douglas (1872–1939), a member of the Yellowstone Ranger Force who played a role in Yellowstone's transition from military to civilian control.

**Druid Peak**  This peak (9,583 feet) was named in 1885 for unknown reasons, perhaps for its relative aloneness or from a Stonehenge-like rock formation on its eastern slopes.

**Dunanda Falls**  This 150-foot waterfall on Boundary Creek was named in 1921 from the Shoshone Indian word meaning "straight down."

**Dunraven Pass**  This pass (8,859 feet) was named for Windham Thomas Wyndham-Quin, the Fourth Earl of Dunraven (1841–1926). He visited Yellowstone Park during a hunting and sightseeing trip to western America in 1874 and published a book about his travels in 1876. The book played an important role in promoting Yellowstone in Britain and Europe.

**Eagle Peak**  This peak (11,367 feet) of the Absaroka Range on the park's east boundary is the tallest mountain in Yellowstone National Park. It was named in 1885 because it was "shaped like a spread eagle."

**Echinus Geyser**  This geyser of the Back Basin of Norris Geyser Basin was named in 1878 "because the pebbles around the basin" had "some resemblance to the spine-covered sea urchin." *Echinus* is a Latin genus of spiny sea urchins; deposits in the Norris area are often spiny due to the acidity of hot spring waters there.

**Eleanor Lake**  Located near the top of Sylvan Pass, this small lake was named by road engineer Hiram Chittenden for his young daughter, Eleanor, during his first tour of duty in Yellowstone (1891–93). Later, Eleanor Chittenden Cress prepared a number of revised editions of her father's famous book *The Yellowstone National Park*.

**Electric Peak**, MONTANA  This peak (10,969 feet) of the Gallatin Range was named in 1872 from the following incident, described by Dr. A.C. Peale of the second Hayden survey:

> We reached the summit of the peak about 4 o'clock. There was a storm cloud all about us. [Henry] Gannett was a little ahead and we saw him hurrying back to us with his hair standing on end. As he neared us we could hear a crackling noise as though there were a lot of frictional electrical machines all about him. We soon began to feel it ourselves. Gannett said [that when] he got to the summit the electricity was so strong that he was obliged to put down the gradienter and hurry down. [A.E.] Brown tried to go up and get it but got a shock on the top of his head and came back in a hurry also. We then descended about 100 feet, having the noise all about us as though there were a lot of electrical machines about us.

**Emerald Pool**  One of Yellowstone's most famous hot springs, Emerald Pool is located at Black Sand Basin of Upper Geyser Basin. It was named in 1872 for its "beautiful emerald tint with yellow-green basin and ornamented edge."

**Excelsior Geyser Crater**  This large (328 by 276 feet) hot spring in the Midway Geyser Basin was once one of the world's tallest geysers, erupting in the 1880s to 300 feet high and 300 feet wide and, according to some accounts, doubling the amount of the Firehole River and washing away footbridges. It was named Excelsior (Latin for "higher") in 1881 by park superintendent P.W. Norris, who believed that it excelled all other geysers.

**Factory Hill**  This hill (9,607 feet) was named in 1885, probably because of imagery used in previously published magazine articles, one of which, for example, proclaimed its author's vision of "steam jets ascending in more than fifty craters, giving [the hill] much the appearance of a New England factory village."

**Fairy Falls**  This 197-foot waterfall on Fairy Creek was named in 1871 by Capt. J.W. Barlow for "the graceful beauty with which the little stream dropped down a clear descent."

**Fall River**  Flowing west to Henrys Fork of the Snake River from Beula Lake, this river was named by fur trappers in the 1830s. Trapper Osborn Russell journaled in 1838 that the river was called "The falling fork" because of "the numerous Cascades it forms whilst meandering thro. the forest previous to its junction with the main river."

**Ferris Fork (of Bechler River)**  Flowing southwest from Pitchstone Plateau, the Ferris Fork was named in 1921 for Warren Angus Ferris, a fur trapper who produced one of the earliest maps of the American West (1836) and who has been credited with being the first tourist to visit present Yellowstone National Park out of curiosity rather than for business. He also was the first person to provide an accurate description of the Yellowstone country and the first to apply the word "geyser" to a Yellowstone hot spring.

**Firehole River**  This river, one of two streams that join to form the Madison River, was so called by 1850. But the name as applied to the valley containing Old Faithful Geyser was, according to Benjamin Bonneville, in use by fur trappers as early as 1832. The reason for the name is somewhat uncertain but probably came about in one of two ways—the "holes of fire" (hot springs and geysers) mentioned as such by park superintendent P.W. Norris and other early travelers; or a valley burned by a forest fire in the 1830s (as mentioned by fur trapper Warren Ferris), the location of which may have been transferred in confusion to the present Old Faithful area.

**Fishing Cone**  One of the park's legendary geothermal wonders, this hot spring located in Yellowstone Lake at West Thumb was proclaimed early by Yellowstone travelers to be the place where fishermen could catch a fish from the lake and cook it right on the hook, a practice now prohibited.

**Folsom Peak**  This peak (9,326 feet) is located in the Washburn Range. It was named for David E. Folsom (1839–1918), a member of the 1869 Folsom-Cook-Peterson party and the first man to write a report of a tour of the park.

# Yellowstone National Park Place Names (*continued*)

**Fountain Paint Pot** This famous "paint pot," a mud hot spring resembling bubbling paint, owes its name to proximity—by 1914 guidebooks were using the term "Fountain Geyser and Paint-Pots."

**Frank Island** The largest island in Yellowstone Lake was named in 1871 by artist Henry Wood Elliot for his brother Frank, about whom very little else is known.

**Frederick Peak** This peak (9,558 feet), located west of Mount Hornaday, was named for Karl T. Frederick (1881–1963), who actively worked to establish Grand Teton and Mount McKinley (now Denali) national parks.

**Frost Lake** This small lake located two miles northeast of Pyramid Peak was named for Ned Ward Frost (1881–1957), an early Yellowstone packer, guide, and big-game hunter who discovered Frost Cave near Cody, Wyoming, and who for many years conducted guided parties into Yellowstone National Park.

**Gallatin Range** This range of mountains took its name from the river that heads here.

**Gallatin River** This river was named in 1805 by Lewis and Clark for Thomas Jefferson's secretary of the treasury, Albert Gallatin (1761–1849).

**Gardner River** This river was named as early as 1831 by fur trappers for Johnson Gardner. Gardner may have named the river for himself, but the name was in use in 1836 by trapper Osborne Russell.

**Gibbon River** This stream, which flows west from Grebe Lake, was named in 1872 by F.V. Hayden for his friend Gen. John Gibbon (1827–96) who partially explored it that year.

**Grand Canyon of the Yellowstone** Not to be confused with Arizona's Grand Canyon, which might have inspired its name, this canyon was named in 1870 by the Washburn expedition.

**Grand Prismatic Spring** Named by mineralogist A.C. Peale in 1878, who wrote, "I have named it Prismatic on account of the brilliant coloring displayed in it."

**Grant Peak** This peak (10,858 feet), located in the Absaroka Range, was named for President Ulysses S. Grant, who signed into law the 1872 act that made Yellowstone the world's first national park.

**Grants Pass** Located on the trail between Lone Star Geyser and Shoshone Lake, this pass (8,022 feet) of the Continental Divide was named in 1882 for Northern Pacific Railroad surveyor M.G. Grant, who surveyed this route for a proposed railroad that year. Congress never approved the railroad, and it was never built.

**Great Fountain Geyser** Named by geologist W.H. Holmes in 1878 because he thought it was the only geyser in Lower Geyser Basin that compared in size to large geysers in the Upper Geyser Basin.

**Hague Mountain** This peak (10,565 feet), located between Little Saddle Mountain and Saddle Mountain, was named for Arnold Hague (1840–1917), a prominent geologist who gave place names to Yellowstone National Park and studied its geology for more than thirty years. He also was instrumental in the setting aside of present Shoshone National Forest and in the passage of the Lacey Act in 1894.

**Harding Geyser** A geyser of the Porcelain Basin of Norris Geyser Basin, this feature began erupting in 1923 and was given the name to honor the visit of President Warren Harding to the park that summer. This place name is one of several exceptions to the park rule prohibiting the naming of thermal features for people.

**Hayden Valley** This magnificent open valley along the Yellowstone River north of Mud Volcano was named for Ferdinand V. Hayden (1829–87) who conducted three government surveys of Yellowstone (1871, 1872, and 1878) that resulted in the accumulation of much scientific knowledge. As much as any other individual, Hayden was also responsible for the action by Congress to create the park.

**Heart Lake** According to Richard "Beaver Dick" Leigh, this lake was so named by trappers before 1871 for Hart Hunney, who hunted in that area between 1840 and 1850 and was killed by Crow Indians in 1852.

**Hedges Peak** This peak (9,685 feet) of the Washburn Range was named for Cornelius Hedges (1831–1907), a member of the 1870 Washburn expedition that received credit for the white discovery of Yellowstone that year. Hedges has also been partially credited with a portion of the national park idea.

**Hellroaring Creek,** MONTANA According to E.S. Topping, a party of prospectors that included A.H. Hubble named this creek in 1867. Hubble "went ahead . . . for a hunt, and upon his return he was asked what kind of stream the next creek was. 'It's a hell roarer,' was his reply, and Hell Roaring is its name to this day."

**Hellroaring Mountain,** MONTANA Named in 1871 by members of the Hayden Survey from nearby Hellroaring Creek.

**Hoyt Peak** This peak (10,453 feet) of the Absaroka Range was named in 1881 for John W. Hoyt (1831–1912), the territorial governor of Wyoming who traveled across the Absaroka Range on a pack trip that year with park superintendent P.W. Norris. At least one source states that Hoyt selected the very peak that was to bear his name.

**Jones Pass** A pass of the Absaroka Range located two miles south of Mount Chittenden. It was named for U.S. Army Capt. William A. Jones (1841–1914) who received credit for its discovery in 1873. Jones originally called it "Stinkingwater Pass."

**Joseph Peak** This peak (10,494 feet) of the Gallatin Range was named for Chief Joseph

(1840–1904), who led the Nez Perce tribe through Yellowstone National Park in 1877 on their historic flight to escape from the U.S. Army.

**Kepler Cascades** Park superintendent P.W. Norris named these waterfalls on the Firehole River in 1881 to honor Kepler Hoyt, "the intrepid, twelve-year-old son of Governor John Hoyt of Wyoming, who unflinchingly shared in all the hardships, privations, and dangers of the explorations of his father" earlier that year.

**Kingman Pass** This pass on the main road between Terrace Mountain and Bunsen Peak was named for Dan C. Kingman (1852–1916) of the U.S. Army Corps of Engineers, who was detailed to Yellowstone National Park in 1883 to build roads, including the first road through this pass. He claimed to have initiated the idea for the park's present "Grand Loop Road," a figure-eight system that still exists in Yellowstone.

**Knowles Falls,** MONTANA A fifteen-foot waterfall of the Yellowstone River, located just downstream from Crevice Creek's mouth. It was named for John S. Knowles, a miner who lived in a cabin at the mouth of Crevice Creek for nearly twenty years (1880–97).

**Lamar River** One of the park's major rivers, the Lamar River originates at Hoodoo Basin and flows northwesterly for sixty-six miles to the Yellowstone River. It was named for Lucius Quintus Cincinnatus Lamar, secretary of the interior (1885–88). The original name of the river was East Fork of the Yellowstone.

**LeHardy's Rapids** These rapids on the Yellowstone River were named in 1873 by the Jones Expedition for Paul LeHardy, a civilian topographer with that group of explorers. LeHardy and a partner rafted the river to sketch it for mapping and their raft came apart at this point, throwing them into the water.

**Lewis Lake** Yellowstone's third largest lake, Lewis Lake was named in 1872 by the U.S. Geological Survey for Meriwether Lewis, who never saw it. Lewis and Clark passed to the north of present Yellowstone National Park in 1805 by a distance of fifty miles.

**Lone Star Geyser** Named in 1879 because of its isolated and solitary position on the upper Firehole River.

**Madison River** One of the three forks of the Missouri River, this stream begins in Yellowstone National Park from the junction of the Firehole and Gibbon rivers and flows northwesterly. It was named in 1805 by Lewis and Clark, who saw it at present Three Forks, Montana. The name was given for James Madison, secretary of state to President Thomas Jefferson.

**Mammoth Hot Springs** Named in 1871 by settler Harry Horr for the immense size of the spring-area deposits on a huge white hillside, rather than for the size of individual springs.

**Mary Lake** This small lake was named in 1873 by Reverend E.J. Stanley, who wrote: "We passed along the bank of a lovely little lakelet . . . on its pebbly shore some members of our party unfurled the Stars and Stripes, and christened it Mary's Lake, in honor of Miss Clark, a young lady belonging to our party."

**McBride Lake** This lake on Buffalo Plateau near Slough Creek was named for James McBride (1864–1942), Yellowstone's first chief ranger. He had worked and lived near the lake in his capacity as a ranger at the upper Slough Creek station. He arrived in Yellowstone in 1886 with the U.S. Army and stayed until his death.

**Miller Creek** Flowing west from Hoodoo Peak to the Lamar River, this stream was named for prospector and early park guide Adam "Horn" Miller (1825?–1913), who retreated from Indians down it in 1870 and who also was one of the four men who discovered gold at Cooke City, Montana, that year.

**Mirror Plateau** Took its name by 1886 from the plateau's Mirror Lake, which Hayden survey members earlier called "a natural mirror."

**Monument Geyser Basin** Named in 1878 by park superintendent P.W. Norris for a number of strangely formed thermal features that resemble gravestones.

**Moran Point** A point on the north rim of the Grand Canyon of the Yellowstone, named for Thomas Moran (1837–1926), one of the West's most famous landscape artists. He accompanied the 1871 Hayden survey to Yellowstone and produced the large 1872 painting of the canyon that was purchased by Congress to hang in the Capitol building.

**Morning Glory Pool** Perhaps Yellowstone's most famous quiet hot spring, this spring was named in 1883 by park concessioner Mrs. E.N. McGowan because its shape and general appearance reminded her of a morning glory flower.

**Mount Chittenden** Located on the park's east boundary, this mountain (10,177 feet) was named for George B. Chittenden, a member of the U.S. Geological Survey who had little to do with Yellowstone but whose friends wanted him so honored.

**Mount Doane** Located in the Absaroka Range near the park's east boundary, this mountain (10,656 feet) was named for Lt. Gustavus Cheyney Doane (1840–92), author of the first official report on Yellowstone.

**Mount Everts** This high plateau (7,842 feet), located just east of Mammoth Hot Springs, was named for Truman C. Everts (1816–1901), a member of the 1870 Washburn expedition (from which he was lost for thirty-seven days).

**Mount Hancock** This mountain (10,214 feet), located on Big Game Ridge south of Heart Lake, was named for Gen. Winfield Scott Hancock (1824–86), who

issued the orders for the military escort that accompanied the 1870 Washburn expedition. Hancock was also an unsuccessful candidate for president in 1880.

**Mount Haynes** Located in Madison Canyon, this mountain (8,235 feet) was named for Frank Jay Haynes (1853–1921), who served as park photographer for Yellowstone National Park between 1881 and 1921 and also operated stagecoaches, bus lines, and photo shops in the park.

**Mount Holmes** This mountain (10,336 feet), located in the Gallatin Range, was named for William Henry Holmes (1846–1933), a geologist and artist with the Hayden surveys who wrote some of the earliest descriptions of park geology in 1878.

**Mount Hornaday** This mountain (10,036 feet), located west of Pebble Creek, was named for William Temple Hornaday (1854–1937), one of the most famous naturalists of his day. He was best known for his campaign to save the American bison from extinction, especially within Yellowstone National Park.

**Mount Humphreys** Located in the Absaroka Range, this mountain (11,009 feet) was named for Gen. Andrew Atkinson Humphreys (1810–83), an early supporter of F.V. Hayden's geological surveys and a person who helped establish the present U.S. Geological Survey.

**Mount Jackson** This peak was named for photographer William Henry Jackson (1843–1942), whose photos influenced congress to preserve Yellowstone as the world's first national park.

**Mount Langford** Located in the Absaroka Range, this mountain (10,440 feet) was named for Nathaniel P. Langford, a member of the 1870 Washburn party credited with the white discovery of Yellowstone and the first superintendent of the new Yellowstone National Park in 1872.

**Mount Norris** Located in the Absaroka Range just south of the Thunderer, this mountain (9,936 feet) was named for Philetus W. Norris (1821–85), second superintendent of Yellowstone National Park and explorer, nature guide, archaeologist, historian, and writer.

**Mount Schurz** This mountain (11,139 feet) in the Absaroka Range was named for Carl Schurz (1829–1906), U.S. secretary of the interior (1877–81), who "took a deep interest in the welfare of the [Yellowstone] reservation" throughout his years in office.

**Mount Sheridan** Located west of Heart Lake in the Red Mountains, this mountain (10,308 feet) was named for Gen. Philip H. Sheridan (1831–88) a strong advocate for Yellowstone without whom the national park story might have been much different.

**Mount Stevenson** Located in the Absaroka Range just south of Mount Doane, this mountain (10,352 feet) was named for

James Stevenson (1840–88), explorer, scientific surveyor, and principal assistant to F.V. Hayden during the Hayden surveys of Yellowstone.

**Mount Washburn** Part of the Washburn Range, this feature (10,243 feet) was named for Henry Dana Washburn (1832–71), the first known white man to ascend it and the man who led the 1870 Washburn expedition, the party that was credited with the white discovery of the Yellowstone region.

**Mud Volcano** In 1870 the Washburn Party named this hot mud spring because it resembled a cone volcano.

**Natural Bridge** Discovered in 1871 by F.V. Hayden, this natural stone arch got its name from the fact that Hayden called the stream flowing under it Bridge Creek.

**National Park Mountain** This high plateau (7,560 feet), located at present Madison Junction, was named because it was supposedly the legendary site where the 1870 Washburn party discussed around their campfire the possibility of preserving Yellowstone as a great national park. More recent historians believe that this tale is apocryphal, perhaps made up by one of the party members, because the true story is much more complicated.

**Norris Geyser Basin** Superintendent P.W. Norris named this area for himself in 1878, calling it "Norris Geyser Plateau."

**Nez Perce Creek** Flowing west from Mary Lake to the Firehole River, this stream was named for the Nez Perce tribe of Indians, who fled up it in 1877 to escape U.S. Army troops pursuing them.

**Obsidian Cliff** Named in 1878 by park superintendent P.W. Norris for the material of which it is composed.

**Old Faithful Geyser** This, the park's most famous geyser, was named in 1870 by the Washburn party "because of the regularity of its eruptions" that occurred then at intervals of about 65 minutes. It erupts today to heights averaging 130 feet.

**Ouzel Falls** This 230-foot waterfall on Ouzel Creek was named in 1885 by the Hague parties for the water ouzel or American dipper, a small, slate-colored bird that feeds underwater.

**Parker Peak** Located in the Absaroka Range, this mountain (10,203 feet) was named for William Parker (1853–1923), one of the men who accompanied Superintendent P.W. Norris on his explorations of Yellowstone in 1880. Parker later lectured on Yellowstone for and with Henry Bird Calfee.

**Peale Island** This island in the South Arm of Yellowstone Lake was named for Albert Charles Peale, the mineralogist who accompanied F.V. Hayden's surveys of Yellowstone and who wrote the first detailed report on Yellowstone's geysers and hot springs.

**Pebble Creek** This stream was named "White Pebble Creek" in 1872 by Hayden Survey members for the chalky white pebbles and rocks near its headwaters.

**Pelican Creek** This stream, which flows southwesterly to Yellowstone Lake from Mirror Plateau, was named in 1864 by a party of prospectors including Adam Miller and John C. Davis. The party shot a pelican for food, discovered that it could not be eaten, and hung the carcass on a tree.

**Phantom Lake** In 1935, park superintendent Roger Toll gave this small, fishless lake its name because it dries up each autumn.

**Phelps Creek** Flowing south from Sheep Mountain to the Yellowstone River, this stream was named for George Phelps, a hunter, scout, and prospector who lived in this area for many years.

**Phillips Fork (of Bechler River)** This stream, which flows from Pitchstone Plateau to Three River Junction, was named for William Hallett Phillips (1853–97), who worked as a lawyer behind the scenes for more than a decade to prevent the exploitation of Yellowstone National Park.

**Pitchstone Plateau** Members of the Hayden Survey named this area in 1878 due to the presence of pitchstone, a kind of obsidian and one of the park's more common volcanic rocks.

**Plenty Coups Peak** This mountain (10,937 feet) located in the Absaroka Range just north of Atkins Peak was named for Crow Indian chief Plenty Coups (1848–1932), an especially influential tribal leader who lived long in the country just northeast of Yellowstone National Park.

**Potts Hot Spring Basin** This area on the west shore of Yellowstone Lake was named for Daniel T. Potts, a fur trapper who wrote the earliest known description of Yellowstone's thermal features (and possibly this very area) in an 1827 letter to his brother. It was subsequently published in the September 27 issue of that year's *Philadelphia Gazette & Daily Advertiser.*

**The Promontory** Extending into Yellowstone Lake from the south, this feature was named in 1871 by members of the Hayden Survey, probably from Lt. G.C. Doane's report of the previous year that stated, "We traveled across a high promontory running into the lake."

**Red Mountains** The 1871 Hayden Survey gave this name in singular form to present-day Mount Sheridan; in 1872, the survey applied that name to the entire range. A traveler to the area in 1890 noted, "In the light of the setting sun, the prevailing tint is a reddish purple, which doubtless gives the name to the range."

**Reese Creek** Forming a portion of the park's north boundary, this stream, which flows from Electric Peak to the Yellowstone River, was named for George Reese (1837–

1913). Arriving in the Yellowstone country as a prospector in 1867, Reese became one of the earliest settlers in this part of the region and settled on this stream in 1875.

**Reservation Peak** This peak (10,629 feet) of the Absaroka Range on the park's east boundary was named in 1895 because it marks the boundary of the great "reservation" of Yellowstone National Park, forever "reserved and withdrawn from settlement, occupancy, or sale."

**Rescue Creek** Flowing east to Blacktail Deer Creek from Mount Everts, this stream was named because it was (mistakenly) thought to be the stream on which explorer Truman C. Everts was found after his thirty-seven days of being lost in Yellowstone in 1870 (the true spot of his finding was farther east some eight miles).

**Richards Creek** This stream, which flows west to Gneiss and Duck creeks from Madison Valley, was named for Alonzo Van Ness Richards (1841–91), who surveyed the park west boundary in 1874.

**Riddle Lake** Nineteenth-century maps of the West long showed a "Lake Riddle" somewhere near the Continental Divide that supposedly drained to both oceans. In 1872, the Hayden survey attached the name to this present small lake located just south of Yellowstone Lake to finally place the fugitive name on an exact spot. The original "riddle" referred to the mystery of which way the lake drained.

**Roaring Mountain** Rising some 400 feet, this mountainside of steaming fumaroles near Twin Lakes was named in 1885 "from the shrill, penetrating sound of the steam constantly escaping from one or more vents located near the summit."

**Robinson Creek,** IDAHO Flowing south and west to the Warm River (Idaho) from near Buffalo Lake, this stream was probably named for Jim Robinson, a thief and rustler who stole cattle in nearby Teton Valley, Idaho, and inhabited this area during the 1880s.

**Rock Creek** This stream, which flows southwest of Robinson Creek from Robinson Lake, was named for Richard W. Rock, a hunter and poacher who lived near Henrys Lake, Idaho, during the 1880s and who was killed by one of his own pet buffalo in 1902.

**Rosa Lake** Located at the head of Indian Creek west of Mount Holmes, this lake was named for Rose (or Rosa) Park Marshall, supposedly the first white child born in Yellowstone National Park on January 30, 1881. A visiting governor named the little girl because "roses were scarce in the Park and she was born in same."

**Rowland Pass** This pass of the Washburn Range located east of Mount Washburn was named for R.B. "George" Rowland, early prospector and explorer who served as one of Superintendent P.W. Norris's assistants.

**Sepulcher Mountain** Located near the

park's north entrance, this mountain (9,652 feet) was probably named from a tombstone-like rock formation on its northern slopes.

**Sheepeater Cliff** Not to be confused with Sheepeater Cliffs (plural) farther downstream, this cliff of columnar basalt on the Gardner River was named around 1903, probably by employees of the Shaw and Powell Camping Company, whose nearby "Willow Park Camp" was the headquarters for many visits to this place.

**Shoshone Lake** The park's second largest lake, Shoshone Lake was named in 1872 because it drained south to Snake River and the geographers thus wanted to adopt "the Indian name of the Snake River" for the lake (Shoshone Indians were also known as Snake Indians).

**Signal Point** This point on the east shore of Yellowstone Lake was so named in 1871 because Hayden survey members used it as a point from which to signal others by building fires.

**Slough Creek** This stream was named in 1867 by a party of prospectors. According to E.S. Topping, one member of their party was sent on ahead and upon his return was asked what kind of stream was ahead. "'Twas but a slough," the scout is said to have replied, and so it remains.

**Snake River** Flowing west and south from the corner of Yellowstone National Park to the Columbia River, this stream was named as early as 1812 for the Snake (Shoshone) Indian tribe, who lived along its banks. The name probably originated from sign language—a serpentine movement of the hand with the index figure extended that referred to the weaving of baskets or grass lodges of the Snake (Shoshone) Indians.

**Soda Butte Creek** This stream was named in 1870 by a party of prospectors that included A. Bart Henderson, who wrote in his diary, "We gave the cone the name of Soda Butte and the creek the name of Soda Butte Creek." Soda water is an effervescent beverage consisting of water charged with carbon dioxide; the water from springs in this area was thought to resemble that drink.

**Specimen Ridge** This name was in use before 1870 and was probably given by prospectors like A. Bart Henderson, Adam "Horn" Miller, and James Gourley. The long ridge that borders the Lamar River was so called from its being the location for specimens of amethyst, opal, chalcedony, petrified wood, and other sought-after souvenirs. This type of collecting is now prohibited.

**Steamboat Geyser** This geyser at Norris Geyser Basin was named in 1878 by the Hayden Survey because the eruptions "reminded one of the sound of an old style paddle wheel steamboat" as its water was dashed out of the vent.

**Steamboat Point** A point on the eastern shore of Yellowstone Lake that was so called in 1871 by members of the Hayden

Survey. Mineralogist Albert Peale wrote that it was "so named from a number of steam jets which are situated on the shore of the lake. From two of them there escaped a vast volume of steam with a continuous noise, exactly resembling that made by some large steamboat when the escape valve is open."

**Stephens Creek** This stream, which flows north to Yellowstone River from Sepulcher Mountain, was named for Clarence Stephens, assistant to park superintendent P.W. Norris, who at one time owned the land through which the creek flows.

**Stevenson Island** This island on Yellowstone Lake was named for James Stevenson (1840–88), explorer, scientific surveyor, and principal assistant to F.V. Hayden during the Hayden surveys of Yellowstone.

**Surprise Creek** Flowing southwest from the Continental Divide to Heart River, this stream was named in 1885 because its course, as mapped that year, was surprisingly different from the "discoveries" of earlier explorations.

**Sylvan Pass** This pass (8,541 feet) of the Absaroka Range was named from its forested setting. The word sylvan means "forested."

**Tendoy Falls** This thirty-foot waterfall of the Ferris Fork (of Bechler River) was named for Tendoy (1834?–1907), a chief of the Lemhi Shoshone Indians, who lived in eastern Idaho near Yellowstone Park for most of his life.

**The Thorofare** A name given early by fur trappers to the entire valley of the Yellowstone River above Yellowstone Lake, so called because the route from Jackson Hole up Pacific Creek and down Atlantic Creek to this valley provided an easy (and the only southern) route onto the Yellowstone plateau. The name also survives today in Thorofare Creek.

**The Thunderer** This mountain (10,554 feet), located north of Mount Norris, was named in 1885 because it was "seemingly a great focus for thunderstorms."

**Topping Point** This point on the west shore of Yellowstone Lake near the outlet of the Yellowstone River was named for Eugene S. Topping, who ran the earliest concessions operation in any national park—a boat for hire on Yellowstone Lake—during the seasonsof 1874, 1875, and 1876.

**Tower Fall** This waterfall of Tower Creek, height 132 feet, was named in 1870 for the rock towers or pinnacles above the falls.

**The Trident** Geologist Arnold Hague named this mountain in 1885—the plateau had three "fingers," like a trident or pitchfork.

**Trischman Knob** This hill (8,600 feet) was named for Harry Trischman (1886–1950), a well-known personality associated with the development of Yellowstone National Park who served as U.S. Army scout, park ranger, and chief buffalo keeper.

**Twin Lakes** These two lakes located just south of Roaring Mountain were named in the 1870s for their similarity, although today the north lake is often blue in color while the southern one is green.

**Two Ocean Plateau** This large plateau south of Yellowstone Lake took its name from Two Ocean Pass, a few miles to the south, a place from which waters flow to both the Pacific and Atlantic oceans.

**Undine Falls** This waterfall of Lava Creek, height sixty feet, was named for mythological water spirits who, in German mythology, were female and lived around waterfalls.

**Union Falls** This 250-foot waterfall of Mountain Ash Creek was named around 1885 by the Hague Survey; two streams join to supply the falling water.

**Virginia Cascade** This sixty-foot sloping cascade of the Gibbon River was named in 1886 for Virginia Gibson, wife of Charles Gibson of the Yellowstone Park Association. While at this time the U.S. Geological Survey generally preferred not to affix personal names to scenic places, an exception was made in this case because the name also designated a state, and naming a feature in such a way seemed appropriate for the nation's first national park.

**Vixen Geyser** This name was given by a woman tourist in 1881 who had the misfortune to feel the geyser's water hitting her. "You vixen, you!" she reportedly exclaimed, and the name stuck.

**Wahb Springs** These hot springs at Death Gulch on Cache Creek were named during the period 1900–04, as park tourists fell in love with Ernest Thompson Seton's 1900 book *The Biography of a Grizzly,* about a grizzly bear named Wahb that died by inhaling the poisonous gases of Death Gulch.

**West Thumb** This most westerly bay on Yellowstone Lake was named from the 1870 Washburn party's pronouncement that the lake was shaped "like a human hand, the fingers extended and spread apart as much as possible." The bay represented the "thumb," and park superintendent P.W. Norris showed it as such on his map of 1880 as "West Bay or Thumb."

**Winegar Lake** This small lake located about three miles east of Cave Falls was named for the Winegar family—including Stephen Winegar and his four sons—all of whom lived near the park's southwest corner and who regularly hunted animals illegally in the park.

**Wrong Creek** Flowing west to Shallow Creek from near Mirror Lake, this stream was named because it was wrongly drawn on so many maps after 1878. These mistakes were not corrected until the 1950s.

**Yancey Creek** This stream, which flows northwest to Elk Creek from Lost Lake, was named for "Uncle" John Yancey (1826?–1903). He established the Pleasant Valley

Hotel on the creek and ran it for nearly twenty years.

**Yellowstone River** Although Crow Indians long referred to this stream as "Elk River," the name Yellowstone was given as early as 1797 as "R. des Roche Jaune" (river of the yellow rock). Contrary to local misnomer, the yellow rocks were probably not those of the Grand Canyon in the park but rather yellowish rocks far downstream (possibly the rim rocks at Billings, Montana, which appear bright yellow in the sunshine after a rain).

# Grand Teton National Park Place Names

All place names listed are located in Wyoming.

**Alaska Basin** Located just outside the park's western boundary, this area was probably named by settlers who had seen Alaska during its 1890s gold rush.

**Amphitheater Lake** This lake was named from its setting in an amphitheater-like glacial cirque.

**Arizona Island** Located on Jackson Lake's northeastern side, this island was named for "Arizona George," who trapped in the area in 1888 and who was found dead on nearby Arizona Creek.

**Bivouac Peak** This peak (10,825 feet) was named by its first recorded ascenders in 1930 from the fact that they were forced to spend a night at the foot of the mountain without food or bedding.

**Bradley Lake** F.V. Hayden named this lake in 1872 for Frank H. Bradley, chief geologist of Hayden's 1872 expedition to the Tetons.

**Buck Mountain** This mountain (11,938 feet) was named by its 1899 ascenders and mappers for George A. Buck, the recorder for that party.

**Cascade Canyon** This glacially carved, U-shaped canyon near the middle of the Teton Range was named from its creek, which cascades dramatically into Jenny Lake.

**Cascade Creek** This stream west of Jenny Lake was named in 1931 for its cascades.

**The Cathedral Group** These three peaks, consisting of Mount Teewinot (12,325 feet), Grand Teton (13,770 feet), and Mount Owen (12,928 feet), appear as cathedral-like spires when viewed from the road north of Jenny Lake.

**Cirque Lake** This lake west of Mount Moran was named from the fact that it is located in a rounded enclosure, or cirque, formed by the head of a glacier pulling rock from the surrounding walls.

**Climbers' Ranch** Located three miles south of Jenny Lake, this ranch was granted a concessions permit in 1970 and named from the fact that it provides mountaineers with overnight accommodations.

**Cloudveil Dome** This mountain (12,026 feet) was named on a 1931 map from its dome-shaped summit that is frequently shrouded in clouds.

**Colter Bay** The name was changed in 1948 from "Mackinaw Bay" to honor John Colter, the credited first white traveler (winter of 1807–08) through Jackson Hole.

**Colter Canyon** Located on the west side of Jackson Lake, this canyon was named for John Colter, the credited first white traveler (winter of 1807–08) through Jackson Hole.

**Cottonwood Creek** Three different streams carry this name in Grand Teton National Park: one flows south from Jenny Lake, one flows south into the Gros Ventre River, and one flows into the Snake River south of Mosquito Creek. All take their names from Wyoming's state tree.

**Death Canyon** This large canyon that adjoins Phelps Lake was probably named by the 1899 Bannon survey, a member of which wandered into the canyon and was never seen again.

**Donoho Point** Located at the southeastern corner of Jackson Lake, this point was named for a trapper who lived there around 1900. The point became an island after the Jackson Lake dam was built.

**Eagles Rest Peak** This peak (11,258 feet) between Waterfalls and Snowshoe canyons was probably named for the sighting of a bald eagle perched in a tree.

**Emma Matilda Lake** This lake was named by W.O. Owen for his wife. He was a topographer with the General Land Office who, with his wife, attempted unsuccessfully to climb the Grand Teton in 1891. Seven years later he succeeded.

**Flagg Ranch** Established by rancher Ed Sheffield during the period 1910 to 1916, the resort was probably named from flags that flew over the nearby soldiers' station, where the U.S. Army from nearby Yellowstone had an outpost during the 1890s.

**Granite Canyon** The name was given from rocks that were thought to be granite, but which actually are metamorphic rocks, chiefly gneiss.

**Garnet Canyon** Formerly called "Bradley Canyon," this canyon above Bradley Lake was named by the U.S. Board on Geographic Names in 1931 from imperfect garnets found there.

**Grand Teton** This highest mountain in the Teton Range (13,770 feet) received its name early (probably before 1800) from French-Canadian fur trappers of the Hudson's Bay Company, who named the main three peaks *"Les Trois Tetons,"* meaning "the three nipples."

**Gros Ventre Range, River** The term means "big belly" in French and referred to an Indian tribe. The name has two possible origins. Some historians believe that sign language used by local Indians was misinterpreted by the French to mean big belly. Others think that the name referred to the swollen stomachs of local Indians who had too little to eat.

**Half Moon Lake** When a meander of the Buffalo Fork was cut off from the main stream by a deposit of sediment, a lake in the shape of a half moon was formed.

**Hermitage Point** Located on the eastern side of Jackson Lake, this point was intended to be the setting for a lodge or hermitage by a now-unknown builder who abandoned the idea after the Jackson Lake dam was built.

**Hidden Falls** West of Jenny Lake, this waterfall was named by early settlers who noticed that the falls could not be seen from the main road on the eastern side of the lake.

**Indian Paintbrush Canyon** The U.S. Board on Geographic Names approved this name in 1931 to honor Indian paintbrush, the state flower of Wyoming.

**Inspiration Point** This point on the western shore of Jenny Lake was named because of its panoramic view of the valley of Jackson Hole.

**Jackson Hole** Early fur trappers referred to mountain valleys as "holes." This valley on the east side of the Teton Range was named for David E. Jackson, who trapped in it during the 1820s.

**Jackson Lake** Variously referred to through the years as "Lake Biddle" and "Teton Lake," this lake, the largest in the valley of Jackson Hole, received its present name in 1829 when trapper William Sublette named it for David E. Jackson, a frequenter of the area.

**Jenny Lake** This lake was named by F.V. Hayden in 1872 for the wife of one of his guides, "Beaver Dick" Leigh. Jenny was a Shoshone Indian woman who was sometimes present on that survey with her husband. She died with her children of smallpox in 1876.

**Lake Solitude** Named from its remote location.

**Leigh Lake** Located south of Jackson Lake, this lake was named for F.V. Hayden's 1872 guide, Richard "Beaver Dick" Leigh.

**Leeks Marina** Named for S.N. Leek, who settled in the valley before 1900 and who used photographs and gave lectures to gain support for the idea of establishing the National Elk Refuge.

**Lizard Creek** This stream, which flows into the eastern side of Jackson Lake below Fonda Point, was so named on an 1899 map, likely as a result of salamanders being mistaken for lizards.

**Lupine Meadows** Named for the lavender-blue lupine, a wildflower that carpets the area in summer.

**Menor's Ferry** On the Snake River, this ferry honors its builder, William Menor, who erected it in 1894 near Moose, Wyoming, one of the few places where the river kept to a single channel.

**Mount Hunt** This mountain (10,783 feet) was named to honor Wilson Price Hunt, who led one of the first white parties—the Astorians—through Jackson Hole in 1811.

**Mount Moran** This most prominent peak (12,605 feet) at the northern end of the Teton Range was named in 1872 by F.V. Hayden for the landscape artist and expedition member Thomas Moran.

**Mount Owen** This peak (12,928 feet) was named for W.O. Owen, who climbed the Grand Teton in 1898 with Bishop Spalding, John Shive, and Frank Petersen.

**Mount St. John** Orestes St. John, geologist for Hayden's 1877 survey, is the namesake of this mountain (11,430 feet). His monographs on the Teton and Wind River ranges are now classics.

**Mount Wister** Located north of Buck Mountain, this peak (11,490 feet) was named for the author Owen Wister, whose classic book *The Virginian* was partially set in Jackson Hole and served to interest many easterners in the romance of the West.

**Mount Woodring** This mountain (11,590 feet) was named for Samuel Woodring, first superintendent of Grand Teton National Park, who began his career in Yellowstone as a ranger in 1920.

**Nez Perce Peak** This mountain (11,901 feet) was named for the Indian tribe, whose most famous leader was Chief Joseph. The tribe traveled through Yellowstone National Park just to the north in 1877 in a famous chapter of western history.

**Paintbrush Canyon** See Indian Paintbrush Canyon.

**Phelps Lake** This lake at the mouth of Death Canyon was named in 1872 by F.V. Hayden for George Phelps, a trapper friend of Hayden's guide, "Beaver Dick" Leigh. Phelps frequented the Jackson Hole and Yellowstone areas in the 1860s and 1870s. A pass in Teton park along with a peak and a creek in Yellowstone are all named for him.

**Raynolds Peak** Located west of Bivouac Peak (10,825 feet), this peak (10,910 feet) was named for Captain W.F. Raynolds, who explored the region in 1859 and 1860.

**Rendezvous Peak** The name of this peak (10,927 feet) honors the historically significant rendezvous (meeting) of fur trappers, which took place between 1825 and 1840.

**Rockchuck Peak** This mountain (11,144 feet) took its name from the "rock chucks" or yellow-bellied marmots that abound on its slopes.

**Rolling Thunder Mountain** The name of this peak (10,908 feet) is taken from the atmospheric rumbling often heard there.

**Schoolroom Glacier** So called because it exhibits textbook examples of glacial features, such as bergschrunds, crevasses, moraines, pedestal boulders, and a small lake at its base, colored milky blue by the rock "flour" suspended in its melted waters.

**Signal Mountain** A tragedy accounts for the name of this mountain (7,733 feet). In 1891, searchers looking for Robert Ray Hamilton agreed to light a signal fire on the summit of this mountain when he was found—its commanding location assured that such a fire could be seen from the flats below. Seven days later, Hamilton's drowned body was discovered two miles downriver from the outlet of Jackson Lake.

**Skillet Glacier** Located on the eastern side of Mount Moran (12,605 feet), this glacier took its name from its characteristic shape.

**Snake River** Stretches of this long, important river have carried many names, such as "Lewis River" and the "Accursed Mad River," but the present name dates from at least 1806, when it was *"Gens du Serpent."* The name has nothing to do with either the river's sinuous route or with reptiles along its banks. Instead the name was given from the Snake or Shoshone Indians who inhabited stretches of it and whose sign-language symbol for their people was a hand moving through space like a snake.

**Spalding Bay** This southernmost bay on Jackson Lake was named for bishop Franklin S. Spalding, who made the famed 1898 ascent of Grand Teton (13,770 feet) with W.O. Owen, John Shive, and Frank Petersen.

**Static Peak** This mountain (11,303 feet) was so named because it is so often hit by lightning.

**Surprise Lake** The lake's original name, "Lake Kinnikinic" (for a plant), was suggested by famous Teton guide and geologist Fritiof Fryxell. The U.S. Board on Geographic Names approved the present name in 1938. Its origin is unknown.

**Taggart Lake** Located at the foot of the Teton Range, this lake was named by F.V. Hayden's expedition in 1872 for the party's assistant geologist, W. Rush Taggart.

**Teewinot Mountain** Fritiof Fryxell named this mountain (12,325 feet) in 1929 from the Shoshone Indian word meaning "many pinnacles."

**Teton Glacier** Located between Mount Owen (12,928 feet) and Grand Teton (13,770 feet), the glacier took its name from the range.

**Teton Range** The range received its name early (probably before 1800) from French-Canadian fur trappers of the Hudson's Bay Company, who named the main three peaks *"Les Trois Tetons,"* meaning "the three nipples." These trappers viewed the range from the west (Idaho) side and the peaks in the range served as landmarks for travelers in that direction for many years.

**Thor Peak** This peak (12,028 feet) was named for Thor, the Norse god of thunder.

**Two Ocean Lake** Named from Two Ocean Pass, many miles to the northeast. Applied to this lake, the name is a misnomer, but that distant pass is famed as being the place where creek waters divide and flow to both the Atlantic and Pacific oceans.

**Webb Canyon** This canyon at the north end of the Teton Range was named for Dr. W. Seward Webb, who came to Jackson Hole in 1879 on a military expedition led by General Carrington. He passed his influence and ideas for protecting the area to his son, Vanderbilt Webb, who became John Rockefeller's lawyer and thus helped purchase lands that eventually became part of the extended Grand Teton National Park.

# Afterword, Sponsors, and Sources

# Afterword

## Creating the *Atlas of Yellowstone*

Creating the *Atlas of Yellowstone* was a monumental task. How, after all, does one select which tales to tell from the remarkable body of scholarship and data on the world's first national park and it surrounding region? Understanding the world of Yellowstone requires examining spatial scales ranging from microbial mats in a hot spring to the continent-wide dispersal of coyote and looking at time scales ranging from the hours needed for a wildfire to blow up to the millions of years necessary for geologic processes. Understanding Yellowstone also requires examining the many connections between places and times and, in particular, the profound interaction of humans and nature in this landscape. Making the *Atlas of Yellowstone* therefore required many hard decisions about what to include and what to leave out. This essay explains how the *Atlas of Yellowstone* came to be and describes the large collaborative effort that went into selecting stories, data, and maps to portray Yellowstone and guide readers to a deeper appreciation of this extraordinary place.

## Origins

The *Atlas of Yellowstone* originated from a class project in the Department of Geography at the University of Oregon. Each year, the Advanced Cartography course focuses on developing cartographic products around a central theme. In January 2003, editors Marcus and Meacham decided to have the following year's class develop a series of thematic maps for Yellowstone's Northern Range, which was the location of our research and that of many other scientists. This area straddles Yellowstone park's northern boundary, the Montana-Wyoming state line, and property controlled by many different agencies and private landowners. Scientists working in the Northern Range thus always faced the daunting task of collecting and stitching together data and maps from many different sources. Many a map user has had the maddening experience of what could be called Murphy's Law of Maps: that any area of interest always falls at the boundary of at least four different maps. Thus one of the most frustrating aspects of mapping helped to initiate the *Atlas of Yellowstone*.

The difficulty that students would face in acquiring this multiagency data, however, led us to refocus the class project on a discrete geographic area for which data are widely available: Yellowstone National Park. While discussing the upcoming class in a social setting and emboldened by several microbrews, we decided that as long as we were doing a class on the topic, we might as well create an *Atlas of Yellowstone*. Why not? This naive decision led us on a much more involved and lengthy journey than we ever imagined. Almost ten years, more than 100 expert contributors, many Yellowstone trips, dozens of cartographers (many of them students), and about 300 pages later, it is clear that building the atlas was a mammoth undertaking, but we've never regretted that inspired moment when we toasted the concept.

One of our first decisions required choosing a format for the atlas. This was a relatively easy decision in light of the then-recent completion of the *Atlas of Oregon*, also produced in the Department of Geography at the University of Oregon under the guidance of William G. Loy. Given the success of that volume, it made sense to use the same format and layout. The core content of the *Atlas of Yellowstone* consists of "page pairs" modeled after the *Atlas of Oregon*. Page pairs are facing pages that focus on a particular topic (for example, wolves), with guidance on the storylines and data sources coming from expert contributors.

This facing page layout forces experts and cartographers to tell their story in a condensed and visually accessible manner with integrated maps, data graphics, and interpretive text. Following the page pairs are a series of reference materials. In the *Atlas of Yellowstone* these include new elevation-tinted reference maps for the Greater Yellowstone Area, new land cover-based maps for Yellowstone and Grand Teton national parks, a place names section, and a gazetteer.

In addition to choosing an overall format and page design, we also developed four overarching geographic themes to provide focus and continuity. The first theme is the idea that **Yellowstone is connected** across many scales and locations, a concept that leads to maps of many different scales and spatial extents. The second theme is that **Yellowstone is dynamic,** changing over time periods that range from millions of years to hours, a concept that lends itself to series of time-sequenced maps and graphs of change. A third theme is that the world of Yellowstone results from the **interaction of humans and nature,** with changes in one leading to changes in the other. The fourth theme is the **importance of Yellowstone** as a place, a concept that is highlighted by page pairs specifically targeted to this topic (for example see "The World's First National Park" and "Teton Climbing History.") This theme is also a natural conclusion to the exploration of the first three themes. Generally, these themes are not called out explicitly throughout the atlas; rather, they provided overarching guidance to the editors and the experts as they initiated and developed each page pair. The process of developing these themes also made it apparent we would have to extend the geographic coverage of the atlas to the entire region; connections, changes, and human-nature interactions occur over many different scales and rarely are constrained by political boundaries.

Although the atlas could not be limited to national park boundaries, Yellowstone and Grand Teton national parks are at the core of the region, and it was clear we needed access to National Park Service expertise and data for the atlas to succeed. In spring 2003, we therefore approached Ann Rodman, director of the GIS center at Yellowstone National Park, and John Varley, director of the Yellowstone Center for Resources, to determine if the national park would share their data and expertise. That first spring visit led to a summer visit to further explore the concept and cement the involvement of Ann Rodman in the project—a relation that led to her joining the team as a coeditor. It was surrounding this trip that we created for purposes of demonstration our first page pair: "Fire History," with expert guidance from Cathy Whitlock, Roy Renkin, and Don Despain. That page pair remains in this atlas, although updated to include data through 2010. A third visit in December 2003 included a meeting with approximately twenty-five park personnel in which we developed an initial table of contents and a preliminary list of expert contributors. The willingness of National Park Service personnel to share their ideas and deep knowledge of the region was crucial to initiating and completing the atlas project.

The advanced cartography class that generated the whole Yellowstone atlas concept finally met in winter 2004. That class and subsequent ones like it generated materials that are the basis for many pages in this atlas. Yellowstone park geology, wolves, climate, transportation, and vegetation pages all had their origins in student work from these classes. Accompanying us on those first trips to Yellowstone was Erik Strandhagen, a master's student in the Department of Geography, who developed three page pairs about river flow as part of his master's thesis. Another master's student, Alethea Steingisser, took that first class in 2004;

in it she developed material on glaciers that is included in this volume. More important, Steingisser finished her thesis on human impacts on geysers around the world and in Yellowstone and graduated to become the production manager on this atlas as well as the lead designer on many of its pages. Almost all of the cartographers on this project received training in the Advanced Cartography course. The involvement of students—with their creativity and seemingly infinite enthusiasm—has been one of the great joys of this project.

In the spring and summer of 2004, we embarked on a next critical step, finding funding. Unfortunately, most funding agencies will not fund atlases, which are viewed as syntheses of existing work rather than work that produces new knowledge. We (of course!) argue this point—there should be sources of major funding for significant new syntheses and visual interpretations that make scholarly materials accessible to the public and create new knowledge through such syntheses rather than primary data gathering. But telling granting agencies that their priorities are wrong is rarely a good starting point for acquiring funds. Fortunately, the University of Oregon stepped forward in 2004 with strong support, support that has continued to the completion of the atlas. Over the course of the following seven years we also received major funding from the Yellowstone Park Foundation, Montana State University, and the University of Wyoming, organizations that are all deeply invested in Yellowstone and are willing to back up with resources their commitment that communicating scholarly work is central to their missions of research, education, and interpretation.

## Selecting the Stories and Contributing Experts

By the fall of 2004, a year and a half after conceiving the atlas idea, we had developed a university–park service partnership, identified core themes and a format for the atlas, acquired seed funding, created sample page pairs, loosely defined a table of contents, and identified potential contributing experts. Our aspiration was to create a comprehensive, state-of-the-art reference volume that centered on Yellowstone and Grand Teton national parks. The target audience was to be visitors, educators, resource managers, and scholars—essentially anyone with a deep interest in Yellowstone.

This broad coverage required an expanded vision for determining which stories to include in the volume. Early on, we decided to portray the subjects that Yellowstone scholars suggested, which meant that identifying lead experts was the key determinant in guiding atlas content. This decision meant excluding more populist input such as reports by visitors on their experiences or oral histories of residents. Although such data can be at the heart of scholarly work, it is difficult or impossible to map these personal experiences. The decision to focus on stories from experts also highlights the often unrecognized importance of more than a century of research in Yellowstone to the advancement of science and the humanities. Finally, relying on expert input allowed us to translate a great deal of valuable National Park Service research into a medium that is more accessible to the public.

The selection of experts and their stories required outreach to the Yellowstone research community. Park personnel from Yellowstone and other parks had already provided ideas about content and continued to do so throughout the making of the atlas. Our personal research experience in Yellowstone also helped in identifying potential stories and experts. The additional experts identified by the National Park Service and our professional contacts in turn led us to new sources and more stories. We also formed an advisory committee of regional experts, acknowledged at the start of this volume. Ultimately, a great deal (but not all) of the atlas content came from collaboration with researchers at Yellowstone and Grand Teton national parks, the U.S. Geological Survey, the University of Oregon, Montana State University, the University of Wyoming, the Museum of the Rockies, the Buffalo Bill Historical Center, Headwaters Economics, and the Yellowstone Ecological Research Center.

We met with many more experts and took notes on many more stories than are included in this atlas. Topics fell by the wayside for a variety of reasons. In many cases, the data did not exist in a geographical form that could be mapped. Oral histories, professional knowledge, and historical accounts all are valid sources of knowledge but do not translate readily to atlas format. Some of those stories might have been converted to maps or change-over-time charts with sufficient time, but we already had a huge task on our hands without adding the work of conducting primary research and converting existing data to a spatial format.

We also avoided topics that could be told primarily through photo-essays; there are already many excellent photographic explorations of Yellowstone. In other cases, experts had superb stories and data but did not have the time to contribute, an understandable issue given that all the topic experts volunteered their time to the atlas. In other instances, we simply did not have the time to follow up with potential contributors because of funding limitations and our publication deadline; we often wished that we had the time and resources to prepare the thousand-page atlas that Yellowstone deserves.

In selecting experts and stories, we wanted to avoid having an atlas that was dominated by National Park Service perspectives. On pages that focus on Yellowstone or Grand Teton national parks, we therefore tried to include one park service employee and a non–park service expert. Even on pages not focused on the national parks, we tried to have two sets of experts creating and critiquing the page. Deadline pressures prevented this in several cases, but by and large we held true to this commitment. A list of experts who worked on each page pair appears at the end of the atlas. We are deeply indebted to them for their generosity of spirit, their willingness to share knowledge, and the time they spent contributing to the effort.

## Building Page Pairs and Reference Materials

Once the broad categories of topics and experts were identified, it was time to create the page pairs that portrayed the stories. This typically was a multiyear process for any one topic and involved a number of steps.

The first step was to meet with the experts for several hours to determine the key stories to tell, identify the associated data sets, and develop a general layout for the page. Meetings were in park service, university, and corporate offices; museum collections; field stations; and private homes—wherever the experts could meet. Meetings commenced with our introducing the atlas concept and themes to the experts (the *Atlas of Oregon* example proved extremely valuable), followed by the simple question, "What stories would you like to tell?" This query always prompted one of the magical experiences of atlas-making, that moment when we had the privilege of hearing the leading experts in the world tell a lifetime of stories about . . . wolves, art of the American West, archaeology, economic trends, new discoveries in Yellowstone Lake, and on and on—every topic included in the table of contents. The treasure-trove of fascinating tales and insights left us both exhilarated and overwhelmed at the amount of information. Many of the stories could not be included in the atlas because of length and time limitations; in retrospect, we wish we had video-recorded these sessions so that all the tales were preserved in this very personal form.

Ultimately we had to distill the topic areas down to those that could be told in our atlas format. With a list of key stories in hand, we moved to a white board and mocked-up the page pair, showing the general layout of the two pages and the locations and size of potential graphics and text. Rarely did these initial layouts survive intact after the many iterations of redesign that occurred once real data and graphics were inserted, but they prompted the critical process of thinking in terms of which stories could be told visually and identifying the data sets necessary for that presentation. The experts would guide us as the mock-up was created, giving them a deeper appreciation for the potential and the limitations of the atlas format. As we created the layout on the white board, we also entered information into a spreadsheet about the key story being told, the associated graphics, the data sources, and the people responsible for providing the data.

Once the mockup was created and key data sources identified, the production process required continued collaboration. Cartographic production and design occurred in the InfoGraphics Lab at the University of Oregon's Department of Geography, although the graphics and layouts were shared frequently with experts to solicit guidance. Data and data-quality assurance were generally provided through Yellowstone's Spatial Analysis Center or by other experts, although some widely accessible data sets (U.S. Census Bureau population data, for example) were directly accessed by the cartographers. Our cartography team gathered for weekly production meetings, displaying their progress on individual page pairs and conducting a remarkably open and free-flowing critique of design elements. The willingness of individuals to significantly change their work (and sometimes start over from scratch) was key to the high quality of cartography—designs underwent many, many iterations before they reached their final form.

Each page pair starts with a page-wide, two-column, or three-column layout. This ensures a consistent feel throughout the atlas. Most page pairs have a dominant map that establishes the overarching storyline accompanied by a series of subsidiary graphics that portray a more nuanced perspective on the landscape or the topic. The ESRI Arc suite of geographic information system (GIS) software was used in the compilation, analysis, and exploration of spatial data. Initial layouts were often built using quickly generated maps from the GIS data sets and graphs from a spreadsheet, which provided a sense of what the page pair might look like. This often led to changes in design, after which the spatial and tabular data were imported and turned into graphics using Adobe Illustrator and Photoshop. All finished page pair graphics and text were placed into Adobe InDesign for final layout design and printing.

The text *always* followed the graphics, meaning that the amount of text allowed on a page had to fit in the space left over after graphics were in place. Not once in this atlas was the page adjusted to allow more text, although we sometimes reduced the text to make more room for visualizations. In the large majority of cases, text was provided and reviewed by outside experts, although the editors took the lead on several topics in their areas of expertise. After outside review, our text editor, Ross

West, reworked the text to ensure consistency of voice and style throughout every page of the atlas. This was a trying task because some experts wrote in an overly technical manner for purposes of the atlas, because writers often exceeded by two to three times the length of text that fit in the allotted space, and because much of the text editing occurred on very tight deadlines.

The multiyear process required to finish many of the page pairs meant that they needed updating near the end of the production phase. For example, our first page pair created in 2003 was "Fire History"; a long-term drought and many more fires occurred since that time. We thus revisited all pages in 2011 and, whenever possible, updated charts, maps, and graphs to portray the most recent data to reflect the economic downturn of the past few years, 2010 Census data, change in wildlife populations and ranges, and many other topics in flux. This proved challenging in some circumstances; the 2010 Census data, for example, was only released six week before the delivery deadline for the atlas going to the printer.

In addition to the page pairs, the *Atlas of Yellowstone* contains new reference materials. Allan Cartography, makers of the award-winning Raven wall map series and Benchmark Road and Recreation Atlases, produced new 1:500,000 maps for the Greater Yellowstone Area and 1:100,000 maps for Grand Teton and Yellowstone national parks as well as a gazetteer to accompany them. Stuart Allan, founder of Allan Cartography, also provided important consultative advice on map design, layouts, and atlas organization throughout the entire production period. Other reference materials include a directory of place name origins for the region and the parks developed by Lee Whittlesey, a list of vertebrate species in the Yellowstone and Grand Teton parks based on National Park Service records, reference maps for U.S. Geological Survey topographic quadrangles, and U.S. Census-based names and counties in the tristate region.

## Interpretations and Stories Not Told

Any atlas is a work of interpretation and many, many decisions about which data to include or exclude. The stories portrayed in this atlas are guided by leading experts in their fields, but there are differences of opinion among the experts. Creating the atlas was at times an exercise in navigating among strong-minded, articulate, and dauntingly persuasive individuals, some of whom do not particularly agree with one another.

In very general terms, we found the pages that generated the widest range of divergent opinions involved historic data, whether it was related to natural systems like geology or to cultural systems like archaeology. For example, in the case of the Sheep Eater Indians, oral histories, limited written accounts, and artifacts leave room for debate about whether this group existed as a discrete entity within the larger Shoshone tribe. Other topics can be contentious as well, ranging from the choice of data to portray regional vegetation to interpretations of economic data to explanations of why geysers erupt. In such cases we relied on the experts listed at the back of this atlas to provide their interpretation. Ultimately, however, we as editors were responsible for selecting among their sometimes conflicting assessments to create the pages.

The astute reader may also wonder about the many tales the atlas does *not* show. As with any account, there are more stories untold than told. A major reason for not including topics was because spatial data did not exist to tell the story in map form. One such example was the seasonal migration of elk through the Northern Range of Yellowstone. We wished to replicate the techniques used in Charles Joseph Minard's famous map, hailed as "the best statistical graphic ever drawn," that displays the death of Napoleon's soldiers and the associated weather over time as they marched on and retreated from Moscow. This could have resulted in a compelling graphic showing increases and decreases in elk due to births, weather, and predation as they penetrated and retreated from Yellowstone's interior. To our disappointment, data did not support this ambitious graphical presentation. Although we could have created a conceptual map, we made it a rule to avoid including maps or graphs not based on direct observation.

Another story we wished to include was to be titled "The Park That Isn't There"; we envisioned it as a page pair mapping the many amusement parks, mines, ski areas, railroads, and other developments proposed but never built because of Yellowstone's national park status. Again, many anecdotal and written records exist but have not been converted to map form. Perhaps most frustrating to us as editors was the relative lack of spatial data for mapping the long-term presence of American Indians in the landscapes of Yellowstone. We also wanted more page pairs that tracked life at the scale of an individual—whether an early settler, a tourist, or an animal—but spatial data were lacking.

In some cases we had experts and data sets identified and page pairs mocked-up, but ran out of time before our publication deadline. Topics in this realm include "Vegetation Change," "Night Skies" (fish-eye perspectives of light pollution in Grand Teton and Yellowstone parks), "Prehistoric Mammals" (to accompany the "Dinosaurs" page pair in this volume), "Park Management," and "Park Employment." Other topics such as "Amphibians and Reptiles," "Small Mammals," "Path of the Pronghorn," "Energy Development," "Exurban Development," and "Environmental Quality" were on the docket with identified experts but never made it to the mockup stage. About twenty additional topics were considered but set aside due to time and resource constraints.

Other page pairs were limited or never created to avoid revealing too much about a vulnerable animal or resource. The map on American Indian names, for example, provides important evidence of the long-term presence of Indians in the area of Yellowstone National Park long before the park was explored and set aside by the U.S. government. The number of mapped names could have been larger, except that many of the names are so descriptive that they could lead people to sites and resources that are best left undisturbed. Likewise, radio collar data on individual wolves and grizzly bears would have made for fascinating maps but also could lead people directly to den or hibernation sites that need to be protected from all disturbances.

If we are fortunate, interest in this volume will generate a second edition in which we can visit many of the topics not covered in this volume. Even more, we hope that readers upset at the absence of a topic will be inspired to create data sets and maps that portray the stories they want told.

## Final Thoughts

The above synopsis summarizes the ten years of work that led to publication of the *Atlas of Yellowstone*. In truth, however, work contributing to the atlas extends over a far longer time and among many more people than are mentioned here. The data used to make maps in the atlas come from over a century of research in Yellowstone National Park, two centuries of western exploration, and millennia of evidence left by native inhabitants. Their messages are passed on in the maps in this atlas.

It has been our privilege to sit with many of the people who inhabit, study, and love Yellowstone. In these meetings, we came to understand that the true story behind the *Atlas of Yellowstone* is a love story, a tale about the deep and abiding attachment and dedication that Yellowstone inspires. It is also a story about our love of maps and the joy we take in portraying the complexity and beauty of this extraordinary place in ways that bring it to life for people far from its boundaries.

Yellowstone has long been an inspiration to the world. We hope that you, the reader, take as much pleasure and inspiration from this atlas as we experienced in creating it.

*W. Andrew Marcus*
*James E. Meacham*
*Ann W. Rodman*
*Alethea Y. Steingisser*

*June 21, 2011*

# Sponsoring Institutions

UNIVERSITY OF OREGON

**The University of Oregon,** founded in 1876, is the state's flagship institution. The University of Oregon's commitment to academic excellence draws well-rounded students and faculty members to the university's 263 academic programs, many of which rank number one in Oregon and among the top twenty in the nation. The university's academic reputation is highlighted by its stature as Oregon's only member of the prestigious Association of American Universities, one of just thirty-four public universities in the United States so honored. Membership in this select group signifies preeminence in graduate and professional education and basic research.

In addition to a comprehensive College of Arts and Sciences, the university also includes the Graduate School, the Robert Donald Clark Honors College, and six professional schools: the School of Architecture and Allied Arts, the Charles H. Lundquist College of Business, the College of Education, the School of Journalism and Communication, the School of Law, and the School of Music and Dance.

A leader in sustainability, the UO is based in the classic college town of Eugene, in a region known for creativity, innovation, and outdoor adventure. The University of Oregon features a diverse student body hailing from all fifty states as well as eighty-five countries. The campus is situated on 295 lush, tree-shaded acres. The 23,000 students at the UO participate in more than 250 student organizations and cheer on the university's nineteen NCAA Division I sports teams.

## College of Arts and Sciences

The College of Arts and Sciences is the heart and soul of the University of Oregon because it is home to the core academic programs that support the entire university. The college provides a liberal arts foundation to the vast majority of UO undergraduates; for students who wish to pursue a degree in a liberal arts or science discipline, the college offers more than seventy undergraduate degrees in forty-six fields of study in the humanities, social sciences, and natural sciences. The College of Arts and Sciences is also the intellectual hub of the UO, granting more than 75 percent of the UO's doctoral degrees. Among its more than 500 faculty members, the college boasts some of the most accomplished and best-known researchers in the world, including more than fifty Guggenheim fellows, thirty-one fellows of the American Association for the Advancement of Science, six members of the National Academy of Sciences, ten American Academy of Arts and Sciences members, and a Medal of Science recipient.

## Department of Geography

Recognized by the National Research Council as one of the leading geography programs in the country, the University of Oregon's Department of Geography is an important center for research and teaching about the changing spatial organization and material character of the planet. The department's research and educational initiatives are rooted in geography's longstanding concern with exploration, mapmaking, and accounts of differences from place to place, but also reflect the needs, challenges, and technological possibilities of the modern world. Through work grounded in field explorations, mapmaking, geographic information analysis, modeling, and critical assessments of spatial arrangements and ideas, Oregon geographers have made important contributions to efforts to understand the impacts of physical and human processes on Earth's places and landscapes. They have also played a fundamental role in reviving interest in geography and demonstrating the discipline's contemporary relevance for science and society.

## InfoGraphics Lab

The InfoGraphics Lab, founded in 1988, is recognized as a national leader in cartographic and atlas design, geographic information systems (GIS), and mobile and web-based interactive mapping applications. The lab is a center of excellence committed to advancing research and teaching innovations in geospatial data access, visualization and analysis through interdisciplinary and cross-institutional collaborations. These collaborations directly tie undergraduate and graduate student learning to the creation of new methods and techniques that benefit both the campus environment and the academic research in the colleges. Six undergraduate and seven graduate students worked as members of the InfoGraphics Lab staff on the *Atlas of Yellowstone*. Atlas-making has been a foundation of the lab's research since its founding. In 2001, the award-winning *Atlas of Oregon* was published under the direction of legendary cartographer and lab cofounder Professor William G. Loy.

# MONTANA STATE UNIVERSITY

Montana State University, the premier land-grant university in the Rocky Mountain region, delivers undergraduate and graduate educational programs; engages scholars in cutting-edge research and creative projects; and provides service to the state, nation, and globe. The Bozeman campus is the largest, flagship institution of the four-campus MSU system and offers more than 125 degree and program options within its academic colleges: Agriculture; Arts & Architecture; Business; Education, Health & Human Development; Engineering; Letters & Science; Nursing; University College; Gallatin College Programs; and Graduate School.

In addition, MSU encompasses Extension offices and the Montana Agricultural Experiment Stations that give it a physical presence in fifty-six counties and five reservations throughout Montana. This statewide presence facilitates the dissemination of knowledge to individuals and communities and empowers them to improve their quality of life.

The university's innovative research- and inquiry-based curriculum encourages exploration and creativity. Curiosity and collaboration drive research discoveries, and every student pursues a research or creative experience in their chosen discipline with a faculty mentor. Scholarship takes place in the field, classrooms, and natural laboratories to uncover new insights, deepen human understanding, and improve lives. MSU is designated as one of only 108 research universities—out of 4,600 institutions—with "very high research activity" by the Carnegie Foundation for the Advancement of Teaching. This means that MSU is among only 2 percent of institutions nationwide to achieve this level of research prominence. Montana State University is ranked in the top twenty schools nationally for the number of Goldwater Scholarship recipients, the nation's premier scholarship for undergraduates studying math, natural sciences, and engineering.

## The Yellowstone Connection

Montana State University, known as University of the Yellowstone®, claims numerous research and creative connections to the Greater Yellowstone Ecosystem. Studies involve life at the microbial scale, especially in Yellowstone's thermal features, while other projects explore wolves, elk, and fishery science as well as climate research and the examination of policy in the park's history. Montana State University is passionately connected to the Greater Yellowstone Ecosystem through learning, discovery, and engagement.

## Learning

MSU engineering students developed an innovative program for their senior capstone project to recycle bear spray canisters. They created a machine that removed the dangerous chemical, allowing the canister to be recycled with other metals. Yellowstone National Park's environmental protection specialist commended the students for developing a solution to this hazardous waste.

MSU film and photography students spend a week in Yellowstone National Park approaching the natural environment from a new perspective, while MSU architecture students brainstorm solutions on how to protect the historical structures and enhance the experience for visitors to one of America's greatest treasures.

The MSU Library Special Collections and Archives house a significant Yellowstone National Park and Yellowstone ecosystem collection, including documents related to the first expeditions to the area that would become Yellowstone: the 1869 Cook-Folsom-Peterson expedition and the Washburn-Langford-Doane expedition in 1870.

## Discovery

MSU is home to two internationally recognized microbial research entities—the Center for Biofilm Engineering and the Thermal Biology Institute. Both programs are noted for their interdisciplinary collaboration, undergraduate and graduate opportunities, and the quality and quantity of their research, which expands the basic understanding of the microbial world and seeks practical applications from their discoveries, many of which originate in Yellowstone.

Montana State University researchers discovered a rare oasis of life in the midst of hundreds of geothermal vents at the bottom of Yellowstone Lake. Using remote operated vehicles and advanced techniques of DNA analysis, 260 new species were found. Understanding how the colony lives in such extreme conditions may lead to the development of new products and inventions.

The research of MSU faculty members and students will identify and categorize viruses from extreme environments, primarily from very hot and very acidic places. These studies will broaden the understanding of the viral world and its relationship to cellular life.

## Engagement

The Museum of the Rockies at Montana State University houses an unprecedented collection of Yellowstone memorabilia from the Hamilton Povah Collection. Povah is the daughter of Charles A. Hamilton, founder of the Hamilton Store chain, which operated in Yellowstone National Park from 1915 through 2002. The collection includes items from the stores themselves and Native American objects that would have been representational of inventory in the store.

Extended University reaches out to high school teachers, offering summer courses focusing on life in extreme environments. These teachers can bring the world of Yellowstone to life for their students.

The Yellowstone Writing Project provides an intensive professional experience for K–12 teachers seeking to enhance their ability to teach writing. The participants immerse themselves in their own writing as well as explore strategies for teaching students of all grade levels to write well.

## Yellowstone Park Foundation

For more than 135 years, Yellowstone National Park's exceptional beauty has inspired philanthropy with generous individuals, corporations, and foundations contributing to its long-term preservation.

The Yellowstone Park Foundation, the park's official fundraising partner, plays a vital role in expanding this tradition of philanthropic support. It works closely with the National Park Service to identify Yellowstone's immediate needs and long-term funding challenges. Then foundation staff members connect generous donors with opportunities to help address these challenges and participate in park stewardship.

Notable past projects include significant annual support of Yellowstone wolf research, a million-dollar landscape restoration of historic Artist Point, and a completed $15 million capital campaign to help build the Old Faithful Visitor Education Center.

By investing in the following six strategic initiatives, contributions to the Yellowstone Park Foundation are having a lasting and significant impact in the park:

### Visitor Experience

Every trip to Yellowstone should be magical, but heavy annual visitation can take its toll on trails, campgrounds, and other facilities. It can also put a strain on the park's ability to provide educational opportunities. The Yellowstone Park Foundation supports projects that enhance recreation, safety, and accessibility, while bolstering both in-park and online education.

### Wildlife, Wonders, and Wilderness

Yellowstone is home to the largest concentration of wildlife in the lower forty-eight states and more geysers and hot springs than the rest of the world combined. The Yellowstone Park Foundation supports research and conservation projects to preserve the wildlife and other precious natural resources for which Yellowstone is famous.

### Cultural Treasures

Yellowstone—the world's very first national park—is the keeper of stories. The Yellowstone Park Foundation supports projects that protect, preserve, research, and share information about Yellowstone's human past. Projects include support for the park's museum collection, archeological surveys, historic preservation, and more.

### Ranger Heritage

Yellowstone rangers have no small job. They are charged with protecting the 2.2 million-acre park's natural resources as well as the safety of visitors. They need trustworthy equipment, modern technology, reliable transportation, and suitable facilities. The Yellowstone Park Foundation supports projects that promote the effectiveness, safety, and efficiency of rangers.

### Tomorrow's Stewards

Yellowstone is one of the world's premier outdoor classrooms and offers several award-winning, ranger-led educational programs for children. Yellowstone Park Foundation support helps expand the reach of these programs to promote the appreciation and stewardship of Yellowstone in the next generation.

### Greenest Park

Yellowstone has long been a leader in natural resource management but still uses large quantities of fossil fuel and treated water and generates much solid waste in serving millions of annual visitors. The Yellowstone Park Foundation supports projects that aim to reduce Yellowstone's ecological footprint and better preserve environmental resources.

The *Atlas of Yellowstone* was made possible in part by a grant from Canon U.S.A., Inc., the largest corporate supporter of wildlife conservation in Yellowstone National Park. The *Eyes on Yellowstone* program, which supports educational and research projects in Yellowstone, is made possible by Canon.

# UNIVERSITY OF WYOMING

University of Wyoming, located in Laramie, opened its doors in fall 1887 when Wyoming was still a territory. UW aspires to be one of the nation's finest public land-grant research universities. The university serves as a statewide resource for accessible and affordable higher education of the highest quality; rigorous scholarship; technology transfer; economic and community development; and responsible stewardship of our cultural, historical, and natural resources. As Wyoming's only university, UW is committed to outreach and service that extend its human talent and technological capacity to serve the people in our communities, our state, the nation, and the world.

More than 2,800 UW faculty members and staff are dedicated to educating more than 13,000 students. Approximately 180 undergraduate, graduate, and professional programs of study are offered with seven colleges: Agriculture and Natural Resources, Arts and Sciences, Business, Education, Engineering and Applied Science, Health Sciences, and Law. The university also houses two schools, the Haub School of Environment and Natural Resources and the School of Energy Resources.

UW's Outreach School delivers undergraduate and graduate degree programs at the University of Wyoming Casper College Center and through its distance learning delivery systems. The university maintains nine outreach education centers across Wyoming and Cooperative Extension Service Centers in each of the state's twenty-three counties and on the Wind River Indian Reservation. The institution is supported by the UW Foundation, which provided a donation to the *Atlas of Yellowstone* project in honor of Peter K. Simpson, a distinguished UW political science professor and political leader.

## Research and Student Learning in Yellowstone National Park

Many UW departments collaborate within Yellowstone National Park and the Greater Yellowstone Ecosystem through their teaching and research endeavors. The Department of Geography provides expert knowledge and research opportunities in physical geography, human-environment interactions, GIScience, and natural resource management and planning. The Department of Zoology and Physiology provides students with applied field experiences for careers in biology, physiology, ecology, and neuroscience. The Environment and Natural Resources Program is a leader in interdisciplinary learning and research and collaborative approaches to solving natural resource problems. The Wyoming Geographic Information Science Center is dedicated to advancing geographic information science and technology in education, government, and business for better place-based management of natural and human resources. The Wyoming Natural Diversity Database provides an important resource for information on the distribution and ecology of rare plants and animals in Wyoming, and the Department of Geology and Geophysics provides students with diverse field research opportunities in landscape development, water management, paleontology, climate change, crustal evolution, tectonics, and volcanic history in Yellowstone.

## History and Culture of the Rocky Mountain Region

Several social science departments, along with the American Heritage Center, the Rocky Mountain Herbarium, and the University Libraries, house substantial collections and staff members who demonstrate scholarly expertise in this area of distinction. The American Heritage Center is a repository open to the public that includes numerous collections related to Yellowstone. Collections include:

• The Fritiof Fryxell Papers—These papers contain materials related to Ferdinand Hayden's 1871 expedition to the Yellowstone National Park region, which included photographer William Henry Jackson and artist Thomas Moran, among others.

• The John W. Meldrum Papers—John W. Meldrum served as U.S. Commissioner of Yellowstone National Park for forty-one years.

• The Adolph Murie Papers—Adolph Murie was a wildlife biologist with the U.S. Departments of Interior and Agriculture for more than thirty years, serving with the National Park Service and Fish and Wildlife Service. Murie conducted wildlife studies in Yellowstone and Grand Teton national parks.

The Rocky Mountain Herbarium, founded in 1893 by Aven Nelson, contains the largest collection of Rocky Mountain plants and fungi in existence, including species collected in Yellowstone by Nelson and his students in 1899. UW botanist Erwin Evert published a comprehensive book in 2010, *Vascular Plants of the Greater Yellowstone Area: Annotated Catalog and Atlas*.

UW Libraries has comprehensive holdings of current and historical published documents and works related to the Greater Yellowstone Ecosystem, with materials ranging from natural and physical sciences to art. Included are the special collections of Grace Raymond Hebard, a former librarian and western historian. An avid collector, Hebard acquired a number of early works on Wyoming and Western U.S. history, which were placed in the Wyoming Room in the library. Her manuscript materials form the nucleus of the archival collections in the American Heritage Center. George Miles, curator of the Western Americana Collection at Yale's Beinecke Library, described the Hebard Collection as "a major research collection of national importance that provides essential service for scholars working in the history and culture of Wyoming and the Mountain Plains West."

With 96 percent of Yellowstone National Park and 100 percent of Grand Teton National Park located within Wyoming, the University of Wyoming is a proud contributor and steward of scientific, cultural, and artistic works within the region.

# Contributing Organizations

 **Buffalo Bill Historical Center**
Cody, Wyoming

The Buffalo Bill Historical Center's varied and extensive collections enable both in-house and visiting scholars to expand knowledge about the history, culture, peoples, and natural history of the American West. Areas of expertise and emphasis include the natural history of the Greater Yellowstone region; western art from the nineteenth century through today; the life and times of William F. Cody (also known as Buffalo Bill); the lives, art, and cultures of the Plains Indian peoples; and the history, development, and use of firearms in Europe, the United States, and the American West. The Center's research library houses more than 30,000 books, 500,000 photographs and negatives, and 315 manuscript collections, and welcomes hundreds of researchers from around the world every year. Most important, recognizing that it is vital to share scholarship as broadly as possible, the center incorporates this new work into its galleries to ensure interpretation is continually fresh and current.

Headwaters Economics is an independent, nonprofit research group. The staff at Headwaters Economics blends innovative research techniques and extensive on-the-ground experience working with a range of partners across the West for more than twenty years. The mission of Headwaters Economics is to improve community development and land management decisions in the West. Headwaters Economics provides original and effective research to people and organizations that make a difference in the West, working with community leaders, landowners, public land managers, elected officials, business owners, and other nonprofit organizations. Its goal is to give these partners credible information they can use to identify and solve problems.

The Yellowstone Ecological Research Center (YERC) specializes in long-term, large-scale collaborative ecological research and education. YERC brings together academic institutions, private companies, and governmental agencies from around the country to work in the group's major research focus area: the 20 million acre Greater Yellowstone Ecosystem (GYE). As an independent, private, nonprofit organization dedicated to increasing the role of science at the natural resources decision-making table, YERC serves as a source of science-based decision criteria for the diverse stakeholders of the GYE and beyond, hewing to the ideals of rational decision-making based on the common ground of shared information.

# Sources

**Style and Usage Note** The *Atlas of Yellowstone* covers a great deal of material, much of it technical and from fields of study (for example, soil science) abounding in terminology unfamiliar to the general reader. Scores of contributors accustomed to writing in the specialized languages of areas such as archaeology, art history, ecology, economics, geology, geography, history, paleontology, wildlife biology, and numerous other disciplines provided text for this project. Throughout the production process our text editors made decisions balancing a desire for detail and depth with the necessities of clarity and readability as well as the brevity required by space limitations. A number of style guides aided us in this effort: *The Chicago Manual of Style* (Sixteenth Edition), the University of Oregon *Grammar and Style Guide*, *The Yellowstone National Park Publications Style Guide*, the U.S. National Park Service's *Harpers Ferry Center Editorial Style Guide*, and *Editorial Style Guide for* Park Science *and* Natural Resource Year in Review. Occasional variances in punctuation, abbreviation, and capitalization were made to recognize the special requirements of an atlas. The choices we made in producing this volume reflect our best attempt to communicate clearly and accurately.

**Base Data** The following data layers were used to contruct maps throughout the *Atlas of Yellowstone*:

ESRI. *World and United States datasets.* Redlands, 2006.

Natural Earth dataset. Online.

U.S. Department of Commerce. Bureau of the Census. Census of the Population.

U.S. Department of the Interior. National Park Service.

U.S. Geological Survey. National Elevation Data Set.

U.S. Geological Survey. National Hydrography Data Set.

U.S. Geological Survey. U.S. Board on Geographic Names. Geographic Names Information System.

U.S. National Park Service. Natural Resource Information Portal. Online.

## Yellowstone in the World, 2–3
*Contributing experts: atlas editors*

Tom Patterson, U.S. National Park Service.

## Yellowstone in the Region, 4–5
*Contributing experts: atlas editors*

Tom Patterson, U.S. National Park Service.

## Greater Yellowstone Detail, 6–7
*Contributing experts: Robert Crabtree, Stuart Allan*

Homer, C.C. Huang, L. Yang, B. Wylie, and M. Coan. "Development of a 2001 National Landcover Database for the United States." *Photogrammetric Engineering and Remote Sensing* 70, No. 7 (July 2004): 829–40.

"The Greater Yellowstone Ecosystem." Map produced by Allan Cartography for Yellowstone Ecological Research Center, 2008.

## The World's First National Park, 8–9
*Contributing experts: John Varley, Paul Schullery*

U.S. Department of the Interior, National Park Service.

U.S. Department of the Interior, National Park Service Public Use Statistics Office. Online.

WDPA Consortium. World Database on Protected Areas. World Conservation Union (International Union for Conservation of Nature) and United Nations Environment Programme–World Conservation Monitoring Centre, 2004.

## Political Boundaries, 10–11
*Contributing experts: William K. Wyckoff, Paul Schullery*

Haines, Aubrey L. *The Yellowstone Story: A History of Our First National Park, Volume 2*. Boulder, Colorado: Yellowstone Library and Museum Association, Colorado

Associated University Press, 1966.

National Geographic Society. *Historical Atlas of the United States*. Washington, D.C., 1988.

United States Department of the Interior Geological Survey. *The National Atlas of the United States of America*. Washington, D.C., 1970.

The Newberry Library. *Atlas of Historical County Boundaries*. Online.

U.S. National Park Service. *Natural Resource Information Portal*. Online.

## Archaeology, 14–15
*Contributing experts: Ann Johnson, C. Melvin Aikens, Elaine Skinner Hale, Lawrence L. Loendorf*

Davis, Leslie B., Stephen A. Aaberg, James G. Schmitt, and Ann M. Johnson. *The Obsidian Cliff Plateau, Prehistoric Lithic Source, Yellowstone National Park, Wyoming*. Selections series No. 6, Division of Cultural Resources, Rocky Mountain Region, U.S. National Park Service, 1995.

Johnson, A., Reeves, O.K. Brian, and M.W. Shortt. *Osprey Beach: A Cody Complex Camp on Yellowstone Lake*. Lifeways of Canada Limited, 2004. Online.

Wedel, Waldo R. "Mummy Cave: Prehistoric Record from Rocky Mountains of Wyoming." *Science* 160 (1968).

U.S. National Park Service, Yellowstone National Park. Unpublished data.

## American Indians, 16–17
*Contributing experts: Rosemary Sucec, Katie White, Lawrence Loendorf, C. Melvin Aikens*

Weixelman, Joseph O. "Fear or Reverence? Native Americans and the Geysers of Yellowstone," in *People and Place: The Human Experience in Greater Yellowstone—Proceedings of the Fourth Biennial Conference on the Greater Yellowstone Ecosystem, October 12–15, 1997*, ed. Paul Schullery and Sarah Stevenson: 50–66. Yellowstone National Park Center for Resources, 2004.

Nabokov, P., and Larry Loendorf. *American Indians and Yellowstone National Park: A Documentary Overview*. U.S. National Park Service, Yellowstone National Park Center for Resources, 2002.

U.S. Geological Survey. *Indian Land Areas Judicially Established 1978* (1993). Map. Online.

U.S. National Park Service, Yellowstone National Park. Unpublished data.

## Sheep Eaters, 18–19
*Contributing experts: Lawrence Loendorf, Ann Johnson, C. Melvin Aikens*

Eakin, Daniel H. "Evidence for Shoshonean Bighorn Sheep Trapping and Early Historic Occupation in the Absaroka Mountains of Northwest Wyoming," in *University of Wyoming National Park Service Research Center 29th Annual Report 2005*: 74–86. University of Wyoming, 2005.

Schullery, Paul and Lee H. Whittlesey. *The History of Greater Yellowstone Wildlife: An Interdisciplinary Analysis, 1796–1881*. Book manuscript written for U.S. National Park Service, forthcoming in 2013.

Sheep trap image courtesy of Daniel H. Eakin. Office of the Wyoming State Archaeologist, University of Wyoming Department of Anthropology.

Petroglyph images courtesy of Linda Olson. Minot State University Department of Art.

Joaquín, David. *Shoshone Camp* (date unknown). Courtesy of David Joaquin. www.twohawkstudio.com.

Lewis, Samuel and William Clark. *A map of Lewis and Clark's track across the western portion of North America, from the Mississippi to the Pacific Ocean: by order of the executive of the United States in 1804, 5, and 6* (copied

by Samuel Lewis from the original drawing of William Clark, 1814). Library of Congress, Geography and Map Division.

## Catlin and the American Indian, 20–21
*Contributing experts: Paul Schullery, Brian Dippie*

Catlin, George. *Buffalo Chase, Bull Protecting Cow and Calf* (1832–33). Smithsonian American Art Museum, Washington, D.C./Art Resource, New York, New York.

Catlin, George. *Buffalo Bull's Back Fat, Head Chief, Blood Tribe* (1832). Smithsonian American Art Museum, Washington, D.C./Art Resource, New York, New York.

Catlin, George. *Peh-to-pe-kiss, Eagle's Ribs, a Piegan Chief* (1832). Smithsonian American Art Museum, Washington, D.C./Art Resource, New York, New York.

Catlin, George. *Outline Map of Indian Locatlities in 1833*, in *Illustrations of the Manners, Customs, and Condition of the North American Indian*. London: H.G. Bohn, 1848. Digital scan from the Beinecke Rare Book and Manuscript Library, Yale University.

## Exploration, 22–23
*Contributing experts: Paul Schullery, Lee Whittlesey, Sarah Bone*

Loy, William G., Stuart Allan, Aileen R. Buckley, and James E. Meacham. *Atlas of Oregon* (Second Edition). Eugene, Oregon: University of Oregon Press, 2001.

Haines, Aubrey L. *The Yellowstone Story: A History of Our First National Park, Volumes 1 and 2*. Boulder, Colorado: Yellowstone Library and Museum Association, Colorado Associated University Press, 1977.

Russel, Osborne. *Journal of a Trapper*. Aubrey L. Haines, ed. Portland, Oregon: Oregon Historical Society, Champoeg Press, 1955.

U.S. National Park Service, Yellowstone National Park. Unpublished data.

## Early Maps, 24–29
*Contributing experts: Lee Whittlesey, Paul Schullery, James Walker*

Bechler, G.R. *Map of the sources of Snake River with its tributaries, reduced from the preliminary map after surveys by Gustavus R. Bechler.*

Clark, William. *A map of part of the continent of North America: between the 35th and 51st degrees of north latitude, and extending from 89° degrees of west longitude to the Pacific Ocean, compiled from the authorities of the best informed travellers by M. Lewis* (copied by Nicholas King, 1805). Library of Congress, Geography and Map Division.

Clark, William. *Clark's Map of 1810*. Yale Collection of Western Americana, Beinecke Rare Book and Manuscript Library.

Doane, Gustavus C. *Map of the Route of the Yellowstone Expedition, 1869*. National Archives (Map No. NWCS-077-CWMF-Q329-30: Yellowstone).

De Lacy, W.W. *Map of the territory of Montana with portions of the adjoining territories: showing the gulch or placer diggings actually worked and districts where quartz (gold andsilver) lodges have been discovered to January 1st 1865* (drawn by W.W. de Lacy for the use of the first legislature of Montana). Library of Congress, Geography and Map Division.

de Smet, Pierre Jean. *Map of the upper Great Plains and Rocky Mountains region, respectfully presented to Col. D.D. Mitchell by P.J. de Smet, 1851*. Library of Congress, Geography and Map Division.

Drouillard, George. *Map of the Big Horn Country, September 1808* (cropped). Missouri History Museum, St. Louis.

Ferris, Warren Angus. *Map of the Northwest fur country, 1836*. Heritage and Research Center, Yellowstone National Park.

Raynolds, W.F., and H.E. Maynaidier. *Map of the Yellowstone River and Missouri Rivers and their Tributaries explored by Captains W.F. Raynolds and H.E. Maynaidier, 1859–1860*. Library of Congress, Geography and Map Division.

Lewis, Samuel and William Clark. *A map of Lewis and Clark's track across the western portion of North America, from the Mississippi to the Pacific Ocean: by order of the executive of the United States in 1804, 5, and 6 (copied by Samuel Lewis from the original drawing of William Clark, 1814)*. Library of Congress, Geography and Map Division.

Washburn, H.D. *Map of the Territory of Montana to Accompany the Report of the Surveyor General, 1869*. National Archives Map No. NWCS-049-OMF-Montana No. 03 (1869).

**Jackson and Moran, 30–31**
*Contributing experts: Christine Brindza, Colleen Curry, Paul Schullery*

Haines, Aubrey L. *The Yellowstone Story: A History of Our First National Park, Volume 1*. Boulder, Colorado: Yellowstone Library and Museum Association, Colorado Associated University Press, 1977.

Jackson, William Henry. *Hot Springs on Gardner River* (1871). U.S. National Park Service, Yellowstone National Park, Yellowstone Digital Slide File, NPS photograph. Online.

Jackson, William Henry. *Hayden Expedition in Camp* (1872). U.S. National Park Service, Yellowstone National Park, Yellowstone Digital Slide File, NPS photograph. Online.

Jackson, William Henry. *Old Faithful* (1872). U.S. Geological Survey Photographic Library. Online.

Moran, Thomas. *The Grand Canyon of the Yellowstone* (1872). Smithsonian American Art Museum, Washington, D.C. Lent by the U.S. Department of the Interior Museum.

Moran, Thomas. *Great Blue Spring of the Lower Geyser Basin, Firehole River, Yellowstone* (1872). Buffalo Bill Historical Center, Cody, Wyoming. Purchased with funds from the William E. Weiss Fund, Mrs. J. Maxwell Moran, Wiley Buchanan III, Nancy-Carroll Draper, Nancy and Nick Petry, Steve and Sue Ellen Klein, William C. Foxley, John F. Eulich, Mary Lou and Willis McDonald IV, and D. Harold Byrd Jr., 24.91.

U.S. National Park Service, Yellowstone National Park. Unpublished data.

**Yellowstone Art, 32–33**
*Contributing experts: Christine Brindza, Colleen Curry*

Pruessl, Carl. *Old Faithful* (1929). Buffalo Bill Historical Center, Cody, Wyoming. Designated purchase with donations from the Arlington Gallery, Dr. and Mrs. Van Kirke Nelson and Family, Thomas and Shannon Nygard, and William E. Weiss Memorial Fund, 3.01.

Bierstadt, Albert *Geysers in Yellowstone* (ca. 1881). Buffalo Bill Historical Center, Cody, Wyoming. Gift of Townsend B. Martin, 4.77.

Coe, Anne. *Out to Lunch* (1990). Buffalo Bill Historical Center, Cody, Wyoming. Gift of the artist and D. Harold Byrd Jr., 2.93.2.

Fell, Olive. *Me and Old Faithful* (date unknown). Buffalo Bill Historical Center, Cody, Wyoming. Gift of Hank and Florence Dais, 18.99.1.

Leigh, William R. *The Lower Falls of the Yellowstone* (1911). Buffalo Bill Historical Center, Cody, Wyoming. William E. Weiss Memorial Fund Purchase, 11.88.

**Science History, 34–37**
*Contributing experts: John Varley, Paul Schullery, Christie Hendrix, Bill Resor*

Allen, E.T., and Arthur Day. *Hot Springs of the Yellowstone National Park*: 466. Carnegie Institute of Washington, 1935.

Hague, Arnold. *Atlas to Accompany Monograph XXXII on the Geology of the Yellowstone National Park*. U.S. Geological Survey Monograph 32, 1904.

Hayden, Ferdinand. *Upper Geyser Basin, Fire Hole River, Wyoming Territory* (1871). Library of Congress, Geography and Map Division.

Preble, Edward Alexander. *Report on Condition of Elk in Jackson Hole, Wyoming, in 1911*. U.S. Department of Agriculture, Biological Survey, 1911.

Thermus aquaticus photograph courtesy of Robert Ramaley. University of Nebraska Medical Center Department of Biochemistry and Molecular Biology.

Wildlife research image courtesy of Frank and John Craighead and the Craighead Environmental Research Institute.

U.S. National Park Service, Yellowstone National Park. Unpublished data.

**Road History, 38–39**
*Contributing experts: Elaine Skinner Hale, Zehra Osman, Lee Whittlesey*

Baldwin, Kenneth H. *Map of the Yellowstone National Park showing roads completed, under construction, and approximately projected* (compiled from map of U.S. Geological Survey and other sources, to accompany Major Charles J. Allen's project dated November 14, 1888). From Kenneth H. Baldwin's *Enchanged Enclosure: The Army Engineers and Yellowstone National Park—A Documentary History*: 87.

Haines, Aubrey L. *The Yellowstone Story: A History of Our First National Park, Volumes 1 and 2*. Boulder, Colorado: Yellowstone Library and Museum Association, Colorado Associated University Press, 1977.

Norris, Philetus W. *Report upon the Yellowstone National Park, to the Secretary of the Interior, for the Year 1877* (1877). Government Printing Office, Washington, D.C, (Map inside back cover)

Norris, Philetus W. *Report upon the Yellowstone National Park, to the Secretary of the Interior, for the Year 1878* (1879), Government Printing Office, Washington, D.C,

Norris, Philetus W. *Report upon the Yellowstone National Park, to the Secretary of the Interior, for the Year 1879* (1880), Government Printing Office, Washington, D.C,

Norris, Philetus W. *Annual Report of the Superintendent of the Yellowstone National Park, to the Secretary of the Interior, for the Year 1880* (1881), Government Printing Office, Washington, D.C,

Norris, Philetus W. *Fifth Annual Report of the Superintendent of the Yellowstone National Park, December 1, 1881* (1881), Government Printing Office, Washington, D.C,

U.S. War Department. *Report of the Chief of Engineers*. Map showing development of Yellowstone National Park road system (drawn to accompany report submitted by A.E. Burns, overseer, March 1, 1900). Library of Congress.

*Yellowstone National Park and North Western Wyoming, U.S. Geological Survey, J.W. Powell, Director*. Surveyed in 1883–85.

*Yellowstone National Park, U.S. Geological Survey, J.W. Powell, Director*. Surveyed in 1884–85.

Yellowstone Digital Slide File. U.S. National Park Service postcards by Frank J Haynes. Online.

**Development at Old Faithful, 40–41**
*Contributing experts: Zehra Osman, Lee Whittlesey*

Byrand, Karl John. "The Evolution of the Cultural Landscape in Yellowstone National Park's Upper Geyser Basin and the Changing Visitor Experience, 1872– 1990." Master's thesis, Montana State University, 1995.

U.S. National Park Service, Yellowstone National Park. Unpublished data.

Yellowstone National Park. *Old Faithful Historic District, Cultural Landscape Inventory*. 2009.

**Roads and Trails, 42–43**
*Contributing experts: John Sacklin, atlas editors*

U.S. National Park Service. Natural Resource Information Portal. Online.

**Traffic, 44–45**
*Contributing experts: John Sacklin, atlas editors*

Federal Highway Administration, Eastern Federal Lands Highway Division. National Park Service 2004 Traffic Data Report, 2005.

Greyhound Lines, Inc. Greyhound Route Map, North America; Public Intercity Transportation, 2010. Online.

Idaho Transportation Department, Road Data and Analysis Section. Traffic Monitoring Stations Graphs. Online.

Montana Department of Transportation, Traffic Data Collection Section. Montana's Automatic Traffic Counters, 2009. Online.

U.S. Department of Transportation, Federal Highway Administration, Office of Operations. Tonnage on Highways, Railroads, and Inland Waterways, 2007. Online.

U.S. Department of Transportation, Research and Innovative Technology Administration, Bureau of Transportation Statistics. National Transportation Atlas Databases, 2010. Online.

Wyoming Department of Transportation, Planning Program. Automatic Traffic Recorder Report, 2009. Online.

**Park Visitation, 46–47**
*Contributing experts: John Sacklin, atlas editors*

U.S. Department of the Interior, National Park Service Public Use Statistics Office. Online

U.S. Department of the Interior, National Park Service Social Science Program. *Yellowstone National Park Visitor Study*. Report No. 178. University of Idaho, Park Studies Unit, Visitor Services Project, 2006.

U.S. National Park Service, Yellowstone National Park and Grand Teton National Park. Unpublished data.

**Tetons Climbing History, 48–49**
*Contributing experts: Alexander M. Tait, atlas editors*

Easterbrook R. Vegetation Mapping Program: Vegetation Coverage for Grand Teton National Park and the John D. Rockefeller Jr. Memorial Parkway, Wyoming, 20070131. U.S. Department of the Interior, National Park Service, Grand Teton National Park, Science and Resource Management, GIS Office. Geospatial Data Set-1043079, 2007.

Ortenburger, Leigh N., and Reynold G. Jackson. *A Climber's Guide to the Teton Range*. Seattle, Washington: Mountaineers, 1996.

Turiano, Thomas. "Teton Range," in *Select Peaks of Greater Yellowstone: A Mountaineering History and Guide*: 395–430. Jackson, Wyoming: Indomitus, 2003.

**The Economy, 50–51**
*Contributing experts: Ray Rasker, John Sacklin*

Headwaters Economics, Economic Profile System Socioeconomic Profiles, 2007 version. Online.

U.S. Department of Commerce. Bureau of the Census. Census of the Population.

U.S. Department of Commerce. Bureau of Economic Analysis. Online.

# Sources *(continued)*

U.S. Department of Labor. Bureau of Labor Statistics, Quarterly Census of Employment and Wages.

**Labor and Employment, 52–53**
*Contributing experts: Anne Fifield*

U.S. Department of Commerce. Bureau of the Census. American Community Survey. Online.

U.S. Department of Labor. Bureau of Labor Statistics. Local Area Unemployment Statistics. Online.

U.S. Department of Commerce. Bureau of Economic Analysis. Online.

**Income, 54–55**
*Contributing experts: Anne Fifield, Bruce A. Blonigen*

U.S. Department of Commerce. Bureau of the Census. American Community Survey. Online.

U.S. Department of Commerce. Bureau of Economic Analysis. Local Area Personal Income. Online.

**Agriculture, 56–57**
*Contributing experts: Stuart Allan, Gene Martin, Rick Wallen*

U.S. Department of Agriculture. National Agricultural Statistics Service, Census of Agriculture.

**Market Access, 58–59**
*Contributing experts: Ray Rasker, John Sacklin*

Federal Aviation Administration. Passenger Boarding and All-Cargo Data for U.S. Airports, 2000–2009. Online.

Headwaters Economics, Economic Profile System Socioeconomic Profiles, 2007 version. Online.

Rasker, R., P.H. Gude, J.A. Gude, and J. van den Noort. 2009. "The Economic Importance of Air Travel in High-Amenity Rural Areas." *Journal of Rural Studies* 25 (2009): 343–53.

Rasker, R., B. Alexander, J. van den Noort, and R. Carter. *Prosperity in the 21st Century West: The Role of Protected Private Lands.* Sonoran Institute. 2004.

U.S. Department of Commerce. Bureau of the Census. Census of the Population.

U.S. Department of Labor. Bureau of Labor Statistics, Quarterly Census of Employment and Wages.

**Wildland Economies, 60–61**
*Contributing experts: Ray Rasker, John Sacklin*

Headwaters Economics, Economic Profile System Socioeconomic Profiles, 2007 version. Online.

National Parks Conservation Association. *Gateways to Yellowstone: Protecting the Wild Heart of our Region's Thriving Economy.* 2006. Online.

U.S. Department of Commerce. Bureau of the Census. Census of the Population.

U.S. Department of Commerce. Bureau of Economic Analysis. Online.

U.S. Department of the Interior. *The National Atlas of the United States of America.* Online.

U.S. Department of Labor. Bureau of Labor Statistics, Quarterly Census of Employment and Wages.

**Protected Areas, 62–63**
*Contributing experts: Story Clark, John Varley*

Interagency Wild and Scenic Rivers Coordinating Council.

US Geological Survey, National Biological Information Infrastructure, Gap Analysis Program. May 2010. Protected Areas Database of the United States (PAD-US) Version 1.1.

**Land Ownership, 64–65**
*Contributing experts: John Varley, Tom Olliff*

U.S. Geological Survey, National Biological Information Infrastructure, Gap Analysis Program. May 2010. Protected Areas Database of the United States (PAD-US) Version 1.1.

U.S. Department of the Interior. *The National Atlas of the United States of America.* Online.

**Population, 66–67**
*Contributing experts: atlas editors, Susan W. Hardwick*

U.S. Department of Commerce. Bureau of the Census. Census of the Population, Decennially 1950–2010.

**County Population, 68–69**
*Contributing experts: William K. Wyckoff, atlas editors*

U.S. Department of Commerce. Bureau of the Census. Census of the Population, Decennially 1870–2010. Online.

The Newberry Library. *Atlas of Historical County Boundaries.* Online.

**City Population, 70–71**
*Contributing experts: William K. Wyckoff, atlas editors*

U.S. Department of Commerce. Bureau of the Census. Census of the Population, Decennially 1880–2010. Online.

**Education, 72–73**
*Contributing experts: Susan W. Hardwick, Donald G. Holtgrieve*

U.S. Bureau of the Census. Educational Attainment by State, 1990–2008. Washington, DC: Statistical Abstract of the United States, 2011.

U.S. Bureau of the Census. "Education," in *Census Atlas of the United States: Census 2000 Special Reports*: 158–75. Washington, D.C.: U.S. Department of Commerce, Economics and Statistics Division.

U.S. Department of Commerce. Bureau of the Census. Census of the Population, 2000 and 2010. Online.

U.S. Department of Education. National Center for Education Statistics. Online.

Idaho High School Activities Association. Online.

Montana High School Association. Online.

Wyoming High School Activities Association. Online.

**Race and Ethnicity, 74–75**
*Contributing experts: Susan W. Hardwick, Donald G. Holtgrieve*

Gibson, C. and K. Jung. *Historical Census Statistics on Population Totals by Race, 1790 to 1990, and by Hispanic Origin, 1970 to 1990, for the United States, Regions, Divisions, and States.* U.S. Bureau of the Census, Working Paper Series No. 56. Online.

U.S. Department of Commerce. Bureau of the Census. Census of the Population, 2000 and 2010. Online.

U.S. Department of Commerce. Bureau of the Census. American Community Survey Five-Year Estimates, 2005–9. Online.

U.S. Department of Commerce. Bureau of the Census. 2010 Census Redistricting Data Summary File. Online.

U.S. Department of Education, Institute of Education Sciences, National Center for Education Statistics. Online.

**Religion and Politics, 76–77**
*Contributing experts: Stuart Allan, Alexander B. Murphy*

Jones, Dale E., Sherri Doty, James E. Horsch, Richard Houseal, Mac Lynn, John P. Marcum, Kenneth

M. Sanchagrin, and Richard H. Taylor. *Religious Congregations and Membership in the United States 2000: An Enumeration by Region, State, and County based on Data Reported by 149 Religious Bodies.* Nashville, Tennessee: Glenmary Research Center, 2002.

Glenmary Research Center. Atlanta, Georgia.

*Dave Leip's Atlas of U.S. Presidential Elections.* Online.

**Elevation, 80–81**
*Contributing experts: atlas editors*

U.S. Geological Survey. National Elevation Data Set. Online.

**Cross Sections, 82–83**
*Contributing experts: atlas editors, Cheryl Jaworowski*

U.S. Geological Survey. National Elevation Data Set. Online.

U.S. Geological Survey. U.S. Board on Geographic Names. Geographic Names Information System. Online.

**Landforms, 84–93**
*Contributing experts: Kenneth L. Pierce, David R. Lageson, Grant A. Meyer, Cheryl Jaworowski, Greg Pederson, Lisa A. Morgan*

ESRI StreetMap North America.

U.S. Geological Survey. U.S. Board on Geographic Names. Geographic Names Information System. Online.

U.S. Geological Survey. National Elevation Data Set. Online.

**Grand Teton Geology, 94–95**
*Contributing experts: David R. Lageson*

Love, J.D., J.C. Reed, and A.C. Christiansen. *Geologic Map of Grand Teton National Park, Teton County, Wyoming.* U.S. Geological Survey Miscellaneous Investigations Series, Map I-2031, scale 1:62,500 (1992).

Love, J.D., A.C. Christiansen, and A.J. Ver Ploeg. *Stratigraphic Chart Showing Phanerozoic Nomenclature for the State of Wyoming.* The Geological Survey of Wyoming, Map Series 41 (1993).

**Yellowstone Geology, 96–97**
*Contributing experts: Kathryn E. Watts, David R. Lageson*

Christiansen, Robert L. *The Quaternary and Pliocene Yellowstone Plateau Volcanic Field of Wyoming, Idaho, and Montana.* U.S. Geological Survey Professional Paper 729-G (2001).

**Yellowstone Volcano, 98–99**
*Contributing experts: Kathryn Watts, David Lageson, Cheryl Jaworowski*

Smith, R.B., and L.J. Siegel. *Windows into the Earth.* New York, New York: Oxford University Press, 2000.

Morgan, L.A., K.L. Pierce, and W.C.P. Shanks. "Track of the Yellowstone Hotspot: Young and Ongoing Geologic Processes from the Snake River Plain to the Yellowstone Plateau and Tetons," in *Roaming the Rocky Mountains and Environs: Geological Field Trips*, ed. R.G. Raynolds. Geological Society of America (2008).

U.S. Geological Survey. *Steam Explosions, Earthquakes, and Volcanic Eruptions–What's in Yellowstone's Future?* Fact sheet 2005-3024 (2005). Online.

Pierce, K.L., D.G. Despain, L.A. Morgan, and J.M. Good. "The Yellowstone Hotspot, Greater Yellowstone Ecosystem, and Human Geography," in *Integrated Geoscience Studies in the Greater Yellowstone Area—Volcanic, Tectonic, and Hydrothermal Processes in the Yellowstone Geoecosystem*, ed. Lisa A. Morgan. U.S. Geological Survey Professional Paper 1717, Chapter A: 1–39 (2007).

Pierce, K.L., and L.A. Morgan. "The Track of the Yellowstone Hotspot: Volcanism, Faulting, and Uplift," in *Regional Geology of Eastern Idaho and Western Wyoming*, ed. P.K. Link, M.A. Kuntz, and L.W. Platt. Geological Society of America Memoir 179 (1992).

Christiansen, R.L., J.B. Lowenstern, R.B. Smith, Henry Heasler, L.A. Morgan, Manuel Nathenson, L.G. Mastin, L.J.P. Muffler, and J.E. Robinson. *Preliminary Assessment of Volcanic and Hydrothermal Hazards in Yellowstone National Park and Vicinity*. U.S. Geological Survey Open-file Report 2007-1071 (2007).

## Geothermal Activity, 100–101
*Contributing experts: Stephen G. Custer, Henry P. Heasler, Carrie Guiles*

Friedmna, I., and D.R. Norton. "Is Yellowstone Losing Its Steam? Chloride Flux Out of Yellowstone National Park," in *Integrated Geoscience Studies in the Greater Yellowstone Area—Volcanic, Tectonic, and Hydrothermal Processes in the Yellowstone Geoecosystem*, ed. Lisa A. Morgan. U.S. Geological Survey Professional Paper 1717, Chapter I: 271–97 (2007).

U.S. National Park Service, Yellowstone National Park. Unpublished data.

## Hydrothermal Areas, 102–103
*Contributing experts: Cheryl Jaworowski, Henry P. Heasler, atlas editors*

Jaworowski, Cheryl, Henry P. Heasler, Christopher M.U. Neale, and Saravanan Sivarajan. "Using Thermal Infrared Imagery and Lidar in Yellowstone Geyser Basins." *Yellowstone Science* 18, No. 1 (2010): 8–19.

## Geysers, 104–105
*Contributing experts: Stephen G. Custer, Alethea Y. Steingisser*

Bryan, T.S.. *The Geysers of Yellowstone* (Fourth Edition). Boulder, Colorado: University Press of Colorado, 2008.

Geyser Observation and Study Assocation. Recent geyser activity data. Online.

Steingisser, A.Y., and W.A. Marcus. "Human Impacts on Geyser Basins." *Yellowstone Science* 17, No. 1 (2009): 7–17.

U.S. National Park Service, Yellowstone National Park. Unpublished data.

## Earthquakes, 106–107
*Contributing experts: David R. Lageson, Kathryn Watts, Henry P. Heasler*

U.S. Geological Survey. Earthquakes Hazards Program Data. Online.

University of Utah. Regional and Urban Seismic Network Station Map. Online.

## Glaciation, 108–109
*Contributing experts: Kenneth L. Pierce, Lisa A. Morgan, Greg Pederson*

Good, J.M. and K.L. Pierce. *Interpreting the Landscape of Grand Teton and Yellowstone National Parks*. Wyoming: Grand Teton Natural History Association, 1996.

Pierce, K.L., D.G. Despain, L.A. Morgan, and J.M. Good. "The Yellowstone Hotspot, Greater Yellowstone Ecosystem, and Human Geography," in *Integrated Geoscience Studies in the Greater Yellowstone Area—Volcanic, Tectonic, and Hydrothermal Processes in the Yellowstone Geoecosystem*, ed. Lisa A. Morgan. U.S. Geological Survey Professional Paper 1717, Chapter A: 1–39 (2007).

Paleogeography and Laurentide Ice Sheet Data (courtesy of Patrick J. Bartlein). University of Oregon Department of Geography.

## Yellowstone Lake, 110–111
*Contributing experts: Lisa A. Morgan, Kenneth L. Pierce, Patrick Shanks, Pat Bigelow*

Morgan, Lisa A., Wayne C. Shanks III, Kenneth L. Pierce, David A. Lovalvo, Gregory K. Lee, Michael W.

Webring, William J. Stephenson, Samuel Y. Johnson, Stephen S. Harlan, Boris Schulze, and Carol A. Finn. "The Floor of Yellowstone Lake is Anything but Quiet—New Discoveries from High-Resolution Sonar Imaging, Seismic-Reflection Profiling, and Submersible Studies," in *Integrated Geoscience Studies in the Greater Yellowstone Area—Volcanic, Tectonic, and Hydrothermal Processes in the Yellowstone Geoecosystem*, ed. Lisa A. Morgan. U.S. Geological Survey Professional Paper 1717, Chapter D: 91–126 (2007).

## Drainage Basins, 112–113
*Contributing experts: Paul A. Caffrey, Richard A. Marston, W. Andrew Marcus*

Natural Resources Conservation Service. Watershed Boundary Data Set. Online.

U.S. Geological Survey. National Hydrography Data Set. Online.

## Rivers, 114–115
*Contributing experts: Mark Story, Peter Bengeyfield, Erik R. Strandhagen*

Strandhagen, Erik R. "Views of the Rivers: Representing Streamflow of the Greater Yellowstone Ecosystem." Master's degree thesis, University of Oregon, 2005.

U.S. Geological Survey. National Hydrography Data Set. Online.

## Streamflow, 116–117
*Contributing experts: Mark Story, Peter Bengeyfield, Erik R. Strandhagen*

Strandhagen, Erik R. "Views of the Rivers: Representing Streamflow of the Greater Yellowstone Ecosystem." Master's degree thesis, University of Oregon, 2005.

U.S. Geological Survey, National Atlas. Realtime USGS Streamflow Stations, 2005. Online.

U.S. Geological Survey. National Water Information System. Annual Water Data Reports; Real-Time Water Data for the Nation, 2010. Online.

U.S. Geological Survey. Real-Time Water Data for the Nation. Online.

U.S. Geological Survey. National Hydrography Data Set. Online.

## Flow Regimes, 118–119
*Contributing experts: Erik R. Strandhagen, Richard A. Marston, W. Andrew Marcus*

Strandhagen, Erik R. "Views of the Rivers: Representing Streamflow of the Greater Yellowstone Ecosystem." Master's degree thesis, University of Oregon, 2005.

## Waterfalls, 120–121
*Contributing experts: Lee Whittlesey, Paul Rubinstein, Mike Stevens, Mark A. Fonstad*

Rubinstein, Paul, Lee H. Whittlesey, and Mike Stevens. *The Guide to Yellowstone Waterfalls and their Discovery*. Englewood, Colorado: Westcliffe Publishers, 2000.

U.S. National Park Service, Yellowstone National Park. Unpublished data.

## Precipitation, 122–123
*Contributing experts: Jacqueline J. Shinker, Patrick J. Bartlein*

National Oceanic and Atmospheric Administration, National Environmental Satellite, Data, and Information Service, National Climatic Data Center. Climatography of the United States No. 20, Monthly Station Climate Summaries, 1971–2000 (February 2004).

PRISM (Parameter-elevation Regressions on Independent Slopes Model) Climate Group, Oregon State University.

## Temperature, 124–125
*Contributing experts: Jacqueline J. Shinker, Patrick J. Bartlein*

National Oceanic and Atmospheric Administration, National Environmental Satellite, Data, and Information Service, National Climatic Data Center. Climatography of the United States No. 20, Monthly Station Climate Summaries, 1971–2000 (February 2004).

PRISM Climate Group, Oregon State University.

## Climate Change, 126–127
*Contributing experts: Patrick J. Bartlein, Sarah L. Schafer, Jacqueline J. Shinker*

Daly, C., M. Halbleib, J.I. Smith, W.P. Gibson, M.K. Doggett, G.H. Taylor, J. Curtis, and P.A. Pasteris. "Physiographically Sensitive Mapping of Temperature and Precipitation across the Conterminous United States." *International Journal of Climatology* 28 (2008): 2031–64.

Mitchell, T.D., and P.D. Jones. "An Improved Method of Constructing a Database of Monthly Climate Observations and Associated High-Resolution Grids." *International Journal of Climatology* 25 (2005): 693–712.

New, M., D. Lister, M. Hulme, and I. Makin. "A High-Resolution Data Set of Surface Climate over Global Land Areas." *Climate Research* 21 (2002): 1–25.

PRISM 30 arc second data, acquired 2009–10. PRISM Climate Group, Oregon State University.

Collins, W.D., C.M. Bitz, M.L. Blackmon, G.B. Bonan, C.S. Bretherton, J.A. Carton, P. Chang, S.C. Doney, J.J. Hack, T.B. Henderson, J.T. Kiehl, W.G. Large, D.S. McKenna, B.D. Santer, and R.D. Smith. "The Community Climate System Model Version 3 (CCSM3)." *Journal of Climate* 19 (2006): 2122–43.

*We acknowledge the modeling groups, the Program for Climate Model Diagnosis and Intercomparison (PCMDI) and the WCRP's Working Group on Coupled Modelling (WGCM), for their roles in making available the WCRP CMIP3 multimodel data set. Support of this data set is provided by the Office of Science, U.S. Department of Energy.*

## Wetlands, 128–129
*Contributing experts: Mary Hektner, Jennifer Whipple, Miles C. Barger*

U.S. Fish and Wildlife Service. National Wetlands Inventory. Online.

Elliott, C.R., and M.M. Hektner. *Wetland Resources of Yellowstone National Park*. Yellowstone National Park, 2000.

## Soils, 130–131
*Contributing experts: Ann Rodman*

Rodman, A., H.F. Shovic, and D. Thoma. *Soils of Yellowstone National Park*. Yellowstone National Park Center for Resources (YCR-NRSR-96-2), 1996.

U.S. National Park Service, Yellowstone National Park. Unpublished data.

## Ecoregions, 132–133
*Contributing experts: James M. Omernik, Daniel G. Gavin, Mary Hektner*

U.S. Environmental Protection Agency. Level III and IV Ecoregions of the Continental United States. Online.

## Vegetation, 134–139
*Contributing experts: Daniel G. Gavin, Mary Hektner*

U.S. Geological Survey. Gap Analysis Program. Online

## Landscape Change, 140–141
*Contributing experts: Ramesh Sivanpillai, Mark A. Fonstad*

U.S. Geological Survey, Earth Resources Observation and Science Center.

U.S. Department of Agriculture, Aerial Photography Field Office, Farm Service Agency.

U.S. Geological Survey, 2002 Digital Orthophoto Quadrangle Imagery.

# Sources *(continued)*

**Fire History, 142–143**
*Contributing experts: Don G. Despain, Cathy Whitlock, Roy Renkin, Mitchell J. Power*

Cook, E.R., C. Woodhouse, M.C. Eakin, D.M. Meko, and D.W. Stahle. "Long-Term Aridity Changes in the Western United States." *Science* 306 (5698): 1015–18.

National Oceanic and Atmospheric Administration. Climate Services and Monitoring Division.

U.S. National Park Service, Yellowstone National Park. Unpublished data.

**1988 Fires, 144–145**
*Contributing experts: Roy Renkin, Don G. Despain*

Despain, Don, Ann Rodman, Paul Schullery, and Henry Schovic. *Burned Area Survey of Yellowstone National Park: The Fires of 1988.* Division of Research and the GIS Laboratory, Yellowstone National Park, 1989.

Franke, M.A. *Yellowstone in the Afterglow.* Yellowstone National Park Center for Resources, 2000.

U.S. National Park Service, Yellowstone National Park. Unpublished data.

**Grizzly Bears, 148–149**
*Contributing experts: Kerry A. Gunther, Chuck Schwartz*

Gunther, K.A. *Food Habits of Grizzly Bears and Black Bears in the Yellowstone Ecosystem.* Information Paper No. BMO-3. Bear Management Office, Yellowstone National Park, March 2003.

Interagency Grizzly Bear Study Team. Annual Reports, 2003–9. Online.

LaRoe, E.T., G.S. Farris, C.E. Puckett, P.D. Doran, and M.J. Mac, eds. *Our Living Resources: A Report to the Nation on the Distribution, Abundance, and Health of U.S. Plants, Animals, and Ecosystems.* U.S. Department of the Interior, National Biological Service, Washington, D.C., 1995.

Schwartz, C.C., S.D. Miller, and M.A. Haroldson. *Wild Mammals of North America: Biology, Management, and Conservation* (Second Edition), ed. G.A. Feldhamer, B.C. Thompson, and J.A. Chapman. Baltimore, Maryland: Johns Hopkins University Press, 2003.

U.S. National Park Service, Yellowstone National Park. Unpublished data.

**Wolves, 150–151**
*Contributing experts: Douglas W. Smith*

Bangs, Ed. *The Reintroduction of Gray Wolves to Yellowstone National Park and Central Idaho: Final Environmental Impact Statement.* Helena, Montana: U.S. Fish and Wildlife Service, 1994.

Demarais, Stephen, and Paul R. Krausman. *Ecology and Management of Large Mammals in North America.* Upper Saddle River, New Jersey: Prentice Hall, 2000.

Phillips, M.K., and D.W. Smith. *The Wolves of Yellowstone.* Stillwater, Minnesota: Voyageur Press, 1996.

Yellowstone Wolf Project, Annual Reports, 2006–9. Yellowstone National Park, U.S. National Park Service. Online.

U.S. National Park Service, Yellowstone National Park. Unpublished data.

**Coyotes, 152–153**
*Contributing experts: J.W. Sheldon, Robert L. Crabtree, Douglas W. Smith*

Grady, Wayne.. *The World of the Coyote.* San Francisco, California: Sierra Club Books, 1994.

Yellowstone Ecological Research Center. Published and unpublished data.

**Bison, 154–155**
*Contributing experts: Rick Wallen, Chris Geremia, Mary Meagher*

Boyd, Delaney P. "Conservation of North American Bison: Status and Recommendations." Master's degree project, University of Calgary Department of Environmental Design, 2003.

Coder, George D. "The National Movement to Preserve the American Buffalo in the United States and Canada between 1880 and 1920." Doctoral dissertation, Ohio State University, 1975.

Geist, Valerius. *Buffalo Nation: History and Legend of the North American Bison.* Stillwater, Minnesota: Voyageur Press, 1996.

Taper, M.L., M. Meagher, and C.L. Jerde. *The Phenology of Space: Spatial Aspects of Bison Density Dependence in Yellowstone National Park.* Montana State University Department of Ecology, 2000.

U.S. National Park Service, Yellowstone National Park. Unpublished data.

**Bison Movement, 156–157**
*Contributing experts: Rick Wallen*

U.S. National Park Service, Yellowstone National Park. Unpublished data.

**Elk, 158–159**
*Contributing experts: P.J. White*

Barber, S.M., D.L. Mech, and P.J. White. "Yellowstone Elk Calf Mortality Following Wolf Restoration: Bears Remain Top Summer Predators." *Yellowstone Science* 13, No. 3 (2005): 37–44.

Eberhardt, L.L., P.J. White, R.A. Garrott, and D.B. Houston. "A Seventy-Year History of Trends in Yellowstone's Northern Elk Herd." *Journal of Wildlife Management* 71 (2007): 594–602.

Homer, C.C. Huang, L. Yang, B. Wylie, and M. Coan. "Development of a 2001 National Landcover Database for the United States." *Photogrammetric Engineering and Remote Sensing* 70, No. 7 (July 2004): 829–40.

U.S. National Park Service, Yellowstone National Park. Unpublished data. *Yellowstone's Northern Range: Complexity and Change in a Wildland Ecosystem.* U.S. National Park Service, Mammoth Hot Springs, 1997.

**Fish, 160–161**
*Contributing experts: Todd M. Koel, Lynn Keating*

Koel, T.M., J.L. Arnold, P.E. Bigelow, P.D. Doepke, B.D. Ertel, and D.L. Mahony. *Yellowstone Fisheries and Aquatic Sciences: Annual Report, 2004.* U.S. National Park Service, Yellowstone National Park Center for Resources (YCR-2005-04), 2005.

May, B. E., W. Urie, B.B. Shepard, S. Yundt, C. Corsi, K. McDonald, B. Snyder, S. Yekel, and K. Walker. *Range-Wide Status of Yellowstone Cutthroat Trout (Oncorhynchus clarki bouvieri): 2001.* U.S. Forest Service, 2003.

U.S. National Park Service, Yellowstone National Park. Published and unpublished data.

**Potential Wildlife Habitats, 162–163**
*Contributing experts: Gary P. Beauvais, Mark D. Andersen*

Wyoming Natural Diversity Database, University of Wyoming.

**Sagebrush-Steppe Habitat, 164–165**
*Contributing experts: Charles R. Preston, Jeffrey K. Gillan*

Comer, Patrick, Jimmy Kagan, Michael Heiner, and Claudine Tobalske. *Current Distribution of Sagebrush and Associated Vegetation in the Western United States (Excluding New Mexico and Arizon).* Interagency Sagebrush Working Group, 2002. Online.

Schroeder, M. A., C.L. Aldridge, A.D. Apa, J.R. Bohne, C.E. Braun, S.D. Bunell, J.W. Connelly, P.A. Deibert, S.C. Gardner, M.A. Hilliard, G.D. Kobriger, S.M. McAdam, C.W. McCarthy, J.J. McCarthy, D.L. Mitchell, E.V. Rickerson, and S.J. Stiver. "Distribution of Sage-Grouse in North America." *The Condor* 106 (2004): 363–76.

Wolf, S. "2009 Wildlife Monitoring: Greater Sage-Grouse Lek Counts," in *2009 Wildlife Conservation, Management, and Research,* ed. S.L. Cain: 30–31. Moose, Wyoming: U.S. National Park Service, 2009.

**Thermophiles, 166–167**
*Contributing experts: David M. Ward, William P. Inskeep, Timothy R. McDermott, Cristina Takacs-Vesbach*

Grogan, D.W. "Yellowstone's Thermophiles: Microbial Diversity from Harsh Environments." *Yellowstone Science* 18, No. 3: 23-32.

Pace, N.R. "A Molecular View of Microbial Diversity and the Biosphere." *Science* 276 (5313) (1997): 734–40.

Photographs courtesy of David M.Ward. Montana State University Land Resources and Environmental Sciences Department.

U.S. National Park Service, Yellowstone National Park. Unpublished data.

Yellowstone National Park. *Yellowstone Resources and Issues.* Mammoth, Wyoming: Division of Interpretation, 2011.

**Dinosaurs, 168–169**
*Contributing experts: Kelli Trujillo, Molly Ward, James G. Schmidt*

Paleogeography imagery courtesy of Ron Blakey, Northern Arizona University Department of Geology.

Sloan, Chris. *How Dinosaurs Took Flight.* Washington D.C.: National Geographic Society, 2005.

U.S. Geological Survey, Preliminary integrated geologic map databases for the United States, Open File Report (2005-1305). Online.

**Vertebrate Species, 170–174**
*Contributing experts: Ann Rodman, National Park Service*

U.S. National Park Service. Natural Resource Information Portal. Online.

**Reference Maps, 178–223**
*Contributing experts: Stuart Allan, atlas editors*

ESRI StreetMap North America.

U.S. Geological Survey. National Elevation Data Set.

U.S. Department of Agriculture, Forest Service Topographic Map Data. Online.

U.S. Department of the Interior. National Park Service.

U.S. Department of Transportation, Federal Aviation Administration.

U.S. Department of Transportation, National Transportation Atlas Database, 2008.

U.S. Geological Survey. National Hydrography Data Set.

U.S. Geological Survey. U.S. Board on Geographic Names. Geographic Names Information System.

**Gazetteer, 224–231**
*Contributing experts: Stuart Allan, atlas editors*

Allan Cartography

**U.S. Geological Survey Map Index, 232–235**
*Experts: atlas editors*

U.S. Geological Survey. Online.

**Counties, 236–237**
*Contributing experts: atlas editors*

U.S. Department of Commerce. Bureau of the Census. Census of the Population, 2010. Online.

**Place Names, 238–249**
*Contributing experts: Lee Whittlesey*

Boone, Lalia. *Idaho Place Names: A Geographical Dictionary.* Moscow, Idaho: University of Idaho Press, 1988.

Cheney, Roberta Carkeek. *Names on the Face of Montana.* Missoula, Montana: Mountain Press Publishing Company, 1984.

Urbanek, Mae Bobb. *Wyoming Place Names.* Missoula, Montana: Mountain Press Publishing Company, 1998.

Whittlesey, Lee H. *Yellowstone Place Names.* Wonderland Publishing, 2006.

# Index

Cities and place names are not listed in this index; their locations on reference maps are listed in the gazetteer (pages 224–231). The origins of place names are provided in the place names pages (pages 238–249).

## A

age distribution (demographic data), 67
agriculture, 56–57
    bison, relationship with regional, 156
    economics of, 51
    ecoregions, effect on, 132–133
    land cover of, 6
air travel
    economic growth, effect on, 50
    flights, numbers of, 45
    market access, 58–59
Allen and Day survey of geothermal areas, 35
alpinism, 48–49
amenities, environmental, 58, 60–61
American bison. See bison
American Indian reservations
    boundaries, contemporary, 64–65
    establishment and history of, 17
    populations of, 74
American Indians, 16–17
    Catlin, George, 20–21
    early explorers and, 24
    obsidian, use and trade of, 14–15
    Sheep Eaters, 18–19
    trails used by, 28
amphibians, list of, 170
ancestry (demographic data), 75
Anderson, Jack, 36
archaeology, 14–15
    See also American Indians
Army Corps of Engineers, U.S., 39, 40
art, 30–33
arrowheads
    archaeology, 14, 15
    Sheep Eaters, 19
artists
    Bierstadt, Albert, 32
    Catlin, George, 20–21
    Coe, Anne, 33
    Fell, Olive, 33
    Jackson, William Henry, 23
    Joaquin, David, 18
    Leigh, William, 33
    Moran, Thomas, 23, 30–31
    Pruessl, Carl, 32
ashfall, Yellowstone volcano, 99
Astorians, The, 22
automobiles
    early roads, role in development of, 39
    infrastructure and visitation, effect on, 40–41
    traffic, flow, and volume of, 44–45

## B

backcountry visitation, 47
    trails, location and elevation profiles of, 42–43
Bannock Trail, 28
barley, 56, 57
Barlow, John, 23
Barlow-Heap Expedition, 23
bathymetry of Yellowstone Lake, 111
bears, grizzly, 148–149
    and elk, 158, 159
Bechler, Gustavus, 29
beetle kill, satellite imagery of, 141
Bierstadt, Albert, 32
bighorn sheep
    American Indians and historic sightings of, 18–19
    in Mummy Cave, 15
birds
    list of, 171–175
    Yellowstone Lake, as habitat for, 111
bison, 154–157
    American Indians and, 19
    early humans and, 14–15
    grizzly bears and, 149
    wolves and, 150
BLM. See Bureau of Land Management
boundaries, political. See political boundaries

Bridger, Jim, 24, 26, 27
brucellosis, 156
buffalo. See bison
Buffalo Bull's Back Fat (Stu-mick-o-súcks), painting of, 21
*Buffalo Chase, Bull Protecting a Cow and Calf* (painting), 21
buildings, development of, 40–41
Bureau of Land Management (BLM), land managed by, 64–65
Bureau of Public Roads, 39
Bureau of Reclamation land, 64–65
burns. See fire

## C

caldera, Yellowstone. See Yellowstone caldera
campground users, number of, 47
carbon dioxide and climate change, 126–127
cars. See automobiles
Carver, Jonathan, 114
Catlin, George, 20–21
cattle, 56–57
charcoal, use of for reconstructing fire history, 143
cheatgrass, 164
chemistry of geothermal areas, 100–101, 166–167
Chittenden, Hiram, 39
Chittenden map of 1895, 38
chloride flux of rivers, 101
Clark (Drouillard) map of 1808, 24
Clark and King's map of 1805, 24
Clark, William, 22, 24–25
    See also Lewis and Clark Expedition
Clark's 1814 map, 19
Clark's manuscript map of 1806–1811, 25
climate
    ecoregions and, 133
    fire and, 143
    glaciation and, 108–109
    of last glacial period in Yellowstone, 109
    precipitation, as function of, 122–123
    stream flow and, 118
    temperature and, 124–127
climate change, 126–127
climbing, 48–49
Cody Complex peoples, 15
Coe, Anne, 33
collars, radio, 36
colleges, 72–73
Colter, John, 22, 25
Colters Hell, 24, 26
concessioners, role of in development, 40–41
conservation (environmental)
    of bison, 154–155
    of grizzly bears, 148
    privately owned land and, 62–63
    science, use of to manage, 36
conservation areas. See protected areas
construction
    economics of, 51
    of park infrastructure, 40–41
Continental Divide and drainage basins, 112–113
Continental Divide National Scenic Trail, 42
continents, movement over time, 169
Cook, Charles, 23
Cook-Folsom-Peterson Expedition, 23
counties
    boundaries, development of, 11
    boundaries, modern, 236–237
coyotes, 152–153
crops, 56–57
cross-sections. See elevation profiles
cutthroat trout, Yellowstone, 160–161
    in Yellowstone Lake, 111

## D

dairy, 56–57
dams and flow regimes, 116–119
DeLacy map of 1865, 27
DeLacy party, 23, 27
DeLacy, Walter, 23, 27, 28
DeSmet, Father Pierre-Jean, 26
DeSmet's map of 1851, 26
demographic data
    age distribution, 67
    ancestry, 75
    education, 72–73
    employment, 52–53
    ethnicity, 74–75

immigration, 74
income, 50–51, 54–55
labor, 52–53
market access, 58–59
minorities, 74–75
politics, 77
population
    cities, 70–71
    counties, 68–69
    density and pyramids, 66–67
poverty, 55
race, 74–75
religion, 76–77
sex, 67
unemployment, 52–53, 61
development
    flow regimes, effect of, 118–119
    near Old Faithful, 40–41
    rivers and, 114
    of roads, 38–39
    and sagebrush-steppe, 164
    satellite imagery of, 141
diet
    of coyotes, 153
    of grizzly bears, 149
dinosaurs, 168–169
discharge (streams), 116–117
Doane map of 1870, 28
Doane, Gustavus C., 28, 34
dogs, importance to Sheep Eaters of, 18–19
domes, volcanic, 96–97
drainage basins, 112–113
Drouillard, George, 24–25
Durrance, Jack, 48, 49

## E

Eagle's Ribs (Peh-tó-pe-kiss), painting of, 21
early maps, 24–29
earthquakes, 106–107
    and volcanism, 98
economics
    of agriculture, 56–57
    employment and labor, 52–53
    freight tonnage, 45
    income, 54–55
    overview of, 50–51
    market access and, 58–59
    of wildland complexes, 60–61
ecoregions, 132–133
education, 72–73
    race and ethnicity and, 75
    and wildland economies, 61
elections, 77
elevation, 80–81
    ecoregions, role in defining, 133
    vegetation, effect of, 135
    See also elevation profiles
elevation profiles
    of last Yellowstone glaciation, 109
    of the Greater Yellowstone Area, 82–83
    of roads and trails, 43
    of Yellowstone Lake, 111
elk, 158–159
    radio collaring of, 36
    survey of, historic, 35
Ellingwood, Albert, 48, 49
Elliott, Henry Wood, 30, 111
employment, 52–53
    and wildland economies, 60–61
Endangered Species Act
    and greater sage-grouse, 165
    and grizzly bears, 148
    and wolves, 150, 151
energy development
    and flow regimes, 118–119
    geothermal, 105
    satellite imagery of, 141
environmental amentities, 58, 60–61
environmental monitoring
    history of, 35
    park management, role in, 36
eruptions
    of geysers, 105
    of volcanoes, 98–99
establishment of national parks, 8–9
ethnicity (demographic data), 74–75
expeditions

Allen and Day survey of geothermal areas, 35
Astorians, The, 22
Barlow-Heap, 23
DeLacy party, 23, 27
Lewis and Clark, 22, 24–25
Hague survey of geology, 34
Hayden survey. *See main entry*
Huston-Wyant, 23
Raynolds, 23, 24
Washburn, 23, 28
exploration, 22–23

**F**

farming. *See* agriculture
Father DeSmet's map of 1851, 26
faults (geology), 107
    in Grand Teton National Park, 94–95
    of Yellowstone Lake, 111
    in Yellowstone National Park, 96–97
Federal Highway Administration, 39
Federal Land Act of 1796, 65
federal research areas, 62–63
Fell, Olive, 33
Ferris, Warren, 26
Ferris's manuscript map of 1836, 26
fire
    history of, 142–143
    1988 fires, 142–145
    satellite imagery of, 141
fish, 160–161
    species list of, 171
    and Yellowstone Lake, 111
floods, 116, 118–119
flow regimes, 118–119
Folsom, David, 23
freight tonnage, 45
Fryxell, Fritiof, 48, 49

**G**

gauging stations (streamflow), 116, 118
gazetteer, 224–231
General Land Office map, Henry
    Washburn (1869), 28
geology
    Grand Teton National Park, 94–95
    Hague survey of, 34
    plate tectonics, 169
    soils and, 130
    Yellowstone National Park, 96–97
    Yellowstone Lake, 111
geothermal areas and features, 100–105
    American Indians and, 15
    art featuring, 30–33
    bison use of, 156
    chemistry of, 100–101, 166–167
    early accounts of, 24–25, 26–27, 28
    energy development in, 105
    infrastructure near, development of, 40–41
    locations of, 100, 104–105
    Sheep Eater religion, role in, 19
    surveys of, early, 34–35
    temperatures of, 100–103
    thermal infrared imagery of, 102–103
    thermophiles in, 166–167
    wetlands and, 128–129
    of Yellowstone Lake, 111
    *See also* geysers
Geothermal Resource Areas, 105
geysers, 100–101, 104–105
    eruption intervals of, 105
    list of named, 104
    paintings of, 32–33
    survey of, Hayden, 34
    thermal infrared imagery of, 102–103
*Geysers in Yellowstone* (painting), 32
glaciers, 108–109
    Wind River Range, erosion of, 93
    Yellowstone Lake, role in forming, 111
    *See also* last glacial period
global warming. *See* climate change
gold prospecting, 23
grains, 56–57
*Grand Canyon of the Yellowstone* (painting), 30–31
Grand Loop Road
    development of, 39
    elevation profiles and overview of, 42–43

Grand Teton National Park
    climbing in, 48–49
    establishment and boundaries of, 11
    geology of, 94–95
    glaciers in, 108
    landforms of, 87
    place names in, 248–249
    sagegrouse and, 165
    *See also* main entries (e.g., elevation)
gray wolves. *See* wolves
*Great Blue Spring of the Lower Geyser Basin,*
    *Firehole River, Yellowstone* (painting), 31
greater sage-grouse, 165
greenhouse effect, 126
Grinnell, George Bird, 34
grizzly bears, 148–149
    and elk, 158–159

**H**

habitats, 162–163
    grizzly bears, seasonal preferences for, 149
    sagebrush-steppe, 164–165
    wetlands, 128–129
Hague survey of geology, 34
hay, 56–57
Hayden Survey map by Bechler, 1872, 29
Hayden Survey, 29, 30–31, 34, 48
    and Yellowstone Lake, 111
Hayden, Ferdinand, 23, 27, 30–31, 34
Haynes, Frank J., 39
Heap, David, 23
Henderson, A. Bart, 23
Heney, Hughes, 24
high schools, 72–73
    race and ethnicity in, 75
hiking trails. *See* trails
historical maps, 24–29
Hopewell culture, 14
hot springs. *See* geothermal areas and features
*Hot Springs on the Gardner River* (photograph), 30
hotspot, Yellowstone, 84, 98, 106
hunting
    by American Indians, 18–19
    of bighorn sheep, 18–19
    of bison, 154
    by early humans, 14–15
    of elk, 158–159
    of grizzly bears, 148
Huston, George, 23
Huston-Wyant Expedition, 23
hydric soils, 128–129, 131
hydrology
    drainage basins, 112–113
    flow regimes, 118–119
    rivers, 114–115
    waterfalls, 120–121
hydrothermal areas
    temperatures of, 102–103
    *See also* geothermal areas
hydrothermal vents, 96–97

**I**

Ice Age. *See* last glacial period
ice sheets (glaciation), 108–109
immigration, 74
income, 50–51, 54–55
    and wildland complexes, 60–61
Indian reservations. *See* American Indian reservations
Indians, American. *See* American Indians
industry, 52
infrared imagery
    of hydrothermal areas, 102–103
    landscape change, use of to observe, 141
infrastructure, development of, 40–41
Intermountain Seismic Belt, 106
internet access, economic effects of, 58
irrigation, 56

**J**

Jackson, William Henry, 23, 30–31
Joaquin, David, 18
John D. Rockefeller Jr. Memorial Parkway
    establishment and boundaries of, 11

**K**

King, Nicolas, 24
Kurz, Frederick, 18

**L**

labor, 52–53
land ownership, 64–65
land cover, 6
landforms
    Cascade Corner (Bechler region), 90
    Grand Canyon of the Yellowstone, 88
    Grand Teton National Park, 85
    headquarters, Grand Teton, 87
    headquarters, Yellowstone, 86
    northeast entrance, 90
    overthrust belt, 92
    Wind River Range, 93
    Yellowstone Lake, 89
    Yellowstone National Park, 84
LANDSAT imagery, 140–141
landscape change, 140–141
Langford, Nathaniel P., 39, 48
languages, 75
last glacial period, 108–109
    bison migration during, 154
    human habitation, effect on, 14
    Yellowstone Lake during, 111
    *See also* glaciers
Laurentide ice sheet, 109
lava flows, 99
Legend Rock State Petroglyph Site, 19
Leigh, William R., 33
leks (sage-grouse) in Grand Teton, 165
Leopold Report, 41
Letters and Notes on the Manners, Customs, and
    Condition of North American Indians (book), 21
Lewis and Clark Expedition, 22
    early maps from, 24–25
Lewis and Clark's map of 1814, 25
Lewis, Samuel, 25
logging, satellite imagery of, 141
*The Lower Falls of the Yellowstone* (painting), 33
Lower Geyser Basin
    geysers, named, 104
    temperature and chemistry of, 101
    thermal infrared imagery of, 102
Ludlow, William, 34

**M**

mammals, list of, 170
Mammoth Hot Springs
    geothermal features of, 100
    landforms surrounding, 86
management
    of bison, 154–155
    of elk, 158
    of fire, 142–145
    of fish, 160–161
    of grizzly bears, 148–149
    science, role of in, 36
    of wolves, 150–151
mantle plume, Yellowstone volcano, 98
manufacturing, 51
market access, 58–59
*Me and Old Faithful* (drawing), 33
microorganisms in geothermal areas, 35, 36, 166–167
Midway Geyser Basin
    geysers, named, 104
    temperature and chemistry of, 101
    thermal infrared imagery of, 102
migration
    of bison, 156–157
    of elk, 158
mining, 50, 51
    rivers and, 114
minorities, 74–75
Mission 66
    Old Faithful, development of during, 41
    road development during, 39
monitoring stations
    seismic, 107
    streamflow, 116–117
    traffic, automobile, 44–45
Moran, Thomas, 23, 30–31

# Index (continued)

mountaineering, 48–49
mudpots, 100
Mummy Cave (archaeological site), 15

## N

national forests, 62–65
    establishment and boundaries of, 11
    logging in, 141
    1988 fires and, 144–145
National Land Cover Database (NLCD), 6–7
national parks
    economics effects on region of, 60–61
    in the U.S. *See* National Park System
    global, 8
National Park Service
    boundaries of units, 64–65
    establishment and units of, 8–9
    infrastructure development, role in, 40–41
    Mission 66, 39, 41
National Parks Omnibus Management Act, 36
national wildlife refuges, 62–63
nature preserves. *See* protected areas
1988 fires, 142–145
    satellite imagery of, 141
Norris Geyser Basin
    chemistry and temperatures of, 101
    geysers in, 105
    thermophiles in, 166–167
Norris, Philetus (P. W.), 34

## O

obsidian, and American Indians, 14–15
Old Faithful (area)
    infrastructure, development of, 40–41
    thermal infrared imagery of, 103
Old Faithful (geyser)
    drawing of, 33
    eruption process, description of, 105
    paintings of, 32–33
    photograph of (1872), 30
    thermal infrared imagery of, 103
*Old Faithful* (painting), 32
*Old Faithful (1872)* (photograph), 30
Old Faithful Inn, 40
Organic Act of 1916, 9
*Out to Lunch* (painting), 33
*Outline Map of Indian Localities in 1833*, 20
over-snow vehicles (OSVs), 47
Owen, William, 48, 49

## P

paleoclimatology
    glaciers, 109
    fire, 143
paleontology, 168–169
parabola, Snake River, 106
*Peh-tō-pe-kiss, Eagle's Ribs, a Piegan Chief* (painting), 21
per capita income. *See* income
permits, scientific research, 37
Peterson, William, 23
petroglyphs, 19
Petzoldt, Eldon, 48, 49
Petzoldt, Paul, 48, 49
pH levels in geothermal areas, 100–101
Pinedale glaciation, 108
place names
    American Indian, 16
    Grand Teton National Park, 248–249
    Greater Yellowstone Area
        cultural features of, 238–239
        physical features, 240–241
    Yellowstone National Park, 242–247
plants. *See* vegetation
plate tectonics, 169
    landforms, role in creating, 84
    and Grand Teton geology, 95
    and Yellowstone geology, 97
    and the Yellowstone volcano, 98–99
political boundaries, 10–11
    of American Indian tribes and reservations, 17
    of counties (modern), 236–237
politics, 77
population
    cities, 70–71
    counties, 68–69

density and structure, 66–67
    race and ethnicity and, 74–75
    wildland complexes and, 60–61
potatoes, 56, 57
potential wildlife habitat, 162–163
pottery
    and archaeology, 14–15
    Sheep Eaters, use of, 19
poverty, 55
Preble's map of elk winter range, 35
precipitation, 122–123
    agriculture, effect on, 57
    bison movement, effect on, 156–157
    climate, as function of, 126–127
    ecoregions and, 132–133
    elk migration, effect on, 158–159
    fire and, 142–145
    streamflow and, 116–117
    vegetation and, 135
preserves, nature. *See* protected areas
presidential elections, 77
privately owned land, 63–65
profiles, elevation. *See* elevation profiles.
protected areas, 62–65
    economics, effects on, 60–61
    in the United States, 9
    in the world, 8
Pruessl, Carl, 32
public lands. *See* protected areas
public transportation, 45

## R

race, 74–75
radio collars, 36
radiocarbon dating, 14–15
    fire history, use of to understand, 143
railroads, development of, 39
rainfall. *See* precipitation
ranching. *See* agriculture.
Raynolds Expedition, 23, 34
Raynolds map of 1860, 27
Raynolds, William, 23, 27
recreation, economics of, 60–61
reference maps
    Greater Yellowstone Area, 178–191
    Grand Teton and Yellowstone
        national parks, 192–223
regional population, 66–67
reintroduction of wolves, 150–151
    coyotes, effect on, 152–153
religion
    contemporary, 76–77
    of Sheep Eaters, 19
remote sensing imagery, 140–141
reptiles, list of, 170
research. *See* scientific research
reservations, Indian. *See* American Indian reservations
reserves, nature. *See* protected areas
resource extraction, 52, 53
rivers, 114–115
    chloride flux of, 101
    drainage basins and, 112–113
    exploration and, 24
    fish in, 160–161
    flow regimes, 118–119
    streamflow, 116–119
    waterfalls, 120–121
    wild and scenic (federal designation), 62–63
roads
    development and history of, 38–39
    location and elevation profiles of, 42–43
    near Old Faithful, development of, 40–41
    traffic patterns on, 44–45
Rubinstein, Paul, 121
runoff, 117
Russel, Osborne
    bighorn sheep, observations of by, 18
    explorations of Yellowstone by, 22

## S

sagebrush-steppe habitat, 164–165
sage-grouse, greater, 165
satellite imagery, 140–141
school districts, 72–73
schools, 72–73
scientific research

history of
    contemporary, 36–37
    early, 34–35
    *See also* specific research subjects (e.g. thermophiles)
seismic activity. *See* earthquakes
seismic monitoring stations, 107
Setchell, William A., 35
sex distribution (demographic data), 67
Sheep Eaters, 17, 18–19, 22
sheep, bighorn, 56–57
    American Indians and historic sightings of, 18
sheep, domestic, 57
Shoshone, Sheep Eater, 17, 18–19
Snake River parabola, 106
snow. *See* precipitation
snow coaches, 47
snow water equivalent, 126–127
snowmobiles, 47
soils, 130–131
    ecoregions and, 132–133
    vegetation and, 135
Spalding, Franklin, 48, 49
spear points, 14
species, list of, 170–175
state boundaries, 10–11
state parks, 62–63
state land, 65
Steamboat Geyser, 105
Stegner, Wallace, 9
Stevens, Mike, 121
Stevenson, James, 48
streams. *See* rivers
streamflow, 116–119
*Stu-mick-o-súcks, Buffalo Bull's Back Fat,*
    *Head Chief, Blood Tribe* (painting), 21
surveys. *See* expeditions

## T

temperature, 124–127
    of geothermal areas, 100–103
territories
    American Indian, 17
    U.S., political boundaries of, 10–11
Teton Range
    climbing of, 48–49
    geology of, 94–95
    glaciers and, 108
thermal features. *See* geothermal features
thermophiles, 36, 166–167
    early study of, 35
topography. *See* elevation
tourism
    economics, effect on, 60–61
    infrastructure, effect on development of, 40–41
    visitation, 46–47
traffic flow and volume, 44–45
trails, 42–43
    development of, 39
    near Old Faithful, development of, 40–41
transects. *See* elevation profiles
transportation
    economics, effect on, 58–59
    roads, development of, 39
    traffic flow and volume, 44–45
    visitation, effect on, 46–47
tribal land, 64–65
tribes, American Indian. *See* American Indians
Tukadika (Sheep Eaters), 18–19

## U

Underhill, Robert, 48, 49
unemployment, 52–53, 61
United States Army, 40
United States Army Corps of Engineers, 39, 40, 42
United States Department of Agriculture land, 64–65
United States Fish and Wildlife Service land, 64–65
United States Forest Service
    early history of, 11
    land managed by, 62–65
United States Geological Survey
    early surveys by, 23, 30–31
    index of maps by, 232–235
universities, 72–73
Upper Geyser Basin
    geysers, named, 104

infrastructure near, development of, 40–41
temperature and chemistry of, 101
thermal infrared imagery of, 103
USGS. *See* United States Geological Survey

**V**
vegetation
ecoregions and, 132–133
Greater Yellowstone Area, 134–139
habitat and, 162–163
human use of for food, 14–15, 18
land cover of, 6
soils, effect on, 130
wetlands and, 128–129
vents, hydrothermal, 96–97
vertebrate species, list of, 170–175
visitation, 46–47
infrastructure, impact on development of, 41
volcanic activity. *See* volcanism
volcanic domes, 96–97
volcanism, 84–85, 98–99
and soils, 131
eruptions, 98–99

**W**
Washburn Expedition, 23, 28
Washburn, Henry, 28
water spirits (Sheep Eater religion), 19
waterfalls, 120–121
weather
bison migration and, 155, 156–157
elk migration and, 158–159
1988 fires and, 144–145
streamflow and, 116–117
visitation and, 46–47
*See also* climate
wetlands, 128–129
hydric soils, 131
wheat, 56–57
whirling disease, 161
Whittlesey, Lee, 121
wilderness (federal land designation), 62–65
wild and scenic rivers, 62–63
wildfires. *See* fire
wildlife
coyotes, 152–153
bison, 154–155
early account of, 22
early survey of, 35
elk, 158–159
fish, 160–161
grizzly bears, 148–149
habitat of, 162–163
prehistoric, 168–169
research of, 36
sagebrush-steppe, importance to, 164
species, list of, 170–175
wolves, 150–151
*See also* individual species listings
wildlife refuges, national, 62–63
wolves, 150–151
and elk, 158–159
coyotes, interrelationship with, 152–153
"The Wonders of the West" (article), 31
Wyant, H. W., 23

**Y**
Yellowstone region, early accounts of, 22
Yellowstone Act (1872), 9
Yellowstone caldera, 84, 96–97, 98–99
and earthquakes, 106–107
and Yellowstone Lake, 110–111
Yellowstone cutthroat trout, 160–161
Yellowstone hotspot, 84, 98, 106
Yellowstone ice cap, 109, 111
Yellowstone Lake, 110–111
and climate change, 127
and fish, 161
Yellowstone National Park
boundaries, development of, 10–11
establishment of, 10–11
and national park idea, 8–9
role of art in, 31, 32–33
role of science in, 34–35
*See also* main entries (e.g., elevation)
Yellowstone volcano, 98–99
Yellowstone Volcano Observatory, 107